International Review of **Cytology** A Survey of **Cell Biology**

VOLUME 179

Cumulative Subject Index

Volumes 138–176

International Review of Cytology

A Survey of Cell Biology

Edited by

Kwang W. Jeon
Department of Zoology
The University of Tennessee
Knoxville, Tennessee

VOLUME 179

Cumulative Subject Index
Volumes 138–176

ACADEMIC PRESS
San Diego London Boston New York Sydney Tokyo Toronto

This book is printed on acid-free paper. ∞

Academic Press
a division of Harcourt Brace & Company
525 B Street, Suite 1900, San Diego, California 92101-4495, USA
http://www.apnet.com

Academic Press Limited
24-28 Oval Road, London NW1 7DX, UK
http://www.hbuk.co.uk/ap/

International Standard Book Number: 0-12-364583-2

PRINTED IN THE UNITED STATES OF AMERICA
97 98 99 00 01 02 BB 9 8 7 6 5 4 3 2 1

CONTENTS

CONTENTS OF VOLUMES 138–176

VOLUME 138

Photoreceptor Rescue in the Dystrophic Retina by Transplantation of Retinal Pigment Epithelium 1
Harold John Sheedlo, Linxi Li, Vinod P. Gaur, Richard W. Young, Anthony D. Seaton, S. V. Stovall, C. D. Jaynes, and James E. Turner

Polyamine Cytochemistry: Localization and Possible Functions of Polyamines 51
David M. Hougaard

Neuropeptides in the Amphibian Brain 89
Ann C. Andersen, Marie-Christine Tonon, Georges Pelletier, J. Michael Conlon, Aldo Fasolo, and Hubert Vaudry

Mitogen-Activated Protein Kinase and Cytoskeleton in Mitogenic Signal Transduction 211
Eisuke Nishida and Yukiko Gotoh

Calcium and Mitosis 239
Peter K. Hepler

New Insights into the Development of Neural Crest Derivatives 269
Nicole M. Le Douarin, Elisabeth Dupin, Anne Baroffio, and Catherine Dulac

VOLUME 139

Cellular Effects of Early Exposure to Sex Hormones and Antihormones 1
Taisen Iguchi

Biogenesis of the Vacuole in *Saccharomyces cerevisiae* 59
Christopher K. Raymond, Christopher J. Roberts, Karen E. Moore, Isabelle Howald, and Tom H. Stevens

The Cell Biology of Pattern Formation during *Drosophila* Development 121
Teresa V. Orenic and Sean B. Carroll

Assays of Random Motility of Polymorphonuclear Leukocytes *in Vitro* 157
Leon P. Bignold

Molecular Origins of Cellular Differentiation in *Volvox* and Its Relatives 189
Rüdiger Schmitt, Stefan Fabry, and David L. Kirk
Actin Matrix of Dendritic Spines, Synaptic Plasticity, and Long-Term Potentiation 267
Eva Fifková and Marisela Morales
Role of Signal Transduction Systems in Cell Proliferation in Yeast 309
Isao Uno

VOLUME 140

PART I. POLLEN GRAIN AND TUBE
Pollen Gene Expression: Molecular Evidence 3
Joseph P. Mascarenhas
Identification and *in Situ* Localization of Pollen-Specific Genes 19
S. P. Davies, M. B. Singh, and R. B. Knox
Pollen Wall and Sporopollenin 35
R. Wiermann and S. Gubatz
Cytoskeleton and Cytoplasmic Organization of Pollen and Pollen Tubes 73
Elisabeth S. Pierson and Mauro Cresti
PART II. GAMETES
The Male Germ Unit: Concept, Composition, and Significance 129
H. Lloyd Mogensen
PART III. PROGAMIC PHASE AND FERTILIZATION
A Dynamic Role for the Stylar Matrix in Pollen Tube Extension 297
Luraynne C. Sanders and Elizabeth M. Lord
Double Fertilization in Nonflowering Seed Plants and Its Relevance to the Origin of Flowering Plants 319
William E. Friedman
Double Fertilization 357
Scott D. Russell
PART IV. MANIPULATION IN POLLINATION AND FERTILIZATION MECHANISMS
In Vitro Pollination: A New Tool for Analyzing Environmental Stress 391
Isabelle Dupuis
In Vitro Fusion of Gametes and Production of Zygotes 407
Erhard Kranz, Horst Lörz, Catherine Digonnet, and Jean-Emmanuel Faure
Plant Transformation Using the Sexual Route 425
Patricia Roeckel, Maurice M. Moloney, and Joël R. Drevet
PART V. SELF-INCOMPATIBILITY BETWEEN POLLEN AND STIGMA
Gametophytic Self-Incompatibility: Biochemical, Molecular Genetic, and Evolutionary Aspects 449
Anuradha Singh and Teh-Hui Kao

Sporophytic Self-Incompatibility Systems: *Brassica S* Gene Family 485
Martin Trick and Philippe Heizmann
Sporophytic Self-Incompatibility Systems: *S* Gene Products 525
H. G. Dickinson, M. J. C. Crabbe, and T. Gaude
PART VI. POSTSCRIPT
Plant Reproductive Biology: Trends 565
Christian Dumas and Scott D. Russell

VOLUME 141

Mitochondrial Genomes of the Ciliate 1
Donald J. Cummings
Mitochondrial DNA of Kinetoplastids 65
Kenneth Stuart and Jean E. Feagin
Evolution of Mitochondrial Genomes in Fungi 89
G. D. Clark-Walker
Structure and Function of the Higher Plant Mitochondrial Genome 129
Maureen R. Hanson and Otto Folkerts
Animal Mitochondrial DNA: Structure and Evolution 173
David R. Wolstenholme
Transcription and Replication of Animal Mitochondrial DNAs 217
David A. Clayton
The Endosymbiont Hypothesis Revisited 233
Michael W. Gray

VOLUME 142

Biological Problems of Regenerative Cementogenesis: Synthesis and Attachment of Collagenous Matrices on Growing and Established Root Surfaces 1
Hubert E. Schroeder
Immunocytochemical Localization of Proteins in Striated Muscle 61
Marvin H. Stromer
Recent Developments in Vertebrate Cell Culture Technology 145
Satish J. Parulekar, Thomas Hassell, and Satish C. Tripathi
Transdifferentiation in Medusae 213
Volker Schmid
Symplast as a Functional Unit in Plant Growth 263
Kiyoshi Katou and Hisashi Okamoto
Intracellular Ca^{2+} Messenger System in Plants 305
Shoshi Muto

VOLUME 143

NF-κB and Rel: Participants in a Multiform Transcriptional Regulatory System 1
Mariagrazia Grilli, Jason J.-S. Chiu, and Michael J. Lenardo
Molecular Genetic Analysis of Muscle Development, Structure, and Function in *Drosophila* .. 63
Sanford I. Bernstein, Patrick T. O'Donnell, and Richard M. Cripps
Contractile Proteins and Myofibrillogenesis 153
Takashi Obinata
Cytophysiology of Fish Chromatophores 191
Ryozo Fujii
A Review of Recent Studies on Homomorphic Self-Incompatibility 257
Kokichi Hinata, Masao Watanabe, Kinya Toriyama, and Akira Isogai
The Genetics of Nitrogen Assimilation and Amino Acid Biosynthesis in Flowering Plants: Progress and Prospects 297
Robert L. Last

VOLUME 144

Karyosphere in Oogenesis and Intranuclear Morphogenesis 1
Mira N. Gruzova and Vladimir N. Parfenov
Applications of *Arabidopsis thaliana* to Outstanding Issues in Plant–Pathogen Interactions .. 53
Jeffery L. Dangl
Toward a New Concept of Cell Motility: Cytoskeletal Dynamics in Amoeboid Movement and Cell Division 85
Yoshio Fukui
Fine Structure, Innervation, and Functional Control of Avian Salt Glands 129
Rüdiger Gerstberger and David A. Gray
Mitosis: Dissociability of Its Events 217
Sibdas Ghosh and Neidhard Paweletz
The Endosymbiotic Origin of Chloroplasts 259
Jean M. Whatley

VOLUME 145

NMR and the Study of Pathological State in Cells and Tissues 1
Jesús Ruiz-Cabello and Jack S. Cohen
Roles of Urease in Plant Cells 65
Joseph C. Polacco and Mark A. Holland
Aspects of Amphibian Metamorphosis: Hormonal Control 105
Sakaé Kikuyama, Kousuke Kawamura, Shigeyasu Tanaka, and Kazutoshi Yamamoto

Control of Metabolism and Development in Higher Plant Plastids 149
M. J. Emes and A. K. Tobin
Biomineralization and Eggshells: Cell-Mediated Acellular Compartments of Mineralized Extracellular Matrix 217
José L. Arias, David J. Fink, Si-Qun Xiao, Arthur H. Heuer, and Arnold I. Caplan
Regulation of Intracellular Movements in Plant Cells by Environmental Stimuli 251
Reiko Nagai

VOLUME 146

Histogenesis of the Avian Retina in Reaggregation Culture: From Dissociated Cells to Laminar Neuronal Networks 1
Paul G. Layer and Elmar Willbold
Growth-Factor-Induced Retinal Regeneration *in Vivo* 49
Carol M. Park and Martin J. Hollenberg
Cytokine Production by Retinal Pigmented Epithelial Cells 75
Peter A. Campochiaro
Biological Clocks in the Retina: Cellular Mechanisms of Biological Timekeeping 83
G. D. Block, S. B. S. Khalsa, D. G. McMahon, S. Michel, and M. Geusz
Plasticity and Differentiation of Retinal Precursor Cells 145
Ruben Adler
Genetics, Development, and Malignancy of Melanocytes 161
D. C. Bennett
Dynamic Aspects of the Plant Extracellular Matrix 261
G. Paul Bolwell

VOLUME 147

Neuromodulatory Control of Rhythmic Behaviors in Invertebrates 1
Allen I. Selverston
The Sertoli–Germ Cell Communication Network in Mammals 25
B. Jégou
The Bacterial Flagellar Motor 97
S. Roy Caplan and Michael Kara-Ivanov
The Salmon GnRH Encoding Gene in Teleost Fish 165
Øivind Andersen and Helge Klungland
Calcium Channels in the Neuromuscular Junction 193
Bertalan Csillik
Structure, Function, and Regulation of the Chalcone Synthase 233
C. R. Martin

Enzymes Involved in Glutamatergic and GABAergic Neurotransmission 285
Peter Kugler

VOLUME 148

Elicitors of Plant Defense Responses 1
Jürgen Ebel and Eric G. Cosio
Membrane and Cytoskeleton Flow in Motile Cells with Emphasis on the Contribution of Free-Living Amoebae 37
Andrzej Grębecki
Cell Differentiation: An Evolutionary Perspective 81
Reed A. Flickinger
Neuropeptide Receptors and Astrocytes 119
Brigitte Krisch and Rolf Mentlein
Functional Aspects of Cell Response to Heat Shock 171
Vladimir Ya. Alexandrov
Cell Biology of Phototaxis in Flagellate Algae 229
Georg Kreimer

VOLUME 149

Na^+ Requirements for Capacitation and Acrosomal Exocytosis in Mammalian Sperm 1
Lynn R. Fraser
Control of Plant Enzyme Activity by Reversible Protein Phosphorylation 47
Steven C. Huber, Joan L. Huber, and Robert W. McMichael, Jr.
Testicular Macrophages 99
James C. Hutson
Analysis of Microfilament Organization and Contractile Activities in *Physarum* 145
Wilhelm Stockem and Klaudia Brix
Molecular and Cellular Aspects of Neurotransmission and Neuromodulation 217
R. Martínez-Rodriguez and R. Martínez-Murillo

VOLUME 150

Mechanical Stresses in Embryonic Tissues: Patterns, Morphogenetic Role, and Involvement in Regulatory Feedback 1
L. V. Beloussov, S. V. Saveliev, I. I. Naumidi, and V. V. Novoselov
Locomotion of Tissue Culture Cells Considered in Relation to Ameboid Locomotion 35
Albert K. Harris

Role of Mechanical Stimulation in the Establishment and Maintenance of Muscle Cell Differentiation 69
David G. Simpson, Wayne Carver, Thomas K. Borg, and Louis Terracio
Finite Element Methods for Developmental Biology 95
G. Wayne Brodland
Substratum Mechanics and Cell Differentiation 119
Michal Opas
Phase Transitions, Interfaces, and Morphogenesis in a Network of Protein Fibers 139
Gabor Forgacs and Stuart A. Newman
The Interphase Nucleus as a Dynamic Structure 149
Umberto De Boni
Cellular Tensegrity: Exploring How Mechanical Changes in the Cytoskeleton Regulate Cell Growth, Migration, and Tissue Pattern during Morphogenesis 173
Donald E. Ingber, Laura Dike, Linda Hansen, Seth Karp, Helen Liley, Andrew Maniotis, Helen McNamee, David Mooney, George Plopper, John Sims, and Ning Wang
Mechanics of the Cytoskeleton and Morphogenesis of *Acetabularia* 225
B. C. Goodwin and C. Brière
The Chemical Basis of Diatom Morphogenesis 243
Richard Gordon and Ryan W. Drum
Appendix: Dialogue on Embryonic Induction and Differentiation Waves 373
Richard Gordon, Natalie K. Björklund, and Pieter D. Nieuwkoop

VOLUME 151

Inherited Genetic Defects: Analysis and Diagnosis at the Cellular Level in Preimplantation Embryos 1
Audrey L. Muggleton-Harris
Structure and Function of the Cyanelle Genome 29
Wolfgang Löffelhardt and Hans J. Bohnert
Diverse Distribution and Function of Fibrous Microtubule-Associated Proteins in the Nervous System 67
Thomas A. Schoenfeld and Robert A. Obar
Cholinesterases in Avian Neurogenesis 139
Paul G. Layer and Elmar Willbold
Role of Nuclear Trafficking in Regulating Cellular Activity 183
Carl M. Feldherr and Debra Akin
Phenolic Components of the Plant Cell Wall 229
Graham Wallace and Stephen C. Fry

VOLUME 152

The Unique Structure of Lepidopteran Spindles 1
Klaus Werner Wolf
Cellular Mechanisms for the Activation of Blood Coagulation 49
Carolyn L. Geczy
Function and Modulation of Expression of Auxin-Regulated Genes 109
Yohsuke Takahashi, Sarahmi Ishida, and Toshiyuki Nagata
Regulation of Mitochondrial Gene Expression in *Saccharomyces cerevisiae* 145
Carol L. Dieckmann and Robin R. Staples
Dynamics of the Calcium Signal That Triggers Mammalian Egg Activation 183
Karl Swann and Jean-Pierre Ozil
Regeneration of Mammalian Retinal Pigment Epithelium 223
Gary E. Korte, Jay I. Perlman, and Ayala Pollack
Habituation as a Tumorous State That Is Interchangeable with a Normal State in Plant Cells 265
Kunihiko Syōno and Tomomichi Fujita

VOLUME 153

Fluorescent *in Situ* Hybridization for the Diagnosis of Genetic Disease at Postnatal, Prenatal, and Preimplantation Stages 1
Darren K. Griffin
Isolation and Function of Human Dendritic Cells 41
Lisa A. Williams, William Egner, and Derek N. J. Hart
Granulated Lymphoid Cells of the Pregnant Uterus: Morphological and Functional Features 105
Chau-Ching Liu, Earl L. Parr, and John Ding-E Young
The Replication Band of Ciliated Protozoa 137
Donald E. Olins and Ada L. Olins
Whole-Chromosome Hybridization 171
S. D. Bouffler
Neuronal Modulation and Plasticity *in Vitro* 233
Robert A. Smith and Zhi-Gang Jiang

VOLUME 154

Intermediate Filament Proteins: Cytoskeletal Elements with Gene-Regulatory Functions? ... 1
Peter Traub and Robert L. Shoeman
Secretion and Endocytosis in the Male Reproductive Tract: A Role in Sperm Maturation 105
Louis Hermo, Richard Oko, and Carlos R. Morales

Evolution of Mammalian Sex Chromosomes and Sex-Determining Genes 191
Jennifer A. Marshall Graves and Jamie W. Foster
Organization of Replication Units and DNA Replication in Mammalian Cells as Studied by DNA Fiber Radioautography 261
Natalia A. Liapunova
Instability of the Homogeneous State as the Source of Localization, Epigenesis, Differentiation, and Morphogenesis 309
Yoram Schiffmann

VOLUME 155

Phylogeny and Ontogeny of Chemical Signaling: Origin and Development of Hormone Receptors 1
G. Csaba
Growth Factor-Induced Cell Migration: Biology and Methods of Analysis 49
Marianne Manske and Ernesto G. Bade
Physiological and Biochemical Aspects of Cytoplasmic Streaming 97
Teruo Shimmen and Etsuo Yokota
Cell and Molecular Biology of Flagellar Dyneins 141
David R. Mitchell
Morphological and Functional Reorganization of Human Carcinomas *in Vitro* 181
Petra Köpf-Maier, Birgit Kolon, and Markus Bugenings

VOLUME 156

Cell and Molecular Biology of *Rhizobium*-Plant Interactions 1
Michael Schultze, Éva Kondorosi, Pascal Ratet, Michel Buiré, and Ádám Kondorosi
Cellular Resistance to Cancer Chemotherapy 77
Torben Skovsgaard, Dorthe Nielsen, Christian Maare, and Karsten Wassermann
Cellular and Molecular Effects of Thymic Epithelial Cells on Thymocytes during Differentiation and Maturation 159
Yoshihiro Kinoshita and Fumihiko Hato
Actin-Binding Proteins in Cell Motility 199
Sadashi Hatano
The Cell Biology of Nematocysts 275
Glen M. Watson and Patricia Mire-Thibodeaux
The Impact of Altered Gravity on Aspects of Cell Biology 301
Dale E. Claassen and Brian S. Spooner

VOLUME 157

Cold-Stable and Cold-Adapted Microtubules 1
Margareta Wallin and Elisabeth Strömberg
Development and Heterogeneity of Prolactin Cells 33
Sumio Takahashi
Centrifuge Microscope as a Tool in the Study of Cell Motility 99
Yukio Hiramoto and Eiji Kamitsubo
Murine B Cell Development: Commitment and Progression from Multipotential Progenitors to Mature B Lymphocytes 129
Barbara L. Kee and Christopher J. Paige
Signal Transduction by the Antigen Receptors of B and T Lymphocytes 181
Michael R. Gold and Linda Matsuuchi
Neuroendocrine Epithelial Cell System in Respiratory Organs of Air-Breathing and Teleost Fishes 277
Giacomo Zaccone, Salvatore Fasulo, and Luigi Ainis
Structure–Function Relationships in Gap Junctions 315
Hartwig Wolburg and Astrid Rohlmann

VOLUME 158

Cell and Molecular Biology of Self-Incompatibility in Flowering Plants 1
F. C. H. Franklin, M. J. Lawrence, and V. E. Franklin-Tong
Mitosis and Meiosis in Cultured Plant Cells and Their Relationship to Variant Cell Types Arising in Culture 65
Vittoria Nuti Ronchi
Ecto-ATPases: Identities and Functions 141
Liselotte Plesner
Molecular Genetic Approaches to the Study of Human Craniofacial Dysmorphologies 215
Gudrun E. Moore
Effects of Electromagnetic Fields on Molecules and Cells 279
Eugene M. Goodman, Ben Greenebaum, and Michael T. Marron

VOLUME 159

Biology of Hypothalamic Neurons and Pituitary Cells 1
Vladimir R. Pantić
Angiogenesis: Models and Modulators 113
Gillian W. Cockerill, Jennifer R. Gamble, and Mathew A. Vadas
Nuclear Remodeling in Response to Steroid Hormone Action 161
Klaus Brasch and Robert L. Ochs

Effects of Axotomy, Deafferentation, and Reinnervation on Sympathetic Ganglionic Synapses: A Comparative Study 196
Jacques Taxi and Daniel Eugene
Chondrocyte Differentiation 265
Ranieri Cancedda, Fiorella Descalzi Cancedda, and Patrizio Castagnola

VOLUME 160

Sterility in Neonatally Androgenized Female Rats and the Decidual Cell Reaction 1
Yasuhiko Ohta
Mechanistic and Developmental Aspects of Genetic Imprinting in Mammals 53
Keith E. Latham, James McGrath, and Davor Solter
Molecular Mechanisms for Passive and Active Transport of Water 99
T. Zeuthen
The Comparative Cell Biology of Accessory Somatic (or Sertoli) Cells in the Animal Testis 164
Sardul S. Guraya
Factors Controlling Growth, Motility, and Morphogenesis of Normal and Malignant Epithelial Cells 222
Carmen Birchmeier, Dirk Meyer, and Dieter Riethmacher
Dynamics of the Seminal Vesicle Epithelium 268
Lucinda R. Mata
Molecular Organization of Hepatocyte Peroxisomes 304
Takashi Makita

VOLUME 161

The Role of Growth Factors in Vascular Cell Development and Differentiation 1
Marsha A. Moses, Michael Klagsbrun, and Yuen Shing
Sexuality of Mitochondria: Fusion, Recombination, and Plasmids 49
Shigeyuki Kawano, Hiroyoshi Takano, and Tsuneyoshi Kuroiwa
Analyzing Renal Glomeruli with the New Stereology 111
John F. Bertram
Membrane Mechanisms and Intracellular Signalling in Cell Volume Regulation 173
Else K. Hoffmann and Philip B. Dunham
Bacterial Stimulators of Macrophages 263
Sunna Hauschildt and Bernhard Kleine

VOLUME 162

PART A

The Nuclear Matrix: A Structural Milieu for Nuclear Genomic Function 1
Ronald Berezney, Michael J. Mortillaro, Hong Ma, Xiangyun Wei, and Jagath Samarabandu

The Architectural Organization of Nuclear Metabolism 67
Jeffrey A. Nickerson, Benjamin J. Blencowe, and Sheldon Penman

The Structural Basis of Nuclear Function 125
Dean A. Jackson and Peter R. Cook

Nuclear Domains and the Nuclear Matrix 151
Roel van Driel, Derick G. Wansink, Bas van Steensel, Marjolein A. Grande, Wouter Schul, and Luitzen de Jong

The Nuclear Matrix and the Regulation of Chromatin Organization and Function 191
James R. Davie

Contributions of Nuclear Architecture to Transcriptional Control 251
Gary S. Stein, André J. van Wijnen, Janet Stein, Jane B. Lian, and Martin Montecino

Chromatin Domains and Prediction of MAR Sequences 279
Teni Boulikas

Scaffold/Matrix-Attached Regions: Structural Properties Creating Transcriptionally Active Loci 389
J. Bode, T. Schlake, M. Ríos-Ramírez, C. Mielke, M. Stengert, V. Kay, and D. Klehr-Wirth

Origins of Replication and the Nuclear Matrix: The DHFR Domain as a Paradigm 455
P. A. Dijkwel and J. L. Hamlin

The Nuclear Matrix and Virus Function 485
W. Deppert and R. Schirmbeck

The Nuclear Matrix as a Site of Anticancer Drug Action 539
D. J. Fernandes and C. V. Catapano

PART B

Nuclear Matrix Proteins as Structural and Functional Components of the Mitotic Apparatus .. 1
D. He, C. Zeng, and B. R. Brinkley

Nuclear Matrix Isolated from Plant Cells 75
Susana Moreno Díaz de la Espina

The Dynamic Properties and Possible Functions of Nuclear Lamins 141
Robert D. Moir, Timothy P. Spann, and Robert D. Goldman

Intracellular Structure and Nucleocytoplasmic Transport 184
Paul S. Agutter

Toward a Molecular Understanding of the Structure and Function of the Nuclear Pore Complex 225
Nelly Panté and Ueli Aebi

Nuclear Pore Complex Proteins 257
Ricardo Bastos, Nelly Panté, and Brian Burke

Targeting and Association of Proteins with Functional Domains in the Nucleus: The Insoluble Solution 303
Heinrich Leonhardt and M. Cristina Cardoso
Nuclear Matrix Acceptor Binding Sites for Steroid Hormone Receptors: A Candidate Nuclear Matrix Acceptor Protein 337
Andrea Lauber, Nicole P. Sandhu, Mark Schuchard, M. Subramaniam, and Thomas C. Spelsberg
Nuclear Matrix and the Cell Cycle 377
Peter Loidl and Anton Eberharter
Specificity and Functional Significance of DNA Interaction with the Nuclear Matrix: New Approaches to Clarify the Old Questions 405
Sergey V. Razin, Irina I. Gromova, and Olga V. Iarovaia

VOLUME 163

Osmotic Flow in Membrane Pores 1
Adrian E. Hill
The Use of Computers in Understanding How Animal Cells Crawl 43
David R. Soll
Programmed Cell Death in Development 105
E. J. Sanders and M. A. Wride
Biochemistry and Molecular Biology of Chromoplast Development 175
Bilal Camara, Philippe Hugueney, Florence Bouvier, Marcel Kuntz, and René Monéger
Sperm-Binding Proteins 249
Kathleen R. Foltz
The Ultrastructure of Epithelial and Fiber Cells in the Crystalline Lens 306
J. R. Kuszak

VOLUME 164

Multifunctional Proteins in *Tetrahymena*: 14-nm Filament Protein/Citrate Synthase and Translation Elongation Factor-1α 1
Osamu Numata
Some Characteristics of Neoplastic Cell Transformation in Transgenic Mice 37
Irina N. Shvemberger and Alexander N. Ermilov
Integration of Intermediate Filaments into Cellular Organelles 91
Spyros D. Georgatos and Christèle Maison
Developmental Failure in Preimplantation Human Conceptuses 139
Nicola J. Winston

G-Protein-Coupled Receptors in Insect Cells 189
Jozef J. M. Vanden Broeck
Microtubules and Microtubule Motors: Mechanisms of Regulation 269
Catherine D. Thaler and Leah T. Haimo

VOLUME 165

Salt Tolerance in Plants and Microorganisms: Toxicity Targets and Defense Responses 1
Ramón Serrano
Polytene Chromosomes in Mammalian Cells 53
Eugenia V. Zybina and Tatiana G. Zybina
Ependymins: Meningeal-Derived Extracellular Matrix Proteins at the Blood–Brain Barrier ... 121
Werner Hoffmann and Heinz Schwarz
Cadherin Cell Adhesion Molecules in Differentiation and Embryogenesis 159
James A. Marrs and W. James Nelson
Escape and Migration of Nucleic Acids between Chloroplasts, Mitochondria, and the Nucleus 207
Peter E. Thorsness and Eric R. Weber
Cytoplasmic Mechanisms of Axonal and Dendritic Growth in Neurons 235
Steven R. Heidemann

VOLUME 166

Hormones and the Cytoskeleton of Animals and Plants 1
Arnold De Loof, Jozef Vanden Broeck, and Ine Janssen
Twisted Liquid Crystalline Supramolecular Arrangements in Morphogenesis 59
Marie-Madeleine Giraud-Guille
Vitellogenin Receptors: Oocyte-Specific Members of the Low-Density Lipoprotein Receptor Supergene Family 103
Wolfgang Johann Schneider
Follicular Dendritic Cells and Germinal Centers 140
Yong-Jun Liu, Géraldine Grouard, Odette de Bouteiller, and Jacques Banchereau
Cell Biology of Kidney Glomerulus 182
Shinichi Ohno, Takeshi Baba, Nobuo Terada, Yasuhisa Fujii, and Hideho Ueda
Distribution and Functions of Parathyroid Hormone-Related Protein in Vertebrate Cells 231
Patricia M. Ingleton and Janine A. Danks

VOLUME 167

Clathrin-Coated Vesicles in Plants 1
Leonard Beevers

Peptides in the Nervous Systems of Cnidarians: Structure, Function, and Biosynthesis 37
Cornelis J. P. Grimmelikhuijzen, Ilia Leviev, and Klaus Carstensen
M Cells in Peyer's Patches of the Intestine 91
Andreas Gebert, Hermann-Josef Rothkötter, and Reinhard Pabst
pp125^{FAK} in the Focal Adhesion 161
Carol A. Otey
Feedback Inhibitors in Normal and Tumor Tissues 185
E. Marshall and B. I. Lord
The Incidence, Origin, and Etiology of Aneuploidy 263
Darren K. Griffin

VOLUME 168

Cell Biology of the Harderian Gland 1
Giovanni Chieffi, Gabriella Chieffi Baccari, Loredana Di Matteo, Michela d'Istria, Sergio Minucci, and Bruno Varriale
Signal Transduction in Cell–Matrix Interactions 81
Jun-Lin Guan and Hong-Chen Chen
Calcium and Graviperception in Plants: Inhibitor Analysis 123
Ninel A. Belyavskaya
Glucagon and Glucagon-like Peptides in Fishes 187
Erika M. Plisetskaya and Thomas P. Mommsen
Gametic Differentiation of *Chlamydomonas* 259
Christoph F. Beck and Michel A. Haring
Systemic Acquired Resistance in Plants 303
M. Schneider, P. Schweizer, P. Meuwly, and J. P. Métraux

VOLUME 169

Cell Biology of Somatolactin 1
Toyoji Kaneko
The Centrosome in Animal Cells and Its Functional Homologs in Plant and Yeast Cells 25
Ron Balczon
Differentiation Processes in the Amphibian Brain with Special Emphasis on Heterochronies . 83
Andrea Schmidt and Gerhard Roth
Cell Biology of Wound Healing 151
Christopher J. Schaffer and Lillian B. Nanney
Differentiated Properties and Proliferation of Arterial Smooth Muscle Cells in Culture 184
Johan Thyberg

Cortical Structure and Function in Euglenoids with Reference to Trypanosomes, Ciliates, and Dinoflagellates 268
G. Benjamin Bouck and Huân Ngô

VOLUME 170

Tropomyosin Isoforms in Nonmuscle Cells 1
Jim J.-C. Lin, Kerri S. Warren, Dawn D. Wamboldt, Tao Wang, and Jenny L.-C. Lin
Effects of Microgravity on the Structure and Function of Plant Cell Walls 39
Elena M. Nedukha
Pro-Opiomelanocortin-Derived Peptides, Cytokines, and Nitric Oxide in Immune Responses and Stress: An Evolutionary Approach 79
Enzo Ottaviani, Antonella Franchini, and Claudio Franceschi
Mammalian Skeletal Muscle Fiber Type Transitions 143
Dirk Pette and Robert S. Staron
Acute Phase Proteins and Transformed Cells 225
Andrzej Mackiewicz

VOLUME 171

Biology of Plant Cells in Microgravity and under Clinostating 1
Elizabeth J. Kordyum
The Use of Antibodies to Study the Architecture and Developmental Regulation of Plant Cell Walls 79
J. Paul Knox
Biophysical Aspects of P-Glycoprotein-Mediated Multidrug Resistance 121
Randy M. Wadkins and Paul D. Roepe
Normal and Pathological Tau Proteins as Factors for Microtubule Assembly 167
André Delacourte and Luc Buée
Differentiation and Transdifferentiation of the Retinal Pigment Epithelium 225
Shulei Zhao, Lawrence J. Rizzolo, and Colin J. Barnstable
The Role of Endothelins in the Paracrine Control of the Secretion and Growth of the Adrenal Cortex 267
Gastone G. Nussdorfer, Gian Paolo Rossi, and Anna S. Belloni

VOLUME 172

Transport of Glucose across the Blood–Tissue Barriers 1
Kuniaki Takata, Hiroshi Hirano, and Michihiro Kasahara

The Role of Suppressors in Determining Host–Parasite Specificities in Plant Cells 55
Tomonori Shiraishi, Tetsuji Yamada, Yuki Ichinose, Akinori Kiba, and Kazuhiro Toyoda
Distribution and Functions of Platelet-Derived Growth Factors and Their Receptors during Embryogenesis 95
Paris Ataliotis and Mark Mercola
Adaptive Crystal Formation in Normal and Pathological Calcifications in Synthetic Calcium Phosphate and Related Biomaterials 129
G. Daculsi, J.-M. Bouler, and R. Z. LeGeros
The Biogenesis, Traffic, and Function of Cystic Fibrosis Transmembrane Conductance Regulator 193
Tamas Jilling and Kevin L. Kirk
Regulation of Phenylpropanoid Metabolism in Relation to Lignin Biosynthesis in Plants 243
Mark S. Barber and Heidi J. Mitchell

VOLUME 173

A Model for Flagellar Motility 1
Charles B. Lindemann and Kathleen S. Kanous
Basement-Membrane Stromal Relationships: Interactions between Collagen Fibrils and the Lamina Densa 73
Eijiro Adachi, Ian Hopkinson, and Toshihiko Hayashi
The Role of Endoxyloglucan Transferase in the Organization of Plant Cell Walls 157
Kazuhiko Nishitani
Microtubule–Microfilament Synergy in the Cytoskeleton 207
R. H. Gavin
Insulin Internalization and Other Signaling Pathways in the Pleiotropic Effects of Insulin 243
Robert M. Smith, Shuko Harada, and Leonard Jarett

VOLUME 174

Structure of the Insect Head in Ontogeny and Phylogeny: A View from *Drosophila* 1
Bryan T. Rogers and Thomas C. Kaufman
Functional Reconstitution of the Tonoplast Proton-ATPase from Higher Plants 85
Kunihiro Kasamo and Hiroyasu Yamanishi
The Role of Molecular Chaperones in Mitochondrial Protein Import and Folding 127
Michael T. Ryan, Dean J. Naylor, Peter B. Høj, Margaret S. Clark, and Nicholas J. Hoogtenraad
Proteoglycans and Related Components in Plant Cells 195
Eugene A. Nothnagel

VOLUME 175

Cell Model Systems in Plant Cytoskeleton Studies 1
Seiji Sonobe
The Structure, Function, and Assembly of Actin Filament Bundles 29
Ruth Furukawa and Marcus Fechheimer
Nuclear Components with Microtubule-Organizing Properties in Multicellular Eukaryotes: Functional and Evolutionary Considerations 91
František Baluška, Dieter Volkmann, and Peter W. Barlow
Sialic Acids in Molecular and Cellular Interactions 137
Sørge Kelm and Roland Schauer
Embryonic Neural Chimeras in the Study of Vertebrate Brain and Head Development 242
Nicole M. Le Douarin, Martin Catala, and Cesira Batini

VOLUME 176

Role of Rab GTPases in Membrane Traffic 1
Vesa M. Olkkonen and Harald Stenmark
Lignification in Plant Cell Walls 87
A. Ros Barceló
Functional Interactions among Cytoskeleton, Membranes, and Cell Wall in the Pollen Tube of Flowering Plants 133
Yi-Qin Li, Alessandra Moscatelli, Giampiero Cai, and Mauro Cresti
Biology of the Postsynaptic Glycine Receptor 201
Christian Vannier and Antoine Triller
Cell Fate Specification by Localized Cytoplasmic Determinants and Cell Interactions in Ascidian Embryos 245
Hiroki Nishida
Distribution of Na^+,K^+-ATPase in Photoreceptor Cells of Insects 307
Otto Baumann

Subject Index

Boldface numerals indicate volume number

A

A23187
agravitropism induction in plants, **168,** 155–157
calcium transport, **168,** 154
AA4.1, marker of B cell differentiation, **157,** 136, 142
Abalone, sperm-binding proteins, **163,** 262–263
A-band-associated proteins, in striated muscle antibodies
cardiac muscle, **142,** 106
sarcomere, **142,** 93–94, 96
thick-filament proteins, **142,** 62–80, 82
ABC transporters, *see* ATP-binding cassette proteins
Abies balsamea, **140,** 323–326
Abortion, *see* Spontaneous abortions
ABP
ABP-120 mutants, computer-assisted motion analysis, **163,** 91–93
actin-binding proteins, **156,** 240–245; **173,** 226–230; **176,** 143–145
androgen-binding protein, Sertoli cell, **160,** 201, 281–282
Abscisic acid
cytoskeleton, **166,** 39
fibrillin activation, **163,** 203
in higher plants, **165,** 34–36
inhibition of plant nodule development, **156,** 46
intracellular calcium messenger system and, **142,** 315, 320, 327, 337
plant hormone, **166,** 39
Acanthamoeba
actin, **156,** 228
actin polymerization in cells, **148,** 67–69
cytoplasmic fraction, **156,** 237
cytoskeletal dynamics and, **144,** 97–98, 109
locomotion, **150,** 56
myofibrillogenesis, **143,** 163, 165, 173
myosin, **156,** 221–223
in retraction of cytoskeleton, **148,** 64–67
profilin, **156,** 227–229
spectrin, **156,** 241
Acanthamoeba castellanii, **156,** 221–233, 238, 241
ACBP, binding to ARS T-rich strand, **162A,** 304
Acceleration, cell, 2D analysis, **163,** 62–63
Acceptor proteins
bound to DNA, **162B,** 345–348
RBF, **162B,** 338
for steroid/thyroid hormone receptors, **162B,** 348–355
Accessibility, in plant–parasite interactions, **172,** 57–59, 62–65
Accessory cells
in dome epithelium, **167,** 131–133
inducing variability of inhibitor effects
interferon, **167,** 231–232

Accessory cells (*continued*)
MIP-1α, **167,** 201–202
TNF-α, **167,** 230
M cells as, **167,** 143
Accessory factors
androgen receptor, **162B,** 362
definition, **162B,** 359–360
estrogen receptor, **162B,** 363–364
glucocorticoid receptor, **162B,** 361–362
thyroid hormone receptor, **162B,** 360–361
ACC synthase, *see* 1-Aminocyclopropane-1-carboxylate synthase
ACE inhibitors, *see* Angiotensin-converting enzyme inhibitors
Acellular afibrillar cementum, regenerative cementogenesis, **142,** 2–3
Acellular extrinsic fiber cementum, regenerative cementogenesis
established root surfaces *in vitro,* **142,** 31, 33–34, 42
established root surfaces *in vivo,* **142,** 46, 49
function of cementum, **142,** 2–5
growing root surfaces, **142,** 5–17, 19, 23, 25, 27–28
Acellular intrinsic fiber cementum, regenerative cementogenesis
established root surfaces *in vitro,* **142,** 41–42
function of cementum, **142,** 2–5
growing root surfaces, **142,** 15–25
Acer pseudoplatanus, intracellular calcium messenger system and, **142,** 309, 324
Acetabularia
calcium
in cytoskeletal dynamics, **150,** 227, 229–231
kinetics, **150,** 240–241
cell models, **175,** 5
cell wall dynamics, **150,** 231–232
characteristics, **150,** 225, 238
cytoplasmic streaming
analysis using tonoplast-free cells, **155,** 112
microtubule system-supported, **155,** 109–110
environmental stimuli in plant cells and, **145,** 270, 294
growth, **150,** 225–228
life cycle, **150,** 226
morphogenesis
branching, **150,** 272
characteristics, **150,** 225–228
simulation, **150,** 232–237
phenocopying, **150,** 237–238
viscoelastic equations, **150,** 239–240
Acetabularia crenulata, morphology, **150,** 237
Acetate, higher plant plastids and, **145,** 187–189
Acetobacter xylinum, cellulose content, effect of microgravity, **170,** 55
Acetohydroxamate, urease in plant cells and, **145,** 75
Acetohydroxy acid synthase, flowering plants and, **143,** 323–324
Acetolysis method, isolation of exines, **140,** 53
Acetone, intracellular calcium messenger system and, **142,** 334
Acetylation
core histones, **162A,** 203–208; **162B,** 386–387
deacetylation, histone, kinetics, **162A,** 205–206
in posttranslational modification of microtubules, **157,** 12–14, 18
Acetyl-L-carnitine, neurite initiation and elongation role, **153,** 255
Acetylcholine
γ-aminobutyric acid metabolism and, **147,** 286
avian salt glands and, **144,** 161, 165–167, 203–204
calcium channels and, **147,** 194, 199–200
inhibitors, **147,** 206–207, 209, 211, 213
myasthenic syndrome, **147,** 218
postsynaptic cytoplasm, **147,** 217
control of rhythmic behaviors and, **147,** 4, 12, 14
effect on prolactin release, **159,** 65
hormonal control of amphibian metamorphosis and, **145,** 128
neurotransmission, **149,** 236–239
non-occurrence in cnidarian nervous systems, **167,** 43–44
proteins in striated muscle and, **142,** 99
Acetylcholine receptor
and calcium channels
CGRP, **147,** 220–222, 225
inhibitors, **147,** 210–212
location, **147,** 200–206

muscarinic, in control of rhythmic behaviors, **147,** 13–14, 16
nicotinic, calcium channels and, **147,** 205–206, 210, 225
thymic epithelial cells, **156,** 187
Acetylcholinesterase
in avian neurogenesis
brain development
biosynthesis, **151,** 149
early neurogenesis, **151,** 144–145
membrane association, **151,** 146–147
small molecular forms, **151,** 145
subcellular fractions, **151,** 147
cell adhesion molecule function
HNK-1 epitope regulation, **151,** 159–164
neurite growth alteration, **151,** 158, 167–169
pathfinding cues, **151,** 163, 166
characteristics, **151,** 139
embryonic retinal cell study, **151,** 150–152
expression
butylcholinesterase activity related to, **151,** 156–158, 168–169
in differentiation, **151,** 141–143
in migrating cells, **151,** 170
periods, **151,** 141
in proliferation, **151,** 141–143
functional analysis, **151,** 149–151
genes, **151,** 140, 169
migratory cells associated with, **151,** 150
neuromodulatory function, **151,** 141
in avian retina cellular expression, **146,** 15, 42
and cellular processes, **146,** 29–30, 39–40
and normal development, **146,** 4, 8
avian salt glands and, **144,** 159
and calcium channels, **147,** 193, 218
location, **147,** 203–204
motor end plate, **147,** 195–199
effect of electromagnetic field, **158,** 301, 317–318
glutamate metabolism and, **147,** 288
neurotransmission, **149,** 239
in perinatal exposure to sex hormones and antihormones, **139,** 28–29
Acetyl CoA, higher plant plastids, **145,** 187–189, 203
Acetyl CoA carboxylase, higher plant plastids and, **145,** 182–183
N-Acetyl-galactosamine, and M cells
lectin histochemistry in studying, **167,** 103–104
species variations in concentration, **167,** 124–125
N-Acetyl-D-galactosamine, **156,** 172
N-Acetylglutamic acid, **156,** 40
Acheta, karyosphere and, **144,** 18, 21
Acheta domestica
antennae, **174,** 6–7
complex eye, **174,** 5–6
epidermis, **174,** 8
labrum, **174,** 7
mouthparts, **174,** 7–8
Achyla, mitochondrial genomes in, **141,** 92–93
Achyla ambisexualis, mitochondrial genomes in, **141,** 59, 91–92
Acid aerosol, symplast and, **142,** 286
Acid exposure, regenerative cementogenesis and, **142,** 51
α1-Acid glycoprotein, regulation in transgenic mouse, **170,** 249
Acidification
and environmental stimuli in plant cells, **145,** 290, 300
ligand receptor complexes, **167,** 23–27
and NMR, **145,** 42
Acidity, urease in plant cells and, **145,** 83
Acid phosphatase
avian salt glands and, **144,** 144, 147
in macrophage identification, **167,** 132
in reduced Peyer's patch M cell lysosome, **167,** 128
Acipenser transmontanus, animal mtDNA and, **141,** 202–204
Acquired immunodeficiency syndrome
role of testicular macrophage, **149,** 126
transcriptional regulation, **143,** 39, 43–44
Acrocallosal syndrome
chromosomal abnormalities, **158,** 250
clinical features, **158,** 249–250
mouse model, **158,** 258
Acrosin, role in sperm binding, **163,** 268
Acrosome reaction
carbohydrate role, **163,** 281–282
changes concomitant with, **163,** 252
cytoskeletal dynamics and, **144,** 97

Acrosome reaction (*continued*)
in invertebrate sperm, **175,** 37–40
and sperm capacitation, **149,** 2–7
calcium role, **149,** 11, 13, 15–16, 32, 35–36
Na^+-H^+ exchanger role, **149,** 30
potassium role, **149,** 7
sodium requirements, **149,** 20, 23
AcSDKP, *see* Hemoregulatory tetrapeptide
ACTH, *see* Adrenocorticotropic hormone
Actin
Acanthamoeba, **156,** 228
accumulation at dome epithelial bacterial attachment sites, **167,** 127, 129
bacterial flagellar motor and, **147,** 98, 156
characteristics, **149,** 145–150
colocalization with tropomyosin, **170,** 12–13
cortical actin layer, **156,** 208
cortical flow role, **148,** 56, 58
detachment from membranes, **148,** 73
and myosin, interaction in cytoskeletal movement, **148,** 65
polymerization, protein regulation, **148,** 71
response to calcium increase, **148,** 72
retraction in fibroblasts, **148,** 63
cytoplasmic streaming mechanism affected by
calcium sensitivity, **155,** 123–124
identification in plants, **155,** 108
cytoskeleton
catenin, linkage, **165,** 170–172
FAK in assembly of protein complexes connecting integrins to, **167,** 165
MAP kinase activation and, **167,** 174–175
models of cellular tensegrity, **150,** 189–196
molecular links anchoring to focal adhesion, **167,** 162–163
in *Physarum* plasmodium, **156,** 202–213, 253
in regulation of FAK phosphorylation, **167,** 171
tension molding, **150,** 196–201
cytoskeleton dynamics and
current concepts, **144,** 96–98, 100, 102–103
cytoskeletal components, **144,** 103, 105, 108
Dictyostelium, **144,** 110–112
mechanism, **144,** 120
theories, **144,** 87–89, 94
in *Dictyostelium,* **156,** 230
Drosophila muscles and, **143,** 138–139
contractile protein genes, **143,** 94–95, 98–99, 109–114
function, **143,** 133–136
interactions, **143,** 126–127, 131
tropomyosin, **143,** 117
troponin, **143,** 118–119
Z-disc proteins, **143,** 122
encoding gene, **139,** 227, 237
environmental stimuli in plant cells and, **145,** 300–301
light, **145,** 278–280, 283
low temperature, **145,** 289
wounding, **145,** 291–292, 295–296
fibroblast growth factor
effects on
retinal precursor cells, **146,** 176, 178
retinal regeneration, **146,** 56–58, 64, 66–69
expression in
melanocytes, **146,** 240
retinal pigmented epithelium, **146,** 77, 79
fish chromatophores and, **143,** 199, 241, 246
cytoskeleton, **143,** 221, 223–224
iridophores, **143,** 218
in generative cell and sperm
correlative studies, **140,** 162–163
fluorescence and electron microscope evidence, **140,** 161–162
implications, **140,** 163–165
globular, **156,** 227, 244, 255
cytoskeletal dynamics and, **144,** 101–102
gelsolin, **156,** 233–235
myofibrillogenesis and, **143,** 158–160, 164–165, 169
phosphorylated, **156,** 229–230
polymerization, **156,** 214–215, 217
structure, **156,** 213–214
in interphase nucleus, **150,** 165–167
intron, **139,** 248
karyosphere and, **144,** 41
lamellipodia
assembly, **150,** 64

concentration, **150,** 39
retrograde surface transport, **150,** 47, 49, 51–53
treadmilling, **150,** 61–65
as matrix of dendritic spines
description, **139,** 272–277
identification, **139,** 273–277
Mg-polymer, **156,** 217–218
modulators, **139,** 279–280
as motor of polymorphonuclear leukocytes, **139,** 163
myofibrillogenesis and
assembly, **143,** 157–159, 174–175, 177–178
capping proteins, **143,** 172–173
differential assembly, **143,** 179, 181
isoform change, **143,** 153–155
isoforms, **143,** 159–160
polymerization, **143,** 160–172
sarcomere formation, **143,** 181–184
–myosin system
motive force generation
actin identification in plants, **155,** 108
Characeae, **155,** 100–103
myosin identification in plants, **155,** 103–108, 130
regulation, **170,** 2, 17–18
network
in cytoskeletal transport, **148,** 72
in membrane-cytoskeletal linkage, **148,** 51–52
motor pattern in, **148,** 48
separation from cell membrane, **148,** 60–61
nonpolymerized, **156,** 226–230
organization in transformed cells, **170,** 7
Physarum, **149,** 155–158
in axenically cultured microplasmodia, **149,** 182
caffeine droplets, **149,** 175, 178
contraction, fluorescence analysis, **149,** 190
macroplasmodia, **149,** 173
myxamoebae, **149,** 172
protoplasmic streaming, **149,** 191
polymers, comparison with proteins, **164,** 4
in pseudopod formation, **139,** 173
regenerative cementogenesis and, **142,** 40
in regulation of CFTR, **172,** 212–213
retinal precursor cells and, **146,** 156, 164
Sertoli-germ cell communication network and, **147,** 75
in striated muscle
cardiac muscle, **142,** 107–109, 111, 118
sarcomere, **142,** 94–95, 99
thick-filament proteins, **142,** 65
thin-filament proteins, **142,** 83–84, 88–91
transdifferentiation in medusae and, **142,** 227, 232
treadmilling, as retrograde surface transport, **150,** 51–54, 61–65
vinculin binding to, **167,** 174
α-Actin, proteins in striated muscle and, **142,** 83–84, 95–97, 102
Actin-binding proteins
characteristics, **149,** 146–148
and cytoskeletal dynamics
current concepts, **144,** 103
cytoskeletal components, **144,** 104–108
Dicryostelium, **144,** 110–112
theories, **144,** 88–89
high molecular weight, **156,** 240–241
Physarum
52-kDa, **149,** 156, 163
ABP-36, **149,** 156, 163–164
fragmins, **149,** 156, 161–162
function, **149,** 156, 160–164
high-molecular-weight, **149,** 156, 163
as membrane skeletal component, **149,** 168
prolifin, **149,** 156, 162
physiological significance of, **139,** 287–288
in pollen tube cytoskeleton, **176,** 143–145
polymerization-depolymerization and, **139,** 292–299
regulation, **156,** 243–245
regulation of MT–microfilament interactions, **173,** 226–230
Actin cross-linking proteins
in *Dictyostelium discoideum,* **175,** 50–56
Drosophila, mutations, **175,** 46–50
echinoderm coelomocytes and oocytes, **175,** 41–44
in filament bundle formation *in vitro,* **175,** 57–58
Limulus sperm bundle, **175,** 40
in microvillus, **175,** 34–35
in stereocilia, **175,** 37

Actin depolymerizing factor, **139,** 278–279; **143,** 164–170, 172; **156,** 227
Actin filament bundles
acrosome reaction, **175,** 37–40
bundling, **156,** 239–240, 249–252
contraction, **156,** 252–253
in *Dictyostelium discoideum,* **175,** 50–56
in *Drosophila* development, **175,** 44–50
echinoderm coelomocytes and oocytes, **175,** 40–44
elongation factor-1α–F-actin interactions, **164,** 23–24
future studies, **175,** 20
in intestinal brush border
cross-linking proteins, **175,** 34–35
organization, **175,** 30–33
in stereocilia, **175,** 35–37
in vitro formation
nucleation and growth mechanism, **175,** 64–65
ordered-phase model, **175,** 69–71
phase transition model, **175,** 65–68
role of actin cross-linking proteins, **175,** 57–58
role of electrostatic forces, **175,** 57–58
thermodynamic–kinetic model, **175,** 60–61
thermodynamics, **175,** 56–58
in vivo formation, **175,** 71–75
Actin filaments
52-kDA protein, **156,** 240
biochemical characteristics, **176,** 141–143
and cytoskeletal dynamics
current concepts, **144,** 96–98, 100–101
cytoskeletal components, **144,** 106–108
Dictyostelium, **144,** 111–113
mechanism, **144,** 116, 118
environmental stimuli in plant cells and, **145,** 260–261, 277–279
fragmentation, **156,** 230–236
fragmin, **156,** 230–232
gelsolin, **156,** 233–235
and microfilaments
biochemical characteristics of pollen actin, **140,** 87–88
structure of actin molecules and filaments, **140,** 86–87
visualization in pollen, **140,** 88–90
and MT, interdependence, **173,** 218–219
myofibrillogenesis and, **143,** 158–160, 165, 172, 174
phalloidin, **156,** 209–210
in pollen grains
overview, **140,** 96
pollen grains, **140,** 96–100
pollen tubes, **140,** 100–102
in pollen tube cytoskeleton, **176,** 146–149
role in MT–microfilament interaction, **173,** 231–232
severin, **156,** 234
stabilization, **170,** 16–17
structure, **156,** 214–215, 217
transdifferentiation in medusae and, **142,** 226–227
villin, **156,** 235–236
Actin-fragmin kinase, **156,** 244, 255
α-Actinin
binding partners
integrin β-subunit cytoplasmic tail, **167,** 163
vinculin *in vitro* interaction with, **167,** 162, 174
cytoskeletal dynamics and, **144,** 96, 105–106, 111–112
in *Dictyostelium discoideum,* **175,** 51, 53
Drosophila muscles and, **143,** 97, 122–123, 131
isolation, **156,** 241
localization, **156,** 212
in membrane-cytoskeletal linkage, **148,** 51–53
myofibrillogenesis and, **143,** 153, 155
assembly, **143,** 159, 174–175
sarcomere formation, **143,** 182–183
proteins in striated muscle and, **142,** 129
cardiac muscle, **142,** 108–109, 111, 114–115, 117–118
sarcolemma, **142,** 125–126
skeletal muscle, **142,** 87–88, 91, 101
β-Actinin
myofibrillogenesis and, **143,** 172–173
proteins in striated muscle and, **142,** 91–92
Actinobacillus pleuropneumoniae, M cells and, **167,** 122
Actinogelin, **156,** 214, 242
Actinomycin D, transdifferentiation in medusae and, **142,** 227, 229
Action potentials
compound, biological clocks in retina and, **146,** 86, 88, 92, 101, 108–109

cytoplasmic streaming affected by, **155,** 120–125, 130
neurotransmission, **149,** 227, 230
Activation, higher plant plastids and, **145,** 172–176
Activin
effect on ascidian notochord induction, **176,** 275–276
role in chondrogenesis, **159,** 310
secretion by Sertoli cells, **160,** 201
Sertoli-germ cell communication network and, **147,** 46
Activin A, neural induction in amphibians, **169,** 124
Actomyosin
active actomyosin, **156,** 220
Ca^{2+}-sensitivity, **156,** 220–221
cytoskeletal dynamics and
current concepts, **144,** 94, 102
cytoskeletal components, **144,** 105–109
Dictyostelium, **144,** 110, 112
mechanism, **144,** 119
theories, **144,** 87–88
environmental stimuli in plant cells and, **145,** 264–265, 282, 289
fibrils, **156,** 203–205
inactive actomyosin, **156,** 220
Mg-polymer, **156,** 217–218
Physarum
in axenically cultured microplasmodia, **149,** 181–183
characteristics, **149,** 153–155
contraction, **149,** 157–158
protoplasmic drops, **149,** 178
in retrograde surface transport, **150,** 47
superprecipitation, **156,** 247–253, 254
Acute phase proteins
alteration in serum, **170,** 244
antioxidant activity, **170,** 242–243
biosynthesis and function, regulation
in hepatoma cells, **170,** 251–252
in knockout mouse, **170,** 250
in laboratory animals, **170,** 246–248
in normal hepatocytes, **170,** 250–251
in transformed cells, **170,** 252
in transgenic mouse, **170,** 248–249
classification, **170,** 228–229
effect of intracellular events
early events, **170,** 260–262
immediate early events, **170,** 262–264
related phenomena, **170,** 264
function, effect of glycosylation changes, **170,** 267–268
as glycoproteins, **170,** 264–265
glycosylation
alterations in, regulation
animal models, **170,** 268–269
in nontransformed cells, **170,** 269–270
in transfected and transgenic nontransformed cells, **170,** 270–271
in transfected transformed cells, **170,** 274
in transformed cells, **170,** 271–274
effect of cytokines
nontransformed cell-secreted APP, **170,** 275–276
transformed cell-secreted APP, **170,** 274–275
network of affecting factors, **170,** 276–279
group identification, **170,** 227–228
host defense function, **170,** 236–240
interspecies differences, **170,** 229–235
microheterogeneity, **170,** 265–266
major microheterogeneity, **170,** 266
minor microheterogeneity, **170,** 267
production, **170,** 243–244
neuroendocrine control, **170,** 244–246
proteinase inhibition, **170,** 240–242
sex differences, **170,** 235
synthesis, controlling network
cofactors, **170,** 256–257
cytokines, **170,** 252–255
hepatocyte hematopoietin receptor subunits, **170,** 257–259
soluble cytokine receptors, **170,** 255–256
Acute phase response
activation mechanisms, **170,** 225–226
homeostatic settings, **170,** 226–227
major components, **170,** 227
Acyl carrier protein, higher plant plastids and, **145,** 182–185
Adaptation, avian salt glands and, **144,** 158, 161, 192
Adaptin
α-type, association with CFTR, **172,** 217
β-type, encoding gene, **167,** 18
γ-type, in clathrin binding, **167,** 21

Adaptor proteins
  assembly polypeptides, **167,** 14–18
  in clathrin-coated vesicles
    receptors and, **167,** 19, 21
    uncoating ATPase and, **167,** 23
    unresolved problems in plant systems, **167,** 26
  role in signaling, **164,** 199–200
Adducin, **166,** 4
Adenohypophysial cells
  chromophobes as precursor cells, **159,** 49–50, 66
  immunocytochemistry, **159,** 50–51
  intercellular communication, **159,** 50–51
  morphology, **159,** 50
  origin, **159,** 3, 49–50
  separation, **159,** 51
Adenosine
  fish chromatophores and, **143,** 230–234, 239
  retinal precursor cells and, **146,** 154–155
Adenosine deaminase, rotation, magnetic field effect, **158,** 305
S-Adenosylmethionine
  flowering plants and, **143,** 321
  SAM decarboxylase, **138,** 72, 81
Adenovirus
  application in gene therapy, **158,** 264
  E1B–E4 complex, **162B,** 210–213
  transcriptional regulation and, **143,** 38, 44
Adenylate cyclase
  avian salt glands and, **144,** 200–201
  decidual cell reaction and stimulation, **160,** 20–21
  effect of electromagnetic field, **158,** 282, 316–317
  regulation
    CDC25 genes in, **139,** 313–314
    IRA genes in, **139,** 313–314
    RAS genes in, **139,** 311–313
    SRV2(CAPI) genes in, **139,** 313–314
  transdifferentiation in medusae and, **142,** 245
Adepithelial cells, *Drosophila* muscles and, **143,** 71
ADF, *see* Actin depolymerizing factor
Adherens junction-specific cell adhesion molecule, **142,** 126
Adhesins, in bacterial binding to M cells, **167,** 124–126
Adhesion
  in avian retina histogenesis, **146,** 9, 21–24, 27, 41
  cellular, *see* Cell adhesion
  differential, hypothesis
    avian retina and, **146,** 21
    experimental confirmation, **150,** 147
  effect on membrane flow in amoebae, **148,** 43
  melanocytes, **146,** 216
  plaques, vertebrate cell culture technology and, **142,** 187
  in retinal pigmented epithelium, **146,** 75
  retinal precursor cells, **146,** 179
Adhesion proteins, plant analog, **174,** 272
Adiantum, stimuli in plant cells and, **145,** 269, 273–274, 283, 301
Adipokinetic hormones, insect, characterization, **164,** 228–229
ADP
  avian salt glands and, **144,** 135
  chromoplast, **163,** 207–208
  Drosophila muscles and, **143,** 133–134
  effect on flagellar beat cycle, **173,** 30–31
  glutamate metabolism and, **147,** 295, 302
  higher plant plastids and, **145,** 174–177, 180–182, 190
ADP-ribosylating toxins, Gram-positive bacteria, **161,** 274–275
ADP ribosylation
  histone, **162A,** 212–213
  intracellular calcium messenger system and, **142,** 309–311
  neurofilament protein modulated by, **154,** 43
ADP ribosylation factor
  in CFTR recycling, **172,** 219
  in secretory vesicle formation, **172,** 228
Adrenal cortex
  effect of endothelin
    on cell hypertrophy and steroidogenic capacity, **171,** 286–288
    on cell proliferation, **171,** 288–289
    mechanism, **171,** 289–290
    role of ET receptor subtypes, **171,** 290–291
  nephron differentiation into cells, **148,** 82
Adrenal gland
  avian salt glands and, **144,** 189, 191–192
  blood flow

effect on endothelin release, **171,** 291–292
regulation by endothelin, **171,** 284–285
endothelin synthesis, **171,** 268–269
Adrenal medulla, release of catecholamine, stimulation by ET, **171,** 282–283
Adrenergic system, fish chromatophores and, **143,** 226, 228, 230–231
Adrenoceptors, fish chromatophores and
motility control, **143,** 227–230, 232, 237
signal transduction, **143,** 240–241, 244
Adrenocorticotropic hormone
avian salt glands and, **144,** 192, 198
control of amphibian metamorphosis and, **145,** 105, 113, 116, 119–123, 122–126, 128
evolution, **159,** 52, 56
fish chromatophores, **143,** 241, 250
light exposure effect on plasma levels, **159,** 56
localization, **159,** 58
onset of synthesis, **159,** 49, 52
as POMC component, **170,** 81–82
processing, **159,** 10, 46
production in lymphocytes, **170,** 83–85
release, **159,** 36
role in immune response and stress, **170,** 87–93
secretion
brain neurons, **159,** 46
effect of hypothalamic lesions, **159,** 48
feedback mechanisms, **159,** 56
pituitary cells, **159,** 4, 55–56
stress response, **159,** 56, 91
Tetrahymena, **159,** 46
storage granules, **159,** 87
structure-function relationships, **159,** 47
Adult T-cell leukemia, transcriptional regulation and, **143,** 42–43
Aedes, karyosphere and, **144,** 9, 18, 36
Afferentation, *see* Deafferentation
African clawed toad, microgravity experiments, **156,** 334
Agaricus, mitochondrial genomes in, **141,** 103–105
Aggregation
MIP-1α molecules, **167,** 199–200
secretory vesicle, **176,** 173–174
Aging
effect on muscle fiber-type transitions, **170,** 186–188
incidence curves, melanocytes and, **146,** 225–226
lens epithelial cells, **163,** 312, 323
maternal, aneuploidy and
autosomal monosomy in
preimplantation embryos, **167,** 267–268
effect of, **167,** 276–279, 289
maternal, effect on aneuploidy, **167,** 278
normal, PHF-Tau proteins in, **171,** 197–198
versus number of Peyer's patches, **167,** 94–95
paternal, and aneuploidy
epidiological studies, **167,** 280
molecular studies, **167,** 280–281
overview, **167,** 279–280
plant cell, effect of microgravity and clinostating, **171,** 53–56
regenerative cementogenesis and, **142,** 17
related loss of photoreceptor cells in Fischer rat, **138,** 37–39
related macular degeneration, transplantation studies, **138,** 41–42
Agitation, vertebrate cell culture technology and, **142,** 196–197
Aglycon, as influenza A virus inhibitor, **175,** 165–166
AGM-1470, therapeutic use in vascular disease, **161,** 32
AGP, *see* Arabinogalactan proteins
Agravitropism, *see* Gravitropism, plants
Agressin, *see* Suppressors
Agrobacterium
mediated gene transfer, **140,** 427–428
mediated transformation, **140,** 12, 434–435, 437–438
plant-pathogen interactions in, **144,** 62–63, 66
Agrobacterium rhizogenes, genetic tumors caused by
DNA sequence analysis, **152,** 288–289
interconversion mechanism, **152,** 287
Agrobacterium tumefaciens
cell wall changes under microgravity, **170,** 62
chalcone synthase and, **147,** 263

Agrobacterium tumefaciens (*continued*)
flowering plants and, **143,** 317
tumorous outgrowth induction by, **152,** 265–266
AIDS, *see* Acquired immunodeficiency syndrome
Airlift reactors, vertebrate cell culture technology and, **142,** 169–170
Airplanes, parabolic flight, **156,** 349–352
Albinism, melanocytes and, **146,** 202–205, 209, 212
Albumin gradient, **166,** 149
Aldosterone
avian salt glands and, **144,** 192
control of amphibian metamorphosis end, **145,** 119–120, 122–123, 125–127
tumors producing, role of endothelin, **171,** 293–294
Aleurone, chalcone synthase and, **147,** 254, 267, 271–274
Alfalfa
effect of KNO_3, **156,** 39
Nod factor effects, **156,** 48–49
nodulation
after inoculation, **156,** 9
gene mutations effects, **156,** 31
role of *nod* gene, **156,** 27–28
stages, **156,** 3
supernodulation mutants, **156,** 42
non-nodulating mutants, **156,** 41
Algae
brown, sperm-binding proteins, **163,** 270, 278
cell walls, antibodies to, **171,** 105–107
chloroplasts and, **144,** 259–260, 295
cryptomonads, **144,** 275
dinoflagellates, **144,** 286, 288–289, 291
ciliates and, **141,** 58
cyanelle genome
chloroplast, **151,** 36
plastid, **151,** 37
effects of microgravity, **171,** 25–30
endosymbiont hypothesis and, **141,** 278, 296–297
eukaryotes, **141,** 276–277
mitochondria, **141,** 327
molecular biology, **141,** 265
nuclear genome, **141,** 252
phylogenies, **141,** 273–275
plastics, **141,** 254–255, 261–262, 279
sequences, **141,** 271–272
green, *see* Green algae
heterokont, *see* Heterokont algae
intracellular calcium messenger system and, **142,** 335–336, 338
microgravity experiments, **156,** 343
mitosis and, **144,** 221–222, 224
red, *see* Red algae
volvocine
evolutionary implications, **139,** 246–258
codon usage patterns and, **139,** 251–258
general considerations for, **139,** 246
intron-exon structure and, **139,** 246–251
experimental analysis of sexual induction in, **139,** 218–224
cAMP levels and effects in, **139,** 220–221
general considerations for, **139,** 218–219
inducer action and, **139,** 219, 223–224
pheromone target and, **139,** 219–220
roles of extracellular matrix in, **139,** 221–223
general considerations for, **139,** 258
genetic and experimental analysis of germ/soma differentiation in, **139,** 204–218
asymmetric division and cell size in, **139,** 208–211
cell type-specific genes in absence of asymmetric division in, **139,** 214–215
cytological control of division symmetry in, **139,** 216–217
differential gene expression in, **139,** 211–214
model for genetic control in, **139,** 206–208
mutations in, **139,** 204–206
molecular analysis of gene structure and expression in
general considerations for, **139,** 224
genes encoding ubiquitous proteins in, **139,** 225–243
genes encoding *Volvox*-specific proteins in, **139,** 243–246
ribosomal genes in, **139,** 224–225
as ontogenetic and phylogenetic model

general considerations for, **139,** 189–191
lineage hypothesis and, **139,** 199–204
organizational and reproductive patterns in, **139,** 191–200
tree diagram in, **139,** 201, 203
Alkaline phosphatase
effect of retinoic acid, **159,** 305
expression by villus *versus* dome epithelial enterocytes, **167,** 131
inactivation of plant enzymes, **149,** 75–76
induction by ascorbic acid, **159,** 297
in M cell identification, **167,** 102
and regenerative cementogenesis, **142,** 15, 40–41
role in cartilage mineralization, **159,** 283–284, 297
in yeast vacuole, **139,** 71–74
Alkalinization
intracellular calcium messenger system and, **142,** 307
symplast and, **142,** 275
transdifferentiation in medusae and, **142,** 245
n-Alkanes, from different pollen species, composition of, **140,** 40
Alkylation
anthracyclines, **156,** 131–132
associated agents
interaction with nuclear matrix proteins, **162A,** 545–547
role in DNA repair, **156,** 122–123, 129–131
O^6-Alkylguanine, repair, **156,** 123–124, 126
Allantoate, urease in plant cells and, **145,** 75–77
Allantoic acid, urease in plant cells and, **145,** 73–74, 76, 78, 95
Allantoin, urease in plant cells and, **145,** 73, 75–79, 95–96
Allatostatins, insect, characterization, **164,** 227–228
Alleles, *see also* Genes
Arabidopsis thaliana and, **144,** 56, 58, 62–63, 67–68
chalcone synthase and, **147,** 252, 256, 258-261, 266, 275
Drosophila muscles and, **143,** 139–140
actins, **143,** 113–114
contractile protein genes, **143,** 101, 104, 123, 135
development, **143,** 73, 86–87
interactions, **143,** 131–132
tropomyosin, **143,** 116
troponin, **143,** 118, 120–121
flowering plants and, **143,** 307–308, 315, 322–323
higher plant mitochondrial genomes and, **141,** 131–134
inactivation
mating-type switches in yeast, **160,** 61–62
methylation mechanisms, **160,** 70–73
melanocyte
and development, **146,** 216, 218–219
and pigment-cell genetics, **146,** 203, 206–208, 210–211
mitochondrial genomes in fungi and, **141,** 107, 109
self-incompatibility and, **143,** 257–258, 289–290
gametophytes, **143,** 280, 282, 286–287
sporophytes, **143,** 259–260, 264, 269, 273–274, 277–278
Allergic reaction
dendritic cell role in, **153,** 90
in humans, Type 1, **140,** 9
Allium sepa, environmental stimuli in, **145,** 291
Allopurinol, urease in plant cells and, **145,** 77–79, 96–97
ALP, *see* Alkaline phosphatase
Alpha mating factor precursor protein, in yeast, **167,** 81
ALS–PDC, *see* Amyotrophic lateral sclerosis–Parkinsonism dementia complex
Alternaria, intracellular calcium messenger system and, **142,** 308
Alu repeats, attachment to matrix, **162A,** 341–342
Alveolar bone, regenerative cementogenesis and
established root surfaces in *vitro,* **142,** 33, 36, 40–41
established root surfaces *in vivo,* **142,** 44, 46–47, 49–50
function of cementum, **142,** 3–5
Alzheimer's disease
GLUT1 expression in, **172,** 36
Tau protein aggregation, **171,** 191–195

Alzheimer's disease (*continued*)
Tau protein phosphorylation, **171,** 188–191
Amacrine cells
in avian retina, **146,** 2–3, 5, 16, 39
retinal precursor cells and, **146,** 148
Amaranthus tricolor, intracellular calcium messenger system and, **142,** 333
Ambrosia artemisiifolia, **140,** 10
Ambystoma gracile
gonadotropin-releasing hormone, **138,** 97–98
hormonal control of metamorphosis, **145,** 107
Ambystoma mexicanum
hormonal control of metamorphosis and, **145,** 119, 125
lack of carnosine immunoreactivity, **138,** 171
neural plate and head region, **138,** 274
opiate receptors, **138,** 148
somatostatin cell location, **138,** 115
thyroxine secretion, **138,** 111
Ambystoma tigrinum, hormonal control of metamorphosis, **145,** 107, 119, 124, 135
Amelanotic melanophores, cytophysiology of, **143,** 197
Amelogenesis imperfecta
candidate genes, **158,** 252–253
clinical features, **158,** 250–251
crystal structure in, **172,** 152
incidence, **158,** 250–251
X-linkage, **158,** 251–252
Amethopterin, **156,** 108
Amides
higher plant plastids and, **145,** 167
urease in plant cells and, **145,** 74, 77
Amidinohydrolases, urease in plant cells and, **145,** 74–75, 79–80
Amidohydrolases, urease in plant cells and, **145,** 74–76
Amiloride, avian salt glands and, **144,** 176
Amino acids
γ-aminobutyric acid metabolism and, **147,** 311, 317–318
animal mtDNA and, **141,** 180–186, 192, 206
Arabidopsis thaliana and, **144,** 73, 75
aromatic, flowering blants and, **143,** 309–318
bacterial flagellar motor and, **147,** 108, 132
calcium channels and, **147,** 205–206, 221
as cellular osmolytes, **161,** 197–202
chalcone synthase and, **147,** 241, 243, 245, 272
chloroplasts and, **144,** 281
ciliates and, **141,** 36–38, 41
cytoskeletal dynamics and, **144,** 97, 106–108
Drosophila muscles and, **143,** 81, 108, 134, 139
actins, **143,** 109, 112
contractile protein genes, **143,** 94, 98, 101, 103
myosin light chains, **143,** 105, 107–108
tropomyosin, **143,** 115–116
troponin, **143,** 118–120, 122
eggshells and biomineralization and, **145,** 220, 223, 231, 234, 237
endosymbiont hypothesis and, **141,** 249
mitochondria, **141,** 304, 307, 309
plastids, **141,** 268–269, 278
environmental stimuli in plant cells and, **145,** 297
excitatory, stabilization of microtubule-associated proteins, **151,** 126–127
extracellular matrix and, **146,** 267, 269, 273, 279–280
fish chromatophores and, **143,** 237, 249
flowering plants and, **143,** 297–298, 304, 325
aspartate, **143,** 318–321
auxin, **143,** 317–318
biosynthesis, **143,** 309–310
pathways, **143,** 318, 323–324
proline, **143,** 321–323
shikimate pathway, **143,** 310–313
tryptophan, **143,** 313–317
glutamate metabolism and, **147,** 285–286, 322–323
aminotransferases, **147,** 287, 289, 291
glutamate dehydrogenase, **147,** 298–301
glutaminase, **147,** 309, 311
glutamine synthetase, **147,** 304–306
GnRH-encoding gene in teleosts and, **147,** 166, 181, 185
higher plant mitochondrial genomes and, **141,** 149, 159, 163–164
higher plant plastids and, **145,** 204
fatty acid synthesis, **145,** 185

structure, **145,** 158–159
synthesis, **145,** 190–201
translocation, **145,** 167–169, 171
hormonal character, **155,** 23
hormonal control of amphibian metamorphosis and, **145,** 116, 127, 137–138, 139
intracellular calcium messenger system and, **142,** 309–311, 328, 333
kinetoplastid mtDNA and, **141,** 71, 82
light-dependent incorporation into proteins, **146,** 107
mitochondrial genomes in fungi and, **141,** 118–119
myofibrillogenesis and, **143,** 154, 163, 166–167
neurotransmission, **149,** 255–266
nuclear magnetic resonance and, **145,** 37, 45
proteins in striated muscle and, **142,** 91
as regulatory metabolite s of plant enzymes, **149,** 87–88
retinal precursor cells and, **146,** 148, 153, 155, 175, 181
self-incompatibility and, **143,** 258
gametophytes, **143,** 284, 286
sporophytes, **143,** 268–270, 272–276
sequence homology with gonadotropic hormone receptors, **155,** 27
sequences at PEPC phosphorylation site, **149,** 62–63
Sertoli-germ cell communication network and, **147,** 40
substitution rates for various proteins, **140,** 500
transcription, **141,** 220–221
transcriptional regulation and, **143,** 29–31, 33–34, 37–38, 47
transport, role of V-ATPase, **174,** 90–91
urease in plant cells and, **145,** 98
biochemical genetics, **145,** 86, 89, 91
elimination, **145,** 81, 83
metabolic origins, **145,** 67–68, 74, 77
Amino acid sequence
Antho-RFamide precursor
from *Anthopleura,* **167,** 63*f*
from *Calliactis,* **167,** 59–61
from *Renilla,* **167,** 64–65*t*
Antho-RPamide I precursor, from *Anthopleura,* **167,** 68*t*
chalcone synthase, **147,** 238–241
chalcone synthase and stilbene synthase
alignment, **147,** 244
comparison, **147,** 243
in clathrin-coated vesicle proteins, **167,** 20
comparisons, **140,** 538, 539, 541, 542–543, 544
elongation factor-1α, **164,** 21
gonadotropin-releasing hormone variants, **147,** 167
human *versus* murine MIP-1α homology, **167,** 199
insect amine receptors, **164,** 238
MMA precursor, **167,** 71, 72*t*
Pol-RFamide precursor, **167,** 74–75
salmon gonadotropin-releasing hormone prepro-hormone, **147,** 178–180
SLGs in *Brassica* series, **140,** 529–532
Tetrahymena 14nm filament-associated protein, **164,** 11, 21
XRP catalytic site, **173,** 184–186
XRPs, **173,** 182–184
Aminoacridines, **156,** 97, 101
γ-Aminobutyric acid
chromatin reorganization by, **150,** 161–162
control of rhythmic behaviors and, **147,** 14
effect on prolactin release, **159,** 64–65
fish chromatophores and, **143,** 237
input to dendritic spines, **139,** 297–299
metabolism, **147,** 285–287, 311–312, 322–323
GABA transaminase, **147,** 316–321
glutamate decarboxylase, **147,** 312–316
glutamate dehydrogenase, **147,** 299–300
glutaminase, **147,** 309, 311
glutamine synthetase, **147,** 304–306
as neurite inhibitor, **153,** 257
neurotransmission, **149,** 256–257, 261–262, 265
regulation of pars intermedia cells, **159,** 53
retinal precursor cells and, **146,** 153, 155, 164, 181
systemic acquired resistance induction, **168,** 326
γ-Aminobutyric acid receptor
characterization, **164,** 193–195
molecular cloning, **164,** 195
neurotransmission, **149,** 264–266

γ-Aminobutyric acid transaminase, neurotransmission and, **147,** 300, 305, 311–312, 316–321
1-Aminocyclopropane-1-carboxylate synthase, auxin-induced expression, **152,** 132–133
Aminoethoxyvinylglycine, **156,** 39, 46
Aminoglycosides, nematocyst discharge, **156,** 290
δ-Aminolevulinate synthase, expression in Harderian gland, **168,** 44
Aminopeptidase I
  biosynthetic routes of, **139,** 67, 69–70
  transcriptional regulation and, **143,** 12–13
Aminopeptidases, *see also* Dipeptidyl aminopeptidase; Processing enzymes
  in Antho-RFamide biosynthesis, **167,** 59, 64–67
  in Antho-RPamides II-IV biosynthesis, **167,** 69
  cnidarian neuropeptide resistance to
    L-3 phenyllactyl group providing, **167,** 49
    pyroglutamyl N-terminal group providing, **167,** 48
    X-Pro sequences on N-terminals, **167,** 50
  in cnidarian preprohormone processing, **167,** 78–80
Aminopyridines, calcium channels and, **147,** 206
Amino terminus
  cnidarian G-protein-coupled receptors, **167,** 56
  cnidarian neuropeptides
    Antho-RFamide, **167,** 59–62, 64–67
    Antho-RPamides II-IV, **167,** 70*t*
    copies of immature MMA, **167,** 71
    Pol-RFamides, **167,** 76*t*
    preprohormone processing, **167,** 78, 81–82
    protective residues on, **167,** 48–50, 68
  collagen V, molecular structure, **173,** 96–97
  FAK, **167,** 165–167
  variability in Rab GTPase, **176,** 15–16
Aminotransferases, glutamate metabolism and, **147,** 287–294, 322
Amitosis
  amoebas, **158,** 77
  cultured plant cells, **158,** 76
Amitrole, induction of chromoplast differentiation, **163,** 204
Ammonia
  γ-aminobutyric acid metabolism and, **147,** 286, 301–304, 306–307
  higher plant plastids and, **145,** 191–201
  urease in plant cells and, **145,** 66, 74, 83–84, 89
  vertebrate cell culture technology and, **142,** 164–165, 173, 195
Ammonium
  flowering plants, **143,** 298
    assimilation, **143,** 303–309
    nitrate, **143,** 299, 301–303
  higher plant plastids and, **145,** 197, 199
  nuclear magnetic resonance and, **145,** 3
Amniocentesis
  miscarriage risk, **158,** 262
  prenatal diagnosis of craniofacial dysmorphologies, **158,** 262
Amniotes, neurulation
  morphogenesis types, **175,** 249
  studies, **175,** 248–249
Amniotic fluid, nuclear magnetic resonance and, **145,** 37
Amoeba, *see also specific amoebae*
  actin network in interior, **148,** 46
  actin polymerization in, **148,** 67–69
  capping in, **148,** 43
  cell motility, *see* Cell motility, amoeboid
  chloroplasts and, **144,** 270
  cortical contraction, **148,** 60, 63–64
  cortical cytoskeleton, centripetal movement, **148,** 53–55
  cortical flow in, **148,** 56–57
  cytoskeleton
    detachment, **148,** 61–63
    dynamics, regulation, **148,** 70–72
    membrane linkage to, **148,** 50–52
    membrane protein hauling, **148,** 47–50
    network contraction, **148,** 72
    retraction, myosins in, **148,** 64–67
  gelation and contraction of cytoplasmic fraction, **156,** 236–239
  gel–sol transformation, **156,** 241–243
  giant amoeba, **156,** 199, 201
  granular cytoplasm, **156,** 210
  hormone effects, **155,** 2
  locomotion, **148,** 37
    rolling membrane, **150,** 56–57

shelled amoebae, **150,** 58–59
tissue culture cell locomotion comparisons, **150,** 35–36, 61–62
membrane flow in, **148,** 73
early theories, **148,** 38–40
and in tissue cells, theories, **148,** 41–43
motor responses to relaxation, **148,** 59–60
myosin, **156,** 221–223
related movement of polymorphonuclear leukocytes *in vitro,* **139,** 157–182
release of contractile cortex from cells, **148,** 70
Amoeba proteus
calcium distribution in, **148,** 70
cortical flow, **148,** 56
cytoplasmic fraction, **156,** 236–239
cytoskeleton
cortical cytoskeleton, **148,** 53–55
detachment, **148,** 61
dynamics, regulation, **148,** 71–72
–membrane interactions, **148,** 50–52
retraction, myosins in, **148,** 64–67
locomotion, **150,** 54–56, 61
membrane, **148,** 39–40
phosphorylated actins, **156,** 230
profilin, **156,** 230
relaxation by anesthetics, **148,** 59–60
Amorpha coffaeformis, template formation, **150,** 260
Amorphin, myofibrillogenesis and, **143,** 182
Amphenone, control of amphibian metamorphosis and, **145,** 120
Amphibians, *see also specific amphibians*
avian salt glands and, **144,** 129
biological clocks in retina, **146,** 84–85
brain, **138,** 90–91
neuropeptides, *see* Neuropeptides, in amphibian brain
development, comparison, **176,** 291–292
differentiation systems, **154,** 61–62
egg surface sperm-binding proteins, **163,** 263–264
embryonic tissue, mechanical stress in
cell surface angle measurement, **150,** 3–7
gastrulation, **150,** 29–31
morphogenetic effects, **150,** 8–11
neural tubes affected by relaxation of stress, **150,** 10–11
neurulation, **150,** 29–31
relaxation of tensile stresses, **150,** 9
tissue dissection as detection method, **150,** 2–5
tissue stretching, **150,** 8
growth factor-induced retinal regeneration, **146,** 49, 51, 54–55
hormonal control of metamorphosis, *see* Hormones, control of amphibian metamorphosis
karyosphere and, **144,** 26–27
microgravity experiments, **156,** 334
mitosis and, **144,** 247
parathyroid hormone-related protein, **166,** 263–264
retinal precursor cells, **146,** 161, 180, 182
Amphidinium
chloroplasts and, **144,** 288–290
mitosis and, **144,** 223
Amphotericin
osmotic conductance (P_{os}), **163,** 3–5
osmotic flow, bimodal analysis, **163,** 30–34
Amplification, *see* Gene amplification
AMSA, *see* Aminoacridines
Amsacrine, drug resistance, **156,** 79, 81, 97, 101–103
Amyloids
as cofactor in Tau protein aggregation, **171,** 193–194
transcriptional regulation and, **143,** 18
Amylopectin, chloroplasts and, **144,** 265
Amyloplasts
higher plant
metabolism, **145,** 169, 171, 203
starch synthesis, **145,** 177, 179–180
structure, **145,** 151, 161, 163
microgravity experiments, **156,** 329
Amyotrophic lateral sclerosis–Parkinsonism dementia complex, **171,** 196–197
Anabaena azollae, effect of microgravity, **156,** 342; **171,** 30
Anacystis nidulans
endosymbiont hypothesis and, **141,** 269, 297
mitochondrial genomes of, **141,** 43, 50, 58
Anacystis tumefaciens, mitochondrial genomes of, **141,** 43, 58
Anaphase-telophase, **140,** 174–175
Anas platyrhynchos, avian salt glands and, **144,** 130, 133
Anastomoses, avian salt glands and, **144,** 145–146, 148

Anatomical abbreviations, **138,** 179
Anchoring, cytoplasmic, **166,** 12
Androgen, Sertoli-germ cell communication network and, **147,** 61
Androgen-binding protein, Sertoli cell, **160,** 201, 281–282
Androgenization, neonatal rats
  decidual cell reaction, **160,** 5
  fertility, **160,** 3–5
Androgen receptor
  accessory factors, **162B,** 362
  Harderian gland, **168,** 41, 43, 52–53
Anemia, aplastic, increased MIP-1α mRNA in, **167,** 238–239
Anemones
  hair bundles, **156,** 287–289
  nematocyst discharge, **156,** 284, 292
Aneuploidy
  chromosomal mosaicism in humans
    incidence and origin, **167,** 287–288
    mechanism of formation, **167,** 288–289
    overview, **167,** 286–287
  conclusions, **167,** 289–291
  etiology
    aberrant genetic recombination, **167,** 282–283
    environmental factors, **167,** 283
    genetic factors, **167,** 283–286
    maternal age, **167,** 276–279, 289
    mechanism of nondisjunction, **167,** 275–276
    overview, **167,** 274–275
    paternal age, **167,** 279–282
  human syndromes, **158,** 222
  incidence
    at conception, **167,** 270–272
    in livebirths, **167,** 265
    in oocytes, **167,** 269
    in preimplantation embryos, **167,** 267–268
    in sperm, **167,** 269–270
    in spontaneous abortions, **167,** 265–267
    in stillbirths, **167,** 265
  maternal age effect in, **167,** 278
  mechanisms in plants, **158,** 117–118
  origin, **167,** 272–274
  overview, **167,** 263–265
ANF, *see* Atrial natriuretic factor
Angell theory, precocious separation
  compatibility with Hawley's first hypothesis, **167,** 289–291
  description, **167,** 275–276
Angelman syndrome
  chromosomal deletions, **160,** 84
  description, **167,** 289
  gene imprinting role, **160,** 85
  prenatal screening with FISH, **153,** 30, 32–33
Angiogenesis
  adult processes, **159,** 320
  avian salt glands and, **144,** 151
  effectors
    angiogenin, **159,** 134–135
    fibroblast growth factor, **159,** 129, 135–136
    gangliosides, **159,** 139
    haptoglobin, **159,** 138
    hyaluronic acid fragments, **159,** 138–139
    interleukin 8, **159,** 137
    phosphatase inhibitors, **159,** 137–138
    transforming growth factor β, **159,** 134
    tumor necrosis factor α, **159,** 129
    vascular endothelial growth factor, **159,** 136–137
  extracellular matrix role, **161,** 3
  inhibition in angiogenic disease, **161,** 28
  melanoma and, **146,** 240
  models
    aortic ring model, **159,** 131, 133
    chicken chorioallantoic membrane assay, **159,** 115–116
    coculture models, **159,** 131
    corneal neovascularization model, **159,** 116–117
    human amnion, **159,** 133
    mesenteric window assay, **159,** 118
    pouch assay, **159,** 117–118
    spontaneous angiogenesis, **159,** 119–120
    three-dimensional gel assay, **159,** 121–129
  role
    actin binding, **159,** 129
    fibroblast growth factors, **146,** 56
    heat shock protein, **159,** 90, 119
    integrins, **159,** 128–129
    laminin, **159,** 128
    protein kinase C, **159,** 121, 125
  stages, **159,** 113–114, 319
  stimulation in angiogenic disease, **161,** 27–28
  tumor, **159,** 114, 145

Angiogenesis inhibitors
cartilage factors, **159,** 141, 320
collagen synthesis inhibitors, **159,** 140
cyclosporin A, **159,** 140–141
estrogen metabolites, **159,** 144–145
fumagellin, **159,** 139
genisten, **159,** 145
heparin, **159,** 141
hyaluronic acid, **159,** 142–143
linomide, **159,** 142
oligosaccharide processing inhibitors, **159,** 143
platelet factor 4, **159,** 141–142
protamine, **159,** 140
ribonuclease inhibitors, **159,** 142
steroids, **159,** 141
thrombospondin, **159,** 143–144
tumor suppressor genes, **159,** 143
Angiogenic modulator 1470, inhibition of cell growth, **161,** 21–22
Angiogenin
binding protein, **159,** 135
ribonuclease activity, **159,** 134
role in angiogenesis, **159,** 134–135
stimulation of vascular cells, **161,** 8–9
Angiosperms
cinnamoyl-CoA reductase substrate specificity in, **172,** 263
cinnamyl alcohol dehydrogenase polymorphism in, **172,** 265
4-coumarate-CoA ligase activity in, **172,** 259–260
division of generative cell in, **140,** 149–150
double fertilization in, **140,** 319–320
effects of microgravity
biochemical content and enzyme activity, **171,** 38–39
cells of generative organs, **171,** 41–42
cell and tissue cultures, **171,** 42–44
intracellular calcium balance, **171,** 39–40
leaf photosynthetic cells, **171,** 36–38
microviscosity and lipid peroxidation, **171,** 40–41
organ culture, **171,** 42
protoplast culture, **171,** 44–46
respiration–heat discharge, **171,** 40
root elongation and differentiation, **171,** 34–35
root meristem, **171,** 32–34
seed germination and seedling growth, **171,** 32
shoot apical meristem, **171,** 35–36
shoot primary structure, **171,** 36
vegetative propagation, **171,** 41
embryonal sac cell, **165,** 104–105
embryo sacs, isolation, **140,** 272–279
evolutionary relationships of double fertilization in, **140,** 350–354
and gymnosperms, differences, **176,** 91–93
higher plant plastids and, **145,** 154, 158, 161
lignin types in, **172,** 245
major patterns of megagametophyte development found in, **140,** 234, 235
synergic-lacking, pollen tube arrival in, **140,** 370
Angiostatin, inhibition of endothelial cell growth, **161,** 22–23
Angiotensin, **166,** 25
Angiotensin II
astrocyte reaction to, **148,** 127–128, 131
avian salt glands and, **144,** 189, 192–196, 199
control of amphibian metamorphosis and, **145,** 127, 129
stimulation of vascular cells, **161,** 14
Angiotensin-converting enzyme inhibitors, effect on smooth muscle cells, **161,** 24–25; **169,** 228
Angiotensinogen, transcriptional regulation and, **143,** 18
Animal cells
clathrin-coated vesicles in
cDNA sequencing, **167,** 13
cross-reactivity to plant clathrin proteins, **167,** 10
internal acidity of, **167,** 25
triskelions, **167,** 11
microtubule-organizing center, **175,** 94–97
Animal models
extracellular matrix in, **140,** 299–300
gametic fusion in, **140,** 372–375
fertilization in animals, **140,** 372–373
mechanism of membrane fusion, **140,** 373–375
microgravity effects, **156,** 320, 334–339, 344–348, 352, 354–356
oscillator mouse, **176,** 231–232

Animal models (*continued*)
recognition and gamete fusion in, **140,** 408–414
spasmodic mouse, **176,** 231–232
spastic mouse, **176,** 233–234
sperm capacitation in, **140,** 409–411
Anionic site, polyethylenimine administration, **166,** 214–215
Anise, microgravity experiments, **156,** 327, 328
Anisomycin, biological clocks in retina and, **146,** 106–107, 110, 118, 121, 132
Ankyloglossia
association with cleft palate, **158,** 255
linkage analysis, **158,** 255–256
Ankylosis, regenerative cementogenesis and, **142,** 44–47, 50
Ankyrin, transcriptional regulation and, **143,** 2, 25–26, 30–31
Ankyrin repeat proteins, transcriptional regulation and, **143,** 36–39
Anlage, early neural, neuroepithelium, transplantation experiments
regional specifications, **175,** 271–272
tectal and cerebellar structures, **175,** 266–271
transplantation effects, **175,** 264–266
Anlagen, central body, and karyosphere, **144,** 35
Annelid cuticle, extracellular matrix, **166,** 73–74
Annuli, karyosphere and, **144,** 9
Anolis, fish chromatophores and, **143,** 243
Anolis carolinensis, fish chromatophores and, **143,** 240
Anopheles gambiae, karyosphere and, **144,** 9–10, 18
Anoxia, symplast and, **142,** 271–273, 275–276, 286–287
Antennae, insect, form diversity, **174,** 6–7
Antennal cuticle, genetic control, **174,** 37
Anther tapetum, significance of for biosynthesis of pollen wall phenolic, **140,** 49–50
Anthocyanidins, chalcone synthase and, **147,** 245
Anthocyanin, chalcone synthase and control of activity
accumulation in flower, **147,** 254–255
control of gene, **147,** 271–276
enzymatic activity, **147,** 241
functions of products, **147,** 245–247
production inhibition, **147,** 257–261
Antho-KAamide neuropeptide, **167,** 53–54
Antho-LWamide neuropeptide, **167,** 71–74
Anthopleura elegantissima, see also Sea anemones
in cloning of G-protein-coupled receptors, **167,** 55–56
isolation of neuropeptides in
Antho-RFamide, **167,** 58, 61–62
Antho-RPamide I, **167,** 68–69
Antho-RPamides II-IV, **167,** 69–71
C-terminal sequences, **167,** 50
metamorphosis-inducing extract in *Hydractinia,* **167,** 55, 71–74
Antho-RFamide neuropeptides
biosynthesis
in sea anemones, **167,** 58–64
in sea pansies, **167,** 64–67
effects on muscle groups, **167,** 54
purification from sea pansy, **167,** 48
Antho-RIamide neuropeptide, **167,** 53–54
Antho-RNamide neuropeptide
L-3-phenyllactyl N-terminal, **167,** 49
effects on muscle groups, **167,** 53–54
Antho-RPamide neuropeptides
biosynthesis
neuropeptide I, **167,** 68–69
neuropeptides II-IV, **167,** 69–71
diversity of action, **167,** 77–78
effects on muscle groups, **167,** 53–54, 68
Antho-RWamide neuropeptides
immunoreactivity in RFamide radioimmunoassay, **167,** 49
neuromuscular transmission in sea anemones, **167,** 51
as transmitters at neuromuscular junctions, **167,** 54
Anthozoa, *see also* Corals; Sea Anemones; Sea pansies
Antho-RFamide occurrence in, **167,** 64
centralization and complexity of nervous system, **167,** 47
life cycle, **167,** 38
Anthracenedione derivatives, **156,** 79–80
Anthracyclines
alkylating anthracyclines, **156,** 131–132
in MDR, **156,** 79–80, 100, 101
in MDR reversal, **156,** 96
ME2303, **156,** 92

MX2, **156,** 92
targeting of DNA topoisomerases, **156,** 97–98
Anthranilate, flowering plants and, **143,** 313–316
Anthranilate synthase, flowering plants and, **143,** 316–317
Antiauxins, habituation in tumorous plant cells, induction, **152,** 269
Antibiotic resistance, ciliates and, **141,** 2–3
Antibiotics
higher plant plastids and, **145,** 160
and mitosis, **144,** 235
sensitivity, endosymbiont hypothesis and, **141,** 267, 306
spaceflight experiments, **156,** 340–342
vertebrate cell culture technology and, **142,** 148
Antibodies
as AGP probe, **174,** 210–214
γ-aminobutyric acid metabolism and, **147,** 315
anti-growth factors in vascular disease control, **161,** 30–31
avian retina and, **146,** 15–16, 19, 24, 31, 36
avian salt glands and, **144,** 171, 196
bacterial flagellar motor and, **147,** 97, 126–127
calcium channels and, **147,** 213, 220
chalcone synthase and, **147,** 253
chloroplasts and, **144,** 260
control of rhythmic behaviors and, **147,** 4, 9
CREST, **162B,** 37–38,41–47
cytoskeletal dynamics and, **144,** 95, 99, 108–109
Drosophila muscles and, **143,** 137
contractile protein genes, **143,** 107, 122–124, 136
development, **143,** 71, 79, 81, 85
eggshells and biomineralization and, **145,** 220
environmental stimuli in plant cells and, **145,** 263, 285–286, 295
extracellular matrix
characterization, **146,** 264, 267, 270–271
immunology, **146,** 284
synthesis, **146,** 272–273, 276, 279–280
fish chromatophores and, **143,** 242, 245
flowering plants and, **143,** 313, 320
glutamate metabolism and, **147,** 290
in glycine receptor purification, **176,** 205–207
GnRH-encoding gene in teleosts and, **147,** 186
higher plant mitochondrial genomes and, **141,** 131
higher plant plastids and, **145,** 169, 174
hormonal control of amphibian metamorphosis and, **145,** 133
immunoglobulins, *see* Immunoglobulins
to integrin in studies of FAK upregulation, **167,** 169–170
intermediate filaments, **162B,** 100–102
markers in identification of testicular macrophages, **149,** 107
matrin cyp, **162A,** 33–34
MDR reversal, **156,** 91
melanocytes and, **146,** 207–208, 240, 245, 250
mitosis and, **144,** 237, 246, 252
monoclonal, *see* Monoclonal antibodies
MPM-2, **164,** 273
myofibrillogenesis and, **143,** 154, 166, 172, 180, 183
against nuclear matrix proteins, **162A,** 27
nuclear protein import affected by, **151,** 199–200
NuMA
microinjections, **162B,** 13–15
pattern in interphase nucleus, **162B,** 8–11
polyamine, **138,** 64–65
proteins in striated muscle and, **142,** 61, 127–129
cardiac muscle, **142,** 103–111, 113–119
detection of abnormalities, **142,** 100–102
myosin, **142,** 62–79
sarcolemma, **142,** 124–125, 127
sarcomere, **142,** 93–98
sarcoplasmic reticulum, **142,** 120–124
thick-filament proteins, **142,** 79–83
thin-filament proteins, **142,** 83–88, 90–92
retinal pigmented epithelium and, **146,** 76–80
retinal precursor cells and, **146,** 154, 160, 171, 173, 175, 179
in retrograde surface transport, **150,** 47, 49

Antibodies (*continued*)
self-incompatibility and, **143,** 265, 284
Sertoli-germ cell communication network and, **147,** 36, 44, 72
transcriptional regulation and, **143,** 36
transdifferentiation in medusae and, **142,** 224, 227, 250
vertebrate cell culture technology and, **142,** 147, 184
Anticholinesterase, in avian neurogenesis, **151,** 166–169
α1-Antichymotrypsin, inhibition of serine proteinase, **170,** 241
Antidiuretic hormone, avian salt glands and, 178, **144,** 180, 182, 184
Antifolates, **156,** 108–116
Antigen-presenting cells, *see* Accessory cells
Antigen recognition activation motif
consensus sequence, **157,** 232–233
kinase binding, **157,** 238–241
phosphorylation, **157,** 233–234, 238, 240
role in lymphocyte signaling, **157,** 184, 188, 231–233
Antigens, *see also* Epitopes
4B3, **138,** 295–297
adherence to dome epithelium
M cell cytoskeleton reorganization induced by, **167,** 127
preference for M cells *versus* enterocytes, **167,** 122–123
specific mechanisms, **167,** 124–126
unspecific mechanisms, **167,** 123–124
γ-aminobutyric acid metabolism and, **147,** 312
artificial carriers of, in oral vaccines, **167,** 141
associated with cell differentiation, **171,** 84–85
B1C8, associated with spindle poles, **162B,** 57–59
bullous pemphigoid, intermediate plaque linkage in hemidesmosomes, **164,** 106–107
CD4
coreceptor with T cell receptor, **157,** 214–215
helper T cell differentiation, **157,** 201–203
interaction with T cell receptor, **157,** 199–202
kinase binding, **157,** 238
MHC class restriction of T cells, **157,** 190–191, 199–200
related proteins, role in sperm binding, **163,** 269–272
transcriptional regulation and, **143,** 40
CD4⁺, helper T cells in gut wall
in dome epithelium, **167,** 134
intraepithelial lymphocytes, **167,** 96
in lamina propria, **167,** 97–98
CD8
coreceptor with T cell receptor, **157,** 214–215
cytolytic T cell differentiation, **157,** 203–204
interaction with T cell receptor, **157,** 199–201
kinase binding, **157,** 238
MHC class restriction of T cells, **157,** 190–191, 199–200
CD8⁺, helper T cells in gut wall
intraepithelial lymphocytes, **167,** 97
in lamina propria, **167,** 97–98
CD10, marker of B cell differentiation, **157,** 139
CD11/CD18, role in inflammation, **152,** 75–76
CD14 receptor, macrophage, **161,** 292
CD19
coreceptor with B cell receptor, **157,** 215
kinase binding, **157,** 229, 238
CD21
coreceptor with B cell receptor, **157,** 215–216
ligands, **157,** 215
CD22
binding site regulation, **175,** 199–200
binding specificity, **175,** 207–208
biological function, **175,** 198
coreceptor with B cell receptor, **157,** 216
role in B cell signal transduction, **175,** 198–199
structure, **175,** 197–198
CD24, marker of B cell differentiation, **157,** 138, 142
CD28
role in apoptosis, **157,** 253
role in helper T cell development, **157,** 202–203

CD33
binding specificity, **175,** 207–208
in sialoadhesin family, **175,** 202–203
$CD34^{3+}$, TGF-β effects on progenitor, **167,** 211
$CD34^{3+}$/HLA-$DR^{high/low}$ cell subset, **167,** 229–230
CD40, role
in apoptosis, **157,** 209
in B cell activation, **157,** 206–207, 248
in helper T cells, **157,** 202
CD43, marker of B cell differentiation, **157,** 137–138
CD45
marker of B cell differentiation, **157,** 137–138
regulation of tyrosine kinases, **157,** 241–243
$CD45RA^+$ T cells in intestinal lamina propria, **167,** 98
CD45RO, lamina propria lymphocyte expression of, **167,** 98
CD68, *see* Myelomonocytic antigen
CDC25 genes in adenylate cyclase regulation, **139,** 313–314
control of rhythmic behaviors and, **147,** 4
and dendritic cells
processing, **153,** 72, 74–75
stimulation of primary immune response, **153,** 77
Drosophila muscles and, **143,** 122
duct-associated lymphoid tissue contact with, **167,** 137–138
extracellular matrix and, **146,** 269, 279
fA12, in perichromosome region, **162B,** 33–34
higher plant plastids and, **145,** 169
IgA produced in intestinal lamina propria and, **167,** 98
in immunohistochemical identification of carcinomas
gastrointestinal carcinomas, **155,** 233–236
gynecological carcinomas, **155,** 242–244
lung carcinomas, **155,** 240–241
technique, **155,** 230–232, 245
urological carcinomas, **155,** 242–243
interaction with M cell surface
preferential adherence, **167,** 122–123
specific mechanisms, **167,** 124–126
unspecific mechanisms, **167,** 123–124
intermediate filament, **162B,** 130
large tumor, SV40
biochemical activities, **162A,** 498–508
coordination with SV40 DNA, **162A,** 517–518
NLS, **162B,** 259–260
regulation of DNA replication, **162A,** 497–498
transforming potential, **162A,** 532–533
melanocytes and, **146,** 195–196, 235, 242–243
membrane, expression during differentiation, **171,** 235–236
in monocyte/macrophage procoagulant activity, **152,** 60
myofibrillogenesis and, **143,** 168, 172–173
P1, nuclear distribution, **162B,** 34–35
plant cell wall and plasma membrane, **171,** 105
presentation, **157,** 190
processing, **157,** 190
by MHC class I and II molecules, **161,** 299–302
proteins in striated muscle and
cardiac muscle, **142,** 103, 111, 113
skeletal muscle, **142,** 63, 68, 84, 87, 90
quail non-chick perinuclear, development, **175,** 245
recognition of, by T-cell receptors, **167,** 96–97
retinal pigmented epithelium and, **146,** 77
retinal precursor cells and, **146,** 160, 171
S-antigen
retinal precursor cells and, **146,** 173
transcripts, **138,** 21–22
self-incompatibility and, **143,** 264
Sertoli-germ cell communication network and, **147,** 37, 68
superantigens
intraepithelial lymphocytes and, **167,** 97
microbial toxins, table, **161,** 277
transcriptional regulation and, **143,** 12–14, 16, 18, 40–41
transcytosis of, by M cells, **167,** 126–128
transport by dome epithelium transport
as main function of M cells, **167,** 100, 112
M cell maturation stages *versus* transport capacity for, **167,** 117
overview, **167,** 93

Antigens (*continued*)
uptake by gut epithelium outside of GALT, **167,** 142
urease in plant cells and, **145,** 92
vertebrate cell culture technology and, **142,** 197
Antihormones, *see also* Sex hormones
action, **139,** 40–43
early exposure to
general considerations for, **139,** 43–45
genital abnormalities induced by, **139,** 13–26
in female guinea pigs, **139,** 13–14
in female humans, **139,** 21
in female mice, **139,** 17–21
in female rats, **139,** 14–17
in male mice, **139,** 21–26
in male rats, **139,** 21–25
nongenital abnormalities induced by, **139,** 26–39
Antihypertensives, inhibition of smooth muscle cell growth, **161,** 24–25
Antimetabolites
affecting DNA replication, **162A,** 549–556
cytosine arabinoside, **156,** 120–121
definition, **156,** 108
5-fluorouracil, **156,** 116–120
methotrexate, **156,** 108–116
purine antimetabolites, **156,** 121–122
Antimotic drugs, **144,** 225, 229, 237
Anti-Müllerian hormone, secretion by Sertoli cells, **160,** 201
Antioxidants, activity of acute phase proteins, **170,** 242–243
Antirrhinum, chalcone synthase and, **147,** 237, 250, 252
control of activity, **147,** 256, 259–261
control of gene, **147,** 262–267, 269–271, 275–276
Anti-Schwann cell myelin protein, **138,** 289
Anti-slit diaphragm antibody, **166,** 215–216
α1-Antitrypsin
haplotypes, **167,** 283
inhibition of serine proteinase, **170,** 241
Antitumor effects, muramyl dipeptide, **161,** 269
Anurans
brain, diagram, **138,** 177–178
karyosphere and, **144,** 27–36
Anxiolytic molecules, endogenous, neurotransmission, **149,** 266–267
Aortic-pulmonary septation, proteins in striated muscle and, **142,** 113
Aortic ring, angiogenesis assay, **159,** 131, 133
AP, *see* Alkaline phosphatase
Apatite, *see also* Calcified tissue; Calcium phosphates; Carbonate apatite
in biomaterials
from bovine bone, **172,** 162–164
coralline, **172,** 161
medical history and uses of, **172,** 159–161
in bone
crystal structure, **172,** 143–145, 166
dissolution, **172,** 167
molecular substitutions in, **172,** 165–166
organic matrix and, **172,** 167
during osteoporosis, **172,** 152, 154
in bone graft interface, **172,** 176–177
biodissolution and, **172,** 167, 170–171
bone ingrowth and, **172,** 173–174
carbonate apatite deposition and, **172,** 171–173
dislocations in, **172,** 170–171
in calcified tissue, **172,** 130, 175–176
carbonate incorporation, **172,** 154–155
chloride incorporation, **172,** 159
in dentin, **172,** 143
in enamel
crystal structure, **172,** 139–142
dissolution, **172,** 148–149
interactions with protein, **172,** 142
physical characteristics, **172,** 131
in fish enameloids, **172,** 145
fluoride incorporation, **172,** 155–157
magnesium incorporation, **172,** 157–158, 165–166
in pathological calcifications, **172,** 176
phosphate incorporation, **172,** 158
pyrophosphate incorporation, **172,** 159
strontium incorporation, **172,** 159
Aphidicolin
modulation of DNA repair, **156,** 129–130
transdifferentiation in medusae and, **142,** 252–253, 255–256
Aphids, urease in plant cells and, **145,** 77
API, *see* Aminopeptidase I

Apical cells
avian salt glands and, **144,** 140, 147, 171–173
effect of microgravity
on meristem shoot, **171,** 35–36
on moss protonema, **171,** 23–25
Apical initial, *see* Stem cell
Apical junctions
RPE, structure and composition, **171,** 238–240
tight junctions, avian salt glands and, **144,** 138–139
Apical membrane, M cell
antigen interaction with, **167,** 122–126
antigen and tracer adherence to, **167,** 118
disadvantage of light microscopy in defining, **167,** 102
endocytosis of antigens at, **167,** 127
preferential adherence of *Salmonella typhimurium* to, **167,** 123
recycling in endocytosis, **167,** 128
shape *versus* enterocytes, **167,** 111
Apicomplexa, endosymbiont hypothesis and, **141,** 326
Apigenin, chalcone synthase and, **147,** 247, 249, 253
Aplastic anemia, increased MIP-1α mRNA in, **167,** 238–239
Aplysia
biological clocks in retina, **146,** 85
entrainment pathway, **146,** 100–115
opisthobranch retina, **146,** 86, 88–92, 95
output pathway, **146,** 128
pacemaker mechanisms, **146,** 116–118, 122
transmembrane loop, **146,** 128–129, 131–132, 134
control of rhythmic behaviors and, **147,** 18
cytoskeletal dynamics and, **144,** 100
Apocytochrome *b*
ciliates and, **141,** 12, 41
kinetoplastid mtDNA and, **141,** 71
mitochondrial genomes in fungi and
DNA diversity, **141,** 92, 96–98, 100, 102–103, 105
generation of mtDNA diversity, **141,** 107, 109
Apodachyla, mitochondrial genomes in fungi and, **141,** 91–92
Apolipoproteins
apo(a), stimulation of vascular cells, **161,** 14–15
as cofactor in Tau protein aggregation, **171,** 191–193
Apoplast, symplast and, **142,** 264, 268, 271–272, 276, 285–287, 291–300
Apoprotein, flowering plants and, **143,** 300–301, 309
Apoptosis
B cell, **157,** 142, 183, 197, 208–209, 250–256
calcium role, **163,** 114–115
cell death mechanisms, **163,** 112–113
DNA fragmentation, **163,** 112–113
endonucleases in, **163,** 112–114
chondrocytes, **159,** 298
defined, **163,** 106, 107
FAK potential suppression of, **167,** 178
gene activation during, **163,** 116–117
genetic regulation, **163,** 119–120
growth factor regulation, **163,** 117–119
hormonal regulation, **163,** 117–119
during invertebrate development
Caenorhabditis elegans, **163,** 123–126
Drosophila melanogaster, **163,** 127–130
grasshopper, **163,** 123–126
insects, **163,** 126
Manduca sexta, **163,** 130–133
melanoblast clones, **146,** 221
morphological features, **157,** 250; **159,** 298; **163,** 109–111
role of
AGPs, **174,** 263–264
Bax, **157,** 255
bcl-2, **157,** 254–255
CD40, **157,** 209
Fas ligand, **157,** 204
nur 77, **157,** 251, 255–256
PTK, **157,** 213
in sheep ileal Peyer's patches, **167,** 95
signal transduction during, **163,** 115–116
T cell, **157,** 183, 198–199, 250–256
techniques for examining cell death, **163,** 120–123
during vertebrate development
among primordial germ cells, **163,** 139
amphibian metamorphosis, **163,** 146
blastocyst development and implantation, **163,** 134
gastrulation, **163,** 124

Apoptosis (*continued*)
heart, **163,** 139–140
kidney, **163,** 140–141
lens, **163,** 141–142
limb bud, **163,** 143–146
during myogenesis, **163,** 138–139
nervous system, **163,** 148–153
neural crest morphogenesis, **163,** 136–137
neural tube closure, **163,** 135–137
palate, **163,** 146–147
reproductive system, **163,** 141
in somites, **163,** 138–139
tail bud regions, **163,** 137–138
APP, *see* Acute phase proteins
Apparent diffusion coeffcient, nuclear magnetic resonance and, **145,** 27
Appendix, M cells in rabbit, **167,** 100
Aquaporin, *see* CHIP 28
Aquaspirillum serpens, bacterial flagellar motor, **147,** 106–107
Aqueous humor
blood barrier and
glucose transport, **172,** 19–22
structure, **172,** 17–19
glucose level in, **172,** 17
Arabidopsis
auxin-regulated gene expression, **152,** 133
chalcone synthase and, **147,** 247, 250, 252, 265, 276
intracellular calcium messenger system and, **142,** 328–329
plastids and
amino acid synthesis, **145,** 193–194, 199
fatty acid synthesis, **145,** 185–186
metabolism, **145,** 170, 173
starch synthesis, **145,** 181–182
structure, **145,** 164
root statocytes, **171,** 21–22
self-incompatibility and, **143,** 277–278, 288
urease in plant cells and, **145,** 98–99
Arabidopsis thaliana
biochemistry, **144,** 72–76
effect of microgravity, **156,** 321–323, 329–332; **171,** 37, 41–42
endosymbiont hypothesis and, **141,** 250, 278
flowering plants and
amino acid biosynthesis, **143,** 311, 313–318
amino acid pathways, **143,** 321–322, 324
ammonium, **143,** 303–305
nitrate, **143,** 299–301
genetic model, **144,** 60–64
intracellular calcium messenger system and, **142,** 309–311
models
plant defense, **144,** 76–77
resistance genes, **144,** 76
systemic acquired resistance, **144,** 77–78
pathogens, **144,** 54–55
biochemistry, **144,** 59–60
genetics, **144,** 55–59
phenotypes
bacteria, **144,** 66–69
fungi, **144,** 64–66
nematodes, **144,** 71–72
viruses, **144,** 69–71
salt tolerance, **165,** 3
urease in plant cells and, **145,** 82, 90
Arabinans, extracellular matrix and, **146,** 264, 292
Arabinofuranosylcytosine, *see* Cytosine arabinoside
Arabinofuranosyl residue, as cell wall-bound protein, **176,** 169
Arabinogalactan proteins
action at molecular level, **174,** 264–266
anatomic distribution, **174,** 215–217
antibodies to, development, **171,** 101–102
antibodies as probe, **174,** 210–214
biosynthesis and degradation
AGP turnover, **174,** 238–239
carbohydrate component, **174,** 236–238
hydrolytic enzymes, **174,** 239–240
polypeptide component, **174,** 235–236
carbohydrate component, **174,** 232–234
carbohydrate–polypeptide linkages, **174,** 230–232
as cell wall-bound protein, **176,** 169
definition, **174,** 201–202
developmentally regulated, detection, **171,** 103–104
developmental regulation, **171,** 102–103
electron microscopy, **174,** 234
extracellular matrix and, **146,** 264, 269, 277, 285, 292, 295, 303
localization to cell surface sites, **174,** 240–245

other functions, **174,** 266–267
plant
analogs of animal extracellular matrix
adhesion proteins, **174,** 272
fibronectin, **174,** 271–272
integrins, **174,** 270–275
short peptides, **174,** 271
vitronectin, **174,** 271–272
comparison to animal proteoglycans and mucins, **174,** 268–270
polypeptide component
cDNA-deduced sequence, **174,** 218–230
conformation, **174,** 230
general features, **174,** 217–218
role in cell death, **174,** 263–264
role in cell division, **174,** 260–262
role in cell expansion, **174,** 262–263
role in plant cell development, **171,** 100–101
role in plant development
early studies, **174,** 245–246
pattern formation in roots and coleoptiles, **174,** 255–258
reproductive phase of development, **174,** 247–255
somatic embryogenesis, **174,** 258–260
self-incompatibility and, **143,** 283
taxonomic distribution, **174,** 214–215
use in medicine, **174,** 267–268
Yariv phenylglycoside as probe, **174,** 204–210
Arabinose, extracellular matrix and, **146,** 265, 268, 277, 285, 311
Arabinosyl transferase, extracellular matrix and, **146,** 274–275
Arabitol, **165,** 11
Ara-C, *see* Cytosine arabinoside
Arachidonic acid
decidual cell reaction induction, **160,** 16–21
effect on regulatory volume increase, **161,** 186
effect on swelling-activated Cl^- and taurine effluxes, **161,** 201–202
metabolism, stimulation by endothelin, **171,** 281–282
pathways, **161,** 221
Arachis arenosa, self-incompatibility and, **143,** 263
Ara-CMP, *see* Ara-cytidine monophosphate
Ara-cytidine monophosphate, **156,** 120
ARAM, *see* Antigen recognition activation motif
Arbacia lixula, avian retina and, **146,** 9
ARBP
binding MAR DNA, **162A,** 303–305
localized to inner nuclear matrix, **162B,** 28
mediated DNA loop formation, **162A,** 543–544
Archaebacteria, endosymbiont hypothesis
future prospects, **141,** 331
lineages of life, **141,** 238, 240–246
mitochondria, **141,** 280, 307–309, 313, 327
nuclear genome, **141,** 247–251
organelles, **141,** 329
plastids, **141,** 253, 265, 267–269
sequences, **141,** 270
Ardisia, urease in plant cells and, **145,** 97
ARF, *see* ADP ribosylation factor
Arg-Gly-Asp synthetic peptide, **167,** 169
Arginase, urease in plant cells and, **145,** 66–68, 71–72, 82–83
Arginine
flowering plants and, **143,** 321
role in glycine receptor activation, **176,** 214
urease in plant cells and, **145,** 98
elimination, **145,** 80–83
metabolic origins, **145,** 66–72, 78–79
Arginine–serine-rich domain, **162B,** 312–313,324–325
Arginine vasopressin
cellular localization, **159,** 34
processing, **159,** 10
release, **159,** 36
role in ACTH release, **159,** 36
Arginine vasotocin
avian salt glands and, **144,** 189, 198–199
control of amphibian metamorphosis and, **145,** 125–126
control of rhythmic behaviors and, **147,** 14
Arg-Phe-NH_2, *see* RFamide
Armadillo gene, **139,** 151
Arogenate, flowering plants and, **143,** 311
Aromatase inhibitors, and sex hormones, **139,** 8
Aromatic hydrocarbons, imprinting of steroid receptors, **155,** 32–33

ARS, *see* Autonomously replicating sequences
Artemia, microgravity experiments, **156,** 336, 356
Artemisia pycuocephala, pollen of, **140,** 190
Artemisia vulgaris, **140,** 12
Arthrin, *Drosophila* muscles and, **143,** 112–113, 136, 138
Arthritis, immune, experimentally induced, and testicular macrophages, **149,** 125
Arthropods
cuticle, extracellular matrix, **166,** 71–73
insects as, **174,** 63–65
microgravity experiments, **156,** 335–336
Articulins, epiplasm components in euglenoids, **169,** 283–284, 287
Artifacts
avoidance by electroelution, **162B,** 424–425
fiber attachments, **162A,** 127–128
MARs as, **162A,** 201
preparational, **162A,** 491–495
Ascaris suum
animal mtDNA and, **141,** 188, 191, 202
nucleotide bias, **141,** 204–205
tRNA, **141,** 197–198
cytoskeletal dynamics and, **144,** 91
endosymbiont hypothesis and, **141,** 290
Ascaris univalens, holokinetic chromosomes, **152,** 33
Ascidia ahodori, embryogenesis, **176,** 249–250
Ascidians, sperm-binding proteins, **163,** 258–259
Ascomycetes
endosymbiont hypothesis and, **141,** 282, 290, 300
mitochondrial genomes in fungi and, **141,** 121
DNA diversity, **141,** 93–103
generation of mtDNA diversity, **141,** 116–118
L-Ascorbate oxidase, in lignin biosynthesis, **176,** 111–114
Ascorbic acid, *see* Vitamin C
Asexual development, volvocacean, **139,** 197–198
Asialoglycoprotein receptor
in galactose residue masking, **175,** 144–145
hepatic galactose receptor, **175,** 145–146
Asian horseshoe crab, axonemal pattern, **173,** 5
Asparagine, flowering plants and, **143,** 303
Asparagine synthetase, flowering plants and, **143,** 306–308
Aspartate
flowering plants and, **143,** 318–321
neurotransmiss.ion, **149,** 259–260
Aspartate aminotransferase
glutamate metabolism and, **147,** 288–294, 305, 322
neurotransmission, **149,** 259, 261–262
Aspartate kinase, flowering plants and, **143,** 320
Aspartate receptor, and neurotransmission, **149,** 260–262
Aspergillus, mitosis and, **144,** 243, 245, 252
Aspergillus nidulans
mitochondrial genomes in fungi and
DNA diversity, **141,** 93–98
generation of mtDNA diversity, **141,** 107–108, 110, 113, 117
mitosis and, **144,** 226, 228–229, 232, 248, 251
Aspergillus oryzae, self-incompatibility and, **143,** 287
Aspirin, effects on peroxisomes, **160,** 303, 315–316
Assembly–disassembly model, **162B,** 206–207
Assembly driven model, **162B,** 191
Assembly proteins, clathrin, **167,** 13–18
Astasia cryptochromes in, **148,** 288
Asterina miniata, sperm-binding proteins, **163,** 279–280
Astrocytes
connexin staining, **157,** 343, 345
gap junction
cell coupling, **157,** 342–343, 347
coupling, **157,** 347
induction by neuronal stimuli, **157,** 343
phosphorylation, **157,** 344
quantitation, **157,** 345
size, **157,** 345, 347
staining, **157,** 344
GLUT1 expression in, **172,** 11
glutamate metabolism and, **147,** 286, 319, 322
aminotransferases, **147,** 287, 290–291
glutamate dehydrogenase, **147,** 295, 298–300

glutaminase, **147,** 307–308
glutamine synthetase, **147,** 302–305
interaction with neurons, **149,** 274–276
neuropeptide receptor-mediated responses, **148,** 119–130, 157, 159–161
angiotensin II, **148,** 127–128, 131
atrial natriuretic peptide, **148,** 127–128, 132–134
bombesin, **148,** 137–138
bradykinin, **148,** 127, 129, 137
brain natriuretic peptide, **148,** 127–128, 132
calcitonin, **148,** 146–147
corticotropin-releasing hormone, **148,** 129, 141–142
C-type natriuretic peptide, **148,** 128, 132–133
endothelins, **148,** 127–128, 133–134
insulin, **148,** 129, 144–146
neuromedin, **148,** 129, 137–138
opioids, **148,** 140–141
oxytocin, **148,** 129, 138–139
somatostatin, **148,** 129, 147–154
substance P, **148,** 128, 135–137
vasoactive intestinal peptide, **148,** 129, 142–144
vasopressin, **148,** 138–139
in regulation of GLUT1 expression, **172,** 34
Sertoli-germ cell communication network and, **147,** 34
Astrocytomas, neuropeptide binding sites, **148,** 127–128, 132–134, 153, 155–158
Atherosclerosis
mouse model, smooth muscle cell changes, **169,** 229–230
pathogenesis role of smooth muscle cells, **169,** 184, 187–190, 226, 231
therapeutic strategies, **161,** 29–30
ATP
γ-aminobutyric acid metabolism and, **147,** 311, 314
animal mtDNA and, **141,** 181
ATPase, retinal precursor cells and, **146,** 147, 160
autocrine glands, **146,** 56
avian salt glands and, **144,** 135, 144, 173
bacterial flagellar motor and, **147,** 156
biological clocks in retina and, **146,** 108
as catecholaminergic neurotransmitter, **149,** 243
chromoplast, **163,** 207–208
conductance by CFTR, **172,** 199, 215
cytoplasmic streaming role
concentration as determining factor, **155,** 112–115
temperature effects, **155,** 127
cytoskeletal dynamics and, **144,** 88, 96, 98, 103, 116–118
Drosophila muscles and, **143,** 66, 98–100, 133–135
dynein outer arm changes caused by, **155,** 159–160, 169
effect on flagellar beat cycle, **173,** 30–31
endosymbiont hypothesis and, **141,** 263, 309
environmental stimuli in plant cells and, **145,** 295, 298
chloroplasts, **145,** 276–277, 282–283
photodinesis, **145,** 259, 263–265
extracellular matrix and, **146,** 281
fish chromatophores and
cytoskeleton, **143,** 222–223
motility control, **143,** 230–232, 239, 242
flowering plants and, **143,** 309
glutamate metabolism and, **147,** 302
half-life in blood stream, **158,** 156
higher plant mitochondrial genomes and, **141,** 158
higher plant plastids and
CO_2 fixation, **145,** 171, 174–176
fatty acid synthesis, **145,** 189–190
metabolism, **145,** 168, 203
starch synthesis, **145,** 177, 180–181
hydrolysis by CFTR, **172,** 195, 197, 200
and instability of homogeneous state
Child's results, **154,** 346–34
colocalization role, **154,** 313
concentration, **154,** 311–312
Drosophila, **154,** 338
localized activity of pumps, **154,** 343- 345
metabolic field, **154,** 356–360
metabolism of proliferation, **154,** 361–362, 364–365
reduction fields, **154,** 347–348, 350, 352, 354
spontaneous endogenous electrophoresis, **154,** 340–342

ATP (*continued*)
Turing biochemical mechanism, **154,** 323–324, 326–327, 330–333
Turing-Child metabolic system, **154,** 310
intracellular calcium messenger system and
calmodulin, **142,** 329, 331
concentration, **142,** 312, 314–317
phosphatidylinositol turnover, **142,** 324, 326–327
protein kineses, **142,** 337
kinetoplastid mtDNA and, **141,** 71
levels, electromagnetic field effect, **158,** 318
mitosis and, **144,** 246
myofibrillogenesis and, **143,** 178
neurotransmitter function, **158,** 165
nuclear magnetic resonance and, **145,** 2, 7, 19, 42–43
Physarum contraction by, **149,** 155
proteins in striated muscle and, **142,** 64–65, 81, 120, 122
as purinergic neurotransmitter, **149,** 254–255
symplast and, **142,** 274
transcriptional regulation and, **143,** 30, 39
uncoating of clathrin-coated vesicles, **167,** 23
urease in plant cells and, **145,** 80
ATPase
activation in *Chlorella,* effect of clinostating, **171,** 28–29
activity of dynein, **173,** 13–16
activity and stoichiometry, in drug pump model, **171,** 132–134
activity of T-antigen, **162A,** 500–501
animal mtDNA and, **141,** 174, 177, 181–183, 207
avian salt glands and, **144,** 147
Ca^{2+}-
ion pumping, **168,** 127–129
muscle fiber type-specific distribution, **170,** 158
and neuromediation, **149,** 227–228
role in cell wall changes under microgravity, **170,** 64
role in plant gravitropism, **168,** 135, 150–151
sperm capacitation regulation by, **149,** 14–16, 39
cell wall-bound
effect of suppressors on, **172,** 77–80
in fungal signal transduction cascade model, **172,** 82–84
in peroxide generation, **172,** 81
ciliates and, **141,** 37
cytoskeletal dynamics and
current concepts, **144,** 94, 97, 103
cytoskeletal components, **144,** 106, 108–109
theories, **144,** 88
Drosophila muscles and, **143,** 133, 139
contractile protein genes, **143,** 98–100, 104
myosin light chains, **143,** 105, 107
troponin, **143,** 117
ecto-, *see* Ecto-ATPases
endosymbiont hypothesis and, **141,** 245, 248, 250, 310
environmental stimuli in plant cells and, **145,** 261, 263, 280, 283, 290
E-type, *see* E-type ATPase
fish chromatophores and, **143,** 199, 222–224, 232, 247
H^+-
effect on pollen tube tip growth, **176,** 179
reversible phosphorylation, **149,** 91
vacuolar, **139,** 81–89
higher plant plastids and, **145,** 174
intracellular calcium messenger system and, **142,** 328, 337
concentration, **142,** 313–314, 316
receptors, **142,** 307–308
K^+-, **138,** 19–21
kinetoplastid mtDNA and, **141,** 72
membrane-bound
cell wall regulation of, **172,** 79–80
effect of suppressors on, **172,** 67–70
functional association with polyphosoinositide metabolism, **172,** 76
Mg^{2+}-, **156,** 253–254; **160,** 305, 307, 313, 321–324
mitochondrial genomes in fungi and, **141,** 92–93, 97, 120
mitosis and, **144,** 234
myofibrillogenesis and, **143,** 180
myosin, muscle fiber types based on, **170,** 144–148

Na^+/K^+-
acrosomal exocytosis role, **149,** 28–29
avian salt glands and
secretory mechanism, **144,** 169–172, 175–176
secretory tissue, **144,** 140–141
stimulus-secretion coupling, **144,** 203, 205
in cockroach, **176,** 325–327
discovery, **158,** 143
effect of electromagnetic field, **158,** 284, 299, 319
in fly, **176,** 329–332
in fly photoreceptors, **176,** 337–338
in honeybee, **176,** 325
in honeybee photoreceptors, **176,** 334–337
in locust, **176,** 325–327
molecular analysis, **176,** 320–323
in moth, **176,** 327–329
in neuromediation, **149,** 227–229
ouabain-sensitive, stimulation by ET, **171,** 281
in photoreceptors, **176,** 323–324
polarity
choroid plexus, **165,** 174–177
retinal pigmented epithelium, **165,** 174–177
redistribution during insect photoreceptor development, **176,** 332–334
sequence homology with adhesion molecules, **158,** 148
sperm capacitation role, **149,** 17–18, 25–26
nematocysts, **156,** 279
proteins in striated muscle and
sarcoplasmic reticulum, **142,** 120–123
skeletal muscle, **142,** 63, 65–68, 81, 102
related activity of dynein, **155,** 146–147
vanadate-dependent, **155,** 151
tonoplast proton, *see* Tonoplast proton ATPase
traffic, **156,** 82
T-tubule
amino terminal sequence, **158,** 180
extracellular organization, **158,** 146, 162
glycosylation, **158,** 163, 179
inhibition by detergents, **158,** 177, 195–196
metal dependence, **158,** 163, 192
phorbol ester receptor function, **158,** 191, 196
purification, **158,** 162–163, 179
solubilization, **158,** 177–179
specific activity, **158,** 155, 163, 179
substrate specificity, **158,** 188
temperature effect on activity, **158,** 162
tissue distribution, **158,** 163–164
ATP-binding cassette proteins
ATP transport hypothesis, **171,** 156
–MDR protein, forms, **171,** 157–158
types, **156,** 82
ATP synthase, cyanelle genes of photosynthetic apparatus, **151,** 48
ATP transport
effect of MDR protein expression, **171,** 155–156
hypothesis for ABC transporters, **171,** 156
Atrazine, environmental stimuli in plant cells and, **145,** 254
Atrial natriuretic factor
ANE7, **144,** 196–198, 203
immunohistochemical localization of, **138,** 133–135
neuroendocrine and neurotransmitter functions, **138,** 135–138
Atrial natriuretic peptide, control of amphibian metamorphosis, **145,** 127
A–T-rich sequences
DNA, **162B,** 414
MAR/ORIs containing, **162A,** 320–322
in MARs, **162A,** 101–102, 542–543; **162B,** 358–359
Atrio-ventricular node, proteins in striated muscle and, **142,** 103, 107
Atropine
avian salt glands and, **144,** 161–162, 165, 203–205
fish chromatophores and, **143,** 197
Attachment sites
in model loop, **162A,** 140–141
nuclear matrix, **162A,** 264
satellite DNA and *Alu* repeats, to matrix, **162A,** 341–342
transcriptionally active chromatin, **162A,** 231–232
Attheya decora, silicalemma formation, **150,** 342
Atypical multidrug resistance, **156,** 100, 107

Aurones, chalcone synthase and, **147,** 246, 248
Autocrine feedback loops
PDGF-stimulated cell production of IFN and inhibitory, **167,** 231
proto-oncogene mutations/deletions and stimulatory, **167,** 232–233
Autocrine motility factor, migration induction by, **155,** 79–80
Autocrine signaling, platelet-derived growth factors and, **172,** 104–105, 112
Autogenous origin of cells, endosymbiont hypothesis and, **141,** 233–234
Autoimmune disease
dendritic cell role in, **153,** 88–89
role of chaperones, **174,** 168–169
and testicular macrophages, **149,** 124–125
Autolysins, **139,** 196–197
Autonomic nervous system
avian salt glands and, **144,** 155
fish chromatophores and, **143,** 226
thymic epithelial cells, **156,** 186–190
Autonomously replicating sequences
applied to mammalian cells, **162A,** 464–466
and ciliates, **141,** 11
T-rich strand, ACBP binding, **162A,** 304
in yeast, **162A,** 462–463
Autophosphorylation
FAK on tyrosine residues, **167,** 168–169, 173
intracellular calcium messenger system and, **142,** 334, 337
Autoradiography
γ-aminobutyric acid metabolism and, **147,** 300
avian retina and, **146,** 10, 15
biological clocks in retina and, **146,** 117
electron microscope, protein secretion dynamics, **160,** 275, 277
endothelin receptor subtypes, **171,** 273–274
glutamate metabolism and, **147,** 291
growth factor-induced retinal regeneration and, **146,** 53, 57, 68
hormonal control of amphibian metamorphosis and, **145,** 135
retinal precursor cells and, **146,** 149, 151, 153–154, 167, 169
Sertoli-germ cell communication network and, **147,** 62
Autoreceptor, catecholaminergic neurotransmission, **149,** 243–244
Autoregulation, nodulation in legumes, **156,** 42, 43
Autosomal recessive retinal dystrophy, history of, **138,** 1–2
Autosomes
imbalances, FISH diagnosis
chromosomes
anomalies, **153,** 28, 30–31
painting, **153,** 28, 30–31
Philadelphia chromosome, **153,** 30
specific repeat probes, **153,** 25–27
trisomy 21 detection, **153,** 27–29
DNA
deletions, **153,** 30–33
duplications, **153,** 33
incidence of disomies in sperm, **167,** 269–270
monosomies, in preimplantation embryos, **167,** 267–268
Auxilin, **167,** 23
Auxin
affected plant cells
elongation tissues, **152,** 111–117
growth, **152,** 109
molecular mechanisms, **152,** 137
suspension culture, **152,** 126, 128–132
tobacco mesophyll protoplasts, **152,** 117–127
chalcone synthase and, **147,** 250
cytoskeleton, **166,** 39
effect on DNA methylation, **158,** 117
environmental stimuli in plant cells and, **145,** 294, 297, 299–300
extracellular matrix and, **146,** 292–294, 297, 301
flowering plants and, **143,** 297, 309, 317–319
habituation in tumorous plant cells, induction
concentration changes, **152,** 267–269
cytokinin interactions with, **152,** 270
inoculum size, **152,** 270–271
mechanisms, **152,** 265–266, 276–277, 285–286, 292
induction of genes, **156,** 37–38; **158,** 125
intracellular calcium messenger system and, **142,** 306–307, 325–326, 338
in nodulation, **156,** 39, 44
plant hormone, **166,** 39

regulated genes, *see* Genes, auxin-regulated
regulation of XRP gene, **173,** 188–189
symplast and
integration of activity, **142,** 286–287, 294–298
Lockhart equations, **142,** 277–279, 283–285
mechanical properties, **142,** 279–282
proton pumps, **142,** 274–275
water relations, **142,** 282–283
transdifferentiation in medusae and, **142,** 216
Avena
intracellular calcium messenger system and, **142,** 308–310
plastids and, **145,** 171
symplast and, **142,** 273
Avena sativa, microgravity experiments, **156,** 321, 324–326, 328, 331–332
Average glomerular values, defined, **161,** 113
Avian pancreatic polypeptide, **138,** 127
Avian salt glands
blood supply
microvasculature, **144,** 148–151
secretion, **144,** 151–154
hormonal control, **144,** 189, 199–200
angiotensin II, **144,** 192–195
arginine vasotocin, **144,** 198–199
atrial natriuretic factor, **144,** 196–198
pituitary, **144,** 191–192
prolactin, **144,** 198
innervation
adrenergic, **144,** 167–168
cholinergic, **144,** 159–162
nerve supply, **144,** 155–159
vasoactive intestinal polypeptide, **144,** 162–167
receptive systems, **144,** 177
neural integration, **144,** 188–190
tonicity, **144,** 177–184
volume, **144,** 185–188
secretory mechanism, **144,** 168–169
current models, **144,** 175–176
early models, **144,** 172–175
Na^+/K^+-ATPase, **144,** 169–172
secretory tissue
duct system, **144,** 145–148
enzymes, **144,** 144–145
fine structure, **144,** 133–140
hypertrophy, **144,** 140–144
zoology, **144,** 130–133
stimulus-secretion coupling, **144,** 200
calcium, **144,** 203–205
cyclic AMP, **144,** 200–202
cyclic GMP, **144,** 201–203
protein phosphorylation, **144,** 205
Avirulence, *Arabidopsis thaliana*
induction, **144,** 75
pathogens, **144,** 54, 56–57
phenotypes, **144,** 67, 69
Avocado, plastids and, **145,** 184–185
AVP, *see* Arginine-vasopressin
Axial organs, symplast and, **142,** 277–279, 287, 297, 300
Axis
anterior–posterior, specification
amphibian development, **176,** 291–292
cytoplasmic determinants and induction, comparison, **176,** 287–291
evolutionary modifications of ascidian development, **176,** 293–294
tilt
computation, **163,** 63
two-dimensional analysis, **163,** 63
Axolemma, **165,** 239
Axonal transport
fast
axonal assembly, **165,** 269–273
dynein, **165,** 271
slow
axonal assembly, **165,** 274–279
cytoskeleton, **165,** 276–278
Axoneme
physical model, **173,** 39–46
role in flagellar beat cycle, **173,** 29–31
structural components, **173,** 2–5
structural variations, **173,** 5–9
Axons
assembly
cytoskeleton, **165,** 274–279
fast axonal transport, **165,** 269–273
kinesin, **165,** 270–271
mechanisms, **165,** 268–279
membrane addition, **165,** 269–273
plasma membrane, **165,** 272
slow axonal transport, **165,** 274–279
characteristics and properties, **165,** 237–238
cytoskeleton, **165,** 239

Axons (*continued*)
  early outgrowth, **165,** 248
  elongation, growth cone, **165,** 256–268
    motility relationship, **165,** 260–266
  giant, **167,** 39
  growth, **165,** 244–245
    mechanisms, **165,** 245–246
    neuron, cytoplasmic mechanisms, **165,** 235–280
  microtubule, **165,** 239–240, 246–247
  microtubule stability, **157,** 17–18
  mitochondrial transport in, **173,** 223–225
  neuronal polarity, **165,** 247
  synthesis, **165,** 238
  transport, by microtubules, **171,** 169
Axoplasm, avian salt glands and, **144,** 159
Axotomy
  effect of animal age, **159,** 207, 213–214
  effect on neurons
    action potential waveform, **159,** 208
    conduction velocity, **159,** 207
    frogs
      perikarya, **159,** 215
      synapses, **159,** 216–220, 248
    input resistance, **159,** 208
    mammals
      perikarya, **159,** 208–210
      synapses, **159,** 210–214
  experimental models, **159,** 196, 207
  inhibition of angiogenesis, **159,** 140
Azacytidine, induction of DNA hypomethylation, **158,** 116
8-Azaguanine
  effect on plant cell culture, **158,** 75
  enzymatic deamination, **158,** 75
Azaserine, urease in plant cells and, **145,** 78
Azide, inhibition of ecto-ATPases, **158,** 189–190
8-Azido-ATP
  ATPase substrate, **158,** 166–167
  photoaffinity labeling, **158,** 167, 181
3′-Azido-3′-deoxythymidine, **167,** 222
Azorhizobium caulinodans
  host-specific modifications, **156,** 25–26
  nod box upstream, **156,** 10–11
  Nod factors, **156,** 22
AZT, *see* 3′-Azido-3′-deoxythymidine
Azuki bean, endoxyloglucan transferase
  enzymatic reaction, **173,** 173–174
  localization, **173,** 181–182
  pH dependency, **173,** 176–177
  substrate specificity, **173,** 174–176

B

B3, monoclonal antibody to nuclear matrix proteins, **162A,** 34–36
B29, marker of B cell differentiation, **157,** 141
B220
  marker of B cell differentiation, **157,** 137–138, 141–142
  role in B cell development, **157,** 156, 161–162
Baby hamster kidney cells, cell culture technology, **142,** 150–151, 170, 173
Bacillariophytes, chloroplasts and, **144,** 280, 282, 290
Bacillus Calmette Guérin, **167,** 120*t,* 122
Bacillus subtilis, microgravity experiments, **156,** 339–340
Bacteria
  animal mtDNA and, **141,** 209
  Arabidopsis thaliana and
    biochemistry, **144,** 73–75
    models, **144,** 78
    pathogens, **144,** 56–58
    phenotypes, **144,** 65–69
  binding by sialic acids, **175,** 175
  chloroplasts and, **144,** 259, 270
  cyanobacteria, *see* Cyanobacteria
  eggshells and biomineralization and, **145,** 235
  encapsulated bacteria, **161,** 284–285
  endosymbiont hypothesis and, **141,** 233, 237
    future prospects, **141,** 331–332
    gene structure, **141,** 308–311, 317–318
    lineages of life, **141,** 243
    mitochondria, **141,** 280, 283, 285, 327–328
    nuclear genome, **141,** 248
    plastids, **141,** 267, 271
    rRNA sequences, **141,** 319–320, 322
  extracellular matrix and, **146,** 273–274, 309
  flowering plants and, **143,** 299, 311, 313, 317
  Gram-negative, *see* Gram-negative bacteria

Gram-positive, *see* Gram-positive bacteria
higher plant mitochondrial genomes and, **141,** 132, 163
kinetoplastid mtDNA and, **141,** 66
–macrophage interactions
adhesion strategies, **161,** 293–296
antigen processing and presentation, **161,** 299–302
bactericidal mechanisms
oxygen-dependent, **161,** 306–307
oxygen-independent, **161,** 307–310
chemotaxis, **161,** 304–306
cytokine production, **161,** 310–313
entry strategies, **161,** 293–296
macrophage surface receptors, **161,** 291–293
phagocytosis, **161,** 302–304
signal transduction mechanisms, **161,** 296–299
M cells and
lectin-like adhesins on outer membrane in invasion by, **167,** 124
phagocytosis, **167,** 127–128
preferential adherence to apical membranes, **167,** 123
as targets for enteropathogenic, **167,** 138–139
uptake by, **167,** 117–122
melanocytes and, **146,** 212
microgravity experiments, **156,** 339–342
mitochondrial genomes in fungi and, **141,** 120
Nod signals, **156,** 18–32
regenerative cementogenesis and, **142,** 2, 44
Rhizobium–plant interactions, **156,** 1–51
in suppression of plant defenses, **172,** 61
transcriptional regulation and, **143,** 30
transport proteins, **156,** 83
urease in plant cells and, **145,** 65, 98–99
biochemical genetics, **145,** 88–90, 92
elimination, **145,** 80, 83
metabolic origins, **145,** 66, 74, 76, 79
nickel metabolism, **145,** 92~97
vertebrate cell culture technology and, **142,** 146, 153, 166, 193
Bacterial flagellar motor
bioenergetic viewpoint, **147,** 97–99
dynamic measurements, **147,** 114
frame-by-frame analysis, **147,** 116–117
freely swimming cells, **147,** 114–115
optical tweezers, **147,** 117–118
tethered cells, **147,** 115–116
function, **147,** 110–114
model comparisons, **147,** 156
kinetics, **147,** 146–150
steady-state behavior, **147,** 150–155
models, **147,** 129–130
chemically explicit hypothesis, **147,** 145–146
conformational transitions, **147,** 144–145
cytomembrane streaming, **147,** 141
dielectric motors, **147,** 142–143
diffusive flow of protons, **147,** 142
electrodes, **147,** 143–144
electrostatic interaction, **147,** 131–135
gated channels, **147,** 135–140
intersecting ligand rows, **147,** 130–131
ion flow, **147,** 144–145
muscular contraction, **147,** 156–159
rotation threshold, **147,** 141–142
twistor mechanism, **147,** 140–141
motor energetics
coupling, **147,** 122–125
dissipation function, **147,** 118–122
energy conversion, **147,** 122
freely swimming bacteria, **147,** 120
rotational thresholds, **147,** 129
stall torque, **147,** 126–127
tethered bacteria, **147,** 120–126
torque-velocity curve, **147,** 128–129
unhindered rotation, **147,** 125–126
structure, **147,** 99
basal body, **147,** 99–103
genes, **147,** 102, 104–105
particle rings, **147,** 105–109
switch complex, **147,** 109–110
Bactericides, mechanisms, macrophage
cytokine production, **161,** 310–313
oxygen-dependent, **161,** 306–307
oxygen-independent, **161,** 307–310
Bacteriophages, microgravity experiments, **156,** 303–305
Baculovirus p35 protein, effects on cell death, **163,** 129
BALT, *see* Bronchus-associated lymphoid tissue
Banding techniques, chromosome, **167,** 264

Barley
flowering plants and
amino acid pathways, **143,** 320, 322
ammonium, **143,** 304–306
nitrate, **143,** 300–302
microgravity experiments, **156,** 322, 325
plant-pathogen interactions in, **144,** 58
plastids and
metabolism, **145,** 184–186, 188, 197, 199
structure, **145,** 162–163
urease in plant cells and, **145,** 81
Basal body
associated fibrillar networks
basal foot in ciliated epithelium, **173,** 208–209
ciliated protozoa
centrosomes, **173,** 214
ciliary motility, **173,** 213
MT–microfilament complex, **173,** 211–213
proximal structures, **173,** 211
neurosensory epithelium
cochlea hair cells, **173,** 210
MT–microfilament interactions, **173,** 210
photoreceptor cells, **173,** 209–210
avian retina and, **146,** 17, 19
bacterial flagellar motor and
models, **147,** 108–110, 131, 141, 143
structure, **147,** 99–104, 108–110
complex of volvacacean vegetative cells, **139,** 192–194
–hook complex, *see* Hook-basal body complex
Basal foot, in ciliated epithelium, **173,** 208–209
Basal lamina
avian retina and, **146,** 3, 16, 21
avian salt glands and, **144,** 135, 159
Base excision repair, **156,** 124–125
Basement membranes
associated proteins in retinal pigment epithelial cells
phenotype expression role, **150,** 126–127
transdifferentiation role, **150,** 128–129, 131
avian salt glands and, **144,** 138, 149, 159
effects on retinal pigment epithelium, **152,** 253
glomerular, thickness estimation, **161,** 161–162
as specialized ECM, **173,** 75
Bases, weak, accumulation in cells, **171,** 149–153
Basophils, karyosphere and, **144,** 25
Bat star, sperm-binding proteins, **163,** 279–280
Bax, role in apoptosis, **157,** 255
Bay K-8644, **156,** 289
BB10010, *see also* Macrophage inflammatory protein-1α
hemopoietic progenitor cell mobilization by, **167,** 242–243
protective effects, **167,** 241
B cell receptor
antigen binding subunit, **157,** 182–184
antigen recognition, **157,** 189–191
antigen recognition activation motif, **157,** 184, 188, 231–233, 239–240
B cell activation, **157,** 202–204, 252–256
components, **157,** 185–187
coreceptors, **157,** 183, 215
cross-linking, **157,** 207–208
phosphorylation, **157,** 241
role
apoptosis, **157,** 208, 250–256
tolerance, **157,** 209–210
signal transduction
cell cycle entry, **157,** 248–250
defects, **157,** 183
phosphatidylinositol 3-kinase pathway, **157,** 227–231
phospholipase C pathway, **157,** 216–221
Ras pathway, **157,** 221–226
regulation of gene expression, **157,** 243–248
tyrosine kinases, **157,** 235–236
structure, **157,** 184–188
B cells
activation, **157,** 204–208
apoptosis, **157,** 142, 183, 197, 208–209
CD5 expression, **157,** 133
clonal deletion, **157,** 197
clonal energy, **157,** 198
cytokine inhibition of progenitor growth, **157,** 154
follicular dendritic cell
CD40 ligation, **166,** 157–158
cytokines, **166,** 157–158

proliferation, **166,** 156–157
survival, **166,** 156–157
in vitro study, **166,** 156–158
germinal center cells, **157,** 208–209
lineage commitment, **157,** 160–162
lipopolysaccharide assay, **157,** 147
markers of differentiation
AA4.1, **157,** 136, 142
B29, **157,** 141
B220, **157,** 137–138, 141–142
BP-1/6C3, **157,** 138
CD10, **157,** 139
CD43, **157,** 137–138
CD45, **157,** 137–138
cell size, **157,** 141
γ5, **157,** 140–141
heat-stable antigen, **157,** 138, 142
immunoglobulin proteins, **157,** 135–136
Ly6A, **157,** 136–137, 142
mb-1, **157,** 141
Rag proteins, **157,** 139–120
terminal deoxynucleotidyltransferase, **157,** 140
time of expression, **157,** 134, 163–164
memory B cells, **157,** 187
negative selection, **157,** 196–198
ontogenesis in mouse, **157,** 132–133, 143–144
peripheral tolerance, **157,** 209–210
positive selection, **157,** 192–193
requirements for lineage progression, **157,** 146–147
role in differentiation
B220, **157,** 156
bone marrow microenvironment, **157,** 144, 163
flt-3/flk-2, **157,** 152
IGF, **157,** 153–154
immunoglobulin proteins, **157,** 156–158
interaction with stromal cells, **157,** 143–145, 148–149
interleukin-7, **157,** 145–146, 148, 150–151, 158–159
interleukin-11, **157,** 152–153
mast cell growth factor, **157,** 151–152
signal transduction, role of CD22, **175,** 198–199
systems for antigen-specific stimulation, **157,** 191
tolerance, **157,** 181
transcriptional regulation and function, **143,** 2–7
genes, **143,** 14, 18
molecular cloning, **143,** 26, 29
viruses, **143,** 40
Bcl-2
dimerization, **157,** 255
gene role in cell death, **157,** 254–255; **163,** 119–120
BCNU, *see* 1,3-Bis(2-Chloroethyl)-*N*-nitrosourea
Beans
environmental stimuli in, **145,** 290
plastids and, **145,** 183, 185–186, 189, 194, 203
urease in plant cells and, **145,** 68, 76–77, 89–91
Beat cycle, flagella
effect of nucleotides, **173,** 30–31
mechanics, **173,** 31–34
minimal requirements, **173,** 29–31
role of Ca^{2+}, **173,** 30–31
Beckwith–Wiedemann syndrome
chromosomal deletions, **160,** 86
gene imprinting role, **160,** 86–87
Beetles, karyosphere and, **144,** 18–25
Behavior
associated effects of CRF, **138,** 110–112
cycles, amoeboid cells, **163,** 79–80
effect of brain transplants, **175,** 291
rhythmic, control, **147,** 1–3
chemical modulation, **147,** 18–20
ion channels, **147,** 3
experimental approaches, **147,** 4–5
invertebrate CPGs, **147,** 5–9
neuronal release, **147,** 9–10
physiology, **147,** 10–11
neuromodulation
circuitry, **147,** 13–15
concentration effect, **147,** 11
interaction between networks, **147,** 15–17
sensory feedback, **147,** 17–18
single neurons, **147,** 11–13
Benazeprilat, inhibition of smooth muscle cell growth, **161,** 24–25
Benzodiazepines
anxiolytic effects, **149,** 266–267
GABA affected by, **149,** 263–264
treatment of pituitary tumors, **157,** 47–48

Benzo-1,2,3-thiadiazole carbothioic acid-S-methyl ester, systemic acquired resistance induction, **168,** 327–328
Bergenholtz's system, cell culture technology, **142,** 160, 162
Berg-Khan model, bacterial flagellar motor end, **147,** 146, 150–151, 153–159
Betacellulin, stimulation of vascular cells, **161,** 12–13
Betaines, **156,** 9, 15; **165,** 8, 11–12
Betula verrucosa, **140,** 11
Bezafibrate, effects on peroxisomes, **160,** 303, 316
bFGF, *see* Fibroblast growth factor, basic
Bibenzyl synthase, chalcone synthase and, **147,** 237, 243
Bicarbonate, conductance
  by CFTR, **172,** 199
  in pancreatic duct cells, **172,** 211
Biliary glycoprotein 1
  sequence homology with ecto-ATPases, **158,** 147
  structure, **158,** 148–149
Bimodal theory
  drag coefficients, **163,** 19–20
  empirical, **163,** 23–24
  fixed-section pores
    P_f/P_d ratios, **163,** 26–30
    P_{os} values, **163,** 30–34
  kinetics, **163,** 23
  leaky pores, **163,** 21–26
  osmotic conductance, **163,** 24–26
  pore radius prediction, **163,** 26–30
  reflexion coefficient, **163,** 24–26
  semipermeable pores, **163,** 19–21
  thermodynamics, **163,** 21–22
  variable-section pores, **163,** 34–36
Binding domains, nuclear localization signal, **162A,** 33–34
Biochemical studies
  embryo sacs and their component cells, **140,** 283–284
  GSI
    identification of S allele-associated proteins, **140,** 459–461
    temporal and tissue-specific expression of S alleles, **140,** 461–463
    in vitro assays of S protein function, **140,** 463–465
  isolated cells, **140,** 226–228
    immunological characterization, **140,** 226–227
    polypeptide-protein characterization, **140,** 227–228
  microtubules of GSP, **140,** 159–160
  pollen actins, **140,** 87–88
  sporopollenin biosynthesis, **140,** 55–61
    application of inhibitors, **140,** 57–58
    tracer experiments, **140,** 58–61
Biogenic amine receptors, insect
  characterization, **164,** 214–216
  dopamine, **164,** 218–219
  function, **164,** 218
  histamine, **164,** 223–224
  5-hydroxytryptamine, **164,** 219–220
  localization, **164,** 216–218
  phenolamines, **164,** 222–223
  release, **164,** 215–216
Biological clocks, in retina
  Bulla, **146,** 128–129
    rhythm generation, **146,** 134–139
    transmembrane loop, **146,** 129–133
  entrainment pathway, **146,** 100–101, 115
    efferent regulation, **146,** 108–114
    membrane potential, **146,** 102–103
    photoreceptor, **146,** 101–102
    protein synthesis, **146,** 106–108
    second messengers, **146,** 103–106
  experimental background, **146,** 83–86
  free-running rhythms, **146,** 96–97
  opisthobranch retina
    functional organization, **146,** 88–93
    ocular rhythm, **146,** 86–88
    pacemaker localization, **146,** 93–95
  oscillator, **146,** 99–100
  output pathway, **146,** 123, 126–128
  pacemaker mechanisms, **146,** 115
    intracellular processes, **146,** 116–125
    membranes, **146,** 115–116
  pathway analysis, **146,** 97–99
Biological functions
  pollen wall substances, **140,** 50–51
  stylar matrix in pollen tube extension, active
    latex bead translocation, **140,** 304–305
    model for pollen tube extension via stylar matrix, **140,** 312–313

occurrence of substrate adhesion molecule vitronectin in plants, **140,** 307–312
stylar matrix of *Vicia faba,* **140,** 305–307
Biology
developmental, *see* Developmental biology
evolutionary, goals, **174,** 68–71
Biomineralization, *see* Eggshells, biomineralization and
Biopolymer, cholesteric ordering, *in vitro,* **166,** 84–92
Biopsy
cleavage-stage, **153,** 5
polar body, **153,** 4
Bioreactors, vertebrate cell culture technology
commercial scale, **142,** 167–187
optimization, **142,** 195–197, 199–201
packed-bed reactors, **142,** 187–192
production system, **142,** 162–167
Biosatellite II, microgravity experiments
microorganisms, **156,** 340
plants, **156,** 303–304, 321–322, 324, 326
Biosynthesis
arabinogalactan proteins
AGP turnover, **174,** 238–239
carbohydrate component, **174,** 236–238
hydrolytic enzymes, **174,** 239–240
polypeptide component, **174,** 235–236
cell wall precursors
callose, **176,** 172
cellulose, **176,** 171–172
pectins, **176,** 170–171
cnidarian neuropeptides
of L-3 phenyllactyl group in, **167,** 49*f*
Antho-RFamide
in sea anemones, **167,** 58–64
in sea pansies, **167,** 64–67
AnthoRPamide I in sea anemones, **167,** 68–69
Antho-RPamides II-IV in sea anemones, **167,** 69–71
higher animals *versus,* **167,** 57–58
in hydrozoans, **167,** 74–77
metamorphosis-inducing in sea anemones, **167,** 71–74
lignins, **176,** 101–102
associated enzymes, **176,** 107–108
carbon flow regulation, **176,** 119–122
cell-to-cell cooperation during, **176,** 108–109
monomeric composition regulation, **176,** 122
polymerization step, **176,** 109–111
role of laccase, **176,** 111–114
role of peroxidase, **176,** 114–117
Biotin
-dUTP
suboptimal concentration, **162A,** 134–136
in visualization of permeabilized cells, **162A,** 43–48
higher plant plastids and, **145,** 182, 190
uptake by receptor-mediated endocytosis, **167,** 19
Bipolar cells, avian retina and, **146,** 2–3, 5, 28
Birds
avian salt glands and, **144,** 129–130, 205
blood supply, **144,** 151
hormonal control, **144,** 191, 195, 198
secretory mechanism, **144,** 177–178
secretory tissue, **144,** 131–132, 140–141, 147
biological clocks in retina, **146,** 84–85
bursa of Fabricus in, **167,** 93, 135
karyosphere and, **144,** 25, 37–38
microgravity experiments, **156,** 334
parathyroid hormone-related protein, **166,** 262–263
retinal histogenesis in reaggregation culture, **146,** 1–47
retinal precursor cells, **146,** 173
Blaps, karyosphere and, **144,** 19, 21, 25
Blastocysts
cell death during development, **163,** 134
confined mosaicism arising in, **167,** 288–289
programmed cell death in, **163,** 134
in vitro culture effects on hatching, **164,** 165–166
Blastomeres
development
in ascidian embryo, **176,** 257–263
instructive and permissive induction, **176,** 286

Blastomeres (*continued*)
multinuclear, **164,** 160–161
platelet-derived growth factor activity in, **172,** 112
Blebbing, tissue cell, definition, **150,** 40
Blood
coagulation, cellular mechanisms activating
in pathological conditions, **152,** 89
phospholipid-dependent activity, **152,** 54–55
process, **152,** 49
procoagulants
adhesion receptor regulation of, **152,** 75–77
expression suppression, **152,** 86–89
and malignancy, **152,** 82–86
properties, **152,** 77–82
tissue factor
characteristics, **152,** 50–51
as cytokine growth factor superfamily member, **152,** 51–53
function, **152,** 53–55
localization, **152,** 55, 57–74
dendritic cells
adherence depletion, **153,** 49
contaminating cells, removal, **153,** 51–52
density-gradient separations, **153,** 49, 51
fresh cell isolation, **153,** 52–53
immunophenotypic cell separations, **153,** 52
isolation protocols, **153,** 48–53
panning techniques, **153,** 51–52
phagocyte depletion, **153,** 52
properties, **153,** 53
ependymin, **165,** 147
nuclear magnetic resonance and, **145,** 29
Blood–brain barrier
astrocytes, **159,** 31
avian salt glands and, **144,** 138, 182, 184, 195
basement lamina, **159,** 31–32
development, **159,** 90
endothelium, **159,** 31
ependyrmn, **165,** 121–151
neurotransmission, **149,** 277–278
pericytes, **159,** 31
permeability, **159,** 25, 31–32
structure, **159,** 24, 30–32
Blood flow
adrenal gland
effect on endothelin release, **171,** 291–292
regulation by endothelin, **171,** 284–285
cerebral, nuclear magnetic resonance and, **145,** 32
Blood supply, avian salt glands and
innervation, **144,** 157, 168
microvasculature, **144,** 148–151
receptive systems, **144,** 182, 184–185
secretion, **144,** 151–154
secretory mechanism, **144,** 173, 175–176, 178
secretory tissue, **144,** 131–132, 141
Blood–testes barrier
development, **160,** 182–183
inter-Sertoli cell junctions, **160,** 183–186
nutrient transfer, **160,** 194
permeability, **160,** 185
related proteins, immunocytochemistry, **160,** 207
Blood–tissue barrier
endothelium and epithelium types, **172,** 6–8
glucose transporters in, **172,** 8
Blood vessels
ADPase activity, **158,** 160–161
development, platelet-derived growth factors in, **172,** 116–117
ecto-ATPase levels, **158,** 160–161
supply Peyer's patches, **167,** 95
Blue fluorescent plant mutants, **143,** 315–317
B lymphocytes
in bursa of Fabricus in birds, **167,** 93
excess TGF-β production by malignant, **167,** 237–238
MIP-1α gene expression in activated, **167,** 199
in Peyer's patches
in dome epithelium, **167,** 133–135
in lamina propria, **167,** 98
in lymphoid follicles, **167,** 95
migration through mucosal system, **167,** 93
regulation and, **143,** 22
B lymphocytopenia, **167,** 94–95

Body plan
  Drosophila, regulatory genes, **174,** 30–33
  insect, patterning, **174,** 45–48, 53–54
    Dll and *cnc* expression, **174,** 52–53
    initiation and segment-specific stages, **174,** 48–52
Boergesenia, environmental stimuli in, **145,** 295
Boergesenia forbesii, protoplast wall regeneration, **170,** 44–45
Bombesin, **148,** 137–138; **166,** 20
Bombesin-related peptides, **138,** 161–164; **166,** 20
Bombina bombina, **138,** 161
Bombina orientalis, **138,** 161
Bombina variegata, **138,** 161
Bombyx mori
  male eupyrene meiosis, **152,** 19
  mitosis and, **144,** 225
Bone
  bovine, as bone graft material, **172,** 162–164
  crystal structure and composition, **172,** 143–145, 165–167
  ectopic endochondral bone formation, **159,** 325
  effects of sintering on, **172,** 162–164
  extracellular matrix, **166,** 75–77
  fracture repair
    callus formation, **159,** 323–324
    collagen types, **159,** 323
    role of growth factors, **159,** 324
  innominate, effect of sex hormones and antihormones on, **139,** 31–33
  interactions with bone graft materials, **172,** 134
  osteoporosis in, **172,** 152, 154
  regenerative cementogenesis and, **142,** 1–2, 4
    established root surfaces *in vitro,* **142,** 28–29, 33–34, 36, 40–42
    established root surfaces *in vivo,* **142,** 44, 46, 49–50
    growing root surfaces, **142,** 17, 19, 23
Bone grafts
  calcium phosphate materials in, **172,** 133, 159–161
    from bovine bone, **172,** 162–164
    coralline, **172,** 161
    implant coatings and, **172,** 165
    interactions with bone, **172,** 134
    synthetic, **172,** 164
  ceramic material interface
    bioactive processes and, **172,** 174–175
    bone ingrowth and, **172,** 173–174
    carbonate apatite deposition in, **172,** 167, 171–173, 175, 176–177
    cell colonization of, **172,** 168–170, 174–175
    crystal biodissolution in, **172,** 167, 170–171
    crystal deposition in, **172,** 171–173
Bone marrow
  mitosis and, **144,** 225
  transcriptional regulation and, **143,** 14
Bone marrow cells
  dendritic
    ontogeny, **153,** 69–70
    studies, **153,** 48
  growth regulation
    leukemia-associated failure and, **167,** 234–235, 237
    microenvironmental influences, **167,** 194–196
  inhibitory fraction blocking CFU-S into DNA synthesis, **167,** 191–192
  long-term culture
    AcSDKP in preventing onset of DNA synthesis in, **167,** 221–222
    AcSDKP synthesis in, **167,** 223
    MIP-1α in regulating proliferation of, **167,** 202–204
    TGF-β inhibition of hemopoietic progenitor cell proliferation in human, **167,** 211–212
  MIP-1α and
    colony formation of progentitor, **167,** 201
    suppression in excess production of, **167,** 239
  multilineage colonies formed from, **167,** 191
  protection
    clinical perspectives, **167,** 239–242
    by glutathione against cyclophosphamide treatment, **167,** 226
  repopulating
    pEEDCK effect on, **167,** 218
    in stem cell continuum, **167,** 190
  TGF-β administration to, **167,** 212

Bone morphogenetic proteins
characterization, **159,** 325
effect on chondrocyte differentiation, **159,** 269, 309–310
Bordering elements, nucleosomally organized, **162A,** 400–403
Bordetella pertussis
associated toxin, **161,** 274–275
binding to sialic acids, **175,** 181
Boron, effect on pollen tube tip growth, **176,** 179–180
Botrytis, accessbility induction and, **172,** 62, 64–65
Bovine serum albumin, advanced glycosylation end products, **152,** 70
Boyden chamber assay, **139,** 180–182; **155,** 53, 55–57
BP-1/6C3 protein, marker of B cell differentiation, **157,** 138
Brachydanio revio, retinal precursor cells and, **146,** 171
Bradykinin, astrocyte reaction to, **148,** 127, 129, 137
Bradyrhizobium elkanii, Nod factors, **156,** 21–22
Bradyrhizobium japonicum, **156,** 8, 12, 13, 17
control elements in, **156,** 17–18
host-specific modifications, **156,** 25–27
Nod factors, **156,** 21
Bradyrhizobium parasponium, **156,** 12
Brain
amphibian
complexity, **169,** 84
development
cell differentiation, **169,** 129–130
cell migration, **169,** 130–134, 137
central nervous system segmentation, **169,** 96–97, 125–129
genome size effects, **169,** 87–88, 138
hormonal control, **169,** 134–136, 138
motor system, **169,** 97–106, 136
neural induction, **169,** 89–91, 93–94, 121–125
neural plate, **169,** 94–96
neuronal migration, **169,** 91
olfactory system, **169,** 106–110, 136–137
segregation of brain parts, **169,** 90, 136
visual system, **169,** 110–115, 117–118, 120–121, 137
heterochrony
cell migration effects, **169,** 134
global, **169,** 85, 87
life history modification, **169,** 88–89
local, **169,** 87
role in morphological differences between species, **169,** 84–85, 137
morphology, **169,** 84–85
suitability as developmental model, **169,** 83–84
anuran, diagram, **138,** 177–178
avian
development, molecular polymorphism of cholinesterases during
biosynthesis, **151,** 149
early neurogenesis, **151,** 144–145
membrane association, **151,** 146–147
retina, mature, **151,** 146
retina and, **146,** 1–2
avian salt glands and
hormonal control, **144,** 195, 197
receptive systems, **144,** 180, 182, 184, 187, 189
–blood barrier
glucose transport, **172,** 10–13
structure, **172,** 9
cytoskeletal dynamics and, **144,** 109–110
dendritic spines, *see* Dendritic spines
derived neurotrophic factor
neurite initiation and elongation role, **153,** 247, 251–252
role in apoptotic processes, **163,** 118
FAK isoform specific for human, **167,** 168
fluid, distribution in ependymin, **165,** 144–147
growth factor-induced retinal regeneration and, **146,** 50, 56
hormonal control of metamorphosis and, **145,** 107, 137, 141
in human development, **158,** 221–222
induction in ascidian tissue formation, **176,** 278–279
myofibrillogenesis and, **143,** 166–168
nuclear magnetic resonance and, **145,** 19, 24, 29, 32, 35, 43–45
parathyroid hormone-related protein, **166,** 251–252

quail–chick chimeric, neuronal connections in, **175,** 263–264
sexually dimorphic nucleus, **139,** 26–28
spectrin, **139,** 284–287
structural relationships, **175,** 242–243
suprachiasmatic nuclei, biological clocks in retina and, **146,** 84–85, 132
transplantation
effect on neuronal circuits, **175,** 291–297
role in reflex epilepsy transfer, **175,** 294
structures in neuroepithelium
regional specifications, **175,** 271–272
tectal and cerebellar structures, **175,** 266–271
transplantation effects, **175,** 264–266
urease in plant cells and, **145,** 83
Brain inhibitory protein, **139,** 278–279
Brain natriuretic peptide, astrocyte reaction to, **148,** 127–128, 132
Brain-specific regulators, transcriptional regulation and, **143,** 46
Brainstem, avian salt glands and, **144,** 187–188
Branchial arches
development and Hox *7-msh* family, genes, **138,** 279–281
in head development, **158,** 217–218
Branching enzyme, higher plant plastids and, **145,** 177
Brassica
endosymbiont hypothesis and, **141,** 292
flowering plants and, **143,** 324
genetic tumors, **152,** 280
higher plant mitochondrial genomes and, **141,** 142, 145, 149, 157
recombination, **141,** 152–154
structure, **141,** 136–138, 141, 143
plant-pathogen interactions in, **144,** 65, 69
plastids and, **145,** 185–186
pollen grains, proteins of, **140,** 551–554
self-incompatibility and, **143,** 258, 289
genetics, **143,** 259, 261
glycoproteins, **143,** 265, 269–270
sporophytes, **143,** 262–264, 272–273, 277, 279
S gene family, **140,** 503–509
SLG expression in *Brassica* stigmas, **140,** 512–514
SLR1 subfamily, **140,** 503–506
SLR2 subfamily, **140,** 506–507
S gene family, sporophytic self-incompatibility systems
conclusions, **140,** 521–522
expression of S homologous sequences in anthers
overview, **140,** 517–518
PCR detection of SLG-homologous transcripts in anthers, **140,** 517–521
expression of S sequences
SLG expression in *Brassica* stigmas, **140,** 512–514
SLG gene and promoter constructs in transgenic plants, **140,** 514–517
genomic organization of S gene family, **140,** 496–498
incompatibility systems, **140,** 485–486
isolation of *Brassica S* sequences
isolation of *SLG* and *SRK* sequences, **140,** 487–493
isolation of *SLR* sequences, **140,** 494–496
nomenclature, **140,** 487
molecular evolution of S gene family, **140,** 498–514
S gene structure, **140,** 511–512
stigma, proteins
history, **140,** 527–528
homology modeling of female glycoproteins, **140,** 544–551
other families of glycoproteins homologous of S locus-specific glycoproteins, **140,** 533–543
primary structure of S locus-specific glycoproteins, **140,** 544–551
Brassica campestris
higher plant mitochondrial genomes and, **141,** 141–142, 149, 152, 161
pollen-specific genes from
mRNAs in with clone *Bcpl,* **140,** 26
S-glycoprotein gene promoter, **140,** 33
Brassicaceae
difficulty of pollen germination, **158,** 19
self-incompatibility bioassay, **158,** 18–19
Brassica napus
Bp4 from, **140,** 14

Brassica napus (*continued*)
microgravity experiments, **156,** 320
nucleotide sequence, **140,** 9
pollen, **140,** 38
pollen-specific genes from, **140,** 20
protoplast culture, effect of microgravity, **171,** 44–45
protoplast wall regeneration, **170,** 42–43
Brassica oleracea
higher plant mitochondrial genomes and, **141,** 141–142
intracellular calcium messenger system and, **142,** 328
primary structure, **140,** 533
protoplast wall regeneration, **170,** 46–47
Brassica perviridis, microgravity experiments, **156,** 329
Brassinosteroids, regulation of XRP gene, **173,** 189–191
Breast, parathyroid hormone-related protein, **166,** 259
Breast cancer
cells, NMR, **145,** 37, 39
cells with melphalan resistance, **156,** 129
characteristics, **155,** 197
coiled body content, **159,** 181–182
effects of estrogen, **159,** 181
high-density cultures, **155,** 198–199, 211, 213
mdr1 gene expression, **156,** 88
mortality, **159,** 181
prevalence, **159,** 181
ultrastructure, **155,** 227
Brettanomyces, mitochondrial genomes in fungi, **141,** 101–102
Brettanomyces anomalus, mitochondrial genomes in fungi, **141,** 98, 102–103
Brettanomyces custersianus, mitochondrial genomes in fungi, **141,** 117–119
Brevin, proteins in striated muscle and, **142,** 110
Brine shrimp, microgravity experiments, **156,** 336
Bristles, *Drosophila*
actin cross-linking proteins, mutations
affecting development, **175,** 49–50
affecting oogenesis, **175,** 46–49
development, **175,** 44–46
Brockmann body, *see* Pancreas, fish
Bromocriptine
effect on pituitary cell maturation, **157,** 56
treatment of pituitary tumors, **157,** 47
Bronchus-associated lymphoid tissue, **167,** 136–137
Brush border
in dome epithelium of rats and mice, **167,** 130
enterocytes
Escherischia coli enterotoxin binding to, **167,** 139
versus M cells, **167,** 123
myosin I, **149,** 149–150
Bryopsis
cytoplasmic streaming supported by microtubule system, **155,** 109
environmental stimuli in plant cells and, **145,** 270, 294–295
Btk, activation by B cell receptor, **157,** 212
Bufo calamita, hormonal control of metamorphosis and, **145,** 113
Bufo icterus, **138,** 152
Bufo japonicus
control of metamorphosis and
adrenocortical hormones, **145,** 119–120, 122–123, 127
growth hormone, **145,** 138
prolactin, **145,** 130, 135
thyroid hormone, **145,** 107, 115
localization of arginine-vasotocin and mesotocin, **138,** 120
Bufo marinus, **138,** 101, 131, 145, 147–148, 166, 168–169; **145,** 135
Bufo melanostictus, hormonal control of metamorphosis and, **145,** 113
Bufo vulgaris, **138,** 95
Bufo woodhousei, hormonal control of metamorphosis and, **145,** 140
Bulla, biological clocks in retina and, **146,** 85, 139
concepts, **146,** 100
entrainment pathway, **146,** 101–108, 111–113, 115
opisthobranch retina, **146,** 88–90, 92–93, 95
output pathway, **146,** 123, 128
pacemaker mechanisms, **146,** 116, 118, 121–123
rhythm generation, **146,** 134–139
transmembrane loop, **146,** 129–133
Bullough's ear model, **167,** 187–188
Bullous pemphigoid antigen, intermediate plaque linkage in hemidesmosomes, **164,** 106–107
Bumetanide, and avian salt glands, **144,** 154

Bundling, actin filaments, **156,** 239–240, 249–252
α-Bungarotoxin, **147,** 204, 212; **156,** 187
Burkitt lymphoma, **156,** 103, 131
Bursa of Fabricus, **167,** 93, 135
Bursatella, biological clocks in retina and, **146,** 86, 113
Bursting properties, control of rhythmic behaviors and, **147,** 12, 15
Buthionine sulfoximine, **156,** 96–97, 136
Butyrate, supporting S/MAR functions, **162A,** 423–425
Butyrylcholinesterase, *see* Cholinesterase

C

E-Cadherin
  epithelial cell junctions, **160,** 223, 228, 230
  metastasis role, **160,** 231–232, 251–252
  mutation and mouse embryogenesis, **160,** 228, 230, 251
N-Cadherin
  avian retina and, **146,** 24
  role in chondrocyte aggregation, **159,** 294
T-Cadherin
  properties of superfamily, **158,** 149
  sequence homology with ecto-ATPases, **158,** 149
Cadherins
  catenin
    complex assembly dynamics, **165,** 166–168
    interactions between, **165,** 163–170
    protein-protein interaction hierarchy, **165,** 164–166
  cell adhesion protein, **165,** 160–163
  choroid plexus, expression, **165,** 174–177
  cytoplasmic protein
    complex assembly dynamics, **165,** 166–168
    interactions between, **165,** 163–170
    protein-protein interaction hierarchy, **165,** 164–166
  desmosomal, *see* Desmosomal cadherin
  embryogenesis, **165,** 181–186
    developmentally regulated expression, **165,** 182–183
    selective cell adhesion, **165,** 181–182
  fat-like
    functional diversity, **165,** 161–162
    structural similarities, **165,** 161–162
  fibroblast, expression effects, **165,** 172–174
  function
    cell function change regulation, **165,** 172–179
    cell structure change regulation, **165,** 172–179
  functional diversity, **165,** 163
  keratinocyte, expression, **165,** 177–178
  mediated cell–cell adhesion
    catenin
      consequence, **165,** 170
      function, **165,** 170–172
    embryogenesis, requirements, **165,** 184–186
  membrane-cytoskeleton, **165,** 192
  role in amphibian brain development, **169,** 134
  Sertoli-germ cell communication network and, **147,** 32–33
  structural similarities, **165,** 163
  tumor cell, expression, **165,** 178–179
Caenorhabditis, endosymbiont hypothesis and, **141,** 288
Caenorhabditis elegans
  animal mtDNA and, **141,** 173, 202–206
    genome content, **141,** 176–178, 180
    rRNA, **141,** 182–183, 187–188, 190–191
    tRNA, **141,** 196–198
  Drosophila muscles and, **143,** 97, 140
  endosymbiont hypothesis and, **141,** 282, 289
  LRP gene, **166,** 127–128
  myofibrillogenesis and, **143,** 180, 183
  programmed cell death during development, **163,** 123–126
Caffeine
  cytoplasmic screening, **156,** 208–210
  droplets in *Physarum* microfilament system analysis, **149,** 154, 175–177, 190–191
Caffeoyl-CoA 3-*O*-methyltransferase, in phenylpropanoid pathway, **172,** 258
Cairns structures, intermediate and latest, **162A,** 496–497,509
Calbindin, eggshells and biomineralization and, **145,** 235
Calcification
  articular, crystal structure in, **172,** 147
  decalcification, **145,** 32, 36, 42, 46, 227, 229, 231

Calcification (*continued*)
eggshells and biomineralization and, **145,** 217–218, 223, 226, 229, 234, 236, 241, 243–244
regenerative cementogenesis and, **142,** 35, 47
Calcified tissue
analytical techniques in
crystal composition, **172,** 138–139
crystal localization, **172,** 135
crystal morphology, **172,** 137–138
crystal phase identification, **172,** 135–137
organic phase removal, **172,** 135
bone, crystal structure and composition, **172,** 143–145, 165–167
crystal composition, **172,** 129–130
invertebrates, **172,** 130
pathological, **172,** 130
dentin, crystal structure and composition, **172,** 142–143, 165–166
diseased conditions
amelogenesis imperfecta, **172,** 152
dentin caries, **172,** 149–151
enamel caries, **172,** 148–149
fluorosis, **172,** 151–152
osteoporosis, **172,** 152, 154
enamel, crystal structure and composition, **172,** 139–142, 165–166
fish enameloids, **172,** 145
pathological
dental calculus, **172,** 147–148
mineral phases, **172,** 145, 147
soft tissue calcifications, **172,** 147
urinary calculus, **172,** 148
whitlockites in, **172,** 157–158
Calcineurin
ENAI, **165,** 33
role in T cell receptor signaling, **157,** 218–219
Calcite, eggshells and biomineralization and, **145,** 238–240
Calcitonin
astrocyte reaction to, **148,** 146–147
encoding gene, related peptide, **138,** 164–166
neuromuscular junction and, **147,** 200, 220–227
location, **147,** 201, 204–205
phenotypic expression affected by, **153,** 275
secretion by neuroendocrine cells, **157,** 283–284, 298–299
Calcium
Acetabularia
cytoskeletal dynamics, **150,** 227, 229–231
kinetics, **150,** 240–241
in morphogenesis simulation, **150,** 232–236
phenocopying, **150,** 238
acrosomal exocytosis role, **149,** 9–11, 14–15, 17
channels, **149,** 31–39
ionic changes, **149,** 37–38
sodium requirements, **149,** 31–37
in activation of plant defense response, **148,** 20–22
agonist-induced Ca^{2+}-dependent volume changes, **161,** 218–219
γ-aminobutyric acidergic receptors affected by, **149,** 264
avian retina and, **146,** 23
avian salt glands and, **144,** 184, 200, 203–205
bacterial flagellar motor and, **147,** 152
binding transition, ependymin, **165,** 127–128
biological clocks in retina and *Bulla,* **146,** 128–138
entrainment pathway, **146,** 103–105, 109, 115
opisthobranch retina, **146,** 91
bone source in chick embryo, **159,** 283, 325
caged compounds, **158,** 43
control of rhythmic behaviors and, **147,** 3, 9, 12
cytoplasmic Ca^{2+}, **139,** 273
cytoplasmic concentration, and cytoskeletal dynamics, **148,** 70
cytoplasmic streaming affected by
action potential role, **155,** 120–125
gravity role, **155,** 128–129
inhibition in Characeae, **155,** 118–120, 130
temperature role, **155,** 126–127
cytoskeletal dynamics and, **144,** 86, 88, 102, 106–108

cytosolic concentration, effect of ET, **171,** 279–280
deficiency effect on cartilage phenotype, **159,** 325
detection methods, transmembrane calcium transfer, **168,** 129
in *Drosophila* muscles
contractile protein genes, **143,** 98–99, 133, 135
development, **143,** 66
myosin light chains, **143,** 105, 108
troponin, **143,** 117, 119
Z-disc proteins, **143,** 125
effect on flagella, **173,** 20–21
eggshells and biomineralization and, **145,** 217, 241
fabrication, **145,** 236, 238, 240
structure, **145,** 223, 227, 231, 234
environmental stimuli in plant cells and, **145,** 298, 300–301
chloroplasts, **145,** 279–283, 286
low temperature, **145,** 289–290
nucleus, **145,** 288
photodinesis, **145,** 261, 263–268
wounding, **145,** 294, 296
extracellular, agonists/antagonists effect on cell division
modulation of extracellular calcium, **138,** 248–252
modulation of phosphatidylinositol cycle, **138,** 252–254
extracellular matrix and
modification, **146,** 292–293, 300
pathogenesis, **146,** 307, 312–313
synthesis, **146,** 282
in fish chromatophores, **143,** 250
motility control, **143,** 227–228, 230–231, 235
signal transduction, **143,** 239, 242–245
flux, electromagnetic field effect
low-frequency fields, **158,** 296–299
radiofrequency fields, **158,** 296
free intracellular, in RVD response, **161,** 215–216
in fungal signal transduction cascade model, **172,** 83
and glial cells, **149,** 275
glutamate metabolism and, **147,** 294, 305, 307
GnRH-encoding gene in teleosts and, **147,** 170
gradients, and Ca^{2+}-cytoskeleton interactions in pollen tubes, **140,** 80–83
heat shock-induced increase, **148,** 205
influx in RVD response, **161,** 215
inhibition of microtubules, **157,** 12–13, 17
integrin signal transduction, **168,** 95–96
intracellular, effect of microgravity, **171,** 39–40
in long-term potentiation of dendritic spines, **139,** 289–290
in mitosis
conclusions, **138,** 264
effect of extracellular calcium agonists/antagonists on cell division
modulation of extracellular calcium, **138,** 248–252
modulation of phosphatidylinositol cycle, **138,** 252–254
endogenous calcium transients during mitosis, **138,** 244–248
intracellular modulation of calcium and related metabolism, **138,** 254–261
mitotic apparatus related to calcium regulation, structure and composition
calmodulin and calcium binding proteins, **138,** 241–244
membranes, **138,** 240–241
microtubules, **138,** 239–240
summary, **138,** 244
regulation between calcium and cell-cycle regulatory proteins, **138,** 261–263
myofibrillogenesis and, **143,** 165, 175
neurite outgrowth regulatory role, **153,** 257
and neuromediation, **149,** 227–231
nuclear magnetic resonance and, **145,** 3, 20
in organization of actin cytomatrix, **139,** 277–278
phosphorylation-altered activity, **154,** 345–346
in phototactic signal transduction, **148,** 290–295, 297–300
in plant defense transmembrane signaling, **172,** 70–71

Calcium (*continued*)
plants
functions, **168,** 125
intracellular concentration, **168,** 125–126
transport mechanisms, **168,** 126–129
in polarization of neutrophils, **139,** 173–174
post-translational modification, intermediate filament protein activation by, **154,** 46–47
proteins in striated muscle and, **142,** 127
sarcoplasmic reticulum, **142,** 119–124
skeletal muscle, **142,** 64, 81, 85, 94, 102
in protoplast wall, effect of microgravity, **170,** 43
quantitation, **158,** 16–17, 43, 296–298
regulation of EF-1α–F-actin interactions, **164,** 24
role
in acrosomal exocytosis, **149,** 27–28
in apoptosis, **163,** 114–115
in axonemal motility regulation, **164,** 288–289
in beat cycle, **173,** 30–31
in catecholaminergic neurotransmission, **149,** 243
in cell wall changes under microgravity, **170,** 64–68
in exocytosis of synaptic vesicles, **149,** 220
in flagellar motility, **173,** 25–29
in histamine release, **149,** 253
in lutein-releasing hormone action, **159,** 42–43
in serotonin presynaptic receptors, **149,** 250
in sperm capacitation, **149,** 39
as secondary messenger in pollen tube growth
active Ca^{2+} transport, **176,** 175–176
intracellular free Ca^{2+} concentration, **176,** 175
self-incompatibility and, **143,** 261, 263
Sertoli-germ cell communication network and, **147,** 32–33, 79
signaling in membrane-cytoskeleton complex, **148,** 71
in signal transduction
mammalian egg-activating
analysis of effects, **152,** 200–201
calcium-induced calcium release, **152,** 188–189, 191, 196–200
decoding, **152,** 210–213
fertilization, **152,** 183–189, 213
mechanism, **152,** 183–184, 189, 206–210
monotonic calcium increase, **152,** 202–206
oscillation, **152,** 196–210
signal generation hypotheses, **152,** 189–196
mechanical forces with integrins, **150,** 85
in plants, **158,** 16–17, 43–44, 46–48, 53
sperm capacitation, **149,** 7, 9–11, 14–17
sperm function regulation, **149,** 9–17
in spine organelles, **139,** 271
spines, **139,** 289–290
synapses affected by, **149,** 221
in tissue factor expression on fibroblasts, **152,** 74
transcriptional regulation and, **143,** 7, 14, 30
transdifferentiation in medusae and, **142,** 247
universal mechanisms of action, **168,** 124–125
Calcium-binding proteins
eggshells and biomineralization and, **145,** 234
role in pollen tube growth, **176,** 177
Calcium-capping proteins, **139,** 279–280
Calcium carbonate
activity in bone grafts, **172,** 174
in calcified tissue of invertebrates, **172,** 130
eggshells and biomineralization and, **145,** 244
fabrication, **145,** 236–237
models, **145,** 241, 243–244
structure, **145,** 221, 230
Calcium channel blockers
agravitropism induction in plants, **168,** 164, 166–168
MDR reversal, **156,** 89–91
mechanism of action, **168,** 164, 166
modulation of arterial smooth muscle cell behavior, **169,** 227–228
Calcium channels
intracellular, in plants, **142,** 305–306, 338
calmodulin, **142,** 328–332

concentration
channels, **142,** 318–321
chloroplasts, **142,** 318
endoplasmic reticulum, **142,** 315–316
mitochondria, **142,** 317–318
plasma membrane, **142,** 312–315
vacuoles, **142,** 316–317
G proteins, **142,** 308–311
phosphatidylinositol turnover, **142,** 321–328
protein kineses, **142,** 332–337
receptors, **142,** 306–308
in neuromuscular junction
calcitonin gene-related peptide, **147,** 220–227
classification, **147,** 193–194
electron microscopic equivalents, **147,** 210, 214–217
inhibitors, **147,** 206–213
location, **147,** 200–206
motor end plate, **147,** 194–199
myasthenic syndrome, **147,** 218, 220
postsynaptic cytoplasm, **147,** 217–219
receptor-dependent channels, **147,** 193
type L channels, **147,** 194, 200, 209, 211–212
type N channels, **147,** 194, 200, 211–212
type P channels, **147,** 212–213
type T channels, **147,** 194, 200, 207
voltage-dependent, neuromuscular junction and, **147,** 193–194, 199, 220
inhibitors, **147,** 206–207, 213
location, **147,** 205
plants, **168,** 126, 135–136
role in pollen tube growth, **176,** 176–177
Calcium chloride, eggshells and biomineralization and, **145,** 243
Calcium-dependent protein kinase
control of plant enzyme, **149,** 49–50, 64, 74, 82–83
intracellular calcium messenger system and, **142,** 332–335
Calcium hydroxyapatite, *see* Apatite
Calcium phosphate
in amelogenesis imperfecta, **172,** 152
in biomaterials, **172,** 176–177
bone grafts and, **172,** 133
from bovine bone, **172,** 162–164
coralline, **172,** 161
dental and medical applications, **172,** 159–161
implant coatings, **172,** 165
synthetic, **172,** 164
in bone graft interface
bioactive processes and, **172,** 174–175
biodissolution, **172,** 170–171
carbonate apatite formation and, **172,** 167, 171–173, 175, 176–177
cell colonization of, **172,** 168–170, 174–175
in dental calculus, **172,** 147–148
in dentin caries, **172,** 149, 151
in eggshells, biomineralization, **145,** 244
in fluorotic enamel, **172,** 151–152
minor elements and, **172,** 154–159
in pathological calcifications, **172,** 130, 145, 147
in soft tissue calcifications, **172,** 147
in urinary calculus, **172,** 148
Calcium reserve assembly, eggshells and biomineralization and, **145,** 227, 238
Caldesmon
cross-linking of actin filaments, **139,** 287
proteins in striated muscle and, **142,** 109
regulation of MT–microfilament interactions, **173,** 228–229
role in actin–myosin interaction, **170,** 17–18
role in nonmuscle tropomyosin function, **170,** 13–15
Caldesmon-like protein, in *Physarum,* microfilament system, **149,** 156
Calibration, nuclear magnetic resonance and, **145,** 24
Calliactis parasitica, see also Sea anemones
ganglion cells lacking cilium, **167,** 42
neuropeptides in, **167,** 58–61
Calliphora, karyosphere and, **144,** 42
Callose
antibodies to, generation, **171,** 93–94
biosynthesis and transport, **176,** 172
in cell wall, effect of clinorotation, **170,** 45–47, 59–60
extracellular matrix and, **146,** 276, 303–304, 307, 310
induction in self-incompatibility, **158,** 18, 20

Callose (*continued*)
inner lining
deposition in cell wall, **176,** 167–168
wall thickening and plug formation, **176,** 168
self-incompatibility and, **143,** 260–261, 282
Calmodulin
antagonism, *see* Chlorprozamine
biological clocks in retina and, **146,** 105
and calcium binding proteins, **138,** 241–244
cytoskeletal dynamics and, **144,** 97, 105
dependent protein kinase II, activation of mammalian eggs, **152,** 210–211
Drosophila muscles and, **143,** 98–99
environmental stimuli in plant cells and, **145,** 265, 281–282, 286, 296
fish chromatophores and, **143,** 230, 242–243, 245
intracellular calcium messenger system and, **142,** 305, 328–332
concentration, **142,** 313–315
protein kineses, **142,** 333–334, 337
mediation of RVD response, **161,** 216–217
mitosis and, **144,** 251
plants, **168,** 128
proteins in striated muscle and, **142,** 109
regulation of EF-1α–F-actin interactions, **164,** 24
regulation of MT–microfilament interactions, **173,** 228–229
role in neurogenesis, **159,** 21, 66
role in plant gravitropism, **168,** 134–135, 158
in signal transduction systems, **139,** 325–326
sperm capacitation regulation, **149,** 16
Calmodulin inhibitors, **139,** 42, **156,** 89
Calnexin, in folding of CFTR protein, **172,** 205–206, 207
Calpain, *Drosophila* muscles and, **143,** 136
Calprotectin, macrophage characterization in dome epithelium using, **167,** 132
Calreticulin, proteins in striated muscle and, **142,** 124
Calsequestrin, proteins in striated muscle and, **142,** 121–122
CALT, *see* Conjunctiva-associated lymphoid tissue
Caltrin, sperm capacitation regulation, **149,** 26–27
Camalexin, in *Arabidopsis thaliana,* **144,** 75
CAM plants, *see* Crassulacean acid metabolism plants
Camptothecin, **156,** 97, 105–106
Campylobacter jejuni, M cells
preferential adherence to, **167,** 123
uptake of entire, **167,** 120*t,* 122
Canalicular bile acid transport protein, ATPase activity, **158,** 150–151
Cancer, *see also* Tumor cells; Tumors
breast
cells, NMR, **145,** 37, 39
cells with melphalan resistance, **156,** 129
characteristics, **155,** 197
coiled body content, **159,** 181–182
effects of estrogen, **159,** 181
high-density cultures, **155,** 198–199, 211, 213
mdr1 gene expression, **156,** 88
mortality, **159,** 181
prevalence, **159,** 181
ultrastructure, **155,** 227
chemotherapy resistance, **156,** 77–137
genetic mechanism, **156,** 78–79
mdr1 gene, **156,** 79, 81, 87–88
multidrug resistance, **156,** 79–97
non-glycoprotein mediated, **156,** 92–97
P-glycoprotein, **156,** 79–92
reversal, **156,** 88–92
specific drugs, **156,** 79–81
cytogenetics
clinical studies, **153,** 216–217
comparative genomic hybridization use in examination, **153,** 218–219
drugs, nuclear matrix as site of action, **162A,** 539–565
kidney, *mdr1* gene expression, **156,** 88
lung cancer
MDR, **156,** 100
MDR reversal, **156,** 92–93
ovarian cancer, **156,** 130–131, 134
and procoagulants, **152,** 82–86
testicular cancer, **156,** 128
Cancer borealis, control of rhythmic behaviors and, **147,** 15

Cancer therapy
nuclear magnetic resonance and, **145,** 50
vertebrate cell culture technology and, **142,** 203
Candida albicans, mitochondrial genomes in fungi and, **141,** 103, 105
Candida glabrata, mitochondrial genomes
DNA diversity, **141,** 93, 98, 100–103
generation of mtDNA diversity, **141,** 109–111, 117–119
Candida sloofii, mitochondrial genomes in fungi, **141,** 116
Ca-NFs, *see* Catecholaminergic nerve fibers
CAN/NUP214, characterization, **162B,** 269–270
CAN/NUP214/p250, 3-D localization, **162B,** 238–241
Capacitation, *see* Sperm, capacitation
Capillaries
growth factor-induced retinal regeneration, **146,** 57, 59
pulmonary, hemangiomatosis, **161,** 26
renal glomerular
length, **161,** 154–155
mean length, **161,** 158
number, **161,** 156–157
surface area, **161,** 155–156, 158
volume density of capillary lumina, **161,** 158–159
supply of Peyer's patches, **167,** 95
Capping
in amoeba, **148,** 43–44
concanavalin receptors, **148,** 58
cytoskeletal dynamics and, **144,** 112–113, 118
membrane protein, **148,** 47
Capping proteins
cytoskeletal dynamics and, **144,** 96, 107
myofibrillogenesis and, **143,** 172–174
Capsecum annuum, plastids and, **145,** 161
Cap Z protein
myofibrillogenesis and, **143,** 173, 182
in striated muscle and, **142,** 91–92
Carassius auratus, growth factor-induced retinal regeneration and, **146,** 54
Carausius morosus, microgravity experiments, **156,** 335, 336
Carbachol, avian salt glands and, **144,** 161
Carbamate, cytoskeleton, **166,** 15
Carbohydrates
in arabinogalactan proteins
AGP biosynthesis and degradation, **174,** 236–238
carbohydrate–polypeptide linkage, **174,** 230–232
as glycan chain, **174,** 232–234
eggshells and biomineralization and, **145,** 230, 234–235
higher plant plastids and, **145,** 150, 193, 201–203
hormonal control of amphibian metamorphosis and, **145,** 122–123
metabolism during chromoplast biogenesis, **163,** 209–211
as regulatory metabolites of plant enzymes, **149,** 86–87
self-incompatibility and, **143,** 268–269
specificity of selectins, **175,** 186–189
transdifferentiation in medusae and, **142,** 249–251
Carbon
flowering plants and, **143,** 306
higher plant mitochondrial genomes and, **141,** 135
higher plant plastids and
amino acid synthesis, **145,** 192–193, 196–197
CO_2 fixation, **145,** 174, 176
fatty acid synthesis, **145,** 184–185, 188
starch synthesis, **145,** 177, 179–181
structure, **145,** 160
translocation, **145,** 167–171
mitochondrial genomes in fungi and, **141,** 98
urease in plant cells and, **145,** 66, 74, 76
Carbonate
eggshells and biomineralization and, **145,** 244
fabrication, **145,** 236–238, 240
models, **145,** 241, 243–244
structure, **145,** 221, 230
incorporation in apatite, **172,** 154–155
Carbonate apatite
in bone graft interface, **172,** 167, 171–173, 175, 176–177
in soft tissue calcifications, **172,** 147
Carbon dioxide
flowering plants and, **143,** 303–305

Carbon dioxide (*continued*)
higher plant plastids and, **145,** 150
amino acid synthesis, **145,** 192–193, 196–197, 199
fatty acid synthesis, **145,** 182, 189
fixation, **145,** 171–176
self-incompatibility and, **143,** 262
urease in plant cells and, **145,** 69, 71, 74–76
vertebrate cell culture technology and, **142,** 164, 167, 169, 189, 198, 200
Carbon flow, regulation in lignin biosynthesis, **176,** 119–122
Carbonic anhydrase
eggshells and biomineralization and, **145,** 224, 226, 234
self-incompatibility and, **143,** 262
Carboxyarabinitol bisphosphate, in higher plant plastids, **145,** 174
Carboxyarabinitol 1-phosphate, in higher plant plastids, **145,** 174–175
Carboxylase, urease in plant cells and, **145,** 80
Carboxylation, *see* Decarboxylation
Carboxypeptidase Y
biosynthesis of, **139,** 64–66
overproduction of, **139,** 70
Carboxy terminus
cnidarian neuropeptides
Antho-RFamide, **167,** 59, 61
Antho-RPamides II-IV, **167,** 69–71
copies of immature MMA, **167,** 71
Pol-RFamides, **167,** 76*t*
preprohormone processing, **167,** 80–82
protective residues on, **167,** 48–49, 50
collagen V, molecular structure, **173,** 99–100
FAK, **167,** 165–168
neuropeptides in higher animals, **167,** 57–58
Rab, isoprenylation, **176,** 44–47
variability in Rab GTPase, **176,** 15–16
Carcinogenesis
proliferation inhibition in, **167,** 233
template/antitemplate model in skin, **167,** 187
Carcinogens, melanocytes and, **146,** 196, 233
Carcinoma, *see also* Cancer; Tumor cells; Tumors
adrenocortical, role of endothelin, **171,** 295
bladder
characteristics, **155,** 196
high-density cultures, **155,** 211
colon
drug resistance, **156,** 100, 115, 118, 119
mdr1 gene expression, **156,** 88
MDR reversal, **156,** 91
ultrastructure, **155,** 215
drug resistance, **156,** 115
epithelial differentiation and prognosis, **160,** 231
esophageal, high-density cultures, **155,** 193–194
gastrointestinal
high-density cultures, **155,** 192–193, 200
histological findings, **155,** 203–206
immunohistochemical findings, **155,** 232–236
ultrastructure, **155,** 215, 217–221
gynecological
high-density cultures, **155,** 197–200
histological findings, **155,** 211, 213, 216
immunohistochemical findings, **155,** 237, 242–245
ultrastructure, **155,** 227
head and neck
high-density cultures, **155,** 193–195, 200
histological findings, **155,** 205, 207–210
immunohistochemical findings, **155,** 233, 237–239
ultrastructure, **155,** 217, 222–225
human, cultures
cell culture systems
monolayer, **155,** 182–183
spheroidal aggregate, **155,** 186–187
stem cell, **155,** 184–186
cytostatic drug experiments, **155,** 248–250
high-density
gastrointestinal carcinomas, **155,** 100, 192–193
gynecological carcinomas, **155,** 197–200
head and neck carcinomas, **155,** 193–195, 200
histological findings, **155,** 201–214, 216
lung carcinomas, **155,** 195–196, 198, 200
methods overview, **155,** 187–192

success rates, **155,** 192–201
urological carcinomas, **155,** 196–200
histological findings, **155,** 201–214, 216
gastrointestinal carcinomas, **155,** 203–206
gynecological carcinomas, **155,** 211, 213, 216
head and neck carcinomas, **155,** 205, 207–210
lung carcinomas, **155,** 207, 209, 212–213
urological carcinomas, **155,** 209, 211, 214
immunohistochemical findings
gastrointestinal carcinomas, **155,** 232–236
gynecological carcinomas, **155,** 237, 242–245
head and neck carcinomas, **155,** 233, 237–239
lung carcinomas, **155,** 237, 240–241
methods overviews, **155,** 230–232
urological carcinomas, **155,** 237, 242–245
keratinizing carcinomas, **155,** 205
methods, **155,** 181–182
mouse models, **155,** 247
organoid culture assay, **155,** 248–252
ultrastructure, **155,** 213
gastrointestinal carcinomas, **155,** 215, 217–221
gynecological carcinomas, **155,** 227
head and neck carcinomas, **155,** 217, 222–225
lung carcinomas, **155,** 218–221, 225–226
urological carcinomas, **155,** 226–229
incidence, **160,** 222, 230
keratinizing, human, cultures, **155,** 205
lung
high-density cultures, **155,** 195–196, 198, 200
histological findings, **155,** 207, 209, 212–213
immunohistochemical findings, **155,** 237, 240–241
ultrastructure, **155,** 218–221, 225–226
mdr1 gene expression, **156,** 88
metastasis, regulation by adhesion molecules, **160,** 231–232, 251–252
microgravity experiments, **156,** 317
motility and growth factors, **160,** 233–234
ovarian
characteristics, **155,** 197
high-density culture methods, **155,** 198
histological findings, **155,** 211, 213
organoid culture assay, **155,** 251
ultrastructure, **155,** 227
pulmonary cells, human A549, inability to activate TGF-β precursor, **167,** 210
rectal
characteristics, **155,** 193
high-density cultures, **155,** 193
ultrastructure, **155,** 215
renal cell
high-density cultures, **155,** 211
ultrastructure, **155,** 226
testicular
characteristics, **155,** 196
high-density cultures, **155,** 211
ultrastructure, **155,** 216
thyroid gland
high-density cultures, **155,** 194
ultrastructure, **155,** 224–225
tumor progression, genetic analysis, **160,** 230, 232–233
tyrosine kinase receptors, **160,** 233–234, 236–237, 242, 248–249, 251–252
urological
high-density cultures, **155,** 196–200
histological findings, **155,** 209, 211, 214
immunohistochemical findings, **155,** 237, 242–245
ultrastructure, **155,** 226–229
Cardamine pratensis, self-incompatibility, **143,** 259
Cardiac muscle
myofibrillogenesis and, **143,** 155, 176, 183
proteins
antibody use, **142,** 101, 115–119
sarcolemma, **142,** 125–126
sarcomere, **142,** 98–99
sarcoplasmic reticulum, **142,** 122, 124
structural proteins, **142,** 110–115
thick-filament proteins, **142,** 79, 83, 102–107
thin-filament proteins, **142,** 84–88, 91–92, 107–110

Cardiovascular system
avian salt glands and, **144,** 188–189, 192
images, nuclear magnetic resonance and, **145,** 31
parathyroid hormone-related protein, **166,** 242–244
Caries
in dentin, **172,** 149–151
in enamel, crystal dissolution and, **172,** 148–149
Carnitine, higher plant plastids and, **145,** 182, 187–189
Carnitine acetyltransferase
bezafibrate effect on activity, **160,** 321–322
peroxisome, cytochemical localization, **160,** 305, 319–321
reaction catalyzed, **160,** 321
Carnosine
immunohistochemical localization of carnosine, **138,** 171–172
possible effects of carnosine, **138,** 172–174
ζ-Carotenic maize mutants, chromoplast-like plastids, **163,** 187
Carotenoids
biosynthesis in chromoplasts, **163,** 211–219
chloroplasts and
dinoflagellates, **144,** 286, 288, 290–291
euglenoids, **144,** 282, 284
green algae, **144,** 265
heterokont algae, **144,** 277–278, 280–282
red algae, **144,** 261, 265
in chromoplasts, **163,** 188–189
fish chromatophores and, **143,** 198, 241
flavonoids and phenolic conjugates
biological functions of pollen wall substances, **140,** 50–51
carotenoids, **140,** 41–43
flavonoids, **140,** 43–45
phenolic conjugates, **140,** 45–48
significance of anther tapetum for biosynthesis of pollen wall phenolic, **140,** 49–50
higher plant plastids and, **145,** 153
in photoreceptors, **148,** 275–276, 283
Carrageenan, in algal cell walls, detection, **171,** 105–106
Carrot
chromosome disjunction, **158,** 96–97
clathrin yield from cells, **167,** 8–9
embryogenesis, **158,** 70–71, 91–92, 103
growth factor, **158,** 91
indefinite growth, **158,** 78
intracellular calcium messenger system and, **142,** 308, 315, 320, 323, 328, 337
meiosis
commitment, **158,** 91–92, 94
RAPD analysis, **158,** 109
RFLP analysis, **158,** 109
variant phenotypes, **158,** 100
microgravity experiments, **156,** 319–320, 323, 328, 333
proembryogenic masses, **158,** 103–104
prophase chromosome reduction, **158,** 77
reductional grouping, **158,** 82–83
Cartilage, *see also* Chondrocytes
angiogenesis inhibition by factors, **159,** 141, 320
calcification, **159,** 283–284
classification, **159,** 265–266
derived inhibitor, inhibition of endothelial cell growth, **161,** 18
differentiation, time of, **148,** 81–82
extracellular matrix
mineral deposition, **159,** 295, 297, 300
turnover, **159,** 295
formation in embryo
mesenchymal cell condensation, **159,** 272–273
ossification, **159,** 256, 283–284
positional signaling, **159,** 271–272
role
chondromodulin 1, **159,** 318
dexamethasone, **159,** 315–316
estrogen, **159,** 316
growth hormone, **159,** 312–314
Hox genes, **159,** 275–277
insulin-like growth factor, **159,** 312–314
Msx-1 gene, **159,** 277–278
parathyroid hormone, **159,** 318–319
platelet-derived growth factor, **159,** 318
prostanoids, **159,** 319
Prx-1 gene, **159,** 278
retinoids, **159,** 274–275, 277, 303, 305–306
testosterone, **159,** 316
vitamin D_3 metabolites, **159,** 316–317
inhibition of angiogenesis, **159,** 141
organ culture, **159,** 300–301

Cascade model, steroid hormone action, **162B,** 341–344
Casein kinase 11, transcriptional regulation and, **143,** 37
Catalase
crystallization in peroxisome, **160,** 314
relationship with oxidases, **160,** 317
CATCH
incidence, **158,** 242
role of TUPLE1 gene, **158,** 243
spectrum of disorders, role in plant cell division, **158,** 122–123, 242
Catecholamine
fish chromatophores and, **143,** 226, 230–231, 237, 240, 243
hormonal control of amphibian metamorphosis and, **145,** 122–123
release by adrenal medulla, stimulation by ET, **171,** 282–283
unsuccessful attempts to isolate in *Hydra* neurons, **167,** 43–44
Catenin
actin-based cytoskeleton, linkage, **165,** 170–172
cadherin
complex assembly dynamics, **165,** 166–168
interactions between, **165,** 163–170
protein-protein interaction hierarchy, **165,** 164–166
cadherin-mediated cell-cell adhesion, consequence, **165,** 170
desmosomal catenin, **165,** 169–170
epithelial cell junctions, **160,** 223
molecular characterization, **165,** 168–169
phosphorylation, **160,** 232
role in development, **165,** 186
β-Catenin, *Wnt/wingless* signaling pathway, **165,** 187–191
Catharanthus roseus, intracellular calcium messenger system and, **142,** 322, 325
Cations
divalent, in adhesion, **139,** 170
uptake system, salt tolerance, **165,** 17–21
Caulerpa
cytoplasmic streaming supported by microtubule system, **155,** 109
environmental stimuli in plant cells and, **145,** 270, 294–295
Cauliflower mosaic virus, *Arabidopsis thaliana,* **144,** 62–63, 67
Causal loop, biological clocks in retina and, **146,** 122
Cavalieri principle, for estimating total volume, **161,** 127–128
CCCP, intracellular calcium messenger system and, **142,** 312–315
CCK, *see* Cholecystokinin
Cdc31, spindle pole body functions, **169,** 66–67
cdk2, at DNA replication sites, **162B,** 317,324
Cecidomyiidae, karyosphere and, **144,** 11
Cecropia, instability of homogeneous state, localized activity, **154,** 344
Cell adhesion
inhibition with sialyated compounds, **175,** 183
Sertoli-germ cell communication network and, **147,** 28, 33–37
Cell adhesion molecules
adherens junction-specific, proteins in striated muscle and, **142,** 126
avian retina and, **146,** 23
C-CAM
glycosylation, **158,** 149
sequence homology with ecto-ATPases, **158,** 148–149
structure, **158,** 148–149
tissue distribution, **158,** 149
cholinesterases as, in avian neurogenesis
function, **151,** 158–159
HNK-1 epitope regulation, **151,** 159–164
neurite growth alteration, **151,** 166–167
pathfinding cues, **151,** 163, 165–166
and dendritic cells, costimulation of T cells, **153,** 78–82
neural, *see* Neural cell adhesion molecule
neurite initiation and elongation role, **153,** 246
Sertoli-germ cell communication network and, **147,** 32–33, 37, 44, 72
Cell capacitance, effects of cyclic AMP on, **172,** 224
Cell columnarization models, mechanical stress in, **150,** 23–24
Cell communication, cytoskeleton, **166,** 3
Cell cultures
anchorage-dependent
bioreactors for, **142,** 166, 172–173, 186–189
traditional cultures for, **142,** 147–151

Cell cultures (*continued*)
S/MAR role, **162A,** 431–435
technology, *see* Vertebrate cell culture technology
Cell cycle
conservation of controlling genes, **158,** 72–73
control of microtubule behavior
dynamics and assembly, **164,** 277–278
$p34^{cdc2}$ effects, **164,** 277–278
severing proteins, **164,** 279–282
cytotoxic agents in targeting S-phase, by non-S-phase-specific drugs, **167,** 240–241
dependence
nuclear matrix-associated proteins, **162B,** 387–396
nuclear matrix constituents, **162B,** 382–387
electromagnetic field effects, **158,** 314, 322–323
embryonic cycles, **175,** 116–117
G_2/M transition, centromere role, **162B,** 40–45
G_0 stage
mitogen and inhibitor concentrations, **167,** 193
self-renewal capacity of stem cell populations in, **167,** 196
G_1 stage
inhibitory regulators in, **167,** 194
variations in duration, **167,** 193
G_2 stage, **167,** 192
human preimplantation conceptuses
cdc kinase activity, **164,** 162–163
developmental arrest after micronuclei formation, **164,** 164–165
maturation promoting factor, **164,** 162
phosphorylation state of proteins during, **164,** 163–164
marker genes, *Rhizobium*–plant interactions, **156,** 37–38
NuMA functions during, **162B,** 19–20
oral morphogenesis, **164,** 4–6, 14nm filament protein role
Physarum polycephalum, synchronicity, **162B,** 381–382
plant, culture
deregulation, **158,** 71–72
endoreduplication, **158,** 74–76
possible functions of MAP kinase during, **138,** 232–234
regulated histone genes, **162A,** 253–263
regulation
current concepts, **162B,** 380–381
description, **167,** 192–194
hemopoietic progenitor, **167,** 200–206
link with DNA replication, **162B,** 322–323
regulatory proteins, regulation between calcium and, **138,** 261–263
specificity of interferon-induced inhibition for, **167,** 231
S-phase, and DNA replication, **162B,** 314–315
stage-specific expression of histone genes, **162A,** 263–264
temperature-sensitive mutants, **158,** 72, 79
transdifferentiation in medusae and, **142,** 254–257
tumor suppressor genes in, **167,** 233
Cell death, *see* Apoptosis
Cell development
ascidian, stage of determination, **176,** 281–286
MT–microfilament interaction
centriole migration, **173,** 216–217
centrosome movement, **173,** 214–215
cortical movement, **173,** 215–216
cytoplasmic streaming in ring canals, **173,** 220–221
role of actin filament arrays, **173,** 218–219
role of growth cones, **173,** 221–222
role of nuclear migration, **173,** 219–220
spindle orientation, **173,** 216
spindle positioning and cytokinesis, **173,** 217–218
Cell differentiation
dome epithelium, **167,** 108–111
evolutionary perspective, **148,** 81, 111–112
AT-rich sequences in, **148,** 89–90
attachment to nuclear matrix, **148,** 94–97
late replication, **148,** 91–94
positive regulation by, **148,** 90–91
cell division in, and H1 histone
H1 accumulation role in unbalanced growth, **148,** 97–100
H1 causal role, **148,** 100–101

expression of genomic domains
cell division and, **148,** 110–111
H1 histone role, **148,** 101–103
hierarchical, **148,** 109–110
homeotic genes in, **148,** 103–105
relation to multigene families, **148,** 107–108
role of attachment to nuclear matrix, **148,** 108–109
heterochrony, **148,** 106–107
more conserved genes, time of expression, **148,** 83–86
temporal sequence of differentiation, **148,** 107
time of differentiation of more conserved cells, **148,** 81–83
tissue-specific features, restriction, **148,** 105–106
tissue-specific genes
coordinate transcription, **148,** 86–88
quantitative regulation, **148,** 88–89
intraepithelial lymphocytes, **167,** 97
loss of inhibitory effect of leukemia cell lines during, **167,** 234
plant cells
angiosperm root, effect of microgravity, **171,** 34–35
associated antigens, **171,** 84–85
effect of microgravity and clinostating, **171,** 53–56
space flight studies, **171,** 3
proto-oncogene participation in, **167,** 232–233
retinal pigment epithelium, molecular markers, **171,** 235–237
and substratum mechanics
cell type conversion, **150,** 119
cellular mechanics of transdifferentiation process, **150,** 128–132
cytomechanics, **150,** 120–125
retinal pigment epithelial cells
cell fate choice regulation, **150,** 125–128
transdifferentiation process, **150,** 119–120
Cell division
arrest, mitosis and, **144,** 228
and differentiation, **140,** 187–188
generative cells, *see* Generative cells, division
MT–microfilament interaction
centriole migration, **173,** 216–217
centrosome movement, **173,** 214–215
cytoplasmic streaming in ring canals, **173,** 220–221
role of actin filament arrays, **173,** 218–219
role of growth cones, **173,** 221–222
role of nuclear migration, **173,** 219–220
spindle orientation, **173,** 216
spindle positioning and cytokinesis, **173,** 217–218
Rhizobium–plant interactions, **156,** 3
role of AGPs, **174,** 260–262
in yeast, cAMP in regulation of, **139,** 320–322
Cell expansion, role of AGPs, **174,** 262–263
Cell fate, specification in ascidian embryogenesis, **176,** 281–286, 297–299
Cell fractionation
induced insolubility, **162A,** 528–531
in situ, in preparation of nuclear matrix, **162A,** 492–495
Cell fusion, mitosis and, **144,** 252
Cell junctional molecules, Sertoli-germ cell communication network, **147,** 32
Cell lines
32D myeloid precursor, Shc protein, **173,** 269–272
ascidian
development, **176,** 251
features, **176,** 251–253
CEM, topoisomerase II-active drug effects, **162A,** 559–565
CHO
DHFR domain, **162A,** 466–476
pp120, role in signal transduction, **173,** 267–269
vertebrate cell culture technology, **142,** 147, 150, 176, 186, 189
COS cells, proteins in striated muscle and, **142,** 126
expression cloning of G-protein-coupled receptors, **164,** 233
fetal, mosaicism, **167,** 287
insect, for expression of mammalian G-protein-coupled receptors, **164,** 246–250
LLC-MK_2, virus–host interaction model, **162A,** 531–532
P19, lamin expression, **162B,** 162

Cell lines (*continued*)
segregation during neural crest ontogeny
clonal analysis of neural crest potentialities, **138,** 285–293
cephalic NC, **138,** 288–292
development restriction, **138,** 292–293
early heterogeneity of migrating neural crest cells, **138,** 284–285
transfected, expression cloning of G-protein-coupled receptors, **164,** 233
tsBN2, endogenous NuMA, **162B,** 16
Cell migration
during CNS development
in cerebellar cortex, **175,** 277–280
in forebrain, **175,** 274–277
in hindbrain, **175,** 280–285
marker techniques, **175,** 272–274
in optic tectum, **175,** 280
culture system analysis, **155,** 52–53
extracellular matrix interactions, **155,** 52
FAK in focal adhesion assembly and, **167,** 176
lymphocytes
through high endothelial venues in gut wall, **167,** 95
through mucosal system, **167,** 93–94
microgravity experiments, **156,** 319
Cell mixing, during neural crest formation, **175,** 253–255
Cell model systems
cortical microtubules
isolated CMTs, **175,** 9
membrane ghosts, **175,** 6–9
perfused cells, **175,** 6
cytoskeletons
permeabilized cells, **175,** 2
plant cells, roles in cytoplasmic streaming
cytoplasmic droplets, **175,** 5
other plants, **175,** 5–6
perfused cells, **175,** 4
permeabilized cells, **175,** 4–5
structure isolation, **175,** 2–3
in vitro functions, **175,** 3–4
future studies, **175,** 18–20
microtubule-organizing center, **175,** 13–14
miniprotoplasts
cytokinesis, **175,** 15–16
for studies of cytoskeletons, **175,** 16–18
phragmoplast, **175,** 9–11
isolated phragmoplasts, **175,** 12–13
permeabilized cells, **175,** 11–12
spindle, **175,** 14–15
Cell motility
actin-binding proteins, **156,** 199–255
amoeboid
computer-assisted analysis
ABP-120 mutants, **163,** 91–93
behavioral cycles, **163,** 79–80
coronin-lacking cells, **163,** 93–94
discoidin-lacking cells, **163,** 93
history, **163,** 47–48
myosin II heavy chain mutants, **163,** 86–89
myosin I mutants, **163,** 89–91
ponticulin mutants, **163,** 92–93
pseudopod formation, **163,** 83, 94–97
pseudopod retraction, **163,** 83–85, 94–97
recent evolution, **163,** 48–50
turning, **163,** 80–83
velocity, instantaneous, **163,** 77–79
velocity cycles, **163,** 79–80
three-dimensional analysis systems
centroid, **163,** 75
contour, **163,** 76
digitization, **163,** 74–75
imaging, **163,** 72–73
location tracking, **163,** 76–77
reconstruction, **163,** 74–75
recording, **163,** 72–73
volume, **163,** 75
two-dimensional analysis systems
cell contour editing, **163,** 54–56
cell shape dynamics, **163,** 68–71
centroid computation, **163,** 56–57
contour measurements, **163,** 67–68
difference pictures, **163,** 69–70
digitization, **163,** 53–54
direction, **163,** 62
directional change, **163,** 62
expansion-contraction dynamics, **163,** 68–71
image processing, **163,** 53
imaging, **163,** 50–53
persistence, **163,** 63
recording, **163,** 50–53
sectoring, **163,** 68
size measurements, **163,** 64–67

velocity, **163,** 58–60
windowing, **163,** 68
definition, **155,** 50
Physarum, **149,** 151–155
Cell phenotype, desmosomal cadherin, **165,** 179–180
Cell polarity, role of Rab proteins, **176,** 42–44
Cell proliferation
CD45RA+ T cells in intestinal lamina propria, **167,** 98
effect of endothelin, **171,** 288–289
inhibitors of hemopoietic stem
AcSDKP, **167,** 221–224
G_1 state and, **167,** 193
interferons, **167,** 230–232
iron binding factors, **167,** 227–228
MIP-1α, **167,** 198–208
negative regulatory protein, **167,** 225–226
oligopeptides, **167,** 226–227
other inhibitors, **167,** 224–232
pEEDCK, **167,** 216–220
suicide assays in confirming effect of, **167,** 197–198
TGF-β, **167,** 208–216
tumor necrosis factor, **167,** 228–230
types, **167,** 191–192
microenvironmental influences and self-renewal, **167,** 194–196
proto-oncogene participation in, **167,** 232–233
regulation of
cell cycle, **167,** 192–194
CFC-S hemopoeitic stem cells, **167,** 191–192
epithelial, **167,** 186, 187–188
myelocyte, **167,** 188, 191
status, **162B,** 433–435
stimulation of hemopoietic, by TNF-α, **167,** 230
Cells
anchorage-dependent, NMR analysis, **145,** 14–15, 37
biological organization, **166,** 59–60
cholesteric supramolecular order, **166,** 70–84
contour
2D analysis systems, **163,** 54–56, 67–68
3D analysis systems, **163,** 76
extracellular matrix, **166,** 70–81
fluidity, **166,** 60–61
function
desmosomal cadherin, **165,** 179–180
salt sensitive, **165,** 13–17
human trophoblast, **165,** 68–69
intercellular communication, **157,** 316, 320
invertebrate
cytokines, **170,** 102–104
nitric oxide, **170,** 113–116
proopiomelanocortin-derived peptides, **170,** 86–87
isolation and culture, RPE, **138,** 10–13
mechanical properties
protoplasmic viscosity, **157,** 112–113
tension at the surface, **157,** 111–112
nontransformed, APP glycosylation alterations, **170,** 269–270
nucleus
architecture, cytoplasmic IF effects, **164,** 100–102
components, *in vitro* binding of cytoplasmic IFs, **164,** 99–100
order, **166,** 60–61
organization and quality, pollen interactions, **140,** 569–570
polyamine content and cytochemical fluorescence yield, **138,** 60–64
separation, associated pectin epitopes, **171,** 88
shape
dynamics, 2D analysis, **163,** 68–71
role of tropomyosin, **170,** 19–20
size, measurements, **163,** 2D systems, 64–67
structure in *Dictyostelium discoideum,* **175,** 50–56
transformed
actin organization, **170,** 7
for acute phase protein studies, **170,** 252
regulation of APP glycosylation alterations, **170,** 271–274
transfected, regulation of APP glycosylation alterations, **170,** 274
tropomyosin expression, **170,** 7–9
transgenic nontransformed, APP glycosylation alterations, **170,** 270–271
trophoblast, **165,** 68–69
vertebrate
cytokines, **170,** 102

Cells (*continued*)
vertebrate (*continued*)
nitric oxide, **170,** 110–113
proopiomelanocortin gene, **170,** 82–85
Cell surface
AGP localization, **174,** 240–245
animal, plant analogs
adhesion proteins, **174,** 272
fibronectin, **174,** 271–272
integrins, **174,** 270–275
proteoglycans and mucins, **174,** 268–270
short peptides, **174,** 271
vitronectin, **174,** 271–272
binding by influenza A virus, **175,** 163
chaperone expression, **174,** 170–171
effects of cyclic AMP on, **172,** 224
glycan, resialylation, **175,** 150–151
plant
cell wall, **174,** 198–199
cell wall synthesis sites, **174,** 200
cuticle, **174,** 200
middle lamella, **174,** 199–200
molecular probing with antibodies, **171,** 83–85
plasma membrane–cell wall interface, **174,** 197–198
plasma membrane–cytoskeleton interaction, **174,** 196–197
Cell-to-cell interaction
during lignin biosynthesis, **176,** 108–109
pattern formation in *Drosophilia* and, **139,** 121–151, *see also Drosophilia,* pattern formation in
Cellular intrinsic fiber cementum, regenerative cementogenesis
established root surfaces *in vitro,* **142,** 35, 41–42
function of cementum, **142,** 2–5
growing root surfaces, **142,** 15–25, 27–28
Cellular mixed stratified cementum, regenerative cementogenesis, **142,** 2–4, 17, 42
Cellular suspension, nuclear magnetic resonance and, **145,** 14, 19, 35
Cellular tensegrity, *see* Tensegrity, cellular
Cellulase
moss, effect of clinostating, **171,** 31
role in cell wall changes under microgravity, **170,** 61–63
Cellulose
biosynthesis and transport, **176,** 171–172
cholesteric ordering, **166,** 84–86
content in pea plant, effect of microgravity, **171,** 39
extracellular matrix
characterization, **146,** 265–266
immunology, **146,** 284–285
modification, **146,** 293–294, 307
synthesis, **146,** 273–274, 276
microfibrils, **173,** 162
distribution and orientation, **176,** 165–167
interaction with xyloglucans, **173,** 195–196
fucosylation, **173,** 165–166
pectic polysaccharide network, **173,** 167
synthesis, **173,** 168–169
in plant cell wall
effect of clinorotation, **170,** 57–58
production, effect of space flight, **170,** 54–55
protoplast wall
effect of clinorotation, **170,** 45–47
effect of microgravity, **170,** 43
polymers based on, cholesteric ordering, **166,** 84–85
–xyloglucan framework, **173,** 162–165
Cell walls
algal, monoclonal antibodies to, **171,** 105–107
effects of suppressors on
ATPase and, **172,** 77–80
peroxide generation and, **172,** 80–81
extensibility, **176,** 178
and filiform apparatus of female germ unit, **140,** 250–256
gram-negative bacteria
lipopeptides, **161,** 281–284
lipoproteins, **161,** 281–284
lipoteichoic acids, **161,** 270–272
model, **161,** 278
peptidoglycans, **161,** 267–269
structure, **161,** 266–267, 276–278
surface proteins, **161,** 272–274
teichoic acids, **161,** 270–272
gram-positive bacteria
ADP-ribosylating toxins, **161,** 274–275
encapsulated, **161,** 284–286
exotoxins, **161,** 274–275
leucocidins, **161,** 274
lipopteichoic acids, **161,** 270–272

peptidoglycans, **161,** 267–268
surface proteins, **161,** 272–274
in host-parasite specificity, **172,** 77–80, 84
lignification, model systems, **176,** 98–101
lignin deposition, **176,** 117–119
microgravity experiments, **156,** 319–320, 332–333
middle lamella, **174,** 199–200
peroxidase activity and, **172,** 272
plant
antigens, **171,** 105
architectural models, **173,** 160–162
biological effects of microgravity, **170,** 41–42
classification, **173,** 159–160
components, taxonomic significance, **171,** 85
construction
cellulose microfibril synthesis, **173,** 168–169
cleavage of load-bearing crosslinks, **173,** 192–194
dynamic aspects, **173,** 167–168
EXGT-mediated molecular grafting, **173,** 194–195
xyloglucan–cellulose microfibril interaction, **173,** 195–196
effect of environmental factors, **170,** 39–41
epidermal, external structure, effect of weightlessness, **170,** 51–54
interaction with microbes, **171,** 107–109
loosening, cleavage model, **173,** 169–171
meristem, microgravity effect on regeneration, **170,** 47
molecular grafting model, **173,** 171–172
polymers, probing with antibodies
enzymes, **171,** 100
extensin, **171,** 95–99
glycine-rich proteins, **171,** 99–100
lignin, **171,** 94–95
matrix polysaccharides, **171,** 92–94
pectin, **171,** 85–92
proline-rich proteins, **171,** 99
polysaccharide content
effect of clinorotation, **170,** 57–60
effect of space flight, **170,** 54–56
regeneration by protoplasts, **170,** 42–47
spatial architecture, **171,** 83–84
structural and functional changes under microgravity
role of calcium, **170,** 64–68
role of enzymes, **170,** 60–64
structure and components, **171,** 79–80
vascular bundle, effect of weightlessness, **170,** 50–51
in plant defenses, **172,** 76–77
–plasma membrane interface, **174,** 197–198
pollen tube, structure, **176,** 163–164
bound proteins, **176,** 168–170
inner callosic lining, **176,** 167–168
outer pectocellulosic layer, **176,** 164–167
porosity, **176,** 177–178
precursor biosynthesis and transport
callose, **176,** 172
cellulose, **176,** 171–172
pectins, **176,** 170–171
in regulation of membrane-bound ATPase, **172,** 79–80
synthesis sites, **174,** 200
vegetative, in pollen and pollen tube, **140,** 192, 193, 194, 196, 198–200, *see also* Freeze fracture, male reproductive cells
volvocacean, organization of, **139,** 195–197
Cementoblasts, regenerative cementogenesis
established root surfaces *in vitro,* **142,** 35, 40–42
growing root surfaces, **142,** 13, 15, 19, 21–23, 25
Cementocytes, regenerative cementogenesis and, **142,** 3–4, 21–22
Cementogenesis, regenerative
dentine, **142,** 25–28
established root surfaces *in vitro*
collagenous matrices, **142,** 31–37
culture material, **142,** 28–31
factors, **142,** 37–43
established root surfaces *in vivo*
biological problems, **142,** 49–51
clinical conditions, **142,** 43–44
guided regeneration, **142,** 46–49
spontaneous regeneration, **142,** 44–46
function of cementum, **142,** 2–5

Cementogenesis (*continued*)
growing root surfaces, **142,** 5
acellular extrinsic fiber cementum, **142,** 5–17
cellular/acellular intrinsic fiber cementum, **142,** 17–25
CENP-E protein, regulated behavior, **164,** 306-307
Central cell
female germ unit, **140,** 264–269
micromanipulation of, **140,** 414–419
Central nervous system
aging, related gonadal changes, **160,** 27–28, 40
γ-aminobutyric acid metabolism and, **147,** 285, 303–304, 308–309, 312, 315, 319
amphibian development
anterior–posterior patterning, **169,** 125–126
dorsoventral patterning, **169,** 126–128
factors regulating gene expression, **169,** 128–129
segmentation, **169,** 96–97
avian retina and, **146,** 35–36
avian salt glands, **144,** 155, 178, 193
biological clocks in retina and, **146,** 112
calcium channels and, **147,** 194, 212–213
cell migration during development
in cerebellar cortex, **175,** 277–280
in forebrain, **175,** 274–277
in hindbrain, **175,** 280–285
marker techniques, **175,** 272–274
in optic tectum, **175,** 280
control of rhythmic behaviors and, **147,** 1–2, 13, 20
Drosophila, **139,** 134–137
Drosophila melanogaster, programmed cell death in, **163,** 129–130
Drosophila muscles and, **143,** 79, 109
fish chromatophores and, **143,** 225
GnRH-encoding gene in teleosts and, **147,** 165
growth factor-induced retinal regeneration and, **146,** 58
in *Halocynthia roretzi* larva, **176,** 248
hormonal control of metamorphosis, **145,** 112, 122, 127
origin of oligodendrocytes, **175,** 285–291
parathyroid hormone-related protein, **166,** 251–252
platelet-derived growth factors in
activity, **172,** 113–114
distribution, **172,** 108, 109
proopiomelanocortin-derived peptides, **170,** 83
retinal precursor cells and, **146,** 171
sexual differentiation, **160,** 2, 39
Tau protein structure in, **171,** 171–172
transplant
history, **138,** 3–5
studies, **138,** 4–5
study of development of ganglion cell connectivity, **138,** 5
vertebrate neuronal cell models, **153,** 236–237
Central pattern generators, control of rhythmic behaviors
chemical modulation, **147,** 18–20
ion channels, **147,** 3, 5–11
neuromodulation, **147,** 12, 14–17
Central plug, in NPC central core, **162B,** 233–236
Central zone, lens epithelial cells, **163,** 312–313, 323, 326
Centrifugal force, environmental stimuli in plant cells and
chloroplasts, **145,** 271, 276, 279, 281
photodinesis, **145,** 257, 259–360
Centrifugal modulation, biological clocks in retina and, **146,** 109
Centrifugation, density gradient, principle, **157,** 100
Centrifuge, spaceflight experiments, **156,** 356
Centrifuge microscopy
determination
endoplasmic viscosity, **157,** 118–120
mechanical structure of echinoderm egg cytoplasm, **157,** 123
motive force
cytoplasmic streaming, **157,** 100, 113–114
force–velocity in actin-myosin system, **157,** 124
Paramecium propulsion, **157,** 123
protoplasmic viscosity, **157,** 112–113
stiffness changes in procaine-activated sea urchin eggs, **157,** 122–123
tension at the cell surface, **157,** 111–112
electrical stimulation effects on cytoplasmic streaming, **157,** 120–122

environmental stimuli in plant cells, **145,** 258
imaging, **157,** 125
instrument
commercial availability, **157,** 124–125
design, **157,** 101–109
light sources, **157,** 101, 104, 106, 108–109, 125
maximal centrifugal acceleration, **157,** 101, 104, 106–108
motility behavior under centrifugation
Paramecium, **157,** 116–117
Spirogyra organelles, **157,** 118
Tetrahymena, **157,** 116–117
Vallisneria chloroplasts, **157,** 117
observation of motile protoplasmic fibrils, **157,** 114–115
rotation of cytoplasmic protrusions in characean cells, **157,** 115–116
stratification of eggs, **157,** 109–110
types
B type, **157,** 103–104
HL type, **157,** 101–103
H type, **157,** 107–109
KK type, **157,** 104–106, 125
K type, **157,** 106–107
Centrin
functions, **169,** 41–42
isolation, **169,** 41
Centriole
cell type distribution, **169,** 27, 59
doubling during cell cycle, **169,** 28
migration in cell division and development, **173,** 216–217
mitosis, **144,** 224, 237–238
relationship to nuclear envelope, **175,** 94–95
Centripetal movements, cytoplasmic, **148,** 42
Centroid, cell motility, computation
in 2D analysis systems, **163,** 56–57
in 3D analysis systems, **163,** 75
Centromere-binding proteins, **162B,** 38–48; **167,** 291
Centromeres
aneuploidy and
in meiotic recombinant errors, **167,** 282–283
premature division of, **167,** 275–276
size, **167,** 285–286
–kinetochore region
chromatin anchoring role, **162B,** 39
and prekinetochore, characterization, **162B,** 36–38
role in microtubule-based movement, **162B,** 45–46
lepidopteran
apyrene spindle organization, **152,** 35
DNA analysis, **152,** 33–34
kinetic organization, **152,** 11
mitosis, **144,** 222, 230, 235, 239–240, 244
Centrosomes
in basal body-associated fibrillar networks, **173,** 214
basal feet, **169,** 28
control of microtubule behavior
phosphorylation effects, **164,** 272–273
γ-tubulin effects, **164,** 271–272
cytoplasmic intermediate filament associations, **164,** 112–114
cytoplasmic organelle localization, **169,** 35
cytoskeletal dynamics and, **144,** 94
inheritance, **169,** 56–59
lepidopteran, in male eupyrene meiosis, **152,** 20
localization, **169,** 28, 30
microtubule
disassembly, **169,** 35–36
organization, **169,** 25–26, 31
release, **169,** 36–37
shearing, **169,** 36
migration, **169,** 37
mitosis and, **144,** 217
dissociation of events, **144,** 231, 237–238, 240–241, 243, 250
mitotic events, **144,** 219–220, 224
morphology, **169,** 26–28
movement in MT–microfilament interactions, **173,** 214–215
nuclear role in assembly, **175,** 94–95
protein components
autoimmune antibody binding, **169,** 40–41
centrosome induction in cell lines, **169,** 38
immunofluorescence localization, **169,** 42–43
isolation of centrins, **169,** 41
microtubule affinity chromatography, **169,** 42

Centrosomes (*continued*)
monoclonal antibody binding, **169,** 39–40, 42
mutant analysis, **169,** 42
purification of centrosomes, **169,** 38–39
related structures in plant cells, **175,** 97–101
replication
centrosome protein localization, **169,** 53
effects of cell cycle arrest, **169,** 54–55
eggs, **169,** 50–51
embryonic cells, **169,** 49–51, 53
polar organizers, **169,** 51–52
signaling pathways, **169,** 55–56
somatic cells, **169,** 50, 52–55
timing during cell cycle, **169,** 48–49, 54–55
as spindle poles, **162B,** 56–59
Cephalic neural crest
clonal analysis, **138,** 288–292
and patterning of vertebrate head, **138,** 273–274
Cephalosporium acremonium, mitochondrial genomes in fungi and, **141,** 93–94
Ceramic reactors, vertebrate cell culture technology and, **142,** 190–192
Ceratodon purpureus, apical cells, effect of microgravity, **171,** 24–25
c-ErbB receptor
carcinoma role, **160,** 248–249
ligands, **160,** 245–246
structure, **160,** 244–245
Cerebellar cortex
cell migration in, **175,** 277–280
glutamate metabolism, **147,** 301, 304, 308
Cerebellum
development, **175,** 262–263
glutamate metabolism, **147,** 291, 294, 298, 314–315
transcriptional regulation, **143,** 46
c-foe, transcriptional regulation and, **143,** 43
Cerebrospinal fluid
avian salt glands, **144,** 182, 184, 195
blood barrier
glucose transport, **172,** 13–14
structure, **172,** 13
ependymin, **165,** 144–146
NMR analysis, **145,** 37
Ceruloplasmin
antioxidant activity, **170,** 242
secretion by Sertoli cells, **160,** 201
Sertoli-germ cell communication network and, **147,** 50, 76–77
CFC-S, *see* Colony-forming cells-spleen
c-*fos,* role in chondrocyte differentiation, **159,** 321
CFTR, *see* Cystic fibrosis transmembrane conductance regulator
CFU-A, *see* Colony-forming unit-A cells
CFU-S, *see* Colony-forming unit-S cells
CGRP, *see* Calcitonin gene-related peptide
Ch21 protein
marker of chondrocyte differentiation, **159,** 290–291
tissue distribution, **159,** 290
Chaetoceros diadema, sintering, **150,** 298, 300
Chagas' disease, gap junction role, **157,** 340
Chalcone isomerase
control of gene, **147,** 274
enzymatic activity, **147,** 235
extracellular matrix and, **146,** 281
functions of products, **147,** 252
gene expression, **156,** 38
Chalcone-naringenin, **147,** 235, 244–246, 252–253
Chalcone synthase
control of activity, **147,** 252–261
control of gene, **147,** 261–262
elicitor induction, **147,** 265
enhancers, **147,** 269–271
light inducibility, **147,** 263–265
tissue specificity, **147,** 266–268
trans-acting factors, **147,** 271–277
transcription, **147,** 268–269
DNA content in meiosis, **158,** 106
enzymatic activity
deoxyflavanones, **147,** 244–245
mechanism of action, **147,** 234–236, 238–241
mutation, **147,** 237, 241, 243
stilebene synthase, **147,** 236–237, 242–244
extracellular matrix and, **146,** 280–281
functions of products, **147,** 245–252
gene expression, **156,** 38
and PAL, activities after pollen-tapetum fractionation, **140,** 50

plant
encoded by gene families, **148,** 26
phytoalexin biosynthesis, **148,** 27
stimulation by anthracene-9 carboxylate, **148,** 21
Channel complex, *see* Central plug
Chaos carolinensis
granular cytoplasm, **156,** 210
locomotion, **150,** 54–56
Chaos chaos
cytoplasmic streaming, **156,** 199, 201
cytoskeletal dynamics and, **144,** 115
Chaperones
assisted proteases, *Escherichia coli*
Clp protease, **174,** 155–157
FtsH protease, **174,** 157
La protease, **174,** 155
assisted proteases, in mitochondria
Clp homologs, **174,** 158–159
ClpP proteolytic subunit homologs, **174,** 158–159
FtsH homologs, **174,** 159–160
La protease homologs, **174,** 157–158
cell surface expression, **174,** 170–171
cytosolic, *see* Cytosolic chaperones
encoding gene, stress response, **174,** 161
gene organization, **174,** 163–166
immune response to, **174,** 167–168
levels in autoimmune lesions, **174,** 169
matrix, *see* Matrix chaperones
mitochondrial, *see* Mitochondrial chaperones
molecular, *see* Molecular chaperones
protein folding role, **160,** 332–333
relationship to disease, **174,** 166–167
roles
in autoimmune disease, **174,** 168–169
in mitochondrial protein degradation
stress proteins as proteases, **174,** 152–154
various studies, **174,** 154
in prion disease, **174,** 171–173
Chara
cytoplasmic streaming in
action potential as factor, **155,** 121
calcium-regulated inhibition, **155,** 118
light as factor, **155,** 125
microtubule system-supported, **155,** 109
reconstruction experiments, **155,** 116
environmental stimuli, **145,** 261, 264–165
Characeae
cell models, **175,** 4–5
cells, cytoplasmic streaming, **156,** 200
cytoplasmic streaming in
actin-myosin system role, **155,** 100–103
demembered cell models
calcium-regulated inhibition, **155,** 118–120, 130
energy requirements, **155,** 112–114
longitudinally cut model, **155,** 112–113
magnesium requirement, **155,** 114–115
pH dependence, **155,** 115
plasma membrane permeabilized cells, **155,** 110–111
preparation, **155,** 110–113
reconstruction experiments, **155,** 116–117
tonoplast-free cells, **155,** 111–112, 115
gravity role, **155,** 127–129
mechanism, **155,** 129–130
Chara globularis, rhizoids, effect of microgravity, **171,** 23
Charcot–Marie–Tooth disease, gap junction role, **157,** 340
Charged-coupled device, in whole-chromosome hybridization microscopy, **153,** 181
Chasmichthys, fish chromatophores and, **143,** 194, 198, 230, 243
ChAT, *see* Cholin-acetyltransferase
Chelonia, avian salt glands and, **144,** 188
Chelonia mydas, avian salt glands and, **144,** 147
Chemical cleavage mismatch analysis, detection of gene mutations, **158,** 235
Chemicals
environmental stimuli in plant cells and, **145,** 251–252
isolated as pollen surface substances, **140,** 43
Chemical shift, in NMR analysis, **145,** 7–8, 23, 43–44
Chemical synapses, cnidarian
bidirectionality of, **167,** 42–43
contacts with multiple neurons, **167,** 40–41
Chemoattractant, gradients, assessment, **163,** 94–97

Chemokines
bone marrow colony formation and, **167,** 202
hemopoietic progenitor cell mobilization by, **167,** 242–243
human *versus* murine, **167,** 199*t*
macrophage secretion of, **167,** 132
Chemokinesis, **139,** 161–162
Chemotaxis
associated peptides and mast cell degranulation, **138,** 174–175
bacterial flagellar motor and, **147,** 104, 111, 114, 116
in bacteria–macrophage interactions, **161,** 304–306
cytoskeletal dynamics and, **144,** 85, 88, 110–111, 117
in migration analysis, **155,** 53, 55–57
retinal pigmented epithelium and, **146,** 76–78
rhizobia, **156,** 2
role of POMC products
in invertebrates, **170,** 88–93
in vertebrates, **170,** 88
sperm, **163,** 251
vertebrate cell culture technology and, **142,** 154
Chemotherapy
hemopoietic growth factors enabling increase in dose intensity, **167,** 239–242
hemopoietic stem inhibitors in, **167,** 205
nondisjunction in men who have undergone, **167,** 283
potential benefit of combining with inhibitor protocols, **167,** 236
resistance, cellular, **156,** 77
Chiasmata, chromosomal
in aberrant genetic recombination, **167,** 282
maternal age in aneuploidy, **167,** 277, 278–279
Chicken, parathyroid hormone-related protein, gene structure, **166,** 234–236
Chicken chorioallantoic membrane assay
angiogenesis quantitation, **159,** 115–116
multiple grafting, **159,** 116
in vitro assay, **159,** 115–116
Chimeras
Fepi/JA, construction, **175,** 294
neural, *see* Neural chimeras
CHIP28
evolution, **160,** 113
hydraulic permeability, **160,** 122–124
osmotic conductance (P_{os}), **163,** 5–6
osmotic flow, bimodal analysis, **163,** 34–36
in seminal vesicle epithelium, demonstration, **160,** 279
structure, **160,** 114–115
tissue localization, **160,** 112–114
water transport, **160,** 112, 114, 145
Chironomus
karyosphere, **144,** 41
RNP particles, **151,** 190–191
Chitin
cholesteric ordering, **166,** 84–86
extracellular matrix, **146,** 285, 295, 304, 310
Chitinase
Arabidopsis thaliana, **144,** 74
extracellular matrix, **146,** 269, 304, 309–311
Chitosan
as exogenous elictor in plants, **148,** 11–12
in phytoalexin signal transduction, **148,** 21
systemic acquired resistance induction, **168,** 324
Chlamydomonas, see also Algae, volvocine
axonemal dynein diversity, **173,** 19–21
dynein ATPase, **173,** 13–14
endosymbiont hypothesis and, **141,** 296
flagellar dyneins
characteristics, **155,** 141
cross-bridge cycle, **155,** 169
flagellar motility *in vivo,* **155,** 166–167
gene modification and expression, **155,** 173
heavy chains, **155,** 150, 152–154
inner row *in situ* localization, **155,** 164
intermediate chains, **155,** 148–151
light chains, **155,** 147–148
microtubule translocation *in vitro,* **155,** 167–168
outer arm, **155,** 155–160
outer row, **155,** 147, 161–162
structure, **155,** 146
gametogenesis
activation by staurosporine, **168,** 292–293
assay systems, **168,** 262–263, 282

blue light signaling, **168,** 271–277, 295–296
cell culture, **168,** 259–260
gamete division and growth, **168,** 281
gene characterization, **168,** 286–290, 293
inhibitors, **168,** 290, 292
model, **168,** 264–265
mutants, **168,** 283–285
nitrogen deprivation signaling, **168,** 265–271
sexual life cycle, **168,** 260–261
synchronization of cell cycle, **168,** 278
ultrastructural changes during gametogenesis, **168,** 263–264
phototaxis
calcium ion requirements, **148,** 292–295
eyespot reflection in, **148,** 259, 261, 265
photoreceptive apparatuses, components, **148,** 270–273
retinal pigments, **148,** 278–284, 286
signal generation, **148,** 253
signal transduction, **148,** 297
plastids, **145,** 159
Chlamydomonas eugametos, gametogenesis, **168,** 293–295
Chlamydomonas reinhardtii
cyanelles in, **151,** 36, 39
dynein assembly, **155,** 146
effect of microgravity, **171,** 29–30
endosymbiont hypothesis and
gene expression, **141,** 305, 307
gene structure, **141,** 314
genome organization, **141,** 289, 295–297
mitochondria, **141,** 282, 326–327
organelles, **141,** 331
plastids, **141,** 267, 272
rRNA sequences, **141,** 320–321, 323–324
gametogenesis, **168,** 261–262, 272, 274, 278–281
intracellular calcium messenger system and, **142,** 309–310
microgravity experiments, **156,** 342
mitochondrial genomes of, **141,** 12, 45, 50, 52–53
phototaxis
eyespot apparatus, **148,** 239
eyespot reflection, **148,** 258–259, 261, 266, 268
photoreceptive apparatuses, components, **148,** 272
retinal pigments in, **148,** 279, 281–282
Chloramphenicol
ciliates, **141,** 4, 7
spaceflight experiments, **156,** 341
Chlorarachnion, chloroplasts and, **144,** 276
Chlorate, flowering plants and, **143,** 300–301
Chlorella
endosymbiont hypothesis, **141,** 296–297, 324
intracellular calcium messenger system, **142,** 329
microgravity experiments, **156,** 343
Chlorella pyrenoidosa, effect of microgravity
on cell ultrastructure, **171,** 26–29
on chloroplasts, **171,** 25–26
Chlorella vulgaris
ATPase activation, effect of clinostating, **171,** 28–29
effect of microgravity
on cell ultrastructure, **171,** 26–29
on chloroplasts, **171,** 25–26
Chloride
avian salt glands and, **144,** 129, 205
blood supply, **144,** 154
hormonal control, **144,** 196–197
innervation, **144,** 155, 157
secretory mechanism, **144,** 173, 175–176
secretory tissue, **144,** 131–133, 139
stimulus-secretion coupling, **144,** 200–201
biological clocks in retina and, **146,** 116, 126
conductance in cystic fibrosis, **172,** 211
incorporation in apatite, **172,** 159
in insect retina, **176,** 316–319
Chloride channels
Cl^-,HCO_3^- exchanger, **161,** 194–196
in cystic fibrosis, **172,** 195–196
in endosome acidification, **172,** 230
inhibitors, **161,** 185–186
maxi, **161,** 185
medium, **161,** 185
mini, **161,** 184–185
swelling-activated efflux, **161,** 201–202
volume-sensitive
cell types affected by, **161,** 181–182
properties, **161,** 184–185

Chloride transport, **165,** 23
Chlormequat, induction of chromoplast differentiation, **163,** 204
Chlorochromoplasts
characteristics, **163,** 185
representative, table, **163,** 186
Chloroethylnitrosoureas, cytotoxic activity, **162A,** 545–547
2-(Chlorophenylthio) triethyl-ammonium chloride, induction of chromoplast development, **163,** 204
Chlorophyceae
eyespot reflection in, **148,** 259, 261, 266, 268
photoreceptive apparatuses, components, **148,** 273
Chlorophyll
chloroplasts
cryptomonads, **144,** 273, 275
dinoflagellates, **144,** 285, 288–289, 292
endosymbiotic origin, **144,** 259–260
in euglenoids, **144,** 284
green algae, **144,** 265, 270, 272
heterokont algae, **144,** 277, 282
red algae, **144,** 261, 263, 265
endosymbiont hypothesis, **141,** 254–255, 257, 261, 276
higher plant plastids, **145,** 164, 166–167, 190
Chlorophyll a, content in pea plant, effect of microgravity, **171,** 38–39
Chlorophyll b
content in pea plant, effect of microgravity, **171,** 38–39
endosymbiont hypothesis, **141,** 269, 276
Chlorophyta
endosymbiont hypothesis, **141,** 295–296
eyespot apparatus, **148,** 235
Chlorophytes
endosymbiont hypothesis
mitochondria, **141,** 282, 324, 327
plastids, **141,** 263, 272, 275–276, 278
endosymbiosis, **144,** 265, 280
Chloroplast endoplasmic reticulum, endosyrnbiont hypothesis, **141,** 258–259
Chloroplasts
animal mtDNA and, **141,** 192
chalcone synthase, **147,** 276–277
ciliates and, **141,** 4, 55, 58–59
Paramecium aurelia, **141,** 12, 36–38, 41–43
rRNA genes, **141,** 45, 50
DNA, higher plant mitochondrial genomes and, **141,** 148–149
effect of microgravity
in angiosperm leaf, **171,** 37–38
in *Chlorella,* **171,** 25–26
endosymbiont hypothesis and, **165,** 208
mitochondria, **141,** 304, 315–316, 318, 323
plastids, **141,** 258, 266–267, 269, 278
endosymbiotic origin, **144,** 259–260, 293–295
cryptomonads, **144,** 272–273
biochemistry, **144,** 276
Chlorarachnion, **144,** 276
envelope membranes, **144,** 275–276
pigments, **144,** 273
structure, **144,** 273–275
dinoflagellates, **144,** 285
alien algal endosymbionts, **144,** 286, 288–289
ancestry, **144,** 291–293
biochemistry, **144,** 291
eyespot, **144,** 290–291
fucoxanthin, **144,** 289–290
phycobilins, **144,** 290
pigments, **144,** 285–286
structure, **144,** 286–287
euglenoids, **144,** 282
biochemistry, **144,** 284–285
eyespot, **144,** 284
pigments, **144,** 282–283
structure, **144,** 283
green algae, **144,** 265
ancestry, **144,** 270–272
biochemistry, **144,** 272
envelope membranes 266, **144,** 268–270
pigments, **144,** 265
structure, **144,** 266–267
heterokont algae, **144,** 277
biochemistry, **144,** 281
chloroplast relationships, **144,** 282
eyespot, **144,** 278, 280–281
pigments, **144,** 277
structure, **144,** 277–279
red algae, **144,** 261–264
biochemistry, **144,** 263–264

cyanelles, **144,** 264–265
glaucophytes, **144,** 264
pigments, **144,** 261
structure, **144,** 261–263
environmental stimuli in plant cells and, **145,** 252, 300–301
chemicals, **145,** 297–298
light, **145,** 268–275
nucleus, **145,** 287
photodinesis, **145,** 252–253, 257–259, 263
transduction chain, **145,** 275–287
wounding, **145,** 295
flowering plants
amino acids, **143,** 311–314, 324
ammonium, **143,** 304–309
nitrate, **143,** 302–303
genomes
algae containing chorophyll b, **151,** 36
higher plants, **151,** 35–37
higher plant
amino acid synthesis, **145,** 191, 194, 196–199
carbohydrate oxidation, **145,** 201, 203
CO_2 fixation, **145,** 172–173
evolution, **145,** 164–167
fatty acid synthesis, **145,** 183–189
genome, **145,** 153–164
starch synthesis, **145,** 177, 179–182
structure, **145,** 151
translocation, **145,** 167–171
higher plant mitochondrial genomes and, **141,** 130, 160–161
abnormal phenotypes, **141,** 133, 135
gene location, **141,** 157, 159
structure, **141,** 147, 156
intracellular calcium messenger system, **142,** 308, 318, 334
calmodulin, **142,** 329–332
G proteins, **142,** 309–310
kinetoplastid mtDNA and, **141,** 69, 71
light radiation effect on centrifugal motility, **157,** 117
mitochondrial genomes in fungi and, **141,** 90
nucleic acid
escape, **165,** 207–228
escape consequences, **165,** 226–227
migration, **165,** 207–228
migration consequences, **165,** 226–227
Chloroplast stromal enzyme, phosphorylation, **149,** 49–53
Chlorosulfuron, flowering plants and, **143,** 323–324
Chlorotetracycline
in analysis of Na+ requirements
in acrosomal exocytosis, **149,** 33–35
in sperm capacitation, **149,** 22–25
environmental stimuli in plant cells and, **145,** 280–281, 289
intracellular calcium messenger system and, **142,** 315
sperm function analysis with, **149,** 3–7
calcium regulation, **149,** 11–16
potassium regulation, **149,** 11–16
Chlorpromazine
agravitropism induction in plants, **168,** 157–158, 160–161
calmodulin antagonism, **168,** 157–158
inhibition of ecto-ATPases, **158,** 191
selectivity and dose, **168,** 160–161
Cholecystokinin
control of rhythmic behaviors and, **147,** 4, 18–19
immunoregulation, **149,** 270
secretion by neuroendocrine cells, **157,** 284–285
transport, **159,** 33
visualization, **157,** 284
Cholecystokinin-related peptides
effects of CCK-related peptides, **138,** 154–155
immunohistochemical localization of gastrin- and CCK-related peptides, **138,** 153–154
Cholera, chloride channel activity and, **172,** 209–210
Cholera toxin, structure, **161,** 286
Cholesteric geometry
consequences, **166,** 63–70
descriptions, **166,** 65
properties, **166,** 63–70
twisted plywood model, **166,** 63–65
Cholesteric ordering
biopolymer, *in vitro,* **166,** 84–92
cellulose, **166,** 84–86
cellulose-based polymer, **166,** 84–85
chitin, **166,** 84–86
collagen, **166,** 86–90
crystallite, **166,** 86, 87
DNA, **166,** 90–92

Cholesteric ordering (*continued*)
fibril, **166,** 86, 87
neutral pH, **166,** 90
monomer, acid pH, **166,** 88–90
Cholesteric supramolecular order
chromosome, **166,** 81–83
cytoplasmic inclusion, **166,** 83–84
cytoskeleton, **166,** 83–84
extracellular matrix, **166,** 70–81
gland secretion, **166,** 83–84
nucleus, **166,** 81–83
Choline, NMR analysis, **145,** 43, 45–46
Choline acetyltransferase
avian retina and, **146,** 16
effect of electromagnetic field, **158,** 318
neurotransmission, **149,** 236–237, 240
retinal precursor cells and, **146,** 155
trophic factor-induced neuronal activity, **153,** 271–274
Cholinergic innervation, avian salt glands
innervation, **144,** 159–163, 165, 167
stimulus-secretion coupling, **144,** 201, 203–204
Cholinergic neurotransmission, **149,** 236–241
Cholinergic receptor, fish chromatophores and, **143,** 229–230
Cholinesterase
in avian neurogenesis
acetylcholinesterase expression related to activity, **151,** 156–158
brain development
early neurogenesis, **151,** 144–145
membrane association, **151,** 146–147
retina, mature, **151,** 146
serum BChE, **151,** 147–148
small molecular forms, **151,** 145
cell adhesion molecule function
anticholinesterases, **151,** 166–169
HNK-1 epitope regulation, **151,** 159–163, 159–164
neurite growth alteration, **151,** 167–169
pathfinding cues, **151,** 163, 165–166
regulation, **151,** 158–159
cell proliferation rate, **151,** 155–156
characteristics, **151,** 139–140, 170
expression
in differentiation, **151,** 141–143, 141–144
periods, **151,** 141, 168–170
in proliferation, **151,** 141–143, 141–144
function, **151,** 139–141
functional analysis of actions, *in vitro* models, **151,** 149–152
genes, **151,** 140, 169
molecular polymorphism during brain development
biosynthesis, **151,** 149
early neurogenesis, **151,** 144–145
membrane association, **151,** 146–147
retina, mature, **151,** 146
retinal, downregulation during *in vitro* regeneration, **151,** 151, 153–156
in avian retina, **146,** 16, 29–31, 41
avian salt glands and, **144,** 157–161, 167
perinatal exposure to sex hormones and antihormones and, **139,** 28–29
production by neuroendocrine cells, **157,** 285
Chondrocytes
differentiation
collagen synthesis, **159,** 268–269, 273, 288–290, 299, 302, 326
markers, **159,** 290, 298–299
mesenchymal stem cell precursor, **159,** 268–271
molecular modulators, **159,** 269–271
role
activin, **159,** 310
adhesion molecules, **159,** 294–295
ascorbic acid, **159,** 292–293
bone morphogenetic proteins, **159,** 309–310
fibroblast growth factor, **159,** 308–312
oncogenes, **159,** 321–323
ovotransferrin, **159,** 314–315
retinoic acid, **159,** 271, 274–275, 277, 300, 303, 305–306
retinoic acid-induced heparin-binding factor, **159,** 317
TGF-β, **159,** 306–309
thyroid hormones, **159,** 306
stages, **159,** 268
in vitro studies
cell–cell interactions, **159,** 293–295, 330
cell morphology, **159,** 285–286, 291, 299

cell–substratum interaction, **159,** 285–286, 291, 330
culture in gels, **159,** 286
density gradient separation, **159,** 288
models, **159,** 285
phenotype maintenance, **159,** 285–292
sensitivity to medium, **159,** 292–293
hypertrophic chondrocytes, **159,** 279, 282, 332
apoptosis, **159,** 298
differentiation to osteoblasts, **159,** 298–302, 306
immunostaining, **159,** 302–303
organ culture, **159,** 300–301
osteoarthritic cartilage association, **159,** 326
maturation
chondroitin sulfate synthesis, **159,** 282
collagen synthesis
type II, **159,** 279, 281, 298
type X, **159,** 278, 281–282, 298
hypertrophy, **159,** 279, 282, 332
metalloproteinase expression, **159,** 282–283, 295
Chondrodysplasia
animal models, **159,** 326, 329–330
classification, **159,** 326
COL2A1 mutations, **159,** 327–329
spondyloepiphyseal dysplasia, **159,** 327
Chondroitin sulfate, eggshells and biomineralization and, **145,** 226, 231
Chondromodulin 1, role in chondrogenesis, **159,** 318
Chorionic gonadatropin receptor, **167,** 56
Chorionic villus sampling
in analysis of chromosomal constitution, **153,** 2–4
assessing mosaicism using, **167,** 287
Chorismate, flowering plants and, **143,** 311–313
Choroid, melanocytes, **146,** 192, 194, 243
Choroid plexus
cadherin, expression, **165,** 174–177
ependymin, **165,** 143–144
glucose transport in, **172,** 13–14
GLUT1 expression in, **172,** 11–12
Na^+/K^+-ATPase polarity, **165,** 174–177
Chromatid linkage protein, mitosis, **144,** 244–245
Chromatids
mitosis, **144,** 217
chromosomes, **144,** 238–239, 241, 244–245, 251
dissociation of events, **144,** 235, 237
genetics, **144,** 231, 233
mitotic events, **144,** 220–221, 225
in nondisjunction mechanism, **167,** 275–276
Chromatin
acceptor protein RBF-1, **162B,** 350–354
acceptor sites for progesterone receptor, **162B,** 346–348
anchoring, role in centromere/kinetochore region, **162B,** 39
association with SV40 virions, **162A,** 530–531
bulk nuclear, **162B,** 386–387
cellular, T-antigen accumulation, **162A,** 502–508
ciliated protozoa, replication models, **153,** 161–166
condensed chromatin, **165,** 82
content in plant meiosis, **158,** 107–108
decondensing, **162B,** 169–170, **165,** 82–84
domains
fractionation, **162A,** 221–226
nuclease-sensitive, **162A,** 218–221
organization, **162A,** 457–458
and prediction of MAR sequences, **162A,** 279–366
RNP, **162A,** 6–8
transcriptionally active, formation, **162A,** 232–234
effect of dynamic microtubules, **175,** 117
extended form, **162A,** 197–198
extraction, **162A,** 494
folded, **162A,** 172
granules, in replication band of ciliated protozoa
macronuclei, **153,** 149–150, 152
models, **153,** 164–165
interaction with lamins, **162B,** 163–164
in interphase nucleus
gene expression, **150,** 150, 168
motion, **150,** 150–157
movement by contractile protein, **150,** 164–167
rearrangements, **150,** 160–164
reorganization, **150,** 149

Chromatin (*continued*)
intracellular calcium messenger system and, **142,** 335
karyosphere and, **144,** 8, 15, 21, 23, 39
lepidopteran
elimination, in female meiosis, **152,** 14–15
in male apyrene meiosis, **152,** 27–28
loop
association with nuclear matrix, **162A,** 542–543
histone H1 in, **162A,** 307
and MARs, **162A,** 199–201
model, attachment sites, **162A,** 140–141
and nucleoskeleton, **162A,** 129–131
relaxation prior to chromatin extraction, **162A,** 494
methylation and gene imprinting, **160,** 71–72
as microtubule-organizing center, **175,** 96–97
mitosis and
chromosomes, **144,** 239–240, 246–247, 249–250
dissociation of events, **144,** 235, 238
genetics, **144,** 232–233
mitotic events, **144,** 219–220, 222, 224
–nuclear binding, role of receptor accessory factors, **162B,** 359–364
and nuclear matrix, **162A,** 173
organization
higher order
alteration, **162A,** 206–207
prekinetochore role, **162B,** 39–40
into matrix-attached DNA loops, **162A,** 565
nuclear matrix role, **162A,** 101,191–235,541–545
polyamine association
chromosomes, **138,** 76–77
condensed nuclei, **138,** 73–76
cytochemical evidence for close association with condensed DNA, **138,** 73
interactions with DNA, chemical aspects, **138,** 78
RNP domains, **162A,** 6–8
specific regions, association between, **162A,** 52–53
structure
interferon-β, **162A,** 393–407
and nucleosome organization, **162A,** 263–269
transcriptionally active, **162A,** 226–227, 231–232
transcriptional regulation and, **143,** 2, 5, 11, 25
and transcription factors, **162A,** 228–230
volume in plant cells, effect of microgravity, **171,** 51–52
X chromosome regulation, **160,** 82–84
Chromatography
animal mtDNA and, **141,** 199
avian salt glands and, **144,** 163
chloroplasts and, **144,** 265
high-performance liquid chromatography
intracellular calcium messenger system, **142,** 314, 322
self-incompatibility, **143,** 265, 272
high-pressure liquid chromatography, in analysis of hormonal control, **145,** 115
intracellular calcium messenger system and, **142,** 314, 321–322
karyosphere and, **144,** 41
retinal precursor cell components, **146,** 154, 181, 182
Chromatophores
dermal, cytophysiology, **143,** 193–194
fish, *see* Fish chromatophores, cytophysiology
Chromatosomes, cytophysiology
aggregation, **143,** 247
cytoskeleton, **143,** 221–223
migration, **143,** 194
motility control, **143,** 233–235, 237, 242, 244
protein combination on membrane, **143,** 199
Chromista, chloroplasts and, **144,** 277–282
Chromis viridis, fish chromatophores and' 216, **143,** 233
Chromophores, retinal precursor cells and, **146,** 180
Chromophytes
cryptochromes in, **148,** 288–289
endosymbiont hypothesis
plastids, **141,** 258, 261, 263, 277
sequences, **141,** 269, 271–272

Chromoplasts
biogenesis
acyl lipid metabolism, **163,** 219–220
ATP/ADP sources, **163,** 207–208
carbohydrate metabolism, **163,** 209–211
isoprenoid pathways, **163,** 211–219
nitrogen metabolism, **163,** 220–221
plastid envelope membrane budding, **163,** 206–207
carotenoid compostion, **163,** 188–189
crystalline
with large β-carotene crystals, **163,** 178
with lycopene crystals, **163,** 181
occurrence in nature, table, **163,** 182
with small β-carotene crystals, **163,** 181
differentiation
chemical regulation, **163,** 203–204
genetic factors
Cucurbita pepo, **163,** 194
maize, **163,** 193–194
pepper, **163,** 193
tomato, **163,** 191–193
hormonal regulation, **163,** 201–203
light regulation, **163,** 199
nutritional regulation, **163,** 200–201
in organello translation during, **163,** 197–199
photosynthetic genes
chromoplast DNA organization and topology, **163,** 194–195
transcription, **163,** 195–197
protein import during, **163,** 197–199
reversion phenomena, **163,** 205
stress regulation, **163,** 203
temperature regulation, **163,** 199–200
fibrillar
occurrence in nature, table, **163,** 182
organized into bundles, **163,** 182–184
organized into dispersed structures, **163,** 184
globular
characterization, **163,** 177–178
occurrence in nature, table, **163,** 179–180
higher plant
metabolism, **145,** 175, 190, 197
structure, **145,** 151, 153, 161–163
isolation, **163,** 187–188
membranous, **163,** 184
prenyllipid composition, **163,** 188–189
reticulo-tubular, **163,** 184–185
substructures, biochemical composition, **163,** 188–191
tubular, **163,** 184
Chromosomal protein, with MTOC properties, **175,** 112
Chromosomes
aberrations
FISH diagnosis, **153,** 25–33
formation mechanisms, **153,** 204–208
abnormalities
fertilization *in vitro,* **164,** 143–145
polymorphisms, **167,** 283
structural *versus* numerical, **167,** 264
Arabidopsis thaliana and, **144,** 62–63, 67
arrangement in nucleus, **159,** 164
autosomes, *see* Autosomes
breakage
heterochromatin role, **158,** 112–117
sites, **158,** 113
centromere repetition, **162B,** 37–38
cholesteric supramolecular order, **166,** 81–83
cytoskeletal dynamics and, **144,** 113
deletions, **158,** 225
DNA replication, eukaryotic, *see* DNA replication, in mammals
domains, **162A,** 170–173
Drosophila muscles and, **143,** 63–64
contractile protein genes, **143,** 99, 108, 115, 124, 126, 131
development, **143,** 65, 79–81, 86–87, 94
myosin light chains, **143,** 105, 107
paramyosin, **143,** 118–119, 121
duplications, **158,** 225
endochromosome, polytene chromosome, **165,** 94–95
endosymbiont hypothesis and, **141,** 262, 276
flowering plants and, **143,** 315
gonosomes, *see* Gonosomes
higher plant mitochondrial genomes and, **141,** 136
higher plant plastids, **145,** 165
histone-depleted, **162B,** 410–411
hybridization, *see* Hybridization, whole-chromosome
in interphase nucleus
patterns, **150,** 160–164
topology, **150,** 157–160
inversions, **158,** 225

Chromosomes (*continued*)
karyosphere and, **144,** 1, 42–45, 47
nuclear matrix, **144,** 41
oocytes of insects, **144,** 9, 11, 15, 19, 21, 23, 25
vertebrates, **144,** 25–28, 30, 33, 35–39
lepidopteran spindle structure
experimental systems, **152,** 32–35
holokinetic, **152,** 33
kinetic organization, **152,** 8–12
meiosis
female, **152,** 13–15
male apyrene, **152,** 24–30, 35–36
male eupyrene, **152,** 16–24, 34–36
microtubules, post-translational modifications, **152,** 30–32
mitosis, **152,** 12–13
structure, **152,** 33–34
technical aspects, **152,** 3–8
locus-specific probes, **153,** 27–28
melanocyte
development and, **146,** 217–220
pigment-cell genetics, **146,** 196, 198–203, 205–209, 211
melanoma
clinical evidence, **146,** 227–228, 232
fish, **146,** 233–234
genetics, **146,** 236, 250–251
growth factors, **146,** 239–240
oncogenes, **146,** 244–246
tumor suppressor genes, **146,** 247–248
–microtubular cycles, interrelationships, **175,** 115–119
minichromosomes, *see* Minichromosomes
mitochondrial genomes in fungi and, **141,** 120
mitosis and, **144,** 217–218
dissociation of events, **144,** 227, 229–252
mitotic events, **144,** 218–223, 225
mitotic, **162A,** 141
mosaicism
incidence and origin, **167,** 287–288
mechanism of formation, **167,** 288–289
overview, **167,** 286–289
movements, **159,** 164
role of kinesin-like proteins, **164,** 304
nucleolar organizer region, **162B,** 306–310
painting, **153,** 16–17, 28, 30–31
periphery, **162B,** 31–36
Philadelphia, **153,** 28–30
polyamine localization, **138,** 76–77
polytene, *see* Polytene chromosome
preimplantation diagnosis of genetic defects
c banding technique, **151,** 15
cytogenetic analysis, **151,** 15–16
preimplantation embryo analysis, **153,** 4, 22–25
premature condensation, **153,** 207–208
recombinations in transgenic mice, **164,** 73–74
scaffold components, **162B,** 21–31
self-incompatibility and, **143,** 289
sex, *see* Sex chromosomes, mammalian
specific repeat probes, **153,** 25–27
structure, and DNA replication initiation, **162A,** 473–476
technological advances in analyzing, **167,** 263–264
transcriptional regulation and, **143,** 26–27, 38, 45
transdifferentiation in medusae and, **142,** 252
translocations, **158,** 225–226
trisomy 21 detection, **153,** 27–29
varying rates of nondisjunction for, **167,** 291
vertebrate cell culture technology and, **142,** 148
X, *see* X chromosome
yeast artificial, probes, **153,** 195
Chromulina, phototactic signal generation, **148,** 255
Chronic lymphocytic leukemia, transcriptional regulation and, **143,** 38
Chroomonas, eyespot apparatus, **148,** 240
Chrysiptera cyanea, fish chromatophores and, **143,** 233, 243
Chrysopa, karyosphere and, **144,** 11, 15, 18, 23, 42
Chrysophyceae, eyespot reflection in, **148,** 259, 269
Chrysophytes, chloroplasts and, **144,** 280, 282, 288–290
Cik1, spindle pole body functions, **169,** 68
Cilazapril, inhibition of smooth muscle cell growth, **161,** 25
Cilia
avian retina and, **146,** 11, 17, 28–30, 40–41

beating, regulation, **164,** 285–289
cndidarian neurons, **167,** 40, 42
flagella, *see* Flagella
Ciliary body epithelium
barrier properties, **172,** 18–19
gap junctions in, **172,** 21–22
GLUT1 in, **172,** 19–21
structure, **172,** 17–18
Ciliary neurotrophic factor, and neuronal plasticity
neurite initiation and elongation, **153,** 252–253
phenotypic expression, **153,** 274
Ciliates
chloroplasts, **144,** 293
endosymbiont hypothesis
mitochondria, **141,** 321, 327–328
molecular biology, **141,** 310, 314
mitochondrial genomes, **141,** 1–4, 58–60
genetic code
codon usage, **141,** 55–57
tRNA, **141,** 52–55
Paramecium aurelia sequences, **141,** 12
comparisons, **141,** 41–42
divergence, **141,** 38–41
organization, **141,** 12–37
phylogenetic tree, **141,** 42–45
rRNA genes, **141,** 45–46
large rRNA, **141,** 48, 50–52
small rRNA, **141,** 47–49
structure
initiation and termination repeats, **141,** 9–10
Paramecium aurelia, **141,** 4–7
sequences, **141,** 10–11
Tetrahymena pyriformis, **141,** 7–9
Cinnamate-4-hydroxylase, *see trans*-Cinnamate-4-monooxygenase
trans-Cinnamate-4-monooxygenase
activity in response to stress, **172,** 254
localization, **172,** 253, 254
in phenylpropanoid pathway, **172,** 253–254; **176,** 102–105
Cinnamic acid, extracellular matrix and, **146,** 280–281, 283
Cinnamic acid 4-hydroxylase, extracellular matrix and, **146,** 279, 295, 311
Cinnamoyl alcohol:NADPH oxidoreductase, extracellular matrix and, **146,** 281–282
Cinnamoyl-CoA reductase
effects of inhibitors, **172,** 263–264
expression, **172,** 263
in lignin branch pathway, **172,** 262
role in lignin-specific pathway, **176,** 105–106
substrate specificity, **172,** 263
Cinnamyl alcohol dehydrogenase
expression
antisense gene, **172,** 268–269
correlation with lignification, **172,** 266–267
developmental, **172,** 266
pathogen stress, **172,** 267
inhibitors, **172,** 267–268, 276–277
in lignin branch pathway, **172,** 264
localization, **172,** 264
mutant forms, **172,** 268
polymorphism, **172,** 265–266
role in lignin-specific pathway, **176,** 105–106
structure, **172,** 264–265
Cinnamyl alcohol β-glucosidase, in deglycosylation of monolignol glycosides, **172,** 270–271
Cinnamyl alcohol glucosyltransferase, in glycosylation of monolignols, **172,** 270
Cinnamyl alcohols
in lignin biosynthesis, **176,** 109–111
in lignin network, **176,** 90
storage forms, **176,** 106–107
Ciona intestinalis
embryogenesis, **176,** 249–250
endoderm development, **176,** 264
Circadian rhythm, *see* Biological clocks
Circuits, functional, control of rhythmic behaviors, **147,** 20
Circularization, higher plant mitochondrial genomes and, **141,** 136
Circumventricular organs
avian salt glands, **144,** 182, 195, 197
components, **159,** 24
effects of serotonin, **159,** 25
regulation of neuroendocrine system, **159,** 25
structure, **159,** 25
termination of hypothalamic neurons, **159,** 6
Cisplatin
cytotoxicity of L-PAM, **156,** 130
DNA repair, **156,** 128

Cisplatin (*continued*)
induction of DNA repair, **156,** 122–123
induction of genes, **156,** 79
induction of interstrand crosslinks, **156,** 133–134
cis-splicing, endosymbiont hypothesis and, **141,** 267, 304
Citrate synthase, *Tetrahymena* 14nm filament protein
identification, **164,** 13–16
modulation, **164,** 18–20
Citric acid, regenerative cementogenesis and, **142,** 51
Citrus, plastids and, **145,** 161
Cl-628
effect on enzyme levels, **139,** 29
genital abnormalities in rats and, **139,** 15–16
Cladophoropsis, environmental stimuli in, **145,** 295
Clarkia xantiana, higher plant plastids and, **145,** 181–182
Clathrin
coated vesicles
CFTR recycling and, **172,** 217
in M cells
apical and basolateral membranes, **167,** 111
membrane-bound tracer uptake by, **167,** 127
in plants
acidification, **167,** 23–26
composition of
assembly proteins and adaptors, **167,** 13–18
heavy and light chains, **167,** 10–13
receptors, **167,** 18–21
function
endocytosis, **167,** 2–5
membrane recycling, **167,** 5–6
protein sorting and transport, **167,** 6–8
historical background, **167,** 1–2
isolation, **167,** 8–9
prospects and unresolved problems, **167,** 26–28
uncoating, **167,** 22–23
heavy and light chains in, **167,** 10–13
triskelions, cages formed by
description, **167,** 13–14
reassembly of, **167,** 14–18
Cleavage
endoproteolytic, **167,** 57–58, 78–82
patterns, ascidian embryo, **176,** 268–272
stage biopsy, for preimplantation embryo analysis, **153,** 5
Cleft lip, incidence, **158,** 253
Cleft palate
family pedigree, **158,** 255–256
incidence, **158,** 253
linkage analysis, **158,** 255–256
modes of inheritance, **158,** 253
mouse model, **158,** 259
X-linkage, **158,** 253–256
Clenodinium foliaceum, chloroplasts and, **144,** 288–290, 292
Cleptochloroplasts, endosymbiosis and, **144,** 289, 292
Clinorotation effects
on animal development, **156,** 334–335
in biological research, **156,** 357–358
effect on cell walls
parenchyma wall structure, **170,** 48–50
polysaccharide synthesis, **170,** 57–60
protoplast wall regeneration, **170,** 45–47
growth factor activation, **156,** 318
Clinostating effects
advantages, **171,** 9–10
on angiosperm root elongation and differentiation, **171,** 35
on ATPase activation in *Chlorella,* **171,** 28–29
on cell walls
comparison to microgravity effects, **170,** 41–42
horizontal clinostation
epidermal walls, **170,** 54
protoplast wall regeneration, **170,** 43–44
protoplast wall regeneration, **170,** 45–46
definition, **171,** 7
limitations, **171,** 9
on microviscosity and lipid peroxidation, **171,** 41
on moss cellulosopectolytic enzymes, **171,** 31
on plant cell aging, **171,** 53–56
on plant cell differentiation, **171,** 53–56
on plant cell and tissue cultures, **171,** 43–44

on plant intracellular processes, **171,** 58
on plant organelle structural–functional rearrangements, **171,** 46–53
on root cap statocytes, **171,** 21–22
types, **171,** 7–9
CLL, *see* Chronic lymphatic leukemia
Clofibrate
age-dependent efficacy, **160,** 333–334
effects on peroxisomes, **160,** 303, 316, 326–328
rat strain sensitivity, **160,** 334–335
Clomiphene, *see also* Antihormones; Sex hormones
effect on enzyme levels, **139,** 28
genital abnormalities and
in human females, **139,** 21
in mice, **139,** 22, 220
in rats, **139,** 13–14, 23
history, **139,** 1
Clonal analysis
γ-aminobutyric acid metabolism and, **147,** 312
animal mtDNA and, **141,** 175, 207
Arabidopsis thaliana and
biochemistry, **144,** 73–74
genetic model, **144,** 61–63
models, **144,** 76
pathogens, **144,** 56–57
phenotypes, **144,** 67, 71–72
auxin-regulated genes
arcA, cDNA, **152,** 129
mechanism, **152,** 111
avian retina and, **146,** 34–35
cDNA, tumor-specific genes in habituated plant cells, **152,** 290–291
chalcone synthase and, **147,** 245, 274
chromosome-specific DNA, in whole-chromosome hybridization, **153,** 193–195
ciliates and, **141,** 6, 9, 47–48, 51
control of rhythmic behaviors and, **147,** 4
cytoskeletal dynamics and, **144,** 107–108
from DNA library, **140,** 6
Drosophila muscles and, **143,** 63–65, 138, 140
contractile protein genes, **143,** 117, 126
development, **143,** 71–73, 77, 87–93
endosymbiont hypothesis and, **141,** 299, 308
expression, G-protein-coupled receptors, **164,** 233
extracellular matrix and
characterization, **146,** 267, 270
modification, **146,** 304
synthesis, **146,** 273, 279, 282–283
flowering plants and, **143,** 297–299
amino acids, **143,** 312, 315–317, 320, 324
ammonium, **143,** 305, 307
GABA receptor, **164,** 195
glutamate metabolism and, **147,** 289, 294, 302, 307
L-glutamate receptor, **164,** 192–193
GnRH-encoding gene in teleosts and, **147,** 176, 182
G-protein-coupled receptor
classical approach, **164,** 232–233
expression in transfected cells, **164,** 233
hybridization under reduced stringency, **164,** 234
polymerase chain reaction, **164,** 234
in *Xenopus* oocytes, **164,** 233
growth factor-induced retinal regeneration and, **146,** 52
higher plant mitochondrial genomes and, **141,** 140, 143, 157–159, 163
higher plant plastids and, **145,** 165, 168, 185, 192
intracellular calcium messenger system and, **142,** 319, 328–329, 333, 335
markers, **138,** 290, 291
melanocytes and
biology, **146,** 193, 195–196
development, **146,** 215–216, 218–219, 221, 224
genetics of melanoma, **146,** 233, 235, 242, 248
pigment-cell genetics, **146,** 204, 207–209, 211, 213
myofibrillogenesis and, **143,** 163, 166
neural crest cells *in vitro,* method, **138,** 287
neural crest potentialities, **138,** 285–293
cephalic NC, **138,** 288–292
development restriction of differentiation and proliferation potentialities of trunk-derived NC cells, **138,** 292–293
nicotinic acetylcholine receptor, **164,** 191–192
random, MARs, **162A,** 359–362

Clonal analysis (*continued*)
regenerative cementogenesis and, **142,** 38, 40–41
retinal precursor cells and, **146,** 175
self-incompatibility and, **143,** 257
gametophytes, **143,** 284–287
sporophytes, **143,** 269, 272, 274–275, 280
testis-determining factor *SRY* gene, **154,** 233–237
tonoplast H^+-ATPase cDNA, **174,** 96–99
transcriptional regulation and, **143,** 7, 10, 25–27, 45, 47
ankyrin repeat proteins, **143,** 36–39
cDNA, **143,** 29–33
Drosophila, **143,** 35–36
genes, **143,** 13–14
models, **143,** 34–35
p65 subunit, **143,** 33–34
Rel, **143,** 27–29
transdifferentiation in medusae and, **142,** 214
urease in plant cells and, **145,** 86, 89, 91, 98
Clover
nodulation, **156,** 30, 42, 46
non-nodulating mutants, **156,** 42
Cluster clone theory, higher plant plastids and, **145,** 165
Clusterin, *see* Testosterone-repressed prostate message-2
Clustersome model, **162A,** 41–48
Clypeolabrum
genetic control, **174,** 35–36
role in segmentation, **174,** 26–27
C-metaphase, *see* Colchicine-metaphase
c-Met receptor
epithelial cell expression, **160,** 224
mutations and cancer, **160,** 233, 236–237
scatter factor/hepatocyte growth factor affinity, **160,** 237, 239
signal transduction, **160,** 237–238
structure, **160,** 237
tissue distribution, **160,** 239–240
CML, *see* Chronic myeloid leukemia
CMT, *see* Cortical microtubules
Cnaptor spinimanus, karyosphere and, **144,** 19
Cnidae
nematocysts
chemistry, **156,** 275–278
contents, **156,** 278
discharge, **156,** 278–295
structural components, **156,** 275–278
nervous system
anatomy, **167,** 39–42
neuropeptides, **167,** 43–55
biosynthesis, **167,** 57–77
peptide receptors, **167,** 55–56
and precursor action, **167,** 77–78
preprohormone processing, **167,** 78–82
neurotransmission, **167,** 42–43
perspectives, **167,** 82
tubule formation, **156,** 277, 279
Cnidaria
animal mtDNA, **141,** 199, 208
transdifferentiation in medusae, **142,** 218
CNS, *see* Central nervous system
Coagulation, *see* Blood; Disseminated intravascular coagulation
Coat color, genetics, melanocytes and, **146,** 208, 214, 251
Coated pit, **166,** 110
Coating assays, as migration analysis method, **155,** 60
Cochlea hair cells, in neurosensory epithelium, **173,** 210
Cockroach
Na^+/K^+-ATPase localization, **176,** 325–327
vitellogenin receptor, **166,** 130–131
Codons, usage in volvocalean genes, **139,** 251–258
codon bias and, **139,** 251–254
sequence variation and, **139,** 254–258
Coelomocytes, echinoderm
actin cross-linking proteins, **175,** 41–44
form and function, **175,** 40–41
Coenobiodiscus muriformis, microfilaments, **150,** 334
Coenocytes, in plant cell model, **175,** 104
Coenzyme A, esterification, in phenylpropanoid pathway, **172,** 258
Cofilin, **139,** 278–279; **143,** 165–169, 172; **156,** 202–203, 205, 212, 214
Cohabitation
MARs
with potential ORIs, **162A,** 344–345
with transcriptional enhancers, **162A,** 345–346

S/MARs with enhancer elements, **162A,** 433–435,442
Coiled bodies
containing splicing factors, **162A,** 71–73
content in breast cancer, **159,** 181–182
function, **159,** 166
marker proteins, **159,** 165–166
regulation of expression, **159,** 166–167
role in RNA processing, **162A,** 168–169
snRNP components, **162B,** 311–312
staining, **159,** 165
COL1 domain, collagen V, molecular structure, **173,** 97–99
COL2 domain, collagen V, molecular structure, **173,** 96–97
COL2A1 domain, mutations in chondrodysplasias, **159,** 327–329
Colcemid, mitosis and, **144,** 240, 244, 246
Colchicine
γ-aminobutyric acid metabolism, **147,** 315
cytoskeletal dynamics and, **144,** 109
cytoskeleton, **166,** 15
as drug, **167,** 129–130
effect on microtubule assembly, **157,** 12
environmental stimuli in plant cells, **145,** 283–284, 291–292
fish chromatophores, **143,** 222
induction of MDR, **156,** 79–80, 84–85
induction of somatic crossing over, **158,** 120
metaphase in cultured plant cells, **158,** 79–80, 83
mitosis and, **144,** 230–231, 235–236, 240, 245–246, 259
Coleoptera, karyosphere and, **144,** 8
Coleoptiles
intracellular calcium messenger system and, **142,** 314, 318, 320, 324
pattern formation, role of AGPs, **174,** 255–258
symplast and, **142,** 273–274, 281, 300
Colistin, microgravity experiments, **156,** 340, 342
Collagen
avian salt glands, **144,** 149
cholesteric ordering, **166,** 86–90
effects on arterial smooth muscle cell culture phenotype
type I, **169,** 202
type IV, **169,** 197–198
eggshells and biomineralization and, **145,** 221, 234, 237, 244
fish chromatophores and, **143,** 194
gel, angiogenesis assay, **159,** 121–122
gene mutations in chondrodysplasia, **159,** 327–329
gene splicing, **159,** 290
hormonal control of amphibian metamorphosis and, **145,** 129–130, 133–134, 136
lattices, hydrated, vertebrate cell culture technology, **142,** 159–160
in matrix-driven translocation
concentration, **150,** 141
percolation phenomenom, **150,** 145–146
regenerative cementogenesis and
established root surfaces *in vitro,* **142,** 28–29, 31–38, 40–42
established root surfaces *in vivo* 44-47, **142,** 49, 51
function of cementum, **142,** 2–4
growing root surfaces, **142,** 9, 11–13, 15, 19, 21–23, 25, 27
retinal precursor cells and, **146,** 183
role
bone fracture healing, **159,** 323
cartilage calcification, **159,** 284
self-incompatibility and, **143,** 269
Sertoli-germ cell communication network, **147,** 60
synthesis in arterial smooth muscle cell culture, **169,** 224–225
synthesis in chondrocytes, **159,** 268–269, 273, 278–279, 281, 288–290, 299, 302, 326
transdifferentiation in medusae and, **142,** 222, 236
types, **159,** 266, 268
vertebrate cell culture technology and, **142,** 159–160, 178, 188
Collagenase
calcium channels, **147,** 211
regenerative cementogenesis and, **142,** 38
transdifferentiation in medusae and
DNA replication, **142,** 252, 254–255
initiation, **142,** 237, 244
isolation of tissues, **142,** 222–223
potential, **142,** 224, 232–234
Collagen fibrils
anchoring to lamina densa
anchoring fibrils, **173,** 136–137

Collagen fibrils (*continued*)
beaded fibril role, **173,** 136
microfibril role, **173,** 134–136
microthreads, **173,** 137–138
role in organogenesis, **173,** 132–133
architecture in tissues, **173,** 131–132
diameter
regulation in different tissues, **173,** 109–110
regulation *in vivo,* **173,** 110–111
direct physical connection with lamina densa, **173,** 127–130
fine
collagen V on, **173,** 113
pNcollagen III on, **173,** 111–112
growth *in vitro*
role of collagen V, **173,** 116–120
role of pNcollagen III, **173,** 114–116
–lamina densa connection, light microscopy, **173,** 125–126
–lamina densa interaction, related molecules
collagen I, **173,** 78–86
collagen III, **173,** 86–88
collagen IV, **173,** 88–94
collagen V, **173,** 95–101
collagen VII, **173,** 101–103
collagen XVII, **173,** 103–105
laminins, **173,** 105–108
nidogen, **173,** 108
osteonectin, **173,** 109
perlecan, **173,** 108–109
species, **173,** 77–78
Collagen I
assembly into fibrils, **173,** 83–85
as connective tissue component, **173,** 78–80
disorders, **173,** 85–86
encoding genes, **173,** 81–82
molecular structure, **173,** 80–81
posttranslational modification, **173,** 82–83
Collagen III
encoding genes, **173,** 87–88
during morphogenesis, **173,** 86
role in fibrogenesis, **173,** 87–88
during tissue remodeling, **173,** 86
Collagen IV
α(IV) chains, length polymorphism, **173,** 90–92
and collagen V, codistribution *in vivo,* **173,** 127
gel, **173,** 94
homotypic interaction, **173,** 92–94
interaction with laminins, **173,** 107–108
and lamina densa skeleton, **173,** 120–123
molecular structure, **173,** 88–90
Collagen V
chain structure, **173,** 95–96
and collagen IV, codistribution *in vivo,* **173,** 127
on fine collagen fibrils, **173,** 113
molecular structure
COL1 domain, **173,** 97–99
COL2 domain, **173,** 96–97
C-terminal region, **173,** 99–100
NC2 domain, **173,** 96–97
NC3 domain, **173,** 96–97
role in collagen fibril growth *in vitro,* **173,** 116–120
role in regulation of collagen fibril diameter, **173,** 110–111
significance *in vivo,* **173,** 130–131
subtypes
function, **173,** 101
preparation, **173,** 100–101
Collagen VII
gene structure, **173,** 102–103
molecular structure, **173,** 101–102
roles *in vivo,* **173,** 103
supramolecular structure, **173,** 101–102
Collagen XVII
gene structure, **173,** 104–105
molecular structure, **173,** 103–104
Collateral sensitivity, **156,** 81, 91
Collomia grandiflora, **140,** 195
Colonies
cnidarian, **167,** 38
hemopoietic
interferon suppression of, **167,** 231
suicide assays, **167,** 197–198
TNF-α suppression of, **167,** 229
Colony-forming cells-spleen
localized control of proliferation, **167,** 191–192
reversible inhibition of by pEEDCK, **167,** 188
Colony-forming unit-A cells
assay in purifying MIP-1α, **167,** 198
proliferation inhibition of, **167,** 198
Colony-forming unit-S cells
rapidity of cell cycling between G_0 to S phase, **167,** 193
suicide assays of hemopoeitic, **167,** 197–198

Colony stimulating factor
CSF-1, secretion by granulated metrial gland cells, **153,** 122
role in apoptotic processes, **163,** 118
Color, fish chromatophores and, *see* Fish chromatophores, cytophysiology of
Columella cells, microgravity experiments, **156,** 328–330, 332
Commelina commmunis, intracellular calcium messenger system and, **142,** 309, 313–314, 327
Commissural giant cells, control of rhythmic behaviors and, **147,** 14
Commitment, retinal precursor cells and, **146,** 150, 165–167, 183
Compactin, cytoskeletal dynamics and, **144,** 89–91
Compartment of uncoupling of receptor and ligand, role of V-ATPase, **174,** 91–92
Competence
distribution in ascidian embryos, **176,** 286
embryonic induction, **150,** 380–386
Complementary DNA
AGP polypeptide sequence, **174,** 218–230
animal mtDNA and, transcription, **141,** 221
Arabidopsis thaliana and, **144,** 74
arcA, clone, **152,** 129
calcium channels and, **147,** 213
chalcone synthase and, **147,** 245, 263
clones
isolation from flowering plants, **140,** 21
and localization of pollen-specific transcripts, **140,** 25–27
cytoskeletal dynamics and, **144,** 106–108
Drosophila muscles and, **143,** 63–64, 138
contractile protein genes, **143,** 100, 103, 115, 118–119
development, **143,** 87
endothelins, cloning, **171,** 268
extracellular matrix and, **146,** 267, 273, 276–277, 279–280, 282
flowering plants and, **143,** 298, 302
amino acid biosynthesis, **143,** 311–313, 316–317
amino acid pathways, **143,** 320, 322, 324
ammonium, **143,** 305, 308
glutamate metabolism and, **147,** 294, 307
GnRH-encoding gene in teleosts and, **147,** 166, 172, 175–176, 183
in habituated plant cells
expression, **152,** 279
tumor-specific genes, clones, **152,** 290–291
higher plant mitochondrial genomes and, **141,** 149, 162–163
higher plant plastids and, **145,** 158, 168, 185, 192
hormonal control of metamorphosis and, **145,** 111, 130, 133, 138
intracellular calcium messenger system and, **142,** 319, 328–329, 333, 335
kinetoplastid mtDNA and, **141,** 77–78
melanocytes and, **146,** 203, 205–207, 212, 214, 249
myofibrillogenesis and, **143,** 154
sarcomere formation, **143,** 182
thin filament proteins, **143,** 160, 163, 166–167, 169, 174
NuMA, **162B,** 6–7
proteins in striated muscle and, **142,** 66–67, 91, 123, 126
retinal pigmented epithelium and, **146,** 77
self-incompatibility and, **143,** 269, 272, 285–286
Sertoli-germ cell communication network and, **147,** 59
transcriptional regulation and, **143,** 25, 27–33, 36
urease in plant cells and, **145,** 89
V-ATPase, molecular cloning
16- and 17-kDa subunit gene, **174,** 97–98
55- and 62-kDa subunit gene, **174,** 97
67- to 73-kDa subunit gene, **174,** 96–97
95- to 116-kDa subunit gene, **174,** 98–99
Complementation, melanocytes and, **146,** 206, 219
Complement components
acute phase, host defense function, **170,** 238–239
in assays of PMN random motility, **139,** 160–161
Complement receptor, macrophage, **161,** 292–293
Complete Freund's adjuvant, **161,** 289–291
Complex eye, insect, form diversity, **174,** 5–6
Concanavalin A
cytoskeletal dynamics and, **144,** 99, 113
effect on ecto-ATPase activity, **158,** 163, 189, 191

Concanavalin A (*continued*)
–lectin, for analysis of APP microheterogeneity, **170,** 265–266
microgravity experiments, **156,** 306, 308, 309
neurite initiation and elongation role, **153,** 247
receptor distribution, electric field effect, **158,** 303
response of thymocytes, **156,** 166, 175
Sertoli-germ cell communication network and, **147,** 36
as T-cell mitogen, **156,** 170–172
transcriptional regulation and, **143,** 28
transdifferentiation in medusae and, **142,** 222, 251
Conception
events leading to confined mosaicism soon after, **167,** 288–289
incidence of aneuploidy at, **167,** 267–268, 270–272
Conceptus, preimplantation-stage
chromosomal abnormalities, **164,** 142–143
oocyte, **164,** 143–144
spermatozoa, **164,** 144–145
cytoplasmic maturation effects, **164,** 145–146
developmental potential
cell cycle regulation, **164,** 161–164
cell number/integrity, **164,** 148–155
culture conditions effects, **164,** 165–166
culture media, **164,** 168–170
cytoskeleton role in multinuclear blastomeres, **164,** 160–161
DNA content, **164,** 155–160
endometrial synchronization, **164,** 166–168
markers for, **164,** 173–174
nucleocytoplasmic ratio effects, **164,** 172
and genomic activation, **164,** 172–173
oocyte maturation effects, **164,** 140–142
Condensed nuclei, **138,** 73–76
Conditional oscillators, in control of rhythmic behaviors, **147,** 12
Conductance
bicarbonate
by CFTR, **172,** 199
in pancreatic duct cells, **172,** 211
diffuse nature cnidarian nerve, **167,** 43
membrane, biological clocks in retina and, **146,** 116, 123, 127
osmotic
antibiotic pores, **163,** 3–5
bimodal theory, **163,** 24–26, 30–34
cellular aquapores, **163,** 5–6
Cones
growth factor-induced retinal regeneration and, **146,** 61
retinal precursor cells and, **146,** 147–148, 160, 172
microenvironmental factors, **146,** 178–179, 181
photoreceptor subtypes, **146,** 174–175
Confocal laser scanning microscopy
M cell identification using AP method, **167,** 102
retinal precursor cells and, **146,** 175
in whole-chromosome hybridization, **153,** 187
Conformational transition, ependymin, **165,** 127–128
Coniferin β-glucosidase, role in cinnamyl alcohol synthesis, **176,** 106–107
Conifers, double fertilization in, **140,** 321–327
basic aspects of reproduction, **140,** 321–322
reports of intra-archegonial double fertilization, **140,** 322–327
Coniferyl alcohol, *see* Monolignols
Coniferyl-alcohol glucosyltransferase, role in cinnamyl alcohol synthesis, **176,** 106–107
Conjugation, oral replacement during, **164,** 6
Conjunctiva-associated lymphoid tissue, **167,** 138
Connectin
myofibrillogenesis and, **143,** 176, 182–183
types, **156,** 245–247
Connective tissue
avian salt glands and, **144,** 133, 137, 149, 159–160
collagen I as component, **173,** 78–80
hormonal control of amphibian metamorphosis and, **145,** 129
interactions, role in organogenesis, **173,** 132–133
parathyroid hormonerelated protein, **166,** 246–247

regenerative cementogenesis and, **142,** 13, 29, 37, 44–47
role of fibrillin fibrils, **173,** 134–135
role of fibronectin fibrils, **173,** 135–136
vertebrate cell culture technology and, **142,** 159, 162
Connective tissue growth factor, retinal pigmented epithelium and, **146,** 77
Connexin
assembly of connexon, **157,** 320–321
cellular distribution of types, **157,** 319–320
degradation, **157,** 339–340
distribution
astrocytes, **157,** 343–345, 347
brain, **157,** 341
central nervous system, **157,** 341–342
expression during oncogenesis, **157,** 338–339
gene loci, **157,** 319
in glucose transfer
blood-tissue barriers, **172,** 8
ciliary body epithelium, **172,** 21–22
rat placenta, **172,** 31
identification, **157,** 355
phosphorylation, **157,** 333–335, 339, 344
regulation, **157,** 333–335, 355
role in intercellular communication, **157,** 320
sequence homology, **157,** 318
Sertoli-germ cell communication network and, **147,** 33, 39
structure, **157,** 318–319
translocation, **157,** 333–334
Connexon
assembly, **157,** 320–321
clustering, **157,** 324, 329
cytoplasmic anchoring, **157,** 325–330
effects of cytochalasin D, **157,** 330
hypoxia effect on density, **157,** 327
intermolecular binding interactions, **157,** 322–325
spacing, **157,** 327–330
stabilization by lipid, **157,** 322
structure, **157,** 321
Conserved sequences
animal mtDNA, **141,** 201, 230
regulation, **141,** 227
replication, **141,** 223–224
endosymbiont hypothesis, **141,** 300
in S proteins, **140,** 467
Constitutive secretory pathway
polarized regulation in, **172,** 228–229
trans golgi network in, **172,** 226
Contactinhibin receptor, identification, **175,** 154–155
Continuity matrix system, **162A,** 8–10
Contractile protein
Drosophila muscles and, **143,** 77, 86
encoding gene, *Drosophila* muscles and
actins, **143,** 109–114
development, **143,** 87, 94
function, **143,** 132–137
interactions, **143,** 126–132
muscle protein genes, **143,** 125–126
myofibrils, **143,** 94–99
myosin heavy chains, **143,** 99–104
myosin light chains, **143,** 104–108
paramyosin, **143,** 108–109
tropomyosin, **143,** 114–117
troponin, **143,** 117–122
Z-disc proteins, **143,** 122–125
in interphase nucleus, **150,** 164–167
myofibrillogenesis
differential assembly, **143,** 179–181
isoform change, **143,** 154–157
sarcomere formation, **143,** 181–184
thick filament proteins, **143,** 176–179
thin filament proteins
actin, **143,** 157–159
α-actinin, **143,** 174–175
actin isoforms, **143,** 159–160
actin polymerization, **143,** 160–172
capping proteins, **143,** 172–174
tropomyosin, **143,** 175–176
troponin, **143,** 175–176
Contraction
actin filament bundle, **156,** 252–253
activities in *Physarum* microfilament system
actin, **149,** 157
actin-binding proteins, **149,** 163
actomyosin thread models, **149,** 164, 166–167
ATP, **149,** 155
axenically cultured microplasmodia, **149,** 191–195
caffeine droplets, **149,** 175, 177, 190–191
evaluation methods, **149,** 184, 186–198
microplasmodia, **149,** 178, 181, 183–184

Contraction (*continued*)
regulation, **149,** 199–202
small sandwich plasmodia, **149,** 195–198
cytoplasm, **156,** 247–253
cytoplasmic fraction, **156,** 236–239
cytoskeletal dynamics of systems, **144,** 86–88, 93–96
equatorial, cytoskeletal dynamics, **144,** 93–94
microgravity experiments, **156,** 343, 347
plasmodial strand, **156,** 201–202
–stretching cell response, in embryonic tissue, **150,** 28–29
Contraction hypothesis, nematocyst discharge, **156,** 278–279
Contrast agents, nuclear magnetic resonance and, **145,** 30–31
Convection current, free, in diatom shell morphogenesis, **150,** 263
Copigments, chalcone synthase and, **147,** 245
Copper, melanocytes and, **146,** 197, 204
Coprinus
mitochondrial genomes in fungi and, **141,** 105
mitosis, **144,** 222
Coprinus stercorarius, mitochondrial genomes in fungi, **141,** 93, 105, 119
Copy number dependence, assay principle, difficulties with, **162A,** 427–428
Coral
in bone grafts, **172,** 161, 174
as Cnidarians, **167,** 38
Cord factor, mycobacterial, **161,** 287–289
Core filaments, *see also* Intermediate filaments
nuclear matrix, **162A,** 91–97
NuMA as part, **162A,** 164–165
NuMA localization, **162B,** 17–18
Corn
intracellular calcium messenger system and, **142,** 313–314, 318, 324
microgravity experiments, **156,** 322, 324, 325, 326, 327
Cornea
growth factor-induced retinal regeneration and, **146,** 57
neovascularization
angiogenesis model, **159,** 116–117
image analysis, **159,** 117
quantitation, **159,** 116–117
Coronaviruses, binding to sialic acids, **175,** 171–172
Coronin, mutants, motion analysis, **163,** 93–94
Cortex
cortical inheritance
ciliates, **169,** 304–306
defined, **169,** 268–269
dinoflagellates, **169,** 306–308
euglenoids, **169,** 299–300, 302
microtubule role, **169,** 302–308
trypanosomes, **169,** 302–304
endoplasmic reticulum–alveolar system
calcium binding, **169,** 288–291
ciliates, **169,** 290–291
dinoflagellates, **169,** 291
euglenoids, **169,** 288–289
functions, **169,** 291–292
epiplasm proteins
ciliates, **169,** 284–286
dinoflagellates, **169,** 286–288
euglenoids, **169,** 282–284
evolution in protists, **169,** 268
functions of components, **169,** 309
microtubules
ciliates, **169,** 297
dinoflagellates, **169,** 297–299
euglenoids
assembly, **169,** 292–293
positioning, **169,** 293–294
proteins, **169,** 293–294
trypanosomes
proteins, **169,** 294–297
stability, **169,** 295–296, 308
organization
ciliates, **169,** 272, 274
dinoflagellates, **169,** 274
euglenoids, **169,** 269, 271
trypanosomes, **169,** 271–272
plasma membrane and surface coats
ciliates, **169,** 279, 281
dinoflagellates, **169,** 282
euglenoids
chemical composition, **169,** 274–277
integral membrane proteins, **169,** 277–279
mucus secretion, **169,** 276
trypanosomes, **169,** 279, 281–282
solubilization, **169,** 268
symplast, **142,** 291
Cortexillins, in *Dictyostelium discoideum,* **175,** 52, 56

Cortical actin network, **156,** 208, 254; **165,** 239
Cortical expansion, cytoskeletal dynamics and, **144,** 88–89, 93
Cortical flow
along cell periphery, **148,** 56
in cell development, **173,** 215–216
current theories
actin polymerization, **148,** 67–69
membrane-cytoskeleton dynamics at leading margin of cell, **148,** 70–72
myosins in cytoskeleton retraction, **148,** 64–67
cytoskeletal dynamics, **144,** 99–101
hypothesis, **148,** 55–59
Cortical network, cytoskeletal dynamics and, **144,** 111
Cortical tension, cytoskeletal dynamics and, **144,** 115–117
Corticoids, control of amphibian metamorphosis and, **145,** 119–123, 125–128
Corticosteroids, **145,** 125, 127; **166,** 25
Corticosterone
avian salt glands and, **144,** 191–192
control of amphibian metamorphosis and, **145,** 119–122, 125–127
effect on APP synthesis, **170,** 256
Corticotropin-releasing factor
hormonal control of amphibian metamorphosis and, **145,** 116, 124–125
immunohistochemical localization of, **138,** 106–110
neuroendocrine and behavioral effects of, **138,** 110–112
Corticotropin-releasing hormone
astrocyte reaction, **148,** 129, 141–142
cellular localization, **159,** 35
effect of hypothalamic lesions, **159,** 48
effect on phagocytosis, **170,** 94–95
mRNA levels, **159,** 35–36
precursor, **159,** 35–36
release, **159,** 36
role in immune response and stress, **170,** 89–90
transport, **159,** 33
Corticotropins, astrocyte reaction to, **148,** 140–141
Cortisol, control of amphibian metamorphosis and, **145,** 120–121
Corydalis sempervirens, flowering plants and, **143,** 313
Corylus auellana
pollen from, **140,** 12, 47, 62
sporopollenin from, **140,** 55
Coscinodiscus asteromphalus, morphogenesis
colloidal silica, **150,** 325
vesicles, **150,** 267
Coscinodiscus concinnus, valve formation, **150,** 257
Coscinodiscus wailesii, morphogenesis
scale vesicles, **150,** 267
valve formation, **150,** 257
Costameres, proteins in striated muscle and, **142,** 114
Cosuppression, in control of chalcone synthase expression, **147,** 258–259
Cotyledons
higher plant plastids and, **145,** 151, 153, 177, 179, 189
urease in plant cells and, **145,** 95
elimination, **145,** 82–84
metabolic origins, **145,** 68–69, 71–72, 76–78
4-Coumarate–CoA ligase
expression
developmental, **172,** 260
pathogen stress, **172,** 260–261
wound stress, **172,** 261
isoform substrate specificity, **172,** 259–260, 261–262
localization, **172,** 259, 260
in phenylpropanoid pathway, **172,** 259, 261–262; **176,** 102–105
plant, encoding genes, **148,** 26–27
Coumarate-3-hydroxylase, in phenylpropanoid pathway, **172,** 254–255; **176,** 102–105
Coumarin, inhibition of cell wall formation, **170,** 46
4-Coumaroyl CoA, effect on chalcone synthase, **147,** 234–237, 243–244, 268
p-Coumaryl alcohol, *see* Monolignols
Coumestrol
effects of perinatal exposure to, **139,** 12–13
genital abnormalities in rats and, **139,** 13–14
Counterion condensation theory, actin filament bundle formation, **175,** 62–64
Coupling
decoupling, nuclear magnetic resonance and, **145,** 9, 44

Coupling (*continued*)
electrical
biological clocks in retina and, **146,** 92, 95, 102
control of rhythmic behaviors and, **147,** 3, 14
interocular, biological clocks in retina and, **146,** 112–114
spin, in NMR, **145,** 8–9
spin decoupling, in NMR, **145,** 8–9
stimulus-secretion, avian salt glands, **144,** 200–205
Cowpea
nodulation, **156,** 17, 33, 42, 46
symplast, **142,** 265–266, 270, 280–281, 298
urease in plant cells and, **145,** 77–78
CPM, *see* Confined placental mosaicism
C protein
myofibrillogenesis and, **143,** 153, 155, 176, 178, 181
in striated muscle, **142,** 62, 79–83, 94
CPT-11, development, **156,** 105–106
CPY, *see* Carboxypeptidase Y
Cranial nerve, avian salt glands and, **144,** 155, 158
Crassulacean acid, metabolism plants, **149,** 50, 59–60, 67–68
Creatine, NMR analysis, **145,** 45–46
Creatine kinase
myofibrillogenesis, **143,** 176
in striated muscle, **142,** 81, 117
Creatine phosphate, NMR analysis, **145,** 2, 19
Crepis capillaris, meristem cell wall regeneration, **170,** 47
CRF, *see* Corticotropin-releasing factor
CRH, *see* Corticotropin-releasing hormone
Crickets, karyosphere and, **144,** 18
Crithidia, kinetoplastid mtDNA and, **141,** 67–68, 83
Crithidia fasciculata, kinetoplastid mtDNA, **141,** 69–70, 73–76, 79
Cross-bridge cycle, role in dynein function, **173,** 38–39
Cross-hybridization
ciliates and, **141,** 10, 13
kinetoplastid mtDNA and, **141,** 73–74
mitochondrial genomes in fungi and, **141,** 119
Crosslinks
ciliates, **141,** 6, 11, 13
in DNA–protein, **162A,** 548–549
load-bearing, cleavage in plant cell walls, **173,** 192–194
Cross resistance
to DNA strand-breaking agents, **156,** 132
MDR cells, **156,** 81
Crown gall disease, and habituation of plant tissue, **152,** 266
CR receptors, types 1, 3, and 4, macrophage, **161,** 292
Crypt cells
dome epithelium enterocytes *versus* ordinary, **167,** 131
epithelium of palatine tonsils, **167,** 135–136
M cells originating as undifferentiated, **167,** 108–110
Crypthecodinium cohnii, mitosis and, **144,** 223
Cryptochromes, as photoreceptors in algae, **148,** 286–289
Cryptomonads, chloroplasts and
biochemistry, **144,** 276
dinoflagellates, **144,** 285, 288–290, 292–293
envelope membranes, **144,** 275–276
euglenoids, **144,** 284–285
general considerations, **144,** 260, 264, 272–273, 293–295
heterokont algae, **144,** 277–278, 281
pigments, **144,** 273
structure, **144,** 273–275
Cryptomonas
cyanelles, translation apparatus protein genes, **151,** 43–44
endosymbiont hypothesis, **141,** 276–277
photosynthetic accessory pigments in, **148,** 274–275
Cryptomonas ϕ, chloroplasts and, **144,** 276, 281
Cryptophyceae
eyespot apparatus, **148,** 239
photosynthetic accessory pigments in, **148,** 274–275
Cryptophytes
chloroplasts, **144,** 280
endosymbiont hypothesis and, **141,** 261, 279
Cryptorchidism, testicular macrophage involvement, **149,** 126–127
Crystalline lens, *see* Lens
Crystallite, cholesteric ordering, **166,** 86, 87

Crystallization, eggshells and biomineralization, **145,** 230, 238–239, 241, 243–244
Crystals, biological, *see* Calcified tissue
CSCC hair bundles, **156,** 288, 292–293
Csk
  phosphorylation of protein kinases, **157,** 242
  regulation of CD45, **157,** 242–243
CT28N, **156,** 245
C_0t analysis, carrot genes in meiosis, **158,** 106
CTC, *see* Chlortetracycline
Ctenocephalides felis
  antennae, **174,** 6–7
  complex eye, **174,** 5–6
  epidermis, **174,** 8
  labrum, **174,** 7
  mouthparts, **174,** 7–8
C-type natriuretic peptide, astrocyte reaction to, **148,** 128, 132–133
Cubozoa
  centralization and complexity of nervous system, **167,** 47
  life cycle of, **167,** 38
Cucurbita maxima, **140,** 57, 58, 60–61
Cucurbita pepo
  intracellular calcium messenger system and, **142,** 309
  pollen, **140,** 191–192
Culex pipiens, karyosphere and, **144,** 9
Culture media, effects on conceptus development, **164,** 168–170
Cultures
  cell, *see* Cell cultures
  cocultures, for human conceptuses, **164,** 169–170
  human carcinoma, *see* Carcinomas, human, cultures
  system analysis, **155,** 52–53
Cuphea lutea, plastids and, **145,** 184
Curcurbita pepo, chromoplast differentiation, genetic factors, **163,** 194
Curie-point pyrolysis low-voltage electron impact spectra, pollen wing sporopollenin, **140,** 63, 64
Curvature, instrinsic, DNA
  in α-globin gene MAR, **162A,** 332–333
  in MAR/ORIs, **162A,** 338–339
Cutaneous malignant melanoma–dysplastic nevus syndrome, genetics, **146,** 226–228, 230, 250
Cuticles
  eggshells and biomineralization, **145,** 219, 234–235, 240
  genetic control, **174,** 36–40
  on plant cell surface, **174,** 200
Cutin
  extracellular matrix and, **146,** 283–284
  in plant cell wall, dynamic aspects, **151,** 259–260
Cyanea lamarckii, **167,** 48
Cyanelles
  chloroplasts and, **144,** 264–265, 272, 276, 285
  endosymbiont hypothesis and, **141,** 258, 272–273
  genome
    characteristics, **151,** 29, 50–51
    chloroplast genomes
      algae, **151,** 36
      higher plants, **151,** 35–37
    Cyanophora, **151,** 38
    Cyanophora paradoxa, see Cyanophora paradoxa
    organization
      ATP synthase, **151,** 48
      biosynthetic pathway proteins, **151,** 49–50
      cytochrome *binf6/inf/f* complex, **151,** 46
      DNA, **151,** 37–39
      electron transport chain components, **151,** 46–47
      isoprenoid pathway, **151,** 49–50
      NAD biosynthesis, **151,** 49
      photosynthetic apparatus genes, **151,** 45–49
      photosystem I, **151,** 45
      photosystem II, **151,** 45–46
      phycobilisomes, **151,** 47–48
      ribosomal RNA genes, **151,** 39–41
      rubisco, **151,** 48–49
      transfer RNA genes, **151,** 41–42
      translation apparatus protein genes, **151,** 42–45
    plastid evolution and, **151,** 30, 53–55
    plastid genomes from algae lacking chlorophyll *b,* **151,** 37
    properties, **151,** 29–30

Cyanelles (*continued*)
higher plant plastids and, **145,** 166–167
protein transport
import, **151,** 51–52
routing, **151,** 52–53
Cyanidium caldarium, chloroplasts and, **144,** 264
Cyanobacteria
chloroplasts and, **144,** 259–260, 294–295
cryptomonads, **144,** 273, 276
dinoflagellates, **144,** 286, 292–293
euglenoids, **144,** 283, 285
green algae, **144,** 268, 270
red algae, **144,** 261, 263–265
endosymbiont hypothesis and, **141,** 234
future prospects, **141,** 331
gene transfer, **141,** 278
mitochondrial, **141,** 280
molecular biology, **141,** 265–268
plastids, **141,** 255, 257–258, 278–279
sequences, **141,** 269–270, 272
higher plant plastids and, **145,** 166
Cyanobacterium, microgravity experiments, **156,** 342
Cyanomorpholino doxorubicin, **156,** 131–132
Cyanophora
chloroplasts and, **144,** 264–265, 272
cyanelle genome organization, **151,** 38
Cyanophora paradoxa
chloroplasts and, **144,** 264, 285
cyanelles
biochemical analysis, **151,** 33
DNA size, **151,** 37
genome size, **151,** 30
inner envelope membrane, **151,** 34–35
metabolism, **151,** 31–32
outer envelope membrane, **151,** 33–34
penicillin-binding proteins, **151,** 34
structure, **151,** 35
endosymbiont hypothesis and, **141,** 258, 272–273
plastids, **145,** 166–167
Cycads, potential double fertilization
basic aspects of reproduction, **140,** 327–328
potential for intra-archegonial double fertilization, **140,** 328
Cyclic AMP
avian salt glands and, **144,** 169, 200–201
in axonemal motility regulation, **164,** 286–288
biological clocks in retina and, **146,** 106, 110–111, 128, 131–133
calcium channels and, **147,** 201, 204, 207
cascade
metabolism regulation and, **139,** 319–320
in signal transduction in yeast cells, **139,** 310–319, *see also* Signal transduction systems and yeast, cAMP cascade and
in cell division in yeast, **139,** 320–322
concentrations and intracellular ca2+, regulation by, **138,** 216–217
control of rhythmic behaviors and, **147,** 3, 11
cytoskeletal dynamics and, **144,** 89, 102–103, 110–111
dibutyryl cAMP
calcium channels and, **147,** 204
glutamate metabolism and, **147,** 302
downstream pathway of, **139,** 318
effects on whole cell capacitance, **172,** 224
electromagnetic field effect on levels, **158,** 282, 308, 316–317
elevated levels in cholera, **172,** 209–210
extracellular matrix and, **146,** 312–313
fish chromatophores and, **143,** 227, 231, 239–245, 250
in gene expression, **139,** 318–319
growth cone regulation, **153,** 260, 262
inhibition of *Chlamydomonas* gametogenesis, **168,** 292
and instability of homogeneous state
Child's results, **154,** 346–347
colocalization role, **154,** 313
concentration, **154,** 311–312
Drosophila, **154,** 338–339
localized activity of channels, **154,** 343, 345
metabolic field, **154,** 356–360
metabolism of proliferation, **154,** 361–362, 364–365
reduction fields, **154,** 347–348, 350, 352–353
spontaneous endogenous electrophoresis, **154,** 340–342
Turing biochemical mechanism, **154,** 323–324, 326–328, 331–333

intracellular, **139,** 315–316
intracellular calcium messenger system and, **142,** 305, 335, 338
melanocytes and, **146,** 210
in microtubule motility regulation, **164,** 286–288
modulation of membrane budding, **172,** 228
phosphodiesterases and, **139,** 315
possible regulation of *trans* golgi network polarity, **172,** 229
protein sialylation and, **172,** 228
proteins in striated muscle and, **142,** 123
purinergic neurotransmission role, **149,** 254–255
regenerative cementogenesis and, **142,** 40
regulation of CFTR, **172,** 218–219
regulation of constitutive secretory pathway, **172,** 226
regulation of endocytosis, **172,** 221–222
regulation of exocytosis, **172,** 224
regulation of glycoconjugate secretion, **172,** 225
reverse transformation of malignant cells by, **154,** 44
role in apoptotic processes, **163,** 115
role in flagellar motility, **173,** 25–29
second messenger role, **149,** 232–236
Sertoli-germ cell communication network and, **147,** 39, 61, 64, 66, 71, 79
in sexual induction in *Volvox carteri,* **139,** 220–221
signal transduction, *see* Signal transduction
in sperm capacitation, **149,** 26
in sperm function, **149,** 16
transdifferentiation in medusae and, **142,** 245
trans golgi network acidification and, **172,** 228
Cyclic AMP-dependent protein kinase, **139,** 316–317; **172,** 195, 197, 199
Cyclic AMP receptor, in amoeba, **155,** 2, 8
Cyclic GMP
avian salt glands and, **144,** 200–203
biological clocks in retina and, **146,** 105–106
control of rhythmic behaviors and, **147,** 3
cytoskeletal dynamics and, **144,** 103
second messenger role, **149,** 232–236
Cyclin A
at DNA replication sites, **162B,** 324
localization during S-phase, **162B,** 317
Cyclin B, complex with p34-cdc2, **162B,** 381
Cyclins
mitosis, **144,** 249–251
regulation of cell cycle, **157,** 249
regulation of microtubule-severing proteins, **164,** 279
Cycloheximide
biological clocks in retina and, **146,** 106, 118, 121
enhancement of growth factor-induced mRNA accumulation, **148,** 98–99
hormonal control of amphibian metamorphosis and, **145,** 121
inhibition of cell wall formation, **170,** 46
transcriptional regulation and, **143,** 2–3, 16
urease in plant cells and, **145,** 91
Cyclooxygenase, pathway of arachidonic acid metabolism, **161,** 221
Cyclooxygenase inhibitors, effect on arterial smooth muscle cells, **169,** 227
Cyclophosphamide
AcSDKP protective effect against, **167,** 222–223
glutathione protection of bone marrow against, **167,** 226
Cyclosis, environmental stimuli in plant cells, **145,** 253, 268, 276, 289
Cyclosporin, MDR reversal, **156,** 89–91
Cyclosporin A
blocking of apoptosis, **157,** 251
effect on T cell receptor activation, **157,** 218–219
inhibition of angiogenesis, **159,** 140–141
reduction in M cell number in rabbits after treatment with, **167,** 143
transcriptional regulation, **143,** 13, 16
Cyclotella cryptica
cell wall serine content, **150,** 259
valve formation, **150,** 257
Cyclotriphosphazatriene, urease in plant cells and, **145,** 82
Cylindrotheca fusiformis, intracellular calcium messenger system and, **142,** 336
Cynops pyrrhogaster, **138,** 92, 95
CYRI, gene product, **139,** 310–311

Cysteine
  and chalcone synthase, **147,** 237, 243
  and melanocytes, **146,** 193, 206
  proteins in striated muscle and, **142,** 65
  transdifferentiation in medusae and, **142,** 236
Cysteine-rich protein, as α-actinin binding partner, **167,** 163
Cysteine synthase, in pepper chromoplasts, **163,** 220–221
Cystic fibrosis
  anti-inflammatory drugs and, **172,** 194–195
  clinical manifestations, **172,** 194
    defective apical chloride conductance and, **172,** 211
  compound heterozygotes in, **172,** 203
  defective chloride conductance in, **172,** 195–196
  elevated sweat chloride levels in, **172,** 210–211
  gene therapy, **158,** 263–264
  genetic basis, **172,** 193–195
  mucin secretion in, **172,** 220–221
  mutations of CFTR and, **172,** 200–203
  preimplantation diagnosis, **151,** 5
  respiratory inflammation, CFTR regulatory activity and, **172,** 231
  sodium chloride reabsorption in, **172,** 196
Cystic fibrosis transmembrane conductance regulator
  activity when heterologously produced, **172,** 230
  ATP conductance and, **172,** 199, 215
  ATP hydrolysis and, **172,** 195, 197, 200
  bicarbonate conductance and, **172,** 199
  chloride channel activity
    biophysical properties of, **172,** 197
    in cholera, **172,** 209–210
    in cystic fibrosis, **172,** 195–196
    in epithelial cells, **172,** 208–209
    functional redundancy in, **172,** 200
    in human reabsorptive sweat duct, **172,** 210–211
    ion selectivity and, **172,** 199
    regulation of, **172,** 197–199
  cyclic AMP protein kinase phosphorylation and, **172,** 195, 197, 200
  cytoplasmic nucleotide binding domain, **172,** 195, 197–200
  cytoplasmic regulatory domain, **172,** 195, 197, 199
  effect of glycerol on, **172,** 208
  endoplasmic reticulum and
    model of protein kinetics within, **172,** 206–207
    polyubiquitination, **172,** 204–205
    protein folding in, **172,** 205–206
  localization in epithelial cells, **172,** 208–209, 216–217
  modulation of membrane budding, **172,** 228
  modulation of outwardly rectifying chloride channels, **172,** 214–215
  modulation of sodium channels, **172,** 213–214
  mucin secretion and, **172,** 220–221
  mutations
    activity of D836X, **172,** 200
    activity of DeltaF508, **172,** 201–202, 204–208
    chloride channel activity of, **172,** 206
    classes, **172,** 200–203
    second site mutations, **172,** 203
    temperature-sensitive, **172,** 208
  in organelle acidification, **172,** 227
  and P-glycoprotein, and SUR, homology, **171,** 125–127
  protein sialylation and, **172,** 228
  recycling
    autoregulation, **172,** 222–223
    clathrin-coated vesicles and, **172,** 217
    cyclic AMP regulation of, **172,** 218–219, 223
    endosomes and, **172,** 216–217
    functional significance, **172,** 217–219
    molecular control of, **172,** 219–220
  regulation, **172,** 197–200
    actin and, **172,** 212–213
    cyclic AMP and, **172,** 218–219
    microtubules and, **172,** 212
  in regulation of endocytosis, **172,** 222
  in regulation of exocytosis, **172,** 223–225
  in regulation of membrane traffic, **172,** 220–221
  in regulation of other transporters and receptors, **172,** 230–231
  trans golgi network acidification and, **172,** 228
  transport
    apical endosomes and, **172,** 215–216

effect of epithelial polarization on, **172,** 216
trans golgi network and, **172,** 215
Cystine, proteins rich in, extracellular matrix, **146,** 269
Cystrones, karyosphere and, **144,** 18
Cytochalasin
cytoskeleton, **166,** 15
Cytochalasin B
avian retina, **146,** 25
cytoskeletal dynamics, **144,** 100; **166,** 15
effect on anemone hair bundles, **156,** 287, 294
environmental stimuli in plant cells
chemicals, **145,** 298
light, **145,** 260, 275–277, 284
wounding, **145,** 291–293
fish chromatophores, **143,** 205
and mitosis, **144,** 234
Cytochalasin D
effect on connexon density, **157,** 330
effect on hair bundles, **156,** 292
retinal precursor cells and, **146,** 156–158
Cytochemical probes
for examining cell death, **163,** 120–124
in lignin detection, **176,** 95
Cytochemistry
embryo sacs and component cells, **140,** 281–282
female germ unit, **140,** 258–260
Cytochrome b, animal mtDNA and, **141,** 174, 180–181, 185
Cytochrome b_6/f complex, cyanelle, **151,** 46
Cytochrome c
ciliates and, **141,** 59
endosymbiont hypothesis and, **141,** 310
mitochondrial genomes in fungi and, **141,** 119
Cytochrome c oxidase
animal mtDNA and, **141,** 174, 179–182, 184–185, 187
ciliates and, **141,** 12
electromagnetic field effects, **158,** 319
Cytochrome oxidase
ciliates and, **141,** 37, 41, 45, 55–58
kinetoplastid mtDNA and, **141,** 71, 75, 77–82
mitochondrial genomes in fungi and, **141,** 92
Cytochrome P450
extracellular matrix and, **146,** 279–280, 284
induction by xenobiotics, **160,** 331
sex differences in hepatocytes, **160,** 330–331
Cytochrome P450 epoxygenase pathway, arachidonic acid metabolism, **161,** 221
Cytochromes, kinetoplastid mtDNA and, **141,** 80
Cytodex 1 beads, **156,** 308–309
Cytogenetics
advances in addressing chromosomal abnormalities in, **167,** 263–264
cancer, **153,** 216–217
as chromosome analysis method, **151,** 15–16
clinical, **153,** 214–219
Cytokeratins
filament role in differentiation systems, **154,** 54–57
M cell
confirming epithelial origin of, **167,** 129
in rabbits, **167,** 104
proteins in striated muscle and, **142,** 112
Cytokine growth factor, tissue factor as superfamily member, **152,** 51–53
Cytokine receptor
costimulation by dendritic cells, **153,** 82–83
soluble, effect on APP synthesis, **170,** 255–256
Cytokines, *see also specific cytokines*
in cells, organs, and immune system, **170,** 102–104
characteristics, **170,** 99–101
dendritic cells affected by
costimulation, **153,** 82–83
regulation, **153,** 84–85
dome epithelium antigen-presenting cell secretion of, **167,** 132, 133
effect on acute phase proteins
APP glycosylation
network of factors, **170,** 276–279
nontransformed cell-secreted APP, **170,** 275–276
transformed cell-secreted APP, **170,** 274–275
APP synthesis, **170,** 252–255
interaction with nitric oxide, **170,** 117–118
intestinal lymphocytes producing
intraepithelial lymphocytes, **167,** 97
TH1 and TH2 cells, **167,** 98

Cytokines (*continued*)
macrophage production of, **161,** 310–313
microgravity, **156,** 310, 314
monocyte/macrophage procoagulant activity, modulation by, **152,** 61, 63
origins, **170,** 98–99
production by retinal pigmented epithelium, **146,** 75–80
relationship with nitric oxide, **170,** 107
role in immune response, **170,** 104–107, 121–122
role in inflammatory response, **170,** 118–122
role in stress response, **170,** 107–108, 121–122
secretion by granulated lymphoid cells of pregnant uterus, **153,** 121–123
Sertoli-germ cell communication network and, **147,** 66, 74
signal transduction pathways
early events, **170,** 260–262
immediate early events, **170,** 262–264
TGF-β inhibition of bone marrow $CD34^{3+}$ cells in presence of, **167,** 211
transcriptional regulation and, **143,** 7, 14, 16, 18, 41
Cytokinesis
avian retina and, **146,** 4
cytoskeletal dynamics and, **144,** 91
mitosis and
dissociation of events, **144,** 227–235
mitotic events, **144,** 220–221, 224–226
plant cell secondary MTOCs, **175,** 107–111
role of nonmuscle tropomyosin, **170,** 20–22
role of tropomyosin, **170,** 23–24
Cytokinin
chromoplast reversion, **163,** 202
environmental stimuli in plant cells and, **145,** 287–288
extracellular matrix and, **146,** 297, 299, 301
habituation in tumorous plant cells, induction of
auxin interactions with, **152,** 270
concentration changes, **152,** 268–269
genetic aspects, **152,** 271–272
inoculum size, **152,** 271
mechanisms, **152,** 265–266, 276, 286–287, 292
molecular genetic aspects, **152,** 277–278
temperature effects, **152,** 271
intracellular calcium messenger system and, **142,** 325
role in plant nodule development, **156,** 44–45
Cytology, embryo sacs and component cells, **140,** 281
Cytomegalovirus
human, MIP-1α receptor and, **167,** 207
transcriptional regulation, **143,** 39–40, 44
Cytoplasm
animal mtDNA and, **141,** 190
avian salt glands and, **144,** 138–140, 204
chemicals, **145,** 297–298
chloroplasts, **145,** 268–270, 276–282, 284
low temperature, **145,** 288–290
nucleus, **145,** 287–288
photodinesis, **145,** 252–253, 257–261, 266
wounding, **145,** 290–296
chloroplasts and, **144,** 270, 280
dinoflagellates, **144,** 286, 288, 291–292
euglenoids, **144,** 283–284
red algae, **144,** 261, 263–265
ciliates and, **141,** 1, 12, 53, 55
consistency, **156,** 245
cytoskeletaldynamics and, **144,** 102, 111, 120
cytoskeletal components, **144,** 106, 109
mechanism, **144,** 117, 119
theories, **144,** 87–88
determinants, localization, and metabolic field, **154,** 355–360
determinants, role
in anterior–posterior axis specification, **176,** 287–291
in cleavage pattern generation, **176,** 268–272
in endoderm cell development, **176,** 263–264
in epidermis development, **176,** 264–267
in gastrulation movements, **176,** 267
in muscle development, **176,** 257–263
in ooplasmic segregation, **176,** 253–255
in tissue differentiation, **176,** 255–257
Drosophila muscles and, **143,** 66–67, 109, 112, 127
endosymbiont hypothesis and, **141,** 312

environmental stimuli in plant cells and, **145,** 251–252, 301
female germ unit, features, **140,** 256–258
fibrils
contractility, **156,** 203, 205
differentiation, **156,** 206–207
plasmodium, **156,** 203–207
spatial localization, **156,** 205
fish chromatophores and, **143,** 192, 221, 244, 246–247
cytoskeleton, **143,** 221
iridophores, **143,** 207–209, 213, 218–219
melanophores, **143,** 197
xanthophores, **143,** 198–199
higher plant mitochondrial genomes and, **141,** 129, 159
abnormal phenotypes, **141,** 133–134
structure, **141,** 142, 149, 152–153
higher plant plastids and, **145,** 149
amino acid synthesis, **145,** 191, 200
starch synthesis, **145,** 177, 179–181
translocation, **145,** 167–168, 170
insulin
role in biological effects, **173,** 266–267
translocation, **173,** 258–260
intermediate filaments, *see* Intermediate filaments, cytoplasmic
intracellular calcium messenger system and, **142,** 308, 311
calmodulin, **142,** 330–332
concentration, **142,** 311–312, 316, 318
phosphatidylinositol turnover, **142,** 321, 323, 325–326
protein kineses, **142,** 332, 336
karyosphere and, **144,** 1, 11, 42, 44
macromolecule translocation, mechanism, **173,** 260–261
maturation, effects on fertilization, **164,** 145–146
M cell
attenuated, membranous nature of, **167,** 111
cytokeratin skeletal disk in apical, **167,** 129
disadvantage of light microscopy in defining, **167,** 102
from rabbit caecum, **167,** 113*f*
–microtubules interaction, model, **175,** 104–107
mitosis and
chromosomes, **144,** 242–243, 245–246, 251
dissociation of events, **144,** 230, 238
mitotic events, **144,** 220, 222–223, 226
myofibrillogenesis and
differential assembly, **143,** 179, 181
sarcomere formation, **143,** 183
thin filament proteins, **143,** 158–160, 165, 171–173, 175
proteins
interaction with insulin, **173,** 261–265
in regulating FAK activity, **167,** 170–171
proteins in striated muscle and
cardiac muscle, **142,** 105
sarcolemma, **142,** 124
sarcoplasmic reticulum, **142,** 121
skeletal muscle, **142,** 66, 85, 87, 92, 102
regenerative cementogenesis and
established root surfaces *in vitro,* **142,** 39–40
growing root surfaces, **142,** 9, 12–13, 19, 21–22, 27
self-incompatibility and, **143,** 260–261, 281
spine, **139,** 272–277
streaming, **156,** 247–254
Characeae, actin-myosin system role, **155,** 100–103
Characeae, demembered cell models
calcium-regulated inhibition, **155,** 118–120, 130
energy requirements, **155,** 112–114
longitudinally cut model, **155,** 112–113
magnesium requirement, **155,** 114–115
pH dependence, **155,** 115
plasma membrane-permeabilization, **155,** 110–111
preparation, **155,** 110–113
reconstruction experiments, **155,** 116–117
tonoplast-free cells, **155,** 111–112, 115
classification, **155,** 97–98
effects of electrical stimulation, **157,** 120–122
extracellular factors affecting
action potential, **155,** 120–125, 130

Cytoplasm (*continued*)
gravity, **155,** 127–129
light, **155,** 125
low temperature, **155,** 125–127
mechanism, **155,** 97–98, 129–130
motive force, **157,** 100, 113–114
motive force generation mechanism
actin identification in plants, **155,** 108
actin-myosin system involvement, **155,** 100–108
Characeae, **155,** 100–103
microtubules, **155,** 109–110
myosin identification in plants, **155,** 103–108, 130
site, **155,** 98–100
sliding theory, **155,** 99–100
observation in centrifuge microscopy, **157,** 110–111
in ring canals, **173,** 220–221
streaming, environmental stimuli in plant cells
chemicals, **145,** 297
chloroplasts, **145,** 283
low temperature, **145,** 288–290
photodinesis, **145,** 252, 257, 259–261, 265
wounding, **145,** 295
streaming, role of plant cell cytoskeletons
cytoplasmic droplets, **175,** 5
other plants, **175,** 5–6
perfused cells, **175,** 4
permeabilized cells, **175,** 4–5
symplast and, **142,** 275
transcriptional regulation and, **143,** 48
biochemistry, **143,** 22–23
molecular cloning, **143,** 28–29, 31, 34, 36–39
transdifferentiation in medusae and, **142,** 213–214, 245, 256–257
potential, **142,** 229–230
urease in plant cells and, **145,** 69, 84, 90–91
viscosity, **160,** 102
Cytoplasmic anchoring, **166,** 12
Cytoplasmic inclusion, cholesteric supramolecular order, **166,** 83–84
Cytoplasmic inheritance
higher plant plastids, **145,** 153
influence of fusion in pattern of, **140,** 379–383
Cytoplasmic localization, **138,** 78–83
Cytoplasmic protein, cadherin
complex assembly dynamics, **165,** 166–168
interactions between, **165,** 163–170
protein-protein interaction hierarchy, **165,** 164–166
Cytosine arabinoside
collateral sensitivity, **156,** 81
inducing leukemia cell cytotoxicity with NBME-IV, **167,** 236
modulation of DNA alkylation repair, **156,** 129–130
in study of pEEDCK ability to inhibit CFU-S proliferation, **167,** 327–328
in suicide assays, **167,** 204–205
as therapeutic drug, **156,** 120–121
Cytoskeletons
abscisic acid, **166,** 39
Acetabularia
calcium effects, **150,** 227, 229–231
characteristics, **150,** 225
actin, *see* Actin cytoskeleton
actin-based, catenin, linkage, **165,** 170–172
in amoebae
actin polymerization, **148,** 69
adhesion, effect on movement, **148,** 43
capping studies, **148,** 44
contraction, effect on membrane flow, **148,** 43, 72
directed lipid flow hypothesis, **148,** 45
dynamics
model, **148,** 69
regulation at leading margin of cells, **148,** 70–72
locomotion, **148,** 37
membrane interaction with, **148,** 72–73
lipid flow, **148,** 45–47
membrane flow, **148,** 45–47
membrane proteins, **148,** 47–53
retraction, myosins in, **148,** 64–67
submembrane layer, lateral movement
cortical cytoskeleton, **148,** 53–55
cortical flow hypothesis, **148,** 55–59
detachment from leading edge, **148,** 59–64
association with mRNA, **162B,** 197–200
auxin, **166,** 39
axon, **165,** 239
axonal assembly, **165,** 274–279
carbamate, **166,** 15
cell communication, **166,** 3
cell model systems
permeabilized cells, **175,** 2

plant cells, roles in cytoplasmic streaming
cytoplasmic droplets, **175,** 5
other plants, **175,** 5–6
perfused cells, **175,** 4
permeabilized cells, **175,** 4–5
structure isolation, **175,** 2–3
in vitro functions, **175,** 3–4
chemical research tools, **166,** 15–16
cholesteric supramolecular order, **166,** 83–84
colchicine, **166,** 15
cortical, RPE, structure and composition, **171,** 240–241
cytochalasin, **166,** 15
cytochalasin B, **166,** 15
and cytoplasmic organization of pollen and pollen tubes
conclusions, **140,** 113–115
effects of inhibitors on cytoskeleton, **140,** 112–113
future considerations, **140,** 115–116
hypotheses on functions of in male gametophyte, **140,** 74–76
localization of cytoskeletal elements in, actin filaments in pollen grains, **140,** 96–102
overview, **140,** 96
pollen grains, **140,** 96–100
pollen tubes, **140,** 100–102
microtubules, **140,** 102–111
overview, **140,** 102
pollen grains, **140,** 102–103
pollen tubes, **140,** 103–111
myosin and kinesin, **140,** 111–112
objectives, **140,** 73–74
proteins and methods of visualization
actin, actin filaments and microfilaments, **140,** 86–90
biochemical characteristics of pollen actins, **140,** 87–88
methods to visualize actin filaments and microfilaments in pollen, **140,** 88–90
structure of actin molecules and filaments, **140,** 86–87
myosin and kinesin, **140,** 93–94
biochemical characteristics of myosin and kinesin-like protein from pollen, **140,** 94–96
kinesin, **140,** 95–96
myosin, **140,** 94–95
profilin, **140,** 96
structure of molecules, **140,** 93–94
kinesin, **140,** 94
myosin, **140,** 93–94
tubulin and microtubules, **140,** 90–93
biochemical characteristics of pollen tubules, **140,** 91–92
methods to visualize microtubules in pollen, **140,** 92–93
structure of molecules and microtubules, **140,** 90–91
vegetative cytoplasm of pollen grains and tubes
calcium gradients and calcium-cytoskeleton interactions in pollen tubes, **140,** 80–83
dry pollen, **140,** 76–77
growing pollen tubes, **140,** 78–80
ionic compounds and intracellular pH in pollen tubes, **140,** 83
movements of organelles and nuclei, **140,** 84–86
pollen hydration, activation and germination, **140,** 77–78
cytoplasmic skeleton, **166,** 2, 3
dendrite, **165,** 239
D_2O, **166,** 15
dynamics, **165,** 237
dynamics in *Amoeba,* **144,** 85–86, 120
Amoeba proteus, **144,** 87, 91, 93
current concepts, **144,** 94
contraction, **144,** 94–96
cortical flow, **144,** 99–101
projection, **144,** 97–99
signal transduction, **144,** 101–103
cytoskeletal components, **144,** 103–104, 110
actomyosin, **144,** 105–109
microtubules, **144,** 109–110
Dictyostelium
actomyosin, **144,** 110–113
capping, **144,** 112–114
microtubules, **144,** 113–114
substrate exploration, **144,** 112
mechanism
biased friction, **144,** 117–120
cell motility, **144,** 114–115
motor proteins, **144,** 116–117
theories, **144,** 86
amoeboid movement, **144,** 86–92
cell division, **144,** 91, 93–94

Cytoskeletons (*continued*)
elements, gene-regulatory function, *see* Intermediate filament proteins, gene-regulatory function
environmental stimuli in plant cells, **145,** 261, 278, 281–282, 296, 300
ethylene, **166,** 39
eukaryote, **166,** 3–4
eukaryotic cell
cytoplasmic cytoskeleton, **166,** 5
intermediate filaments, **166,** 5
major cytoskeletal proteins, **166,** 4–6
membrane skeleton, **166,** 4–5
microfilaments, **166,** 5
microtubules, **166,** 5
nuclear skeleton, **166,** 5–6
eukaryotic multicellular organism, cytoskeletal diversity, **166,** 6–7
extended, **150,** 177–179
fish chromatophores, **143,** 195, 199, 220–224, 245
focal adhesion in stabilization of actin, **167,** 161
gibberellin, **166,** 39
herbicide, **166,** 15
hormone
adrenocorticotropic hormone, **166,** 24
androgen, **166,** 27
angiotensin, **166,** 25
behavior, **166,** 31, 34
bombesin, **166,** 20
bombesin-like peptide, **166,** 20
cellular level motility, **166,** 23, 30
chemotaxis, **166,** 23, 30
corticosteroid, **166,** 25
1,25-dihydroxyvitamin D_3, **166,** 25
direct effects, **166,** 16–31
endothelin, **166,** 25
estrogen, **166,** 27
follicle-stimulating hormone, **166,** 26
glucosteroid, **166,** 25
gonadotropin, **166,** 26
growth, **166,** 16–17
growth factors, **166,** 16–17
growth hormone, **166,** 18
human chorionic gonadotropin, **166,** 26
indirect effects, **166,** 16–31
insulin, **166,** 16, 18–20
insulin-like growth factor, **166,** 18
invertebrates, **166,** 31–35, 36
luteinizing hormone, **166,** 26
mevalonate, **166,** 29
motility, **166,** 23–31
muscular level motility control, **166,** 23–31, 32–33
myotropin, **166,** 23–31, 32–33
ooeyte maturation, **166,** 31, 35
organismal level movement, **166,** 31, 34
parathyroid functioning, **166,** 22–23, 28–29
parathyroid hormone, **166,** 29–30
progesterone, **166,** 27
steroidogenesis regulation, **166,** 22, 24–25, 26–27
thyroid, **166,** 22–23, 28–29
thyrotropin, **166,** 28
thyrotropin-releasing hormone, **166,** 28
thyroxine, **166,** 28–29
transepithelial transport, **166,** 17–21
vasopressin-mediated water, **166,** 17–21
interaction with plasma membrane, **174,** 196–197
intracellular calcium messenger system and, **142,** 335
ion transporter regulation, **161,** 206–207
kinetin, **166,** 39
lens epithelial, **163,** 323–326
localization of cytoplasmic determinants, **154,** 355–360
lumicolchicine, **166,** 15
M cell
description, **167,** 104
tracer adherence inducing reorganization of, **167,** 127
in transport of antigens and tracers, **167,** 128–130
mechanisms of ion transporter regulation, **161,** 206–207
melanocytes and, **146,** 211
membrane-bound organelle movement
dynein-like proteins, **176,** 159–160
kinesin-like proteins, **176,** 157–159
motor protein interrelations, **176,** 160–163
myosin heavy chain polypeptides, **176,** 154–157
–membrane connections
fibronectin-like proteins, **176,** 152–154
integrins, **176,** 151–152
spectrin, **176,** 150–151
vitronectin-like proteins, **176,** 152–154
membrane skeleton, **166,** 2, 3

microgravity environment, **156,** 344–346
microtubule, in eukaryotic cells, **175,** 119–121
nocodazole, **166,** 15
nuclear skeleton, **166,** 3
organization in megagametophyte, **140,** 284–287
in origin of living state, **166,** 3–4
phalloidin, **166,** 16
Physarum plasmodium, **156,** 202–213, 253
plant hormone, cell elongation, **166,** 37, 38–39
polarity, **166,** 7–9
pollen tubes, **176,** 140
actin-binding proteins, **176,** 143–145
actin biochemical characteristics, **176,** 141–143
actin filament distribution, **176,** 146–149
microtubule-associated proteins, **176,** 143–145
microtubule distribution, **176,** 146–149
tubulin biochemical characteristics, **176,** 141–143
primordial cell functions, **166,** 3
proteins in striated muscle
cardiac muscle, **142,** 109
sarcolemma, **142,** 125
skeletal muscle, **142,** 66, 88, 93, 98
regulation of cellular tensegrity, *see* Tensegrity, cellular, cytoskeletal regulation
in regulatory volume decrease, **161,** 203–205
in regulatory volume increase, **161,** 205–206
reorganization elicited by growth factors
induction of membrane rufffings as common cellular response to growth factors, **138,** 214–216
protein kinase C-dependent and independent pathways, **138,** 217–219
regulation by intracellular Ca^{2+} and cAMP concentrations, **138,** 216–217
retinal pigment epithelium affected by changes in, **152,** 255
retinal precursor cells and, **146,** 147, 155–157, 160
role in multinuclear blastomere formation, **164,** 160–161
schematic representation, **166,** 3
Sertoli-germ cell communication network and, **147,** 35, 74
signal transduction
actin filament, **166,** 14–15
cytoplasmic anchoring, **166,** 12
desensitization, **166,** 10–11
endocytosis, **166,** 10–11
exocytosis, **166,** 12–14
intracellular signaling, **166,** 9–10
plasma membrane receptor binding, **166,** 10–11
receptor intracellular transport, **166,** 11–12
steroid hormone, **166,** 11
slow axonal transport, **165,** 276, 277–278
in stretch-activated channels, **161,** 211
structure, **165,** 237
studies with miniprotoplasts, **175,** 16–18
subdivisions, **166,** 2
taxol, **166,** 15
transdifferentiation in medusae and, **142,** 213, 236, 249
twisted liquid crystalline supramolecular arrangement, **166,** 95
vasopressin, **166,** 17–21
vertebrate cell culture technology and, **142,** 160
vinblastine, **166,** 15
Cytosol, glutamate metabolism and, **147,** 289–290, 294, 302
Cytosolic chaperones
DnaJ homologs, **174,** 139–140
GrpE homologs, **174,** 140
Hsp70 homologs, **174,** 139
interaction with mitochondrial preproteins, **174,** 138–139
mitochondrial import stimulatory factor as, **174,** 141
presequence binding factor, **174,** 140–141
Cytosolic enzymes, coordinate control
amino acids, **149,** 87–88
carbohydrates, **149,** 86–87
nitrate, **149,** 86
Cytostatic factor
in calcium activation of mammalian eggs, **152,** 210
and mitosis, **144,** 247

Cytotomy, in embryonic tissue, **150,** 27
Cytotoxicity
AcSDKP prevention of CFU-S recruitment induced by, **167,** 222–223
activity of intestinal intraepithelial lymphocytes, **167,** 97
proliferation inhibitors in reducing cumulative toxicity from, **167,** 240–242
role of DNA repair in determination, **156,** 126–131
role of POMC products, **170,** 95–97
suicide assays of hemopoeitic colonies using S-phase
cytosine arabinoside, **167,** 204–205
hydroxyurea, **167,** 204
types, **167,** 197–198
Cytotrophoblasts
assessing mosaicism in, **167,** 287
in human placenta
as blood barrier, **172,** 26
GLUT1 in, **172,** 27
in rat placenta, as blood barrier, **172,** 29
in spontaneous abortions, **167,** 288
Cytovillin, cytoskeletal dynamics, **144,** 106

D

DAG, *see* Diacylglycerol analogs
DAHP synthase, flowering plants and, **143,** 311–312
DALT, *see* Duct-associated lymphoid tissue
Dark-field microscopy, bacterial flagellar motor, **147,** 111, 113
Darkness
adaptation in plant cells, **145,** 257, 276, 280
biological clocks in retina and, **146,** 86, 89, 95–96
retinal precursor cells and, **146,** 161, 164–165
Darwin, Charles, **140,** 449, 525
Datura inoxia, flowering plants and, **143,** 323
Daucus carrota
effect of microgravity, **156,** 319–320, 323, 328, 333; **171,** 43–45
protoplast wall regeneration, **170,** 42–43
Daughter cells
proliferation regulation in number remaining as stem cells, **167,** 196
unequal numbers of chromosomes in, overview, **167,** 264
Daunorubicin, in MDR, **156,** 79
Daylily, microgravity experiments, **156,** 322
Deacetylation, histone, kinetics, **162A,** 205–206
Deadaptation, avian salt glands and, **144,** 140–141
Deafferentation
experimental models, **159,** 196
partial deafferentation effects
frogs
B neurons, **159,** 232–235, 249
C neurons, **159,** 232–235, 249
mammals, **159,** 227–228
reaction to axotomy, **159,** 220–221
total deafferentation effects
frog
physiology of transmission, **159,** 230
synapse morphology, **159,** 230–232, 249
mammals
physiology of transmission, **159,** 222–224, 229
synapse morphology, **159,** 224–227
Decalcification
eggshells and biomineralization, **145,** 227, 229, 231
regenerative cementogenesis, **142,** 32, 36, 42, 46
Decapacitation factors, in sperm, **149,** 5–7, 14–15
Decarboxylation, urease in plant cells and, **145,** 67
Decidual cell reaction
androgen-sterilized female rats
effect on reaction
age of androgenization, **160,** 11–15
aging, **160,** 25–29
androgen dose, **160,** 9–10
duration of androgenization, **160,** 9–10
estrogen, chronic administration, **160,** 29–30
estrogen, neonatal administration, **160,** 23–24
nonaromatizable androgen administration, **160,** 21–22
nonaromatizable androgen administration with estrogen, **160,** 25
ovarian hormones, **160,** 6–9
prostaglandin, **160,** 16–21
pseudopregnant rats, **160,** 5–6

endometrial cell proliferation, **160,** 28
experimental induction, **160,** 2–3, 15–16
uterine, **153,** 105–106
Decoupling, nuclear magnetic resonance and, **145,** 9, 44
Defense mechanisms
plant, *Arabidopsis thaliana* and, **144,** 53–55, 73, 76–77
response elicitors in plants
plants, **148,** 1–2
elicitor-receptor model, **148,** 29
endogenous, **148,** 5, 12–14
galacturonides, **148,** 13
jasmonic acid, **148,** 14
methyl jasmonate, **148,** 14
oligouronides, **148,** 13
salicylic acid, **148,** 14
exogenous, **148,** 4–6
chitosan, **148,** 11–12
general, **148,** 8–12
race-specific, **148,** 6–8
perception, **148,** 15–19
primary signals, **148,** 3
regulation, **148,** 26–29
secondary signals, **148,** 2–3
signal transduction, **148,** 19–25, 29–30
role of lignins, **176,** 99
Degeneration
corticobasal, with Tau 64 and 69, **171,** 200–201
Drosophila muscles, **143,** 118, 120–121
frontal lobe, with pathological Tau triplet, **171,** 199
synergic selection, **140,** 368–370
Degenerin, *Deg-1,* role in neurodegeneration, **163,** 119
Deglycosylation, monolignol glycosides, **172,** 270–271
Degradation
protein, *see* Proteolysis
sporopollenin, **140,** 61–63
Dehydroepiandrosterone-16 alphahydroxylase, perinatal exposure to sex hormones and antihormones, **139,** 28–29
5-Deiodinase, hormonal control of metamorphosis and, **145,** 107
Deiodination, hormonal control of metamorphosis and, **145,** 110
Deletions
animal mtDNA and, **141,** 175, 207
regulation, **141,** 227
transcription, **141,** 218, 221
Arabidopsis thaliana, **144,** 57, 63
Drosophila muscles and, **143,** 81, 117
higher plant mitochondrial genomes and, **141,** 134, 150, 154, 156–157
kinetoplastid mtDNA and, **141,** 66, 83
mitochondrial genomes in fungi and, **141,** 98, 112, 115–116
transcriptional regulation and, **143,** 12, 41
Delta-sleep inducing peptide, **138,** 315–319
Dematin, **166,** 5
Demineralization, regenerative cementogenesis and, **142,** 32–33, 35, 39, 42, 51
Dendraster, sperm-binding proteins, **163,** 279
Dendrites
cytoskeleton, **165,** 237–239
early outgrowth, **165,** 248
fish chromatophores and, **143,** 195, 246–248
cytoskeleton, **143,** 221–224
iridophores, **143,** 209–210, 218–219
leucophores, **143,** 201, 203, 205
motility control, **143,** 225–226, 233, 235, 245
growth
mechanisms, **165,** 245–246
neuron, cytoplasmic mechanisms, **165,** 235–280
theory, in diatom morphogenesis, **150,** 297–301
melanocytes and, **146,** 193–194, 211
microtubule, **165,** 239–240
neuronal polarity, **165,** 247
transcriptional regulation and, **143,** 18
Dendritic antigen-presenting cells, in dome epithelium, **167,** 132–133
Dendritic cells, *see* Follicular dendritic cells; Human dendritic cells
Dendritic spines
cytoplasm composition in
actin-associated proteins in
actin cross-linking proteins, **139,** 280–287
actin depolymerization and polymerization inhibiting proteins, **139,** 278–279
actin modulators, **139,** 279–280
physiological significance of, **139,** 287–288
actin in, **139,** 272–277
Ca^{2+} in, **139,** 277–278

Dendritic spine (*continued*)
description, **139,** 267–269
electrical resistance in, **139,** 296–297
excitable, **139,** 296–297
function, **139,** 269
general considerations, **139,** 272
organelles, **139,** 271–272
stalk resistance in, **139,** 296–297
synaptic plasticity and, **139,** 269–271
actin cytomatrix organization, **139,** 291–295
actin polymerization–depolymerization, **139,** 292–295
GABA synapses, **139,** 297–299
modeling of spine function, **139,** 296–297
morphological changes, **139,** 295–296
phosphorylation, **139,** 291–292
long-term potentiation and, **139,** 288–299
postsynaptic mechanism, **139,** 289–291
calcium, **139,** 289–290
protein kineses, **139,** 290–291
substrates, **139,** 291
Denervation
avian salt glands and, **144,** 158
effect on muscle fiber type transitions, **170,** 176–177
De novo assembly, intermediate filaments, **164,** 119–120
Density gradient centrifugation, principle, **157,** 100
Dental calculus, mineral phases in, **172,** 147–148
Dental implants, calcium phosphate coatings and, **172,** 165
Dentin
caries in, **172,** 149, 151
crystal structure and composition, **172,** 142–143, 165–166
topical fluoride treatment, **172,** 157
Dentine, regenerative cementogenesis and, **142,** 4–5, 51
established root surfaces *in vitro,* **142,** 32–33, 38–39, 41–42
established root surfaces *in vivo* 45-47, **142,** 49
growing root surfaces, **142,** 5, 7, 9, 19, 23, 25–28
Dentoclasts, regenerative cementogenesis and, **142,** 47
5′-Deoxy-5-fluorouridine, mechanism, **156,** 117
Deoxyaglycone, formation, **156,** 80
Deoxychalcone, chalcone synthase and, **147,** 244–245, 261
Deoxycorticosterone, control of amphibian metamorphosis and, **145,** 120–121
Deoxyflavanones, chalcone synthase and, **147,** 244–245
Deoxyisoflavonoid, chalcone synthase and, **147,** 261
Depactin
binding sites for, **156,** 214
myofibrillogenesis and, **143,** 169
role in nonmuscle cells, **156,** 227
Dephosphorylation
intracellular calcium messenger system and, **142,** 324, 337
mitosis and, **144,** 248–249, 252
Depigmentation, melanocytes and, **146,** 219–224, 241, 243
Depolarization
avian salt glands and, **144,** 157
bacterial flagellar motor and, **147,** 113
biological clocks in retina and
Bulla, **146,** 129–132, 134–138
concepts, **146,** 99
entrainment pathway, **146,** 102–106, 114
opisthobranch retina, **146,** 90–91
output pathway, **146,** 123, 126
calcium channels and, **147,** 194, 199, 206
control of rhythmic behaviors and, **147,** 13, 16
intracellular calcium messenger system and, **142,** 320, 325, 327
MDR cells, **171,** 137–139
retinal precursor cells and, **146,** 155
symplast and, **142,** 269, 272, 275–276
Depolymerization
cytoskeletal dynamics and, **144,** 109–110
environmental stimuli in plant cells and, **145,** 279, 285–286, 296
mitosis and, **144,** 243, 245
myofibrillogenesis and, **143,** 164–170, 172
Depression, effect of serotonin, **149,** 251
Depth resolved surface spectroscopy, in pathology, **145,** 21
DES, *see* Diethylstilbestrol

Desensitization
G-protein-coupled receptor, **164,** 212
plasma membrane receptor, **166,** 10–11
Desmin
binding to human erythrocyte inside-out vesicles, **164,** 108
muscle differentiation by, **154,** 49–52
myofibrillogenesis, **143,** 182
in striated muscle, **142,** 62
cardiac muscle, **142,** 112–114, 118
skeletal muscle, **142,** 87, 90, 96–99, 101–102
in vitro binding to nuclear components, **164,** 99–100
Desmoplakin I, linking of intermediate filaments to desmosome, **164,** 104–105
Desmosine, eggshells and biomineralization and, **145,** 220–221
Desmosomal cadherin
cell function, **165,** 179–180
cell phenotype, **165,** 179–180
embryogenesis, morphoregulatory roles, **165,** 180–181
functional diversity, **165,** 162–163
structural similarities, **165,** 162–163
Desmosomal catenin, **165,** 169–170
Desmosomes
avian salt glands, **144,** 135
connection of M cell basolateral membranes to enterocytes, **167,** 131
intermediate filament associations, **164,** 102–105
lens
between elongating fibers, **163,** 338
between epithelial cells, **163,** 337
proteins in striated muscle and, **142,** 112, 126
regenerative cementogenesis and, **142,** 9
Sertoli-germ cell communication network, **147,** 28–29, 32–33, 36
Destabilization, transdifferentiation in medusae and, **142,** 224, 236
Destrin, myofibrillogenesis and, **143,** 166–167
Development
adult *Drosophila,* genetics, **174,** 40–45
amphibians, comparison, **176,** 291–292
ascidian, evolutionary modifications, **176,** 293–294
ascidian cell, stage of determination, **176,** 281–286
asexual, volvocacean, **139,** 197–198
branchial arch, and Hox *7-msh* family, genes, **138,** 279–281
cell, *see* Cell development
cerebellum, **175,** 262–263
cnidarian, **167,** 38
CNS, cell migration during
in cerebellar cortex, **175,** 277–280
in forebrain, **175,** 274–277
in hindbrain, **175,** 280–285
marker techniques, **175,** 272–274
in optic tectum, **175,** 280
and differentiation systems, **154,** 60–62
Drosophila
bristles, mutations affecting, **175,** 49–50
ring canal and bristle formation, **175,** 44–46
effect on muscle fiber type transitions, **170,** 158–162
embryonic, *see* Embryonic development
genetic control, evolution, **174,** 54–56
dorsal ridge evolution, **174,** 60–62
insects as arthropods, **174,** 63–65
mouthpart evolution, **174,** 56–60
insect head, **174,** 65–68
M cell maturation, site-specific variations in, **167,** 115–117
of Peyer's patches
dome epithelium, **167,** 108–111
species compared, **167,** 94
photoreceptor, Na^+/K^+-ATPase redistribution, **176,** 332–334
plant, role of AGPs
early studies, **174,** 245–246
pattern formation in roots and coleoptiles, **174,** 255–258
reproductive phase of development, **174,** 247–255
somatic embryogenesis, **174,** 258–260
plant cell wall
cleavage of load-bearing crosslinks, **173,** 192–194
EXGT-mediated molecular grafting, **173,** 194–195
xyloglucan–cellulose microfibril interaction, **173,** 195–196
postnatal
epithelial clear cells, **154,** 173–175
principal cells, **154,** 169–170
reducibility to molecular genetics, **154,** 362–367

Development (*continued*)
retina, **146,** 149–150
retinal cells, **146,** 166, 182
photoreceptor default hypothesis, **146,** 167, 170
photoreceptor subtypes, **146,** 174–175
Tribolium castanaeum larva, **174,** 43–45
Developmental biology, finite element methods
application, **150,** 100–101
approach, **150,** 98–99
definition, **150,** 95
formulataion
assembly, **150,** 112–114
discretization, **150,** 107–108
element geometries, **150,** 103–105
equations, **150,** 102–103
external driving forces, **150,** 112–113
physical properties, **150,** 109–112
quantities, **150,** 103–104
shape functions, **150,** 104–106
solution techniques, **150,** 114
hourglassing, **150,** 110–111
modeling
evolution, **150,** 100, 102
role of, **150,** 95–98
practical considerations, **150,** 116
recent advances, **150,** 116
time-dependent phenomena, **150,** 114–115
verification of results, **150,** 115
Dexamethasone
effect on acute phase proteins
APP glycosylation, **170,** 276–279
APP synthesis, **170,** 256–257
role in cartilage formation, **159,** 315–316
DFR, *see* DNA fiber radioautography
DG, intracellular calcium messenger system
phosphatidylinositol turnover, **142,** 321–322, 326–328
protein kineses, **142,** 332, 334
DHFR, *see* Dihydrofolate reductase
DHT, effect on mouse pelvis, **139,** 29–30
Diabetes
avian salt glands and, **144,** 198
GLUT1 expression in, **172,** 35
Diabetes mellitus nephropathy, **166,** 209–212
Diacetoxypent-1-yl-DOX, **156,** 92
N-(5,5-Diacetoxypentyl) doxorubicin, **156,** 132
Diacylglycerol
analogs, immunologic studies in microgravity, **156,** 311, 333
avian salt glands and, **144,** 200, 203
cytoskeletal dynamics and, **144,** 102
fish chromatophores, **143,** 244
inhibition of ecto-ATPases, **158,** 191, 196
–inositol triphosphate pathway, in Sertoli-germ cell communication network, **147,** 39, 79, 81
transdifferentiation in, medusae and, **142,** 247–249, 257
vimentin binding stimulated by, **154,** 45
Diagnostic reagents, based on matrix proteins, **162A,** 107–108
Diamagnetic substances, NMR, **145,** 30
cis-Diaminedichloroplatinum, *see* Cisplatin
Diatom morphogenesis, chemical basis
applications to organisms, **150,** 344–345
branching patterns, alternative mechanisms, **150,** 343–344
characteristics of diatom, **150,** 245–247
definition of diatom, **150,** 243
shell, *see* Diatom shell morphogenesis, chemical basis
silica
colloidal properties, **150,** 320–326
diffusion, **150,** 322–323
formation, **150,** 320–322
precipitation, **150,** 271–275, 345
requirements, **150,** 324–326
reserves, **150,** 338–339
transport model, **150,** 341–342
silica antenna, diatom as, **150,** 338–340
silica deposition vesicle, **150,** 253–254
silicalemma
characteristics, **150,** 253–255
formation, **150,** 342–343
sintering
costae number in pennate diatoms, **150,** 301
cross-costae formation, **150,** 301–302
growth of dendrites theory, **150,** 297–301
impurities, role in, **150,** 303–307
instability theory, **150,** 296–297
mechanisms, **150,** 288–292, 346
molecular basis, **150,** 292–295
mother liquor, role in, **150,** 306–307
punctae formation, role in, **150,** 303–305

punctae shapes, role in, **150,** 305–306
silica surface chemistry, **150,** 295–296
sintering, modeling
adsorption potential choice, **150,** 310–312
amorphous precipitate, **150,** 307–310
anisotropic approach, **150,** 312–320
physical simulations, **150,** 315, 320
space-filling branching patterns
costae, **150,** 271
heuristic modeling, **150,** 271–273
nucleating structure, **150,** 273–275
structural considerations, **150,** 243–244
terminology, **150,** 347–348
valve
microfilaments, **150,** 332–335
microtubules, **150,** 329–331
nucleus, **150,** 331–332
plasmalemma, **150,** 326–327
plasmolysis, **150,** 337–338
silica ionophores, **150,** 328–329
silica receptors, **150,** 327–328
silica transport vesicles, **150,** 328, 335–336
subcellular mechanisms, **150,** 336–337
Diatom shell morphogenesis, chemical basis
bubbles, **150,** 257–258
free convection current mechanism, **150,** 263
Golgi apparatus role, **150,** 264–270
models of space-filling branching patterns
computer implementatiation, **150,** 278–286
future considerations, **150,** 286–288
procedure, **150,** 275–278
silica precipitation, **150,** 275–288
silica sources and sink distribution, **150,** 281–284
mold prepattern hypothesis, **150,** 255–258
patterns, **150,** 346
physicochemical mechanisms, **150,** 262–264
prepattern concept, **150,** 260–262
stages, **150,** 247–255
template prepattern hypothesis, **150,** 258–260
Dibutyryl cyclic AMP
calcium channels and, **147,** 204
glutamate metabolism and, **147,** 302
Dicarboxylates, higher plant plastids and, **145,** 167, 170
2,6-Dichloroisonicotinic acid
Arabidopsis thaliana, **144,** 78
systemic acquired resistance induction, **168,** 326–327
2,4-Dichlorophenoxyacetic acid
intracellular calcium messenger system and, **142,** 325–326
messenger RNA response to, **152,** 116
3-(3′4′-Dichlorophenyl) 1,1-dimethylurea, environmental stimuli in plant cells and, **145,** 253–254, 259, 267
5,6-Dichloro-1-β-D-ribobenzimidazole, biological clocks in retina, **146,** 122
Dichotomosiphon tuberosus, environmental stimuli in plant cells and, **145,** 270, 283–287, 301
Dictyosomes
association of coated vesicles with, **167,** 2–4
volume in plant cell, effect of microgravity, **171,** 50
Dictyostelium
actin polymerization in cells, **148,** 68–69
chemotactic response to cAMP, **148,** 70
cortical flow in, **148,** 56
cyclic AMP receptors, **155,** 8
cytoskeletal dynamics, **144,** 85
actomyosin, **144,** 110–113
current concepts, **144,** 96–98, 101, 103
cytoskeletal components, **144,** 103–106, 108–109
mechanism, **144,** 116–117
microtubules, **144,** 113–114
regulation, **148,** 71–72
retraction, myosins in, **148,** 65–66
theories, **144,** 86, 88–89, 94
dynein, outer row heavy chains, **155,** 153
locomotion, **150,** 57–61
membrane-cytoskeleton interactions, **148,** 51–52
mitochondrial genomes, **141,** 55, 59
mutant, ameboid movement of, **139,** 163
myofibrillogenesis, **143,** 173, 178
pseudopod formation and, **139,** 173
Dictyostelium discoideum
actin, **156,** 230
actin cross-linking proteins, **175,** 50–56
cytoskeletal dynamics and, **144,** 95
myosin, **156,** 223–225
spectrin, **156,** 241

DIDS
effect on regulatory volume increase, **161,** 186
effects on swelling-activated Cl^- and taurine effluxes, **161,** 201–202
Diet, sialyated compounds in, **175,** 183
Diethylpyrocarbonate, inhibition of ecto-ATPases, **158,** 191
Diethylstilbestrol effects, *see also* Sex hormones
enzyme inhibitors and, **139,** 8
on enzyme levels, **139,** 29
on female humans, **139,** 21
general considerations for, **139,** 43–44
on gonadal steroidgenesis, **139,** 10–11
on guinea pigs, **139,** 13–14
history, **139,** 1
on mice, **139,** 17–19, 22, 36
molecular changes and, **139,** 11–12
polyovular follicles and, **139,** 8–10
on rats, **139,** 22
on receptor systems, **139,** 6–7
on sexually dimorphic brain and sexual behavior, **139,** 26–28
on *in vitro* cell growth, **139,** 7–8
Differential adhesion hypothesis
avian retina and, **146,** 21
experimental confirmation, **150,** 147
Differentiation
amphibian, **154,** 61–62
ascidian tissue, autonomy, **176,** 255–257
avian retina and, **146,** 40–41
cellular processes, **146,** 25, 27, 30, 36, 39
normal development, **146,** 4–5
retinogenesis *in vitro,* **146,** 8, 13, 15–23
cell, *see* Cell differentiation
characterization of systems, **154,** 48–49, 62–64
cultured cells, **154,** 58–59
cytokeratin filament role, **154,** 54–57
in embryonic induction, **150,** 408
extension of cellular processes, **154,** 53–54
gonadal, *see* Gonadal differentiation
leukemia cell, by vimentin, **154,** 52
mammalian, **154,** 60–61
muscle, by desmin, **154,** 49–52
osteoblast
and gene control, **162A,** 253–263
osteocalcin gene expression during, **162A,** 265–269
premature, hormone-induced, **162B,** 418
and proliferation metabolism, comparison, **154,** 360–362
redundancy of intermediate filament protein function, **154,** 64–66
retinal regeneration and, **146,** 49, 53, 67–69
development, **146,** 50–52
induction by growth factors, **146,** 60, 63–64
ocular tissues, **146,** 55–56, 58–59
S/MAR role, **162A,** 431–435
transdifferentiation, *see* Transdifferentiation
transgenic animals, **154,** 59
waves, **150,** 414
Difflugia, locomotion, **150,** 58, 62
Diffusion
antigens and tracers by M cells, **167,** 117–122
nuclear magnetic resonance and, **145,** 26–29
passive, through NPC, **162B,** 243–244
Diffusionist model
generalizability, **162B,** 192–193
interpretation of pre-mRNA migration, **162B,** 214–215
intracellular transport, **162B,** 186–189
soluble RNA, **162B,** 184–186
DiGeorge syndrome
chromosomal abnormalities, **158,** 239, 241
clinical features, **158,** 239–240, 262
environmental induction, **158,** 239
incidence, **158,** 239
mode of inheritance, **158,** 239
role of TUPLE1 gene, **158,** 243
similarity to velocardiofacial syndrome, **158,** 242
Digitalis purpurea, environmental stimuli in, **145,** 289–290
Digitization, cell motility
in 2D analysis systems, **163,** 53–54
in 3D analysis systems, **163,** 74–75
Diglycosyl diacylglycerides, induction of cortical cell division, **156,** 50
Dihydrocytochalasin B
effect on cartilage phenotype, **159,** 291
transdifferentiation in medusae and, **142,** 232, 249

Dihydrodipicolinate synthase, flowering plants and, **143,** 320–321
7,8-Dihydrofolate, NADPH reduction, **156,** 111
Dihydrofolate reductase
competitive inhibition, **156,** 111–116
locus in CHO cells, **162A,** 466–476
Dihydropyridine
calcium channels and, **147,** 194, 200, 207–209
proteins in striated muscle and, **142,** 127
1,4-Dihydropyridine, intracellular Ca^{2+} messenger system, **142,** 318, 320
Dihydrouridine, animal mtDNA, **141,** 195, 197–199, 201, 208
5α-Dihydroxytestosterone
effect on decidual cell reaction, **160,** 21–22
synthesis, **160,** 199–200
1,25-Dihydroxyvitamin D_3, **166,** 25
Dimethrin, peroxisome proliferation, **160,** 316
Dimethylsulfoxide, myofibrillogenesis and, **143,** 166, 168
6,6′-Dimycolytrehalose, structure, **161,** 287–289
Dinitrophenol, mitosis and, **144,** 246
Dinoflagellates
chloroplasts and, **144,** 275–276, 283, 285–295
cortex
cortical inheritance, **169,** 306–308
endoplasmic reticulum–alveolar system, **169,** 291
epiplasm proteins, **169,** 286–288
microtubules, **169,** 297–299
organization, **169,** 274
plasma membrane and surface coats, **169,** 282
photoreceptive apparatuses, **148,** 247–251
photosynthetic accessory pigments in, **148,** 276
Dinophyceae, phototaxis in
eyespot apparatus, **148,** 235
eyespot reflection, **148,** 259, 267, 269
photoreceptive apparatuses, components, **148,** 270
photosynthetic accessory pigments, **148,** 275–276
Dinophysis, chloroplasts and, **144,** 290
Dinophysoids, chloroplasts and, **144,** 290, 292, 294–295
Dioxigenin, probes for genetic disease diagnosis by FISH, **153,** 10
Dipeptides
immunohistochemical localization of carnosine, **138,** 171–172
possible effects of carnosine, **138,** 172–174
Dipeptidylaminopeptidase, **167,** 59–61
Dipeptidylaminopeptidase A, **139,** 76–77
Dipeptidylaminopeptidase B, **139,** 71–74
Diploid cells, vertebrate cell culture technology, **142,** 149, 151, 174, 176
Diptera
karyosphere and, **144,** 8–11
polytene chromosome, **165,** 53
Diptheria toxin A, self-incompatibility and, **143,** 278
Direct flight muscles, in *Drosophila,* **143,** 64
Direct method transfer, **140,** 436
Direct microscopy, living polymorphonuclear leukocytes, **139,** 177–180
Direct transfer model, **162B,** 190–191, 205–207
Discoidin, mutants, motion analysis, **163,** 93
Disease
collagen I disorders, **173,** 85–86
human, animal mtDNA and, **141,** 207
infectious, dendritic cell role
HIV, **153,** 89
leprosy, **153,** 89–90
regenerative cementogenesis and, **142,** 43
role of chaperones, **174,** 166–169
treatment with human RPE transplants, **138,** 41
Disease resistance
Arabidopsis thaliana and, **144,** 53–54
genetic model, **144,** 61, 64
pathogens, **144,** 54–56, 58–60
phenotypes, **144,** 65–68, 72
and extracellular matrix, **146,** 305, 311, 314
Disector–Cavalieri method, for glomerular volume, **161,** 134–141
Disector–fractionator method, for glomerular volume, **161,** 134–141
Disector method, for numerical density, **161,** 122–125

Dislocations, in apatite crystals, **172,** 170–171
Disomies
FISH-sperm assays in calculating rates of, **167,** 270
incidence in sperm, **167,** 269–270
uniparental, syndromes associated with, **167,** 289
Displacement binding, endothelin receptor subtypes, **171,** 271–273
Disseminated intravascular coagulation
leukemia associated with, **152,** 83–84
tissue factor role, **152,** 57
Dissipation function, bacterial flagellar motor end, **147,** 118–120
Dissociation, mitotic events
chromosomal events, **144,** 238–252
cytokinesis, **144,** 233–235
experimental evidence, **144,** 235–238
genetic evidence, **144,** 226–233
Diuretic hormone receptors, insect
characterization, **164,** 230
cloning, **164,** 241–242
Diversity, glycine receptor subunit isoforms, **176,** 208–210
Division apparatus, *Tradescantia* generative cell
anaphase-telophase, **140,** 174–175
background, **140,** 172
mitotic apparatus, establishment, **140,** 172–174
D-loop region, endosymbiont hypothesis, **141,** 298–299
DNA
acceptor proteins bound to, **162B,** 345–348
Arabidopsis thaliana and, **144,** 57, 63, 67
AT-rich sequences, **148,** 89–91
conserved genes, **148,** 89–90, 112
and DNA loops, relationship, **148,** 109
in homeotic genes, **148,** 104–105
late replication, **148,** 91–94, 112
nuclear matrix attachment, **148,** 94–97, 109
positive regulation, **148,** 90–91
role in hierarchical expression, **148,** 110
autosomal imbalance diagnosis
deletions, **153,** 30–33
duplications, **153,** 33
avian salt glands and, **144,** 141, 143, 158
bendable, in determining nucleosome positioning, **162A,** 348
binding activity in plants, **148,** 28
biolistic delivery, **140,** 429–431
cellular transfer, genes, in habituated cells
expression, **152,** 289–290
homology with transfer DNA of Ri plasmid, **152,** 287–289
role, **152,** 287, 293
chalcone synthase and
control of activity, **147,** 256, 258
control of gene, **147,** 262–263, 272–275, 277
enzymatic activity, **147,** 248
chloroplasts and, **144,** 259, 263, 281
cryptomonads, **144,** 273, 275–276
dinoflagellates, **144,** 288, 290
cholesteric ordering, **166,** 90–92
chromoplast, during differentiation, **163,** 194–195
ciliates and, **141,** 2–4, 6, 9, 55
Paramecium aurelia sequences, **141,** 12–45
rRNA genes, **141,** 47–48
structure, **141,** 5–9
complementary, *see* Complementary DNA
condensed, cytochemical evidence for close association with, **138,** 73
consensus elements, **162B,** 339–340
content of human preimplantation conceptuses, **164,** 155–160
cyanelle genome, organization, **151,** 37–39
cytoplasmic, inheritance, **140,** 577–578
and double fertilization, data, **140,** 587
in cytoskeletal mechanical signaling, **150,** 209–210
direct delivery to microspore-derived embryos
biolistic DNA delivery, **140,** 429–431
microinjection, **140,** 428–429
Drosophila muscles and, **143,** 63–65, 138
contractile protein genes, **143,** 105, 107, 115, 117–118, 124
development, **143,** 72, 81, 86
electromagnetic field effects
repair, **158,** 321
synthesis, **158,** 305–306

encoding S proteins, cloning of, **140,** 465–466
endosymbiont hypothesis and
eukaryotes, **141,** 277
gene expression, **141,** 298–301, 303–305
gene structure, **141,** 308–311, 313–316
gene transfer, **141,** 277, 324–325
genome organization, **141,** 289–298
mitochondria, **141,** 280–282, 286, 288, 327–329
molecular biology, **141,** 263, 265–267
Plasmodium, **141,** 325–326
plastids, **141,** 254, 262, 279
ribosomes, **141,** 272, 276
rRNA sequences, **141,** 32O, 324
sequences, **141,** 269
extracellular matrix and, **146,** 261, 267, 281, 291
fiber, radioautography analysis
basic principles, **154,** 265–269
critical data, **154,** 269–274
double-pulse labeling prior to, **154,** 274–280
method, **154,** 263
replicon model, **154,** 296
flowering plants and, **143,** 311, 323
fragmentation, in apoptosis, **163,** 112–113, 121–123
GnRH-encoding gene in teleosts and, **147,** 166, 176, 186
higher plant mitochondrial genomes and
abnormal phenotypes, **141,** 130, 133–134
gene location, **141,** 157–159
recombination, **141,** 148–149, 151–154, 156–157
structure, **141,** 136–137, 140–141, 146
higher plant plastids and, **145,** 151, 204
evolution, **145,** 165–166
genome, **145,** 153–155, 158–159, 161, 163–164
hormonal control of amphibian metamorphosis and, **145,** 140
in hybridization experiments of coordinate transcription, **148,** 86
hypomethylation, induction by azacytidine, **158,** 116
and instability of homogeneous state
metabolism of proliferation, **154,** 360
reduction fields, **154,** 347
Turing biochemical mechanism, **154,** 324
interaction
with lamins, **162B,** 163–164
with nuclear matrix, **162B,** 405–441
intermediate filament proteins and
cytokeratin filaments, **154,** 54–57
cytoskeleton role in reverse transformation, **154,** 40
double nuclear membrane barrier, **154,** 27–28
evolutionary aspects, **154,** 69–70
loop formation, **154,** 23–24
as matrix elements in gene expression, **154,** 25–26
muscle differentiation, **154,** 51
nuclear functions, **154,** 5–7
nuclear organization, **154,** 7–12
post-translational modification, **154,** 48
redundancy of function, **154,** 65–66
repetitive DNA sequence interactions, **154,** 32–37
transcription factors and similarities, **154,** 13–19
intracellular calcium messenger system and, **142,** 310–311
intrinsically curved
in α-globin gene MAR, **162A,** 332–333
in MAR/ORIs, **162A,** 338–339
karyosphere and, **144,** 8, 41–43, 45
oocytes of insects, **144,** 11, 14, 17–19, 21, 23, 25
vertebrates, **144,** 26, 28, 32, 35–38
kinked, MAR/ORI class harboring, **162A,** 340–341
lepidopteran, **152,** 33–34
loops, binding to nuclear matrix, **148,** 108–109, 112
male heredity components containing, **140,** 131–132
MAR/ORI
methylation level, **162A,** 354–355
prone to mutagens, **162A,** 353–354
sequence composition and genomic position, **162A,** 356–358
melanocytes and, **146,** 194
genetics of melanoma, **146,** 225, 227, 248
pigment-cell genetics, **146,** 207–208, 212
methylation
allelic inactivation, mechanisms
chromatin role, **160,** 71–72

DNA (*continued*)
imprinted gene function, **160,** 72–73
replication timing, **160,** 70
continuity between gametes and embryonic cells, **160,** 63–65
gene inactivation by, **154,** 209, 212
imprinted genes, **160,** 57, 62–63, 67–69, 72–73
in MAR/ORIs, **162A,** 354–355
postfertilization events in imprinted gene regulation, **160,** 74–75
transgenes, **160,** 65–66
mitochondrial, *see* Mitochondrial DNA
mitosis and
chromosomes, **144,** 240–241, 243, 248–249
dissociation of events, **144,** 226–228, 231, 238
mitotic events, **144,** 222, 225
in muscle cell differentiation, gene regulation, **150,** 69–70
myofibrillogenesis and, **143,** 154, 167
organization within nucleus, **162A,** 192–202
pollen grains and vectors of transforming, **140,** 431–440
polyamine interactions with chemical aspects of, **138,** 78
polymorphisms
meiotic recombination errors and trisomy, **167,** 282
paternal age and trisomies of paternal origin, **167,** 280–281
preimplantation embryo analysis, **153,** 4
probes for genetic disease diagnosis by FISH, **153,** 7, 10–11, 13–14
promiscuous, endosymbiont hypothesis and, **141,** 293
proteins in striated muscle and, **142,** 96
in quantitative regulation of tissue-specific genes, **148,** 88
regenerative cementogenesis and, **142,** 37
repair
alkylation repair, **156,** 129–131
excision repair, **156,** 124–125
gene-specific repair, **156,** 132–134
postreplication repair, **156,** 125–127
resistance to anticancer drugs, **156,** 122–137
replication
antimetabolite effects, **162A,** 549–556
discrete foci, **162A,** 69–70
domains, distribution and ultrastructure, **162A,** 165–167
foci
proteins present at, **162B,** 315–318
targeting of proteins to, **162B,** 318–322
visualization, **162B,** 314–315
forks, interaction with drugs, **162A,** 560–562
initiation in mammalian cells, **162A,** 463–476
and lamin B, **162B,** 164–165
link with cell cycle regulation, **162B,** 322–323
machinery, anchored to nuclear matrix, **162A,** 312–314
nuclear matrix-associated, **162A,** 39–53
nuclear structure requirement, **162A,** 460–462
role of lamins, **162B,** 160
SV40, structural topography, **162A,** 495–519
transdifferentiation in medusae and, **142,** 220, 235–246, 249–250, 257
replication, in mammals
chromosomal organization, **154,** 263–265
DNA fiber radioautography analysis
basic principles, **154,** 265–269
critical data, **154,** 269–274
double-pulse labeling prior to, **154,** 274–280
method, **154,** 263
DNA organization, **154,** 301
molecular mechanisms, **154,** 261–252
process, **154,** 261
replication fork rate, **154,** 286–292
replication unit size, **154,** 277, 281–286
replicon model, **154,** 294–301
replicon termination, **154,** 292–294
in replication band of ciliated protozoa
chromatin replication models, **153,** 162–165
cytochemical studies, **153,** 154
function, **153,** 144–145, 147–148
macronuclear models, **153,** 137–138, 142
ultrastructure, **153,** 152

replication sites
nuclear matrix, **162A,** 37–53,549–556
transcriptionally active, **162A,** 139
visualization in 3-D, **162A,** 49–53
retinal precursor cells and, **146,** 168
ribosomal, *see* Ribosomal DNA
satellite, attachment to matrix, **162A,** 341–342
self-incompatibility and, **143,** 257–258, 272, 286, 289
sequence analysis, in interphase nucleus, **150,** 149, 161–162, 167
sequence elements, **162B,** 412–416
Sertoli-germ cell communication network and, **147,** 34
strand breaks, **162A,** 547–549
strand separation and single strands, S/MAR, **162A,** 410–416
supercoiled, **162A,** 220,349
SV40
encapsidated minichromosomal, **162A,** 520–521
intermediates, **162A,** 508–516
virion-contained, subnuclear distribution, **162A,** 523–525
T, integration, **162A,** 435–437
in temporal expression experiments, **148,** 83–84
and testis-determining factor genes, **154,** 230–231
topoisomerase II, biological activities, **162A,** 556–559
transcriptional regulation and, **143,** 1–2, 47–48
biochemistry, **143,** 22–25
function, **143,** 2
inducible activation, **143,** 7, 11
molecular cloning, **143,** 25–26, 28–31, 33–34, 37–39
proteins, **143,** 45–47
structure, **143,** 22
viruses, **143,** 42, 44
transdifferentiation in medusae and, **142,** 214
initiation, **142,** 237, 247–248, 250
isolation of tissues, **142,** 222
potential, **142,** 223–224, 227, 229, 232–235
transfer, tumorous outgrowth induction by, **152,** 266, 278–279
unwinding elements
in MAR/ORIs, **162A,** 322–323
S/MARs, **162A,** 442–443
vertebrate cell culture technology and, **142,** 150, 169, 176, 194
in whole-chromosome hybridization
chromosomal aberration analysis, **153,** 204–207
cloned chromosome-specific DNA probe libraries, **153,** 193–195
detection methods, **153,** 181, 183
genomic probes, **153,** 191–193
polymerase chain reaction, **153,** 196–197
probe labeling methods, **153,** 179–180
probe techniques, **153,** 190–191
in situ hybridization technique, **153,** 172–178
tumor, as whole chromosome hybridization probe, **153,** 192
typical methods, **153,** 188
Z-DNA formation, **154,** 37
DNA-binding proteins
activity on Southwestern blots, **162A,** 29–31
associated with nuclear matrix, **162A,** 230–231
centromeric and telomeric, **162B,** 29–30
nonstringent sequence specificity, **162A,** 303–309
single-stranded, interaction with MAR/ORIs, **162A,** 349–350
DNA escape
detection of nucleic acids from mitochondria in higher eukaryotes, **165,** 217
direct detection from mitochondria of yeast, **165,** 213–216
indirect detection from mitochondria of yeast, **165,** 216–217
intact mitochondrial DNA in isolated HeLa cell nuclei, **165,** 217–218
mitochondrially encoded large rRNA, **165,** 218
organelle, **165,** 213–218
from rho$^+$ mitochondria, **165,** 214–216
from rho- mitochondria, **165,** 213–214
yeast mitochondria, **165,** 222–225
DNA glycosylase, excision of damage, **156,** 125
DNA ligase, excision of damage, **156,** 124

DNA ligase I, characterization, **162B,** 319–322
DNA loops
as basic units of genome evolution, **162B,** 435–438
excision by selective DNA cleavage, **162B,** 425–435
gene distribution specificity, **162B,** 416–425
interface with matrix, **162A,** 457–460
MAR-binding protein role, **162B,** 26–29
size determination, **162B,** 407–411
supercoiling, **162A,** 541–545
DNA methyltransferase, targeting sequence, **162B,** 318–319
DNA polymerase
animal mtDNA and, **141,** 224, 228
attached, **162A,** 131–139
ciliates and, **141,** 7–8
concentration in factories, **162A,** 126–127
kinetoplastid mtDNA and, **141,** 83
mF plasmid, **161,** 90–91
in replicative DNA synthesis, **156,** 125
transdifferentiation in medusae and, **142,** 252
DNA polymerase α, association with isolated nuclear matrix, **162A,** 41–42
DNA primase, target for anticancer drugs, **162A,** 552
DNase I
hypersensitive sites
DNase I-sensitive domains, **162A,** 227–228
interferon-β, **162A,** 393–397
MAR/ORIs, **162A,** 346–347
resistant conformation, **162A,** 219–221
sensitive chromatin fragments, **162A,** 221–226
sensitivity of nuclear filaments, **162A,** 94–97
DNA synthesis
AcSDKP inhibition of onset of, **167,** 221
cell shape effects, **151,** 212–213
hemopoietic inhibitory regulators and, **167,** 194
microtubule disruption and initiation of, **138,** 219–221
MIP-1α and LTBMC, **167,** 203
MIP-1α regulation of, in spermatogenesis, **167,** 208
non-S-phase-specific cytoxics and, **167,** 240–241
platelet-derived growth factors in, **172,** 95–96, 115–116
primed *in situ,* **153,** 20
silica receptor role, **150,** 327
as stage in cell cycle, **167,** 192
suicide assays using cytotoxics, **167,** 197
DNA transfer, direct
microinjection and biolistics, **140,** 435
pollen tube pathway and *Agrobacterium* mediated transformation, **140,** 434–435
as transformation strategy, **140,** 432–434
DNP, karyosphere and, **144,** 42, 45
DNR, **156,** 79
Domains
arginine/serine-rich domain, **162B,** 312–313,324–325
artificial, creating and reusing, **162A,** 437–440
binding
NLS, **162A,** 33–34
nuclear localization signal, **162A,** 33–34
borders, S/MARs as, **162A,** 425–435
chromatin, **162A,** 218–234
chromosome, **162A,** 170–173
COL1 domain, collagen V, molecular structure, **173,** 97–99
COL2 domain, collagen V, molecular structure, **173,** 96–97
DHFR, as paradigm of ORI, **162A,** 455–476
DNA replication, **162A,** 165–167
HTH DNA building domain, **156,** 7
in internal matrix network, **162B,** 105–111
kinase domain, FAK, **167,** 165
lamin rod, **162B,** 143–145
LIM, in zyxin and cCRP, **167,** 163
NC2, collagen V, molecular structure, **173,** 96–97
NC3, collagen V, molecular structure, **173,** 96–97
nuclear
correspondence with nuclear matrix components, **162B,** 123–127
derived insoluble solutions, **162B,** 323–325
and nuclear matrix, **162A,** 151–175
with unknown function, **162A,** 167–170

nucleolar, **162B,** 114
position independence, **162A,** 425–427
prototype, formed by human interferon-β, **162A,** 390–410
RNA, **162A,** 78–79
RNP, **162A,** 6–8
RS, **162B,** 312–313
sequence, Rab GTPase effector region, **176,** 14–15
speckled
bright spots in, **162B,** 311–314
containing splicing machinery components, **162A,** 160–164
function, **162A,** 75–78
location, **162A,** 73–75
location and function, **162A,** 73–78
Src homology 2, ligands, **157,** 211, 222, 229
structural, plant nuclear matrix, **162B,** 88–90
transcript, nuclear distribution, **162A,** 71
transmembrane, G-protein-coupled receptor, **164,** 205–207
Dome epithelium
antigen transport by
as main function of M cells, **167,** 100, 112
M cell maturation stages *versus* capacity for, **167,** 117
overview, **167,** 93
enterocytes in, role of, **167,** 130–131
M cell identification in
alkaline phosphatase, **167,** 102
cross-correlation of labeling patterns, **167,** 104–107
cytoskeleton, **167,** 104
lectins, **167,** 103–104
monoclonal antibodies, **167,** 102
ultrastructure, **167,** 100–102
nonepithelial cells in
lymphocytes, **167,** 133–135
macrophages, **167,** 131–133
origin, differentiation, and development of, **167,** 108–111
overview, **167,** 99–100
types of cells in, **167,** 130
Dopa, melanocytes and, **146,** 193, 196
Dopamine
biological clocks in retina and, **146,** 85
in catecholaminergic neurotransmission, **149,** 243
control of rhythmic behaviors, **147,** 11–13
effect of hypothalamic lesions on levels, **159,** 49
effect on pituitary cell proliferation, **157,** 47–48
effect on release
gonadotropic hormone, **159,** 69
prolactin, **159,** 63–64
functions in insect CNS, **164,** 218–220
hormonal control of amphibian metamorphosis, **145,** 128, 134, 136–137
localization in insect CNS, **164,** 217
regulation of pars intermedia cells, **159,** 53
retinal precursor cells and, **146,** 155, 164
Dopamine receptor
cloning, **164,** 235–236
function, **149,** 243–244
Dopaminergic system, drug interaction with, **149,** 247–249
Dopaquinone, melanocytes and, **146,** 193
Doritis pulcherrima, enzyme activity, effect of microgravity, **171,** 39
Dorsal ridge
Drosophila melanogaster, **174,** 16–17, 40
evolution, **174,** 60–62
segmental organization, **174,** 27
Dorsal root ganglion neurons, plasticity
fibronectin role, **153,** 243–244
growth cones, **153,** 260
laminin role, **153,** 244–245
nerve growth factor role, **153,** 248–252
neurite inhibitors, **153,** 256
neurotrophic factor role, **153,** 255–256
phenotypic expression, **153,** 272, 275–276
substrate role, **153,** 238–239
synaptic connections, **153,** 265
Dosage
avian salt glands, **144,** 167
dependence of growth factor-induced retinal regeneration, **146,** 64
Double fertilization
cytoplasmic DNA inheritance, data, **140,** 587
delivery of male gamete in embryo sac and their condition prior to fusion, **140,** 370–372

Double fertilization (*continued*)
gamete fusion in plants and animals
models of fertilization and membrane fusion, **140,** 372–377
in animal systems, **140,** 372–375
membrane fusion in angiosperms, **140,** 375–377
preferential fertilization, **140,** 377–378
influence of fusion in pattern of cytoplasmic inheritance, **140,** 379–383
paternal cytoplasmic inheritance mechanism, **140,** 382–383
uniparental, maternal cytoplasmic inheritance mechanism, **140,** 379–382
nonflowering seed plants, *see* Nonflowering seed plants, double fertilization
nuclear fusion, **140,** 383–385
process
history, **140,** 357–359
overview, **140,** 359
receipt of pollen tube, **140,** 359–370
pollen tube arrival in synergic-lacking angiosperm, **140,** 370
role of synergic in, **140,** 360–368
selection of degenerated synergic, **140,** 368–370
Down's syndrome
associated craniofacial abnormalities, **158,** 223
immune system in patients, **158,** 322
incidence of, in live births, **167,** 265
maternal age and, **167,** 276–277
PHF-Tau protein in, **171,** 195
DOX, *see* Doxorubicin
Doxorubicin
MDR, **156,** 79–80, 102, 104, 107
MDR reversal, **156,** 91
resistant cells, **156,** 83
DPAP, *see* Dipeptidylaminopeptidase
DPAP A, *see* Dipeptidylaminopeptidase A
DPAP B, *see* Dipeptidylaminopeptidase B
Drag coefficients, in bimodal theory, **163,** 19–20, 26–30
Drosophila
actin cross-linking proteins, mutations
affecting development, **175,** 49–50
affecting oogenesis, **175,** 46–49
animal mtDNA and, **141,** 174–175, 189, 203
genome content, **141,** 176–177, 180
protein genes, **141,** 184, 187
biological clocks in retina and, **146,** 84, 96, 116
cytoskeletal dynamics and, **144,** 97
development, ring canal and bristle formation, **175,** 44–46
direct flight muscle, **143,** 64
effect of homeotic genes, **148,** 103–104
embryo, equivalence group, **176,** 295–296
endosymbiont hypothesis and, **141,** 297
heat shock prompt proteins in, **148,** 201
instability of homogeneous state
dorsoventral system, **154,** 335–339
localization, **154,** 322, 344
metabolic field, **154,** 310, 355, 360
oxygen-driven differentiation balance, **154,** 326
reduction fields, **154,** 347–348, 352, 354
terminal system, **154,** 335–339
intracellular transport, role of nonmuscle tropomyosin, **170,** 18–19
karyosphere and, **144,** 11, 43
kinetoplastid mtDNA and, **141,** 73
maternal meiosis I nondisjunction and *nod*^{DTW} mutation in, **167,** 278–279
melanocytes and, **146,** 191
mitochondrial genomes of, **141,** 12
mitosis and, **144,** 252
dissociation of events, **144,** 226, 230–235, 237, 245
mitotic events, **144,** 225–226
muscles, genetic analysis, **143,** 63–64, 140–143
contractile protein genes
actins, **143,** 109–114
function, **143,** 132–137
interactions, **143,** 126–132
isoforms, **143,** 94
muscle protein genes, **143,** 125–126
myofibrils, **143,** 94–99
myosin heavy chains, **143,** 99–104

myosin light chains, **143,** 104–108
paramyosin, **143,** 108–109
tropomyosin, **143,** 114–117
troponin, **143,** 117–122
Z-disc proteins, **143,** 122–125
development, **143,** 65–66
adult muscles, **143,** 71–77
differentiation, **143,** 76–77
genes, **143,** 78–94
larval muscles, **143,** 66–71
expendable muscle groups, **143,** 64
gene dosage, **143,** 127, 131
genetic programming, **143,** 137
mapping domains, **143,** 139–140
protein components, **143,** 137–138
protein isoforms, **143,** 138–139
transgenic organisms, **143,** 64–65
myofibrillogenesis and, **143,** 175, 181
Na^+/K^+-ATPase, **176,** 320–323
pattern formation in
epidermis
embryogenesis and segmentation and, **139,** 123–125
en gene and, **139,** 128–134
ptc gene and, **139,** 130–134
segment polarity genes and, **139,** 126–128
wg gene and, **139,** 128–134
eye
ommatidia and, **139,** 143–146
R7 cell fate and, **139,** 146–148
general considerations, **139,** 121–122, 148–151
nervous systems, **139,** 134–143
central, **139,** 134–137
neurogenic genes and, **139,** 140–143
peripheral, **139,** 134–136, 137–140
postshock cell repair, **148,** 208–209
proteins in striated muscle and, **142,** 88, 96
retinal precursor cells and, **146,** 174–175, 184
testis-determining factor *SRY* gene-related sex determination, **154,** 242
transcriptional regulation and, **143,** 25, 27, 35–36
transdifferentiation in medusae and, **142,** 215, 256
tropomyosin gene, **170,** 6
zeste gene product
function, **154,** 65
matrix element, **154,** 24–26
Drosophila melanogaster
adult, genetics of development, **174,** 40–45
aging, effect of altered gravity, **171,** 55–56
and animal mtDNA, **141,** 184
antennae, **174,** 6–7
body plan, regulatory genes, **174,** 30–33
complex eye, **174,** 5–6
dorsal ridge, **174,** 16–17
engrailed gene expression, **174,** 11–13
epidermis, **174,** 8
evolution of segmentation, **174,** 28–29
head development, **174,** 1–2
karyosphere, **144,** 8
labrum, **174,** 7
larval epidermis, genetic control
antennal cuticle, **174,** 37
clypeolabrum, **174,** 35–36
dorsal ridge, **174,** 40
intercalary cuticle, **174,** 37–38
labial cuticle, **174,** 39–40
mandibular cuticle, **174,** 38–39
maxillary cuticle, **174,** 39
ocular cuticle, **174,** 36–37
required genes, **174,** 33–35
microgravity experiments, **156,** 335, 352, 355, 356
mouthparts, **174,** 7–8
procephalon, **174,** 13–16
programmed cell death in, **163,** 127–130
segmentation
clypeolabral region, **174,** 26–27
dorsal ridge, **174,** 27
head gap genes, **174,** 23–26
labial segments, **174,** 20–23
maternal components, **174,** 19–20
maternal polarity systems, **174,** 18
maxillary segments, **174,** 20–23
segment-polarity genes, **174,** 18–19
yolkless gene, **166,** 126–127
Drosophila yakuba, and animal mtDNA
control region, **141,** 202
genome content, **141,** 176–178, 180–184
nucleotide bias, **141,** 204–206
protein genes, **141,** 187–188
rRNA, **141,** 190
tRNA, **141,** 193, 196, 198

Drought tolerance, stomatal opening regulation, **165,** 35
Drug accumulation
  rate, effect of membrane potential, **171,** 144–149
  steady-state description, **171,** 141–144
Drug pumps
  ATPase activity and stoichiometry, **171,** 132–134
  coupling, **171,** 130
  function, **171,** 129–130
  kinetics, **171,** 134–135
  photolabeling, **171,** 136
  thermodynamics, **171,** 135
Drug resistance
  antibiotics, spaceflight experiments, **156,** 340–342
  collateral sensitivity, **156,** 81
  cross resistance, **156,** 81
  cytosine arabinoside, **156,** 120–121
  DNA repair, **156,** 122–137
  DNA topoisomerases, **156,** 97–108
  5-fluorouracil, **156,** 118–120
  genetic mechanism, **156,** 78–79
  methotrexate, **156,** 113, 115, 116, 117
  multidrug resistance, **156,** 79–97
  NMR analysis, **145,** 37
  purine antimetabolites, **156,** 121–122
  reversal, **156,** 88–92
  spaceflight experiments, **156,** 340–342
Drugs, *see also specific drugs*
  anticancer, **162A,** 539–565
  biological clocks in retina and, **146,** 121
  cationic, effect of membrane potential
    on membrane flux, **171,** 139–141
    on steady-state drug accumulation, **171,** 141–144
  cytostatic, carcinoma culture experiments, **155,** 248–250
  interactions with neurotransmitters
    γ-aminobutyric acid system, **149,** 266
    cholinergic system, **149,** 240–241
    dopaminergic system, **149,** 247–249
    and neurotransmitter transporters, **149,** 225
    noradrenergic system, **149,** 247–249
    serotoninergic system, **149,** 251–252
  interfering with viral life cycle steps, **162A,** 487,531
  intracellular calcium messenger system and, **142,** 315
  M cells as potential gateway for, **167,** 122
  and mitosis, **144,** 225, 229, 237, 246
  NMR analysis, **145,** 20, 31, 37, 41
  psychotropic, **149,** 217–218, 247, 266
  retinal precursor cells and, **146,** 155
  topoisomerase II-active, **162A,** 559–564
  transdifferentiation in medusae and, **142,** 235, 249–250, 252–253
  vertebrate cell culture technology and, **142,** 203
Drug transport
  advantages of drug pump model, **171,** 132–136
  definitions, **171,** 128–129
  effect of membrane potential, **171,** 137–139, 153–155
    on cationic drug flux, **171,** 139–141
    on drug accumulation rate, **171,** 144–149
    on weak base accumulation, **171,** 149–153
  effect of ΔpH, **171,** 153–155
  historical context, **171,** 130–132
  pump coupling, **171,** 130
  role of pumps, channels, and exchangers, **171,** 129–130
Dry pollen, **140,** 76–77
Duchenne muscular dystrophy, proteins in striated muscle, **142,** 100, 124–126
Ducks, avian salt glands
  blood supply, **144,** 151, 154
  hormonal control, **144,** 193, 196, 199–200
  innervation, **144,** 155, 158–159, 161, 168
  receptive systems, **144,** 177–178, 180, 182, 185–186, 189
  secretory mechanism, **144,** 172
  secretory tissue, **144,** 131–133, 135, 137, 140–141, 143, 145–146
  stimulus-secretion coupling, **144,** 200, 203
Duct-associated lymphoid tissue, **167,** 137–138
Dunaliella, intracellular calcium messenger system and, **142,** 309, 322–323, 326, 335
Duodenum, Peyer's patches in, **167,** 94
Dyes, fluorescent see Fluorescent dyes
Dynactin
  regulation of cytoplasmic dynein, **164,** 291–293
  regulation of MT–microfilament interactions, **173,** 232–233

Dynamin
cross-linking of microtubules, **157,** 20
fish chromatophores, **143,** 222
Dynein
ATPase activity, **173,** 13–16
axonemal
cAMP role in regulating motility, **164,** 286–288
Ca^{2+} role in motility regulation, **164,** 288–289
properties, **164,** 283–285
regulation of ciliary and flagellar beating, **164,** 285–289
axonemal, diversity, **173,** 19–21
cytoplasmic
functions, **164,** 290–291
properties, **164,** 289–290
regulation, **164,** 291–295
cytoskeletal dynamics, **144,** 109–110
fast axonal transport, **165,** 271
fish chromatophores, **143,** 216, 221–223, 242, 247–249
flagellar
characteristics, **155,** 141–142, 173–174
cross-bridge cycle, **155,** 168–169
in flagellar motility *in vivo,* **155,** 166–167
inner row, **155,** 162–166
purification, **155,** 162–164
in situ localization, **155,** 164–166
subunit compostiion, **155,** 162–164
in microtubule motility
doublet sliding, **155,** 142–145
sliding converted to bending, **155,** 145–146
translocation *in vitro,* **155,** 167–168
outer row
heavy chains, **155,** 150–155
intermediate chains, **155,** 148–150
light chains, **155,** 147–148
outer arm structure, **155,** 155–161
purification, **155,** 146–155
structure-function relationships, **155,** 161–162
subunit composition, **155,** 146–155
regulation, **155,** 170–173
mutations, **155,** 171–173
radial spokes, **155,** 170–171
sliding velocity, **155,** 170–171
structure, **155,** 142–145
inner arm, effect of dynein regulatory complex, **173,** 7–9
role of cross-bridge cycle, **173,** 38–39
–tubulin interaction, role of t-force, **173,** 46–47
–tubulin sliding, dynamics, **173,** 17–19
Dynein-like proteins, in cytoskeleton-based motor proteins, **176,** 159–160
Dynein motor
in eukaryotic flagella, **173,** 4
and geometric clutch model, **173,** 51
isolation, **173,** 18
Dynein regulatory complex
definition, **155,** 170
effect on inner arm dynein, **173,** 7–9
mutations, **155,** 171–173
regulation of MT–microfilament interactions, **173,** 233
Dysmorphology
craniofacial abnormality classification, **158,** 216
differentiation from deformation, **158,** 215–216
Dysplastic nevus syndrome, **146,** 226–228, 230, 244, 246, 250
Dystrophin
cytoskeletal dynamics, **144,** 106
Drosophila muscles, **143,** 122
proteins in striated muscle, **142,** 88, 100, 124–126

E

E1B–E4 complex, host transcript immobilization, **162B,** 210–213
E7, localization in nuclear matrix, **162A,** 108–109
Early B cell factor, role in B cell lineage commitment, **157,** 162
Early firing origins, model, **162A,** 474–476
Early pregnancy factor, in Cpn10 bioassay, **174,** 150–151
EBF, *see* Early B cell factor
Ecdysone receptor, characterization, **164,** 204
Echinoderms
coelomocytes and oocytes
actin cross-linking proteins, **175,** 41–44
form and function, **175,** 40–41

Echinoderms (*continued*)
sperm-binding proteins, **163,** 259–262, 270–271
Echoderm, dorsal, from Hensen's node, **175,** 251–253
Echo planar imaging, and NMR, **145,** 29, 32
Echo time, in NMR, **145,** 11–12, 23
Eclosion hormone, insect, characterization, **164,** 229
ECM, *see* Extracellular matrix
Ecto-ATPase
abundance, **158,** 154–155
activators
calcium, **158,** 192
concanavalin A, **158,** 163, 189, 191
fatty acids, **158,** 196
glycoside detergents, **158,** 191
insulin, **158,** 197
magnesium, **158,** 192
saponin, **158,** 191
amino terminal sequences, **158,** 180–181
assay, **158,** 183–185
catalytic mechanism, **158,** 183
discovery, **158,** 142–143
EC classification, **158,** 142, 198
ectoorientation, **158,** 146, 162, 171
glycosylation, **158,** 153, 179, 181
identification criteria, **158,** 183–184
induction by epidermal growth factor, **158,** 174
mercurial-sensitive enzyme, **158,** 174, 178, 190
metal dependence, **158,** 141–143
microvilli localization, **158,** 144
photoaffinity labeling, **158,** 167, 181–182
purification
bovine brain enzyme, **158,** 166
chicken gizzard enzyme, **158,** 165, 180, 182–183
chicken liver enzyme, **158,** 180
chicken oviduct enzyme, **158,** 182
chromatography, **158,** 179
density gradient centrifugation, **158,** 179
difficulty, **158,** 154
immunoprecipitation, **158,** 181
pig pancrease enzyme, **158,** 172
rabbit skeletal muscle, **158,** 182
rat liver enzyme, **158,** 170
solubilization, **158,** 152–153, 176–178
T-tubule ATPase, **158,** 162–163, 179
role
adenosine recycling, **158,** 146–147
bile acid transport, **158,** 150–152
calcium pumping, **158,** 145
cell adhesion, **158,** 148–150, 152
magnesium pumping, **158,** 145
malignant transformation, **158,** 144–145, 172–175, 198
membrane permeability, **158,** 144
nonsynaptic information transfer, **158,** 197–198
platelet aggregation, **158,** 197
purinoreceptor function, **158,** 146
viral function, **158,** 144
secretion in microvesicles, **158,** 145, 168
sequence homology with related proteins, **158,** 147–148
size, **158,** 181–183
specific activity, **158,** 154–155, 179
blood vessels, **158,** 160–161
cancers, **158,** 172–175
cardiac cells, **158,** 159–160
endothelial cells, **158,** 159–160
erythrocytes, **158,** 159
heart, **158,** 164
insects, **158,** 175–176
kidney, **158,** 170–171
liver, **158,** 169–170
lung, **158,** 167
lymphocytes, **158,** 156–157
nervous tissue, **158,** 165–167
oviduct, **158,** 167–168
pancreas, **158,** 171–172
parotid gland, **158,** 172
placenta, **158,** 169
plants, **158,** 176
platelets, **158,** 158
polymorphonuclear leukocytes, **158,** 157–158
prostasomes, **158,** 168–169
skeletal muscle, **158,** 161–164
smooth muscle, **158,** 164–165
sperm, **158,** 168
turnover number, **158,** 155
vascular smooth muscle cells, **158,** 159–160
viruses, **158,** 175
subfamily classification, **158,** 187–188
substrate inhibition, species differences
chicken, **158,** 195
frog, **158,** 195

rabbit, **158,** 195
rat, **158,** 193–195
substrate specificity, **158,** 153, 166–167, 169, 188–190
unfolding difficulty, **158,** 154
Ecto-ATPase inhibitors
azide, **158,** 189–190
chlorpromazine, **158,** 191
detergents, **158,** 152–153, 177
diacylglycerol, **158,** 191, 196
DIDS, **158,** 190–191
diethylpyrocarbonate, **158,** 191
mercury, **158,** 174, 178, 190
suramin, **158,** 157–158
trifluoperazine, **158,** 191
vanadate, **158,** 191
Ectocarpus
eyespot reflection in, **148,** 259, 267–268
phototactic signal generation, **148,** 256
Ectoderm, neural plate derivation, **175,** 255–257
Ectomeninx, **165,** 129
Ectoplasm
cytoskeletal dynamics, **144,** 86–87
environmental stimuli in plant cells, **145,** 276, 300
Edema, nuclear magnetic resonance and, **145,** 46
Editing
kinetoplastid mtDNA and, **141,** 75–82
RNA, *see* RNA, editing
EDL muscle, proteins in, **142,** 70, 74–75, 77, 81, 88
EDTA
cytoskeletal dynamics and, **144,** 109
eggshells and biomineralization and, **145,** 227, 229, 234
environmental stimuli in plant cells and, **145,** 261, 263
karyosphere and, **144,** 21, 45
Edwards' syndrome, associated craniofacial abnormalities, **158,** 223
EEM, *see* Extraembryonic mesoderm
Eeniella, mitochondrial genomes in fungi and, **141,** 101–102
Effector cell protease receptor 1, procoagulant activity in prothrombinase complex, **152,** 80–81
Effector cells, fish chromatophores and, **143,** 225, 229, 238, 244
Efferent ducts, testis, structure and function, **154,** 131, 137–146
Egeria, environmental stimuli in, **145,** 263, 297–298
EGF, *see* Epidermal growth factor
Egg cells
female germ unit, **140,** 260–264
and zygotes, culture, **140,** 418–419
Egg-laying hormones, control of rhythmic behaviors and, **147,** 15
Eggs
activation
general features, **163,** 252
lectin-like binding events, **163,** 268, 281
two-step, **163,** 287–288
ascidian, anterior–posterior specification, **176,** 287–291
chambers, *Drosophila,* actin cross-linking proteins, **175,** 46–49
jelly, sperm-binding proteins, *see* Sperm-binding proteins
mammalian, calcium signal activating, *see* Calcium signal
microgravity experiments, **156,** 334–338, 356
role of cytoplasmic determinants
in cleavage pattern generation, **176,** 268–272
in endoderm cell development, **176,** 263–264
in epidermis development, **176,** 264–267
in gastrulation movements, **176,** 267
in muscle development, **176,** 257–263
in ooplasmic segregation, **176,** 253–255
in tissue differentiation, **176,** 255–257
sperm-binding proteins, *see* Sperm-binding proteins
surface
carbohydrates as binding sites, **163,** 257
sperm-binding proteins, *see* Sperm-binding proteins
sperm receptors, **163,** 280–282
structures, comparative, **163,** 256–257
Eggshells, biomineralization
eggshell types, **145,** 217–218
fabrication, **145,** 235–240
models, **145,** 240–244
structure, **145,** 218–219
cuticle, **145,** 234–235
mamillary layer, **145,** 224–229

Eggshells (*continued*)
palisade, **145,** 229–234
shell membrane layer, **145,** 220–224
EGTA
agravitropism induction in plants, **168,** 153–154
biological clocks in retina and, **146,** 104, 131
calcium channels and, **147,** 214, 216
environmental stimuli in plant cells and, **145,** 276, 290
EHNA, fish chromatophores and, **143,** 205, 222–223
Ehrlich ascites, RVD response, signal transduction pathway, **161,** 225
Eicosanoids
inhibition of smooth muscle cell growth, **161,** 26
insect, characterization, **164,** 230–231
role in volume sensing, **161,** 220–222
Einstein–Smoluchowski model, Brownian motion, **162B,** 186–189
Eisenberg-Hill model, bacterial flagellar motor and, **147,** 157–159
Elaioplasts, higher plant, **145,** 151
Elasmobranchs, avian salt glands and, **144,** 147–148, 177, 196
Elastase, eggshells and biomineralization and, **145,** 220
Elastin
effect on arterial smooth muscle cell culture phenotype, **169,** 202–203
eggshells and biomineralization and, **145,** 220–221
synthesis in arterial smooth muscle cell culture, **169,** 223–224
Electrical coupling
biological clocks in retina and, **146,** 92, 95, 102
control of rhythmic behaviors and, **147,** 3, 14
Electric field
effect on cells
DNA synthesis, **158,** 306
proliferation rate, **158,** 307
induction by magnetic field, **158,** 286, 291–292
resting membrane potential, **158,** 281
Electric potential, symplast, **142,** 265–269, 271–273
Electrocardiogram, and NMR, **145,** 31
Electroelution, preparation of nuclear matrix DNA, **162B,** 424–425
Electrofusion, and micromanipulation of single gametes, **140,** 416–418
Electrogenesis
intracellular calcium messenger system and, **142,** 316, 318
symplast and, **142,** 299
electrophysiological structure, **142,** 267–273
integration of activity, **142,** 286, 298
proton pumps, **142,** 275–276
Electrolytes, NMR analysis, **145,** 19
Electromagnetic field, *see also* Electric field; Magnetic field
biological system selection, **158,** 324–325
cellular effects
acetylcholinesterase, **158,** 317–318
adenylate cyclase, **158,** 316–317
ATP levels, **158,** 318
calcium flux, **158,** 282, 296–299, 304, 322
cell cycle, **158,** 314, 322–323
chemical reaction rates, **158,** 284–285
chick embryo, **158,** 294–295, 305–307, 324
choline *O*-acetyltransferase, **158,** 318
cyclic AMP levels, **158,** 308, 316–317
cyclotron resonance model, **158,** 287
cytochrome-*c* oxidase, **158,** 319
decoupling of triplet states, **158,** 285
DNA synthesis, **158,** 305–306
dose response, **158,** 280, 286
effect of cell culture state, **158,** 294
evidence for effects, **158,** 280–281
experimental controls, **158,** 292
foci formation, **158,** 304–305
frequency response, **158,** 286, 296–299
genotoxicity, **158,** 320–321
immune system, **158,** 321–322
morphology, **158,** 304
myoblasts, **158,** 303
myosin light chain kinase, **158,** 317
neurites, **158,** 303–305
ornithine decarboxylase, **158,** 314–315
phosphatidylinositol pathway, **158,** 282–283
plasma membrane
adenosine deaminase rotation, **158,** 305

concanavalin A receptor distribution, **158,** 303
conductivity, **158,** 299
depolarization, **158,** 304
glucosamine incorporation, **158,** 304
hydrophobicity, **158,** 301
membrane order, **158,** 300
permittivity, **158,** 299–300
potassium efflux, **158,** 304
proliferation rate, **158,** 307, 322
protein kinase C, **158,** 315–316
protein profile, **158,** 312–313
reproducibility of data, **158,** 293–294, 325
respiration rate, **158,** 293, 323
RNA polymerase, **158,** 313–314
sodium, potassium-ATPase, **158,** 284, 299, 319
transcription, **158,** 294, 308–312, 324
exposure systems
design, **158,** 290
electrodes, **158,** 291
harmonics, **158,** 289
magnetic coils, **158,** 290
representative environmental exposures, **158,** 288
sham-exposure, **158,** 290–291
thermal regulation, **158,** 289–290
uniformity of field, **158,** 288–289
tumor hyperthermia therapy, **158,** 280
Electron density
avian salt glands and, **144,** 139, 159
cytoskeletal dynamics and, **144,** 96
karyosphere and, **144,** 21, 38
Electron microscopy
in analysis of calcifications, **172,** 137–138
animal mtDNA and, **141,** 202
arabinogalactan proteins, **174,** 234
in autoradiography of protein secretion dynamics, **160,** 275, 277
avian salt glands and
blood supply, **144,** 148–149
innervation, **144,** 158, 163, 167
receptive systems, **144,** 180, 182
secretory tissue, **144,** 133, 143–144
bacterial flagellar motor and, **147,** 100, 105–106
calcium channels and, **147,** 210, 214–217, 221, 227
inhibitors, **147,** 207
location, **147,** 203
motor end plate, **147,** 195, 199
cell wall, **146,** 305
chloroplasts and, **144,** 268, 286
ciliates and, **141,** 4, 8
cnidarian nervous system anatomy studies, **167,** 40
environmental stimuli in plant cells and, **145,** 266, 277, 284
glutamate metabolism and, **147,** 290
higher plant mitochondrial genomes and, **141,** 135, 147
karyosphere and, **144,** 7
oocytes of insects, **144,** 8, 13–17, 19, 21–25
vertebrates, **144,** 37
melanosomes containing eumelanin and pheomelanin, **146,** 193
microtubule-organizing center, **175,** 96
microtubules, **175,** 106–107
of Peyer's patches
of blood vessels supplying, **167,** 95
M cell identification using, **167,** 99–100
plant cell wall architecture, **173,** 160–162
proteins in striated muscle and, **142,** 128–129
antibody use, **142,** 101
cardiac muscle, **142,** 107, 112
sarcomere, **142,** 97
thick-filament proteins, **142,** 62, 64, 67, 78, 80, 82
thin-filament proteins, **142,** 84, 88, 91
radial Müller glial cells, **146,** 16
regenerative cementogenesis and, **142,** 13, 32
retinal precursor cells and, **146,** 151, 157, 164, 169, 182–183
Sertoli-germ cell communication network and, **147,** 28, 34
studies with, **140,** 407–408
in study of microgravity effects on statocytes, **171,** 11
symplast and, **142,** 270
transdifferentiation in medusae and, **142,** 222
Electron spectroscopy, in analysis of calcifications, **172,** 139
Electron transfer
endosymbiont hypothesis and, **141,** 282, 309

Electron transfer (*continued*)
higher plant mitochondrial genomes and, **141,** 132
kinetoplastid mtDNA and, **141,** 80
Electron transport
higher plant plastids and, **145,** 150, 191, 193
intracellular calcium messenger system and, **142,** 329, 331–332
Electrophoresis
calcium channels and, **147,** 201
ciliates and, **141,** 3, 6, 36
cytoskeletal dynamics and, **144,** 105, 109
eggshells and biomineralization and, **145,** 221
environmental stimuli in plant cells and, **145,** 263
GnRH-encoding gene in teleosts and, **147,** 185
higher plant mitochondrial genomes and, **141,** 136
hormonal control of amphibian morphosis and, **145,** 116, 129, 133, 140
intracellular calcium messenger system and, **142,** 310–311
karyosphere and, **144,** 40
kinetoplastid mtDNA and, **141,** 66
mitochondrial genomes in fungi and, **141,** 90
Sertoli-germ cell communication network and, **147,** 62
spontaneous endogenous, analysis of instability of homogeneous state, **154,** 339–342
transcriptional regulation and, **143,** 2
urease in plant cells and, **145,** 85, 88–89, 89
Electrophysiology
Sertoli-germ cell communication network and, **147,** 39
symplast structure, **142,** 265–274, 297
Electrostatic interactions
in actin filament bundle formation *in vitro,* **175,** 57–58
bacterial flagellar motor and, **147,** 99, 131–135, 142–144
calcium channels and, **147,** 203
Electrostimulation
induction of fast-to-slow muscle fiber type transitions, **170,** 163–170
induction of slow-to-fast fiber type transitions, **170,** 177–178
Elemental characterization, **140,** 282–283
Elicitin, systemic acquired resistance induction, **168,** 324–325
Elicitors, *see also* Inducers
in accessibility induction, **172,** 58
activation of plant defense genes, **172,** 70
chalcone synthase and, **147,** 265, 268–271
effects on cell wall-bound ATPase, **172,** 78
effects on peroxide generation, **172,** 80–81
effects on polyphosphoinositide metabolism, **172,** 71–73
interactions with suppressors, **172,** 65–67
in *Mycosphaerella,* **172,** 62–63
in phytoalexin theory, **172,** 57
pisatin accumulation and, **172,** 65–66
stimulation of membrane-bound protein phosphorylation, **172,** 74–75
in suppression of plant defenses, **172,** 84
Ellipticine, **156,** 79, 81, 97, 101
Elliptio complanatus, dynein–tubulin sliding in, **173,** 17–19
Elodea, environmental stimuli in, **145,** 252, 260–261, 263
Elodea canadensis, environmental stimuli in, **145,** 252–253, 260
Elongation
animal mtDNA and
regulation, **141,** 226
transcription, **141,** 221–222
avian salt glands and, **144,** 135
cytoskeletal dynamics and, **144,** 108, 113
endosymbiont hypothesis and, **141,** 240, 293, 307, 315
regenerative cementogenesis and, **142,** 11–12
retinal precursor cells, **146,** 155–157, 160, 163–165
symplast and, **142,** 265, 299–300
auxin, **142,** 277–280, 283–285
electrophysiological structure, **142,** 265–273
integration of activity, **142,** 286–287, 294, 296–299
proton pumps, **142,** 273–277
Elongation factor-1α
bundle formation
F-actin interactions, **164,** 23–24
regulation by Ca^{2+}/CaM, **164,** 24

in *Euplotes eurystomus* replication band, **164,** 23
microtubule-severing activity, **164,** 279
and molecular evolution, **164,** 29
multifunctionality, **164,** 25–27
properties, **164,** 20–22
and protein synthesis, **164,** 28–29
Elongation factors
cytoskeletal dynamics and, **144,** 107
endosymbiont hypothesis and, **141,** 242, 248, 278, 290, 311–313
structure, **159,** 22
Embryogenesis
ascidian
fate restriction, **176,** 281–286, 297–299
larval structure, **176,** 246–249
stage of determination, **176,** 281–286
time frame, **176,** 249–250
cadherin, **165,** 181–186
developmentally regulated expression, **165,** 182–183
selective cell adhesion, **165,** 181–182
cadherin-mediated cell-cell adhesion, requirements, **165,** 184–186
desmosomal cadherin, morphoregulatory roles, **165,** 180–181
in *Drosophilia,* **139,** 123–125
platelet-derived growth factors in
activity, **172,** 110–117
distribution, **172,** 104–110
somatic, role of AGPs, **174,** 258–260
vertebrate, RPE development during, **171,** 227–228
Embryonic cells
FAK and
induced deficiency of, in mice, **167,** 177–178
Xenopus nervous system, expression in, **167,** 168
FRNK expression in, **167,** 170, 171
sac, angiosperm plant, **165,** 104–105
Embryonic induction
artificial, **150,** 410–411
associated field, definition, **150,** 411–412
competence
definition, **150,** 380–382
development, **150,** 382–386
definition, **150,** 387–389
differentiation waves, **150,** 414
embryonic field, definition, **150,** 411–412
gastrulation movements, **150,** 373, 397–404
and genetics, relationship, **150,** 375–380
homoiogenetic, **150,** 408–409
inducing tissue role, **150,** 406–407
interactions, **150,** 413
mechanism, **150,** 373–374
morphogenetic movements, **150,** 411
morphogens, definition, **150,** 409–410
neurulation movements, **150,** 405–406
occurrence, **150,** 389–391, 414
polarity, definition, **150,** 413
positional information, definition, **150,** 412
pregastrulation movements, **150,** 391–397
regional, **150,** 407–408
regulation, definition, **150,** 409
related to specific differentiation, **150,** 408
responding tissue role, **150,** 407
Embryos
ascidian
anterior–posterior specification, **176,** 287–291
brain induction, **176,** 278–279
fate restriction, **176,** 281–286, 297–299
instructive induction, **176,** 286
notochord induction, **176,** 272–278
permissive induction, **176,** 286
role of cytoplasmic determinants
in cleavage pattern generation, **176,** 268–272
in endoderm cell development, **176,** 263–264
in epidermis development, **176,** 264–267
in gastrulation movements, **176,** 267
in muscle development, **176,** 257–263
in ooplasmic segregation, **176,** 253–255
in tissue differentiation, **176,** 255–257
sensory pigment cell induction, **176,** 279–281
stage of determination, **176,** 281–286
trunk lateral cell induction, **176,** 281
cells, time of differentiation, **148,** 81–82
development
avian retina
cellular processes, **146,** 23, 28, 39

Embryos (*continued*)
normal development, **146,** 2, 5, 9–10
retinogenesis *in vitro,* **146,** 21
extracellular matrix and, **146,** 295, 299
growth factor-induced retinal regeneration and, **146,** 49, 51, 69
fibroblast growth factors, **146,** 67–69
induction by growth factors, **146,** 60–67
ocular tissues, **146,** 57–58
transdifferentiation, **146,** 54–55
mechanical forces, **150,** 175–176
melanocytes and
biology, **146,** 194–196
genetics of melanoma, **146,** 227
pigment-cell genetics, **146,** 199, 214–215, 217, 219–222
photoreceptor default hypothesis, **146,** 167–168, 173
retinal precursor cells and, **146,** 149, 174
chick embryo cells, **146,** 151–155
experimental analysis, **146,** 157, 164–165
microenvironmental factors, **146,** 177–180, 182
tumorigenesis in transgenic mice, **164,** 74–76
distribution of competence, **176,** 286
Drosophila, equivalence group, **176,** 295–296
eggshells and biomineralization and, **145,** 217, 223, 227, 238
higher plant plastids and, **145,** 164, 180–181
hormonal control of amphibian metamorphosis and, **145,** 106, 115, 118, 123, 140–141
incidence of aneuploidy in
preimplantation, **167,** 267–268
spontaneously aborted, **167,** 265–267
karyosphere and, **144,** 11, 38–39, 42
microgravity experiments, **156,** 334–338, 356
microspore-derived, and gene transfer in microspores, **140,** 427–431
preimplantation
diagnostic procedure, **153,** 2
sexing of, **153,** 22–25
preimplantation diagnosis of inherited genetic defects, *see* Genetic defects, preimplantation diagnosis
sac
and component cells, isolation, **140,** 269–280
in angiosperms, methods for, **140,** 276–279
assessment of viability and quality in, **140,** 271–280
characterization, **140,** 281–284
technique and protocols, **140,** 270–271
delivery of male gamete in, **140,** 370–372
and gamete analyses, **140,** 573
isolation and pistil receptivity
embryo sac and gamete analyses, **140,** 573
pistil receptivity and *in vitro* pollination, **140,** 571–572
tissue, mechanical stress
absolute stress value range, **150,** 7–8
amphibians
cell surface angle measurement, **150,** 3–7
morphogenetic effects, **150,** 8–11
neural tubes affected by relaxation of stress, **150,** 10–11
relaxation of tensile stresses, **150,** 9
tissue dissection as detection method, **150,** 2–5
tissue stretching, **150,** 8
cell columnarization models, **150,** 23–24
hyperrestoration of preexisting stress hypothesis, **150,** 24–25
morphogenetic effects
in amphibians, **150,** 8–11
qualitative reconstruction of process, **150,** 29–31
morphogenetic role, **150,** 1–2
morphomechanical situations
intercalation capabilities, **150,** 27–28
intercalation capacities, lack of, **150,** 25–27
stretching–contraction cell relay, **150,** 28–29
superimposed cell sheets, **150,** 28
ontogenetic processes, **150,** 31–32

stress-force feedback
cell columnarization models, **150,** 23–24
definition, **150,** 2
surface tension
cellular level, **150,** 19, 22–23
mechanisms, **150,** 11–12
subcellular level
relaxation of tension, **150,** 12–19
stretching, **150,** 19–21
tissue level, **150,** 19, 22–23
urease in plant cells and, **145,** 98
biochemical genetics, **145,** 84–86
elimination, **145,** 80–84
metabolic origins, **145,** 67, 69, 71–72, 77
nickel metabolism, **145,** 92–93, 95
in vitro fertilized, sampling, **153,** 4
Enamel
amelogenesis imperfecta in, **172,** 152
caries in, **172,** 148–149
crystal/protein interactions in, **172,** 142
crystal structure and composition, **172,** 139–142, 165–166
fluorosis in, **172,** 151–152
topical fluoride treatment, **172,** 157
Enameloids, fish, crystal structure, **172,** 145
Encapsulation, vertebrate cell culture technology and, **142,** 177–178, 197
End labeling
in situ, **163,** 121–123
TdT-mediated dUTP-biotin nick, **163,** 123
Endo-1,4-β-glucanase, role in cell wall changes under microgravity, **170,** 61–63
Endo B, expression by antisense RNA, **154,** 64
Endochromosome, polytene chromosome, **165,** 94–95
Endocrine system
mechanisms, hormonal control of amphibian metamorphosis, **145,** 105, 141
organs, thymic epithelial cell function, **156,** 180–186
tumor, parathyroid hormonerelated protein, **166,** 261
Endocytic compartments, associated Rab, **176,** 23–25
Endocytosis
antigens by M cells
M cell surface irregularities and, **167,** 114
as predominant method of transport, **167,** 117–118
as stage in transcellular transport, **167,** 126–128
CFTR regulation of, **172,** 222
cyclic AMP regulation of, **172,** 221–222
early events, role of Rab proteins, **176,** 36–39
extracellular matrix, **146,** 312–313
insulin and insulin receptors, **173,** 249–250
late pathway, role of Rab7, **176,** 39–40
in plants
coated vesicles, **167,** 2–5
receptor-mediated, **167,** 19
regenerative cementogenesis and, **142,** 40
Sertoli-germ cell communication network, **147,** 58
sperm maturation role
adsorptive endocytosis
epithelial cells of efferent ducts, **154,** 137–141
rete epithelial cells, **154,** 131–134
Sertoli cells, **154,** 117–118
epididymal clear cells, **154,** 170–177
epithelial cells of efferent ducts
quantitative studies, **154,** 145–146
sulfated glycoprotein-1, **154,** 139, 143–145
sulfated glycoprotein-2, **154,** 139, 142
fluid-phase endocytosis
epithelial cells of efferent ducts, **154,** 137–141
rete epithelial cells, **154,** 131–134
Sertoli cells, **154,** 116–118
germ cells, **154,** 128–130
immobilin, **154,** 171, 175–176
process, **154,** 105
in Sertoli cells
receptor-mediated at base, **154,** 120–122
secondary lysosomes, **154,** 118–120
sulfated glycoprotein-1, **154,** 134–135

Endocytosis (*continued*)
sulfated glycoprotein-2, **154,** 134–135
transferrin, **154,** 135–137
vas deferens, **154,** 177–178
transdifferentiation in medusae and, **142,** 232
in vacuolar biogenesis in *Saccharomyces cerevisiae,* **139,** 79–81
in yeast, regulation by Rab GTPase, **176,** 27–28
Endoderm
dorsal, Hensen's node origin, **175,** 255
transdifferentiation in medusae, **142,** 256
isolation of tissues, **142,** 222–223
potential, **142,** 232–236
tissue organization, **142,** 218, 220
Endoderm cells, development in ascidians, **176,** 263–264
Endodesmis, environmental stimuli in, **145,** 263, 296, 300
Endo-1,4-β-glucanase, *see* Cellulase
Endoglycosidase, as inhibitor of fungi growth, **148,** 1
Endolymph, blood barrier and, glucose transport, **172,** 23–24
Endomembrane system, in pollen tube, organization
endoplasmic reticulum, **176,** 138–139
Golgi apparatus, **176,** 139–140
plasma membrane, **176,** 137–138
secretory vesicles, **176,** 140
Endomeningeal fibroblast, ependymin
electron microscopy studies, **165,** 134–135, 137, 138, 139
expression, **165,** 134–135
in situ hybridization studies, **165,** 134, 135
zebrafish embryo transient expression studies, **165,** 134
Endomeninx, **165,** 129
Endometrium
granulocytes, human pregnant uterus
immune-effector function, **153,** 124–125
nonimmunological functions, **153,** 126
properties, **153,** 105, 117, 119–121
rabbit, estrogen RBF, **162B,** 354–355
synchronization during IVF treatment, **164,** 166–168
Endomitosis
associated cycles, **144,** 239
human trophoblast, **165,** 68–69
induced polyploidization, **144,** 225
trophoblast, **165,** 68–69
Endoneurial blood vessesls, GLUT1 in, **172,** 25
Endonucleases
in apoptotic cell death, **163,** 113–114
in lens fiber DNA degradation, **163,** 142–143
Endoplasm
cytoskeletal dynamics, **144,** 87, 110
environmental stimuli in plant cells, **145,** 259, 263, 265, 284, 299
Endoplasmic reticulum
–alveolar system, *see* Cortex
avian salt glands and, **144,** 140, 171, 203
CFTR and
model of protein kinetics within, **172,** 206–207
protein folding in, **172,** 205–206
protein processing and degradation in, **172,** 204–208
retention of mutant protein in, **172,** 201–202
chloroplasts and, **144,** 268, 270, 273, 278, 291
endosymbiont hypothesis and, **141,** 329
environmental stimuli in plant cells and, **145,** 263, 289
epididymis
caput epithelium, **154,** 159, 161–163
initial segment, **154,** 154
fish chromatophores and, **143,** 230, 243
–Golgi interface trafficking, role of Rab, **176,** 30–31
hormonal control of amphibian metamorphosis and, **145,** 116
intracellular calcium messenger system and, **142,** 315–316, 318–319, 325–326
mitosis and, **144,** 219–220, 222, 225, 245
plant cell
protein transport, **167,** 7–8
V-ATPase polypeptides in, **167,** 24
in plant statocytes, effect of microgravity, **171,** 50–51
in pollen tube, **176,** 138–139
proteins in striated muscle and, **142,** 124
role of Rab1, **176,** 20–21
rough ER, **165,** 246
avian salt glands and, **144,** 139, 143

regenerative cementogenesis and, **142,** 9, 11, 19, 21
self-incompatibility and, **143,** 261, 281
Sertoli-germ cell communication network and, **147,** 41, 61
smooth, **139,** 271–272, 277–278
fish chromatophores and, **143,** 243
self-incompatibility and, **143,** 261
SNARE complex formation, role of Ypt1p, **176,** 60–62
Endopolyploidy, mitosis and, **144,** 225, 240–241
Endoprotease cleavage
in cnidarian neuropeptide biosynthesis, **167,** 78–82
in neuropeptide biosynthesis in higher animals, **167,** 57–58
Endoproteinases, *see also* Processing enzymes
in AcSDKP synthesis from thymosin β4, **167,** 223
in Antho-RFamide biosynthesis, **167,** 59, 64–67
in Antho-RPamide I biosynthesis, **167,** 68, 78–79
Endoreduplication
K-252 effect, **158,** 75
and mitosis, **144,** 225, 231, 240
β-Endorphin
analgesic potency, **159,** 47
analogs, **159,** 47
effect on pituitary cell proliferation, **157,** 48
effect on testicular macrophages, **149,** 115
precursor, **159,** 46
processing, **159,** 10
related molecules, production in lymphocytes, **170,** 83–84
role in immune response and stress
in invertebrates, **170,** 89, 93
in vertebrates, **170,** 88
Endosomes
acidification, **172,** 226–227, 230
associated insulin, role in biological effects, **173,** 253–258
CFTR recycling and, **172,** 215–220
compartment
controversy over existence in plant cells, **167,** 5
M cell, transport of endocytic vesicles to, **167,** 127
early Rab, **176,** 23–24
–Golgi communication, role of Rab9, **176,** 32–33
recycling, role of Rab11, **176,** 39
vitellogenin, **166,** 110–111
Endosperm
higher plant plastids and
carbohydrate oxidation, **145,** 201, 203
fatty acid synthesis, **145,** 183, 189
metabolism, **145,** 169, 180
structure, **145,** 151
urease in plant cells and, **145,** 68
Endosymbiont hypothesis
chloroplast, **165,** 208
forms of proof, **141,** 236–237
future prospects, **141,** 331–333
higher plant plastids, **145,** 164–166
lineages of life
archaebacteria, **141,** 240–244
rRNA, **141,** 237–239
universal tree, **141,** 244–246
mitochondria, **141,** 280–281, 326–329; **165,** 208
gene expression, **141,** 298–308
gene structure, **141,** 308–319
gene transfer, **141,** 324–325
genome organization, **141,** 288–298
molecular biology, **141,** 281–288
Plasmodium, **141,** 325–326
rRNA sequences, **141,** 319–324
nuclear genome
eukaryotes, **141,** 251–253
nucleus, **141,** 247–251
organelles, **141,** 329
flagella, **141,** 330–331
hydrogenosomes, **141,** 330
peroxisomes, **141,** 329–330
plastids, **141,** 253–254, 278–279
alagal phylogenies, **141,** 273–275
eukaryotes, **141,** 275–277
gene transfer, **141,** 277–278
hypotheses of origin, **141,** 255, 257–258
molecular biology, **141,** 262–268
phylogenetic diversity, **141,** 254–255
secondary symbioses, **141,** 256, 258–262
sequence comparisons, **141,** 268–273
transfer of genetic material between compartments, **165,** 208–213
Endothelial cell growth factor
migration induction by, **155,** 80
vascular, *see* Vascular endothelial growth factor

Endothelial cell inhibitors
16K prolactin, **161,** 20–21
angiogenic modulator 1470, **161,** 21–22
angiostatin, **161,** 22–23
cartilage-derived inhibitor, **161,** 18
GP140/thrombospondin, **161,** 18–19
interferon-α and interferon-β, **161,** 20
placental ribonuclease inhibitor, **161,** 21
platelet factor 4, **161,** 17–18
thalidomide, **161,** 22
Endothelial cells
average volumes and surface areas, **161,** 151–154
bifunctional vascular growth factors, **161,** 15–16
collagenase content, **159,** 122
cytokine response, **169,** 173–174
development, **161,** 1–5
disease states, therapeutic strategies, **161,** 28
antibodies to growth factors, **161,** 30–31
drugs, **161,** 32
toxins, **161,** 31–32
ecto-ATPase levels, **158,** 159–160
effectors of migration, **159,** 129, 136
FAK potential suppression of apoptosis, **167,** 178
glomerular, quantity estimation in average glomeruli, **161,** 148–151
G protein role in differentiation, **159,** 125
human umbilical vein
spontaneous angiogenesis, **159,** 119–120
tube formation in Matrigel, **159,** 123, 125
kidney glomerulus, **166,** 182
–leukocyte interaction, selectin role, **175,** 183–186
migration, **159,** 114; **169,** 172–173
nitric oxide localization, **170,** 111
origin, **159,** 113–114
protease inhibition, **159,** 122
stimulators
angiogenin, **161,** 8–9
fibroblast growth factors, **161,** 6–7
hepatocyte growth factor, **161,** 13
platelet-derived endothelial cell growth factor, **161,** 14
vascular endothelial growth factor, **161,** 7–8
vascular, cytokine production and, **146,** 80
wound healing role, **169,** 172–174
Endothelial cell tissue factor, regulation, **152,** 64–73
Endothelial leukocyte adhesion molecule-1, in selectin family, **175,** 184–186
Endothelin
cDNA cloning, **171,** 268
effects
on adrenal blood flow, **171,** 284–285
on adrenal cortex growth
cell hypertrophy and steroidogenic capacity, **171,** 286–288
cell proliferation, **171,** 288–289
mechanism, **171,** 289–290
role of ET receptor subtypes, **171,** 290–291
on arachidonic acid metabolism, **171,** 281–282
on catecholamine release by adrenal medulla, **171,** 282–283
on cytosolic Ca^{2+} concentration, **171,** 279–280
on intraadrenal renin angiotensin system, **171,** 284
on intramedullary regulatory peptides, **171,** 283
on ouabain-sensitive Na^+/K^+-ATPase, **171,** 281
on phospholipase A_2, **171,** 281–282
on phospholipase C, **171,** 279–280
on secretory activity
interrenal cells, **171,** 278–279
zona fasciculata reticularis cells, **171,** 277–278
zona glomerulosa cells, **171,** 275–277
on thyrosine kinase, **171,** 280
family members, **171,** 267–268
gene expression studies, **171,** 269
immunohistochemical studies, **171,** 269–270
release, affecting factors
adrenal blood flow, **171,** 291–292
natriuretic peptide, **171,** 293
nitric oxide, **171,** 292–293
roles
in aldosterone-producing tumors, **171,** 293–294

in idiopathic hyperaldosteronism, **171,** 294
in tumors, **171,** 295
secretagogue action, role of ET receptor subtypes, **171,** 285–286
synthesis in adrenal gland, **171,** 268–269
Endothelin-1, effect on arterial smooth muscle cell cultures, **169,** 215–216, 223
Endothelin receptor subtypes
autoradiographic studies, **171,** 273–274
gene expression studies, **171,** 270–271
immunohistochemical studies, **171,** 273
roles
in ET growth action, **171,** 290–291
in ET secretagogue action, **171,** 285–286
in idiopathic hyperaldosteronism, **171,** 294
saturation and displacement binding studies, **171,** 271–273
Endothelins
astrocyte reaction to, **148,** 127–128, 133–134
production by neuroendocrine cells, **157,** 285
Endothelium
avian salt glands and, **144,** 138–139, 149, 151, 160
in blood-tissue barriers, **172,** 6–8
cytokines and, **146,** 77, 80
growth factor-induced retinal regeneration and, **146,** 59
retinal precursor cells and, **146,** 153, 178
Endoxyloglucan transferase
azuki bean
enzymatic reaction, **173,** 173–174
localization, **173,** 181–182
pH dependency, **173,** 176–177
substrate specificity, **173,** 174–176
fluorescence detection, **173,** 177–179
identification, **173,** 172–173
mediated molecular grafting, **173,** 194–195
mediated molecular grafting *in muro,* **173,** 179–180
purification, **173,** 172–173
xyloglucan disproportioning reaction, **173,** 179
xyloglucan endotransglycosylase activity, **173,** 177
Endozepines, **138,** 319–324
End-plate potential, calcium channels and, **147,** 194, 203–206, 209
Energy
cytoplasmic streaming requirements, **155,** 112–114
metabolism, related muscle fiber transitions, **170,** 152–153
Enhancer blocking assays, and domain boundary definition, **162A,** 428–430
Enhancers
cohabitation with S/MARs, **162A,** 433–435,442
Drosophila muscles and, **143,** 65, 80, 117
replication, model, **162A,** 315–316
transcriptional
cohabitation with MARs, **162A,** 345–346
MARs, **162A,** 309–312
transcriptional regulation and
function, **143,** 5–6
genes, **143,** 13
proteins, **143,** 45
viruses, **143,** 41, 43–44
vitamin D responsive element function as, **162A,** 261–263,266–269
Enkephalins
astrocyte reaction to, **148,** 140–141
evolution of precursors, **159,** 10
neurons, **159,** 47
processing, **159,** 9
Enod12 protein, **156,** 34
Enol-pyruvoylshikimate phosphate synthase, flowering plants, **143,** 312
Enterocytes, dome epithelial
brush border inhibiting binding of bacteria to, **167,** 123
controversy over M cell development from, **167,** 108
M cells sharing characteristics of, **167,** 101–102
in rabbit caecum, **167,** 113*f*
role, **167,** 130–131
Enteroendocrine cells, absence from dome epithelia of mice, **167,** 130
Enteroglucagon, fish, **168,** 190
Enteropathogenic microorganisms, M cells as targets for, **167,** 138–139
Enterotoxins
binding to M cells and enterocytes, **167,** 139

Enterotoxins (*continued*)
immune response *vs.* tolerance induction by, **167,** 142
Entrainment, biological clocks in retina and *Bulla,* **146,** 129, 134
concepts, **146,** 96–98
efferent regulation, **146,** 108–115
pathway, **146,** 100–103, 115
protein synthesis, **146,** 106–108
second messengers, **146,** 103–106
Environment
effect on plant cell walls, **170,** 39–41
higher plant plastids and, **145,** 176
signals, in XRP gene regulation, **173,** 191–192
stimuli in plant cells, **145,** 251–252, 298–301
chemicals, **145,** 297–298
light
chloroplasts, **145,** 268–275
nucleus, **145,** 287–288
transduction chain, **145,** 275–287
low temperature, **145,** 288–290
photodinesis, **145,** 252–259
motile systems, **145,** 259–263
signal transduction, **145,** 263–268
wounding
higher plants, **145,** 290–294
siphonous green algae, **145,** 294–296
in vivo effects on RPE differentiated states, **171,** 243–245
Environmental stress, analysis with *in vitro* pollination
conclusions, **140,** 402–403
heat shock response in reproductive tissues, **140,** 400–402
influence of temperature stress at different phases of *in vitro* reproductive process, **140,** 397–400
kernel development, **140,** 399
pollen selection, **140,** 399–400
postpollination, **140,** 397–398
stress on reproductive tissues, **140,** 398–399
stress responses to high temperature, **140,** 391–392
temperature stress on *in vivo* reproductive process, influence of, **140,** 392–394
control of male and female quality, **140,** 396–397
in vitro pollination in maize, **140,** 394–396
Enzymes, *see also specific enzymes*
animal mtDNA and, **141,** 174, 180, 210, 230
regulation, **141,** 226
replication, **141,** 224–225
Arabidopsis thaliana and, **144,** 72–73
associated with lignin biosynthesis, **176,** 107–108
avian salt glands and, **144,** 159, 189, 193
secretory mechanism, **144,** 169, 171–172
secretory tissue, **144,** 138, 144–145
branching, higher plant plastids and, **145,** 177
ciliates and, **141,** 2–3, 6, 9, 55
cnidarian neuropeptide resistance to
L-3-phenyllactyl group providing, **167,** 49
proline and Leu-Pro-Pro N-terminal sequence providing, **167,** 68
pyroglutamyl N-terminal group providing, **167,** 48
X-Pro sequences on N-terminals, **167,** 50
concentration into clathrin-coated vesicles in plants, **167,** 19
dome epithelial enterocytes and activity of digestive, **167,** 131
Drosophila muscles and, **143,** 127
eggshells and biomineralization and, **145,** 231, 241, 244
endosymbiont hypothesis and
mitochondria, **141,** 300–302, 305, 311–312, 315
plastids, **141,** 266
environmental stimuli in plant cells and, **145,** 261, 266
fish chromatophores and, **143,** 241, 243, 246–247
flowering plants and, **143,** 297–298, 325
amino acid biosynthesis, **143,** 311–313, 315–317
amino acid pathways, **143,** 319–320, 322–324
ammonium, **143,** 303–307
nitrate, **143,** 300–302
higher plant mitochondrial genomes and, **141,** 140–141, 161

higher plant plastids and, **145,** 2–4, 149
amino acid synthesis, **145,** 191–194, 196–198, 200–201
carbohydrate oxidation, **145,** 201–203
CO_2 fixation, **145,** 172–176
fatty acid synthesis, **145,** 182–185, 187, 189
starch synthesis, **145,** 177, 179, 182
structure, **145,** 153, 161
translocation, **145,** 170
hormonal control of amphibian metamorphosis and, **145,** 123–124
hydrolytic, in AGP biosynthesis and degradation, **174,** 239–240
inhibitors, sex hormones and, **139,** 8
intracellular calcium messenger system and, **142,** 305, 314, 338
calmodulin, **142,** 329–330, 332
G proteins, **142,** 308–309
phosphatidylinositol turnover, **142,** 321–324, 328
protein kineses, **142,** 332–336
kinetoplastid mtDNA and, **141,** 66, 69, 80
M cell lysosomal, **167,** 128
metabolic, muscle fiber types based on, **170,** 148–150
mitochondrial genomes in fungi and, **141,** 90, 105, 121
mitosis and, **144,** 250–251
nuclear magnetic resonance and, **145,** 33
perinatal exposure to sex hormones and antihormones and, **139,** 8, 28–29
in plant cell wall, **171,** 100
proteins in striated muscle and, **142,** 66–68, 120–121
regenerative cementogenesis and, **142,** 15, 22, 37
role in microgravity effects on plant cell walls, **170,** 60–64
salt tolerance, **165,** 13–14
self-incompatibility and, **143,** 262, 278, 286, 289
transcription, **141,** 222
transcriptional regulation and, **143,** 1
transdifferentiation in medusae and, **142,** 236–238, 244–245, 247, 249–250, 257
urease in plant cells and, **145,** 65, 84, 98–99
biochemical genetics, **145,** 86, 89
metabolic origins, **145,** 66, 72, 76
Eocytes, endosymbiont hypothesis and, **141,** 241, 243, 249
Eoplasts, higher plant, **145,** 150–151
Ependymal cell, ependymin, granular cells, **165,** 140–142, 143–144
Ependymin
adhesive properties, **165,** 149–150
antiadhesive properties, **165,** 149–150
blood, **165,** 147
blood-brain barrier, **165,** 121–151
brain fluid, distribution, **165,** 144–147
calcium-binding transition, **165,** 127–128
cellular analysis, **165,** 128–147
cerebrospinal fluid, **165,** 144–146
choroid plexus, **165,** 143–144
complex localization pattern, **165,** 137–139
blood vessels, **165,** 139, 141
extracellular matrix, **165,** 139, 141
fibroblast, **165,** 139, 140
inner endomeningeal layer, **165,** 139, 140
intermediate layer cells, **165,** 139
macrophages, **165,** 139, 140
conformational transition, **165,** 127–128
endomeningeal fibroblast
electron microscopy studies, **165,** 134–135, 137, 138
expression, **165,** 134–135
in situ hybridization studies, **165,** 134, 135
zebrafish embryo transient expression studies, **165,** 134
ependymal cell, granular cells, **165,** 140–142, 143
extracellular matrix protein, **165,** 148–150
functional aspects, **165,** 147–150
gene structure, **165,** 122–123
glycosylation, **165,** 125–126
leptomeninges, as secretory organ, **165,** 147–148
lymphocyte, granular cells, **165,** 140–142, 143
mammalian homologues, **165,** 128
mature secretory protein characterization, **165,** 125–127
microheterogeneities, **165,** 126–127
modulation of, **165,** 150
molecular mass determination, **165,** 126
neuronal cell, **165,** 144
perimeningeal fluid, **165,** 146

Ependymin (*continued*)
periventricular zone, granular cells, **165,** 140–142
precursor sequence comparison, **165,** 124
precursor structure, **165,** 123–125
Reissner's fiber, **165,** 143–144
saccus dorsalis, **165,** 143–144
saccus vasculosus, **165,** 143–144
teleost fish
inner endomeningeal layer, **165,** 133–134
intermediate endomeningeal layer, **165,** 129–132
meninges, **165,** 128–134
outer endomeningeal layer, **165,** 129
Ephedra
double fertilization in
basic features of reproduction, **140,** 328–333
early reports of intra-archegonial double fertilization in, **140,** 333–337
fate of second fertilization product in, **140,** 339–342
recent studies of fertilization in, **140,** 337–339
evolutionary relationships of double fertilization in, **140,** 350–354
Ephestia kuehniella
kinetic chromosome organization, **152,** 9–11
meiosis
female, **152,** 14
male apyrene, **152,** 24, 26–27
male eupyrene, **152,** 16–19, 21
microtubules, post-translational modifications, **152,** 30–32
structural features, **152,** 3–8
Epidemiology, melanoma, **146,** 225–226
Epidendrum radicans, effect of microgravity
on leaf photosynthetic cells, **171,** 36–39
on shoot primary structure, **171,** 36
Epidermal cells
environmental stimuli in plant cells and, **145,** 275–277
increased mitotic activity in damaged, **167,** 187–188
MIP-1α transcripts in, **167,** 208
weightlessness effect on external wall structure, **170,** 51–54
Epidermal chromatophores, cytophysiology of, **143,** 193, 195, 199, 236
Epidermal growth factor
activation of protein kinase C, **156,** 317
effect on arterial smooth muscle cell culture proliferation, **169,** 212–213
effect on nuclear transport, **151,** 214
effect on pituitary cell proliferation, **157,** 52–53
induction of ATPase, **158,** 174
melanocytes and, **146,** 204, 206, 219, 239–241
migration induction by, **155,** 67–70
neurite initiation and elongation, **153,** 254–255
perinatal exposure to, **139,** 5–7
precursor homology domain, **166,** 115–116
regenerative cementogenesis and, **142,** 40
retinal precursor cells and, **146,** 173, 176–178
role in apoptotic processes, **163,** 118
self-incompatibility, **143,** 269
Sertoli-germ cell communication network and, **147,** 78
synthesis in arterial smooth muscle cell culture, **169,** 222–223
Epidermal growth factor-like growth factor, heparin-binding
gene expression, **161,** 11
stimulation of vascular cells, **161,** 10–12
structure, **161,** 11–12
Epidermal growth factor receptor
family members, **160,** 244
ligands, **160,** 245
Epidermal pentapeptide, inhibition of keratinocyte proliferation, **167,** 27
Epidermis
in ascidian larva, **176,** 264–267
Drosophila larva, genetic control
antennal cuticle, **174,** 37
clypeolabrum, **174,** 35–36
dorsal ridge, **174,** 40
intercalary cuticle, **174,** 37–38
labial cuticle, **174,** 39–40
mandibular cuticle, **174,** 38–39
maxillary cuticle, **174,** 39
ocular cuticle, **174,** 36–37
required genes, **174,** 33–35
Drosophilia
embryogenesis and segmentation and, **139,** 123–125

  en gene in, **139,** 128–134
  ptc gene in, **139,** 130–134
  segment polarity genes and, **139,** 126–128
  wg gene in, **139,** 128–134
 insect, form diversity, **174,** 8
Epidermis bullosa simplex, intermediate filament protein disturbances characterized by, **154,** 54–55
Epididymis, sperm maturation
 cell types and functions, **154,** 146–177
 endocytosis, epithelial clear cell role
  basal cells, **154,** 177
  immobilin, **154,** 171, 175–176
  narrow cells, **154,** 176–177
  postnatal development, **154,** 173–175
 secretion by principal cell 147-156
  caput epididymis, **154,** 159–169
  cauda epididymis, **154,** 159–169
  corpus epididymis, **154,** 159–169
  immobilin, **154,** 157–158
  intermediate zone, **154,** 157, 159
  postnatal development, **154,** 169–170
  SPG-2, **154,** 155, 157, 164, 166
 sperm membrane in, modification during migration, **154,** 178–184
Epifagus virginiana, plastids, **145,** 158, 161
Epifluorescence microscopy
 pseudo-2D images, **162A,** 49–50
 in whole-chromosome hybridization, **153,** 184–187
Epigenesis
 instability of homogeneous state as source, **154,** 309
 and preformation, antithesis between, **154,** 313–322
Epilepsy, avian reflex
 role of neuronal circuits, **175,** 291–293
 symptom dissociation, **175,** 294–297
 transfer, **175,** 294
Epinephrine
 avian salt glands, **144,** 167, 200
 fish chromatophores, **143,** 197, 243
Epiphyseal growth plates, histological regions, **159,** 279
Epiphysis cerebri, structure, **159,** 27
Epiplasm, *see* Cortex
Epipodophyllotoxins, **156,** 79, 81, 97, 100
Episomal theory, higher plant plastids and, **145,** 164–165
Epith, **167,** 78–82
Epithelial cells
 apical polar domains, **165,** 246
 basement membrane contact, **160,** 223
 basolateral polar domains, **165,** 246
 CFTR and
  chloride channel activity, **172,** 208–209
  localization in, **172,** 208–209, 216–217
 creating *Hydra* consisting only of, **167,** 39
 differentiation
  carcinoma prognosis, **160,** 231
  embryogenesis, **160,** 222, 224
  intestine, **160,** 225–226
  liver, **160,** 226
  lung, **160,** 225, 227
  pancreas, **160,** 227–228
 FAK *in vivo* function, **167,** 177
 FAK potential suppression of apoptosis of, **167,** 178
 glomerular
  average volumes and surface areas, **161,** 151–154
  quantity estimation in average glomeruli, **161,** 148–151
 hepatic, liver-regulating protein, in Sertoli-germ cell communication, **147,** 72
 inhibitory effect of TGF-β, **167,** 209
 intestinal, functions, **167,** 92
 ions and cell volume, **160,** 110
 junctions, **160,** 222–223
 kidney glomerulus, **166,** 182
 leaky epithelia and hyperosmolarity, **160,** 105–108
 lens
  central zone, **163,** 312–313, 323, 326
  epithelial–fiber cell interface, **163,** 343
  intercellular junctions, **163,** 337
  polarization, **163,** 326
  pregerminative zone, **163,** 312–313, 326
  transitional zone, **163,** 312–313
  ultrastructural characteristics, **163,** 323–326
 nonubiquitous Rab expression, **176,** 18–20
 polarity
  animal, **166,** 8–9
  role of Rab proteins, **176,** 42–44
 proliferation
  chemotherapy and inhibitors of, **167,** 241–242
  colonic, pGlu-His-Gly as inhibitor, **167,** 227

Epithelial cells (*continued*)
neglect of inhibitory factors in research on, **167,** 186
regulation of
MIP-1α and, **167,** 208
mouse ear model for, **167,** 187–188
rete testis, structure and function, **154,** 131–137
retinal pigment, *see* Retinal pigment epithelial cells
thymic epithelial cells, **156,** 159–191
tyrosine kinase receptor expression, **160,** 224, 230
water transport
homeostasis, **160,** 105–108
ion cotransport system
dark cell epithelia, **160,** 127–129
leaky epithelia, **160,** 125–126
ion reuptake, **160,** 143–144
leaky epithelia
isotonic transport, **160,** 139–140
transport rate *versus* external osmolarity, **160,** 139–140
uphill transport, **160,** 138–139
molecular models, **160,** 137–146
Epithelium
avian retina and, **146,** 11, 25, 29
avian salt glands and, **144,** 149, 163, 176, 200, 205
hormonal control, **144,** 196–197
secretory tissue, **144,** 135, 138–140, 143, 146–148
in blood-tissue barriers, **172,** 6–8
cell division in, H1 histone role, **148,** 1–2
ciliated, basal foot in, **173,** 208–209
cytoskeletal dynamics and, **144,** 101, 113
dome, *see* Dome epithelium
eggshells and biomineralization and, **145,** 219, 236
grease in plant cells and, **145,** 83
growth factor-induced retinal regeneration and, **146,** 51–52, 57, 59
hormonal control of amphibian metamorphosis and, **145,** 118, 131, 133, 140
lymphoepithelium
in BALT, **167,** 136–137
in NALT, DALT, and CALT, **167,** 137–138
melanocytes and, **146,** 196, 233
neural, origin of oligodendrocytes, **175,** 285–291
neuroepithelium
avian retina and, **146,** 41
cellular processes, **146,** 29, 32, 34, 39–40
normal development, **146,** 2, 4
growth factor-induced retinal regeneration and, **146,** 50, 54, 67
melanocytes and, **146,** 194
retinal precursor cells and, **146,** 173–174, 183
neurosensory
cochlear hair cells, **173,** 210
MT–microfilament interactions, **173,** 210
photoreceptor cells, **173,** 209–210
polarized cells, vesicular transport in, **173,** 225
regenerative cementogenesis and, **142,** 2
established root surfaces *in vivo* 44- 45, **142,** 47, 50
growing root surfaces, **142,** 9, 12, 19, 27
retinal pigment, *see* Retinal pigment epithelium
retinal precursor cells and, **146,** 149, 162
Sertoli-germ cell communication network and, **147,** 25, 27, 32, 79–80
function, **147,** 38, 40–41, 57, 70, 73
germ cell depletion, **147,** 63, 65
stage-dependent changes, **147,** 58–60
and sperm maturation
epididymal clear cells, **154,** 170–177
vas deferens, **154,** 177–178
symplast and, **142,** 288–289, 291, 297
transdifferentiation in medusae and, **142,** 217–218, 220, 224, 256
vertebrate cell culture technology and, **142,** 160, 162
Epitopes, *see also* Antigens
Drosophila muscles, **143,** 136
eggshells and biomineralization and, **145,** 220, 237
extracellular matrix, **146,** 275, 285
higher plant plastids and, **145,** 168
HNK-1, regulation by cholinesterases in avian neurogenesis, **151,** 159–164
M cell
monoclonal antibody binding to, **167,** 102

providing immune system access to, **167,** 92
proteins in striated muscle, **142,** 61, 128
cardiac muscle, **142,** 111, 113
myosin, **142,** 63, 65–67, 73
sarcolemma, **142,** 124–126
sarcomere, **142,** 93–96
sarcoplasmic reticulum, **142,** 121
thick-filament proteins, **142,** 79–83
thin-filament proteins, **142,** 88, 90
Epodermolytic hyperkeratosis, **154,** 55
Epoxygenase pathway, arachidonic acid metabolism, **161,** 221
EPSPs, *see* Excitatory postsynaptic potentials
Equatorial contraction, cytoskeletal dynamics and, **144,** 93–94
Equisetum arvense, endosymbiont hypothesis and, **141,** 294
Equivalence group, lateral inhibition in, **176,** 295–296
ER, *see* Endoplasmic reticulum
ERAP160, estrogen receptor-associated protein, **162B,** 364
Eremia velox, karyosphere and, **144,** 37
Eremosphaera, environmental stimuli in plant cells and, **145,** 268, 271
Ernodesmis, cell models, **175,** 6
Erwinia, plant-pathogen interactions in, **144,** 57
Erwinia amylovara, hypersensitive response elicitors in, **148,** 11
Erysiphe cruciferarum, plant-pathogen interactions in, **144,** 65
Erysiphe f scheri, plant-pathogen interactions in, **144,** 58–59
Erythrocytes
aggregation, spaceflight observations, **156,** 320
ecto-ATPase levels, **158,** 159
hormonal control of amphibian metamorphosis and, **145,** 110, 121
intracellular calcium messenger system and, **142,** 333
membrane, intermediate filaments associations, **164,** 107–108
membrane-cytoskeletal linkage, **148,** 50
nuclear magnetic resonance and, **145,** 19
progenitors, **167,** 191
Sertoli-germ cell communication network, **147,** 34
superoxide dismutase in, **167,** 225–226
symplast and, **142,** 289
transcriptional regulation, **143,** 2, 36
Erythroid cells, TGF-β as inhibitor of precursor, **167,** 211
Erythromycin, spaceflight experiments, **156,** 341
Erythropoiesis, inhibitors for, **167,** 191
Erythrytol, **165,** 11
Escherichia coli
animal mtDNA and, **141,** 187, 189–191, 208
bacterial flagellar motor
dynamic measurements, **147,** 115–117
function, **147,** 112, 114
structure, **147,** 100, 106–107, 109
binding by sialic acids, **175,** 175–178
chaperone-assisted proteases
Clp protease, **174,** 155–157
FtsH protease, **174,** 157
La protease, **174,** 155
cyanelles, translation apparatus protein genes, **151,** 43–44
endosymbiont hypothesis and
mitochondria, **141,** 297–298, 311–312, 315
plastids, **141,** 258, 263, 265–267
enterotoxin binding to enterocytes, **167,** 139
flowering plants and, **143,** 308, 316–317, 322
higher plant mitochondrial genomes and, **141,** 132, 158, 164
higher plant plastids and, **145,** 184–185
intracellular calcium messenger system and, **142,** 311
K99-fimbriae expression, **175,** 177–178
M cells and
CS3 adhesin in binding to, **167,** 124
pilus plasmids in adherence to, **167,** 124
preferential adherence to, **167,** 123
uptake of entire, **167,** 120*t,* 122
mitochondrial genomes in fungi and, **141,** 120
mitochondrial genomes of, **141,** 38–39, 43, 47–48, 50
as model of molecular chaperones, **174,** 131–132

Escherichia coli (*continued*)
postshock cell repair, **148,** 208
salt stress, **165,** 25–27
salt tolerance, **165,** 3
S-fimbriae expression, **175,** 176–177
signal transduction, **165,** 25–27
spaceflight experiments, **156,** 303, 304, 339–342
urease in plant cells and, **145,** 92
Esterification, pectin in plant cell walls, **171,** 87–92
Estradiol
control of amphibian metamorphosis and, **145,** 121
effect on enzyme levels, **139,** 29
perinatal exposure to, **139,** 2
Sertoli-germ cell communication network and, **147,** 67–70, 75, 77
Estramustine phosphate, cytotoxic activity, **162A,** 547
Estrogen
dietary, **139,** 12–13
effect on decidual cell reaction, **160,** 6–7, 10–11, 23–24
effect on Harderian gland secretion, **168,** 53–54
effect on peroxisome proliferator response, **160,** 329–330
effect on pituitary cells
maturation, **157,** 55
proliferation, **157,** 46
effect on release
gonadotropic hormone, **159,** 69, 83, 91
growth hormone, **159,** 65–67, 73–74, 91
lutein-releasing hormone, **159,** 82–83, 91
prolactin, **159,** 65–67, 73–74, 91
effect on rooster liver
coiled body content, **159,** 176
gene expression, **159,** 177–178
nuclear morphology, **159,** 174–175, 177
effects on phagocytosis, **149,** 120
GnRH-encoding gene in teleosts and, **147,** 175–177
inhibition of angiogenesis, **159,** 144–145
microtubule-associated protein response, **151,** 125
neonatal exposure to sex hormones and antihormones end, **139,** 10–11
nidatory, role in implantation, **160,** 18–19, 37
RBF, in rabbit endometrium, **162B,** 354–355
RBF-1, in avian oviduct, **162B,** 354
regulation of pars intermedia cells, **159,** 55–55
role in cartilage formation, **159,** 316
seminal vesicle epithelium response, **160,** 288
sensitivity of postnatal rats, **159,** 83–86
uptake in androgenized rats, **160,** 8, 13
Estrogen receptor
accessory factors, **162B,** 363–364
GnRH-encoding gene in teleosts and, **147,** 175–176
Harderian gland, **168,** 41–43
Estrogen responsive element, GnRH-encoding gene in teleosts and, **147,** 175, 186
Estrus, persistent
decidual cell response, **160,** 27–30
induction by androgenization, **160,** 4, 11
interruption by ovariectomy, **160,** 7
ET, *see* Endothelin
Ethacrynic acid, **156,** 136
Ethamoxytriphetol
genital abnormalities and
in mice, **139,** 17–18
in rats, **139,** 16
history, **139,** 1
Ethmoidal gamglion, avian salt glands and, **144,** 158, 163
Ethylene
Arabidopsis thaliana and, **144,** 75
cytoskeleton, **166,** 39
extracellular matrix and, **146,** 301–305
plant hormone, **166,** 39
systemic acquired resistance induction, **168,** 320–321
–vinyl acetate copolymer, **146,** 60–62, 66
Ethylenediaminetetraacetic acid, *see* EDTA
Ethyleneglycol-bis-(β-aminoethylether)-*N,N'*-tetraacetic acid, *see* EGTA
Etioplasts, higher plant, **145,** 150–151, 162–163, 171, 188
Etoposide, *see* VP-16
Ets-1, phosphorylation, **157,** 245
E-type ATPase
adhesion molecule association, **158,** 150
assay, **158,** 185–187
cation inhibition, **158,** 186
classification, **158,** 142, 153

metal dependence, **158,** 141–142, 186
solubilization, **158,** 177–178
substrate specificity, **158,** 153, 188–190
types, **158,** 187, 189–190
Eubacteria, endosymbiont hypothesis
future prospects, **141,** 331
gene expression, **141,** 299, 301–302, 306–308
gene structure, **141,** 308–309, 312–318
gene transfer, **141,** 277
genome organization, **141,** 297
lineages of life, **141,** 238, 240–241, 244–246
mitochondria, **141,** 263, 265–269, 280, 327–328
nuclear genome, **141,** 247–251
organelles, **141,** 329–330
plastids, **141,** 253–255, 258
rRNA sequences, **141,** 321, 323
sequences, **141,** 270–272
Euchromatin
mitosis and, **144,** 222, 230
regenerative cementogenesis and, **142,** 9, 21, 40
Eudorina elegans, **139,** 198–199
Euglena
chloroplasts and, **144,** 284–285, 294
endosymbiont hypothesis and, **141,** 261, 272–273, 275–276
mitochondrial genomes of, **141,** 59
mitosis and, **144,** 222
phototaxis in
cryptochromes, **148,** 287–288
eyespot apparatus, **148,** 245–247
photoreceptive apparatuses, components, **148,** 270–271
retinal pigments, **148,** 284–286
signal generation, **148,** 252, 256
signal transduction, **148,** 298, 300
plastids and, **145,** 157, 167
Euglena gracilis
chloroplasts and, **144,** 264, 280
cryptochromes in, **148,** 288
cyanelles
plastid genome, **151,** 36
translation apparatus protein genes, **151,** 43
effect of microgravity, **171,** 30
endosymbiont hypothesis and, **141,** 252, 272–273, 275, 332
phototactic signal generation, **148,** 256–257
Euglenoids, chloroplasts and, **144,** 264, 282–285, 293–294
Euglenophyceae
eyespot apparatus, **148,** 243, 245, 247
eyespot reflection, **148,** 258–259, 269
photoreceptive apparatuses, components, **148,** 270, 273
Euglenophytes, endosymbiont hypothesis and, **141,** 261–263, 272
Eukaryotes
animal mtDNA and, **141,** 173, 195–196, 217
chloroplasts and, **144,** 259–260, 263, 295
cryptomonads, **144,** 275–276
dinoflagellates, **144,** 286, 292–293
euglenoids, **144,** 283–284
green algae, **144,** 268, 270
heterokont algae, **144,** 278
ciliates and, **141,** 43, 45, 53, 55, 59
DNA replication of chromosomes, *see* DNA, replication in mammals
endosymbiont hypothesis and, **141,** 233–237
acquisition, **141,** 275–277
algae, **141,** 273, 275
future prospects, **141,** 331
gene expression, **141,** 302, 305–308
gene structure, **141,** 310, 312–314, 316–317
gene transfer, **141,** 324
genome organization, **141,** 288, 292, 294, 297
lineages of life 238, **141,** 240–241, 243–246
mitochondria, **141,** 280, 282, 288, 327–328
molecular biology, **141,** 267
nuclear genome, **141,** 247–253
organelles, **141,** 329, 331
Plasmodium, **141,** 326
plastids, **141,** 253, 255, 258–261, 278–279
rRNA sequences, **141,** 300–301
sequences, **141,** 269–270
H^+-ATPase peripheral studies, **174,** 99–100
higher, mitosis and, **144,** 219, 224–226
higher plant mitochondrial genomes and, **141,** 129, 161

Eukaryotes (*continued*)
higher plant plastids and, **145,** 149, 165–166
lower, mitosis and, **144,** 252
dissociation of events, **144,** 226–230, 233
mitotic events, **144,** 219–224
microgravity experiments, **156,** 339, 342, 343
mitochondrial genomes in fungi and, **141,** 120
mitosis and, **144,** 237
urease in plant cells and, **145,** 80
Eukaryotic cells
chromosomal–microtubular cycles, **175,** 115–119
cytoskeleton, **166,** 3–4
cytoplasmic cytoskeleton, **166,** 5
intermediate filaments, **166,** 5
major cytoskeletal proteins, **166,** 4–6
membrane skeleton, **166,** 4–5
microfilaments, **166,** 5
microtubules, **166,** 5
nuclear skeleton, **166,** 5–6
prominent structures, **175,** 119–121
relationship of nucleus and microtubules
in animal and plant cells, **175,** 94–97
lower eukaryotes, **175,** 93–94
Eukaryotic genome, functional organization, **162A,** 456–463
Eukaryotic multicellular organism, cytoskeletal diversity, **166,** 6–7
Eukaryotic secretory pathway, protein sorting in, **139,** 89–93
Eumelanins, biology, **146,** 192–193, 199, 208–211
Eumelanosomes, biology, **146,** 193
Eunotia soleirolii, morphogenesis involving plasmalemma, **150,** 326
Euplotes
mitochondrial genomes of, **141,** 1
replication band
chromatin replication models, **153,** 163, 165
early studies, **153,** 141–142
function, **153,** 144, 146
immunochemical studies, **153,** 156–157, 159–160
ultrastructure, **153,** 149, 152
Euplotes eurystomus
nuclear organization, **154,** 11
replication band
characterization, **153,** 137, 139
chromatin replication models, **153,** 165
cytochemical studies, **153,** 154
function, **153,** 144–145, 148
immunochemical studies, **153,** 156, 158–160
ultrastructure, **153,** 150–152
replication band, elongation factor-1α in, **164,** 23
Euplotes octocarinatus, replication band, **153,** 146–147
EURECA, **156,** 355
European Retrievable Carrier, **156,** 355
Eurycea bislineata, hormonal control of metamorphosis and, **145,** 107
Eustigmatophyceae
eyespot apparatus, **148,** 235
eyespot reflection, **148,** 259, 269
Eustigmatophytes, chloroplasts and, **144,** 277, 280–282
Evolution
animal mtDNA and, **141,** 175–176, 189, 209–210, 230
Arabidopsis thaliana and, **144,** 54, 58–59, 64, 67
aspects of GSI, **140,** 475–477
associated modifications in ascidians, **176,** 293–294
ciliates and, **141,** 41–42, 53
considerations of double fertilization in nonflowering plants, **140,** 349–354
comparison among nonflowering seed plants, **140,** 349–350
evolutionary relationships of double fertilization in *Ephedra* and angiosperms, **140,** 350–354
considerations of GSP, **140,** 165
eggshells and biomineralization and, **145,** 217
endosymbiont hypothesis and, **141,** 233–235, 237
algae, **141,** 273
eukaryotes, **141,** 276
future prospects, **141,** 331–333
gene expression, **141,** 301, 303, 305–306, 308
gene structure, **141,** 308–314
genome organization, **141,** 288, 291–294, 296
lineages of life, **141,** 238–239, 243, 245

mitochondria, **141,** 281, 324–328
nuclear genome, **141,** 247, 250–251, 253
organelles, **141,** 330
plastids, **141,** 257–262, 278–279
rRNA sequences, **141,** 319–320, 323
secondary symbioses, **141,** 265–266
sequences, **141,** 268–269, 271–272
evolutionary biology, goals, **174,** 68–71
gene-regulatory function of intermediate filament proteins, **154,** 66–70
genetic control of development, **174,** 54–56
dorsal ridge evolution, **174,** 60–62
insects as arthropods, **174,** 63–65
mouthpart evolution, **174,** 56–60
GnRH-encoding gene in teleosts and, **147,** 170, 181–184, 186
higher plant plastids and, **145,** 149, 155, 164–167, 174
hormonal imprinting studies, **155,** 37
immune, stress, and inflammatory responses, **170,** 118–122
implications of volvocalean genes, **139,** 246–258
nervous system
from cnidarians, **167,** 39, 77
glycoprotein hormone/receptor couple, **167,** 56
regulatory evolution, **174,** 3
segmentation, **174,** 27–29
sex chromosomes, mammalian, **154,** 202–206
testis-determining factor *SRY* gene organization, **154,** 241–244
X chromosome inactivation, **154,** 211–214
stress response, **170,** 97–98
tRNA, **141,** 199–201
unit period, GnRH-encoding gene in teleosts and, **147,** 184
in whole-chromosome hybridization studies, **153,** 199–200
Excision repair, **156,** 124–125
Excitation, nuclear magnetic resonance and, **145,** 5–6, 45
Excitatory input, biological clocks in retina and, **146,** 99
Excitatory postsynaptic potentials, **167,** 42
Exercise, effect on muscle fiber type transitions
detraining, **170,** 178–180
endurance training, **170,** 171–172
sprint and resistance training, **170,** 172
EXGT, *see* Endoxyloglucan transferase
Exines
cell walls, **140,** 193
isolation and purification
acetolysis method, **140,** 53
by extraction hydrolysis method, **140,** 54
from hydrated and autoclaved pollen of gymnosperms, **140,** 54–55
from separated wings of *Pinus* pollen, **140,** 55
using 4-methylmorpholine N-oxide, **140,** 54
Exo-1,4-β-glucanase, *see* Glucan 1,4-β-glucosidase
Exocytosis
acrosomal
calcium role, **149,** 9–11, 14–15, 17, 39
ionic changes leading to, **149,** 37–39
potassium role, **149,** 17–18
process, **149,** 7–9
sodium requirements, **149,** 21–24, 26, 39
calcium channels, **149,** 31–37
functional evidence, **149,** 27–28
Na^+-Ca^{2+} exchanger, **149,** 29
Na^+-H^+ exchanger, **149,** 30–31
Na^+/K^+-ATPase, **149,** 28–29
antigens to basolateral membrane, by M cells, **167,** 127
biosynthesis and transport of cell wall precursors
callose, **176,** 172
cellulose, **176,** 171–172
pectins, **176,** 170–171
secretory vesicle aggregation, **176,** 173–174
secretory vesicle pathway, **176,** 173
calcium channels and, **147,** 199, 212, 216, 225, 227
constitutive, role of Rab8a, **176,** 35–36
fish chromatophores and, **143,** 230
regulation
and CFTR, **172,** 223–225
and cyclic AMP, **172,** 224
role in peptide hormone secretion, **166,** 12–14
Exo-1,4-β-glucanase, moss, effect of clinostating, **171,** 31

Exons
trapping assay, **158,** 233–234
tropomyosin, role in functional diversity, **170,** 22–25
Exopolygalacturonase, *see* Galacturan 1,4-α-galacturonidase
Exotoxins
Gram-negative bacteria, **161,** 286–287
Gram-positive bacteria, **161,** 274–275
pyrogenic, **161,** 276
Extensin
extracellular matrix and, **146,** 268–269, 277, 283, 285
related glycoproteins, **176,** 169–170
stabilization of microtubules in plants, **157,** 22
structural role in plant cell wall, **171,** 95–99
Extension–compaction–traction, and cytoskeletal dynamics, **144,** 89–91
Extracellular fluid
avian salt glands and, **144,** 185–187
tonicity, avian salt glands and, **144,** 129, 177, 180
volume, avian salt glands and, **144,** 129, 201
hormonal control, **144,** 195–196
receptive systems, **144,** 177, 185–189
Extracellular matrix
in angiogenic process, **161,** 3
animal, plant analogs
adhesion proteins, **174,** 272
fibronectin, **174,** 271–272
integrins, **174,** 270–275
short peptides, **174,** 271
vitronectin, **174,** 271–272
animal egg membrane, **166,** 79, 80
in animal systems, structure function, **140,** 299–300
annelid cuticle, **166,** 73–74
arthropod cuticle, **166,** 71–73
associated molecules, interactions with laminins, **173,** 107–108
avian retina and, **146,** 16, 25
bone, **166,** 75–77
cell, **166,** 70–81
cell migration
coating assay as analytical tool, **155,** 60
composition effects, **155,** 51–52, 55, 57
cellular interaction, **155,** 52
functions, **168,** 81
regulation of gene expression, **168,** 94–95
signal transduction
cellular functions regulated by signaling, **168,** 102–108
crosstalk with other signaling pathways, **168,** 99–102
cytosolic alkalinization, **168,** 96–97
intracellular calcium regulation, **168,** 95–96
phosphatidylinositol 4,5-bisphosphate regulation, **168,** 97–98
phosphatidylinositol 3-kinase role, **168,** 98–99
and cellular tensegrity
actin cytoskeletal models, **150,** 192
focal adhesion, **150,** 207–208
mechanical stress response, **150,** 185–187
microtubules, **150,** 201–203
nucleus, **150,** 206–207
signal transfer, **150,** 209
tension molding, **150,** 196–201
cholesteric supramolecular order, **166,** 70–81
cytomechanics, **150,** 120–124
cytoskeletal force balance alterations, **150,** 179–180, 182–184
development, **173,** 75
dome epithelium, villus basal lamina biochemical composition, **167,** 110
Drosophila muscles, **143,** 85
fish scale, **166,** 77, 78
growth factor-induced retinal regeneration and, **146,** 55–57, 59
growth factors and, TGF-β, **167,** 195
human bone, **166,** 75–77
follicular dendritic cell
albumin gradients, **166,** 149
detailed phenotypic analysis, **166,** 149–153
FACS sortimg, **166,** 151–153
isolation, **166,** 149–153
MACS sorting, **166,** 149
Percoll gradients, **166,** 149
parathyroid hormone, gene structure, **166,** 235
parathyroid hormone-related protein, gene structure, **166,** 234–236
immunology, **146,** 284–290
integrin binding to, FAK phosphorylation and, **167,** 169
integrins as transmembrane receptors, **167,** 162

invertebrate integument, **166,** 71–74
major structures, **173,** 76–77
MAP kinase activation by cell adherence to, **167,** 174
mechanical forces
coupling with extended cytoskeleton, **150,** 178
resistance, **150,** 177
tension, **150,** 176
microgravity environment, **156,** 346–348
mineralization, **162A,** 256–263
modification, **146,** 287, 289, 291
development, **146,** 294–301
growth regulators, **146,** 291–294
pathogenesis, **146,** 305–313
stress, **146,** 301–305
in muscle cell differentiation
components, **150,** 71–72
interaction, **150,** 72, 84–85
role, **150,** 69, 71
plant, **146,** 261–262, 313–314
characterization, **146,** 262–263
phenolics, **146,** 271–272
polysaccharides, **146,** 263–267
proteins, **146,** 267–270
plant cell wall, **166,** 79–81
in plants
cell wall, **140,** 300–301
stylar wall, **140,** 301
role of in pollination, **140,** 303–304
structure and chemical composition, **140,** 301–303
proteins in striated muscle and, **142,** 93, 100
regenerative cementogenesis and, **142,** 32
retinal pigmented epithelium and, **146,** 78, **152,** 253–254
retinal precursor cells and, **146,** 176, 178–180
scute, **166,** 77–78
Sertoli-germ cell communication network, **147,** 25, 45, 53, 72, 80
structural changes, **150,** 174
synthesis
polysaccharides, **146,** 272–276
wall phenolics, **146,** 278–284
wall proteins, **146,** 276–278
tissue, **166,** 70–81
transdifferentiation in medusae and, **142,** 257
isolation of tissues, **142,** 223
potential, **142,** 232–236
tissue organization, **142,** 220–222
vertebrate cell culture technology and, **142,** 159
vertebrate skeletal structure, **166,** 74–77, 78
Extracellular matrix protein
ependymin, **165,** 148–150
in phenotypic expression, **150,** 129
Extracellular matrix-track assay, migration analysis, **155,** 58–60
Extrachromosomal network, nuclear web system, **162B,** 195–197
Extraction
high ionic strength, onion nuclear matrix, **162B,** 80–86
high salt, DNA from nuclear matrix, **162B,** 416–422
LIS
in nuclear isolation, **162A,** 12
preparation of scaffolds, **162B,** 422–424
physiological ionic strength, plant nuclear matrix, **162B,** 87–88
Extraction hydrolysis method, sporopollenin, **140,** 54
Extraembryonic mesoderm, mosaicism in
chorionic villus sampling in assessing, **167,** 287
spontaneous abortions and, **167,** 288
Extreme sequence diversity, **140,** 466–467
Eye
complex, insect form diversity, **174,** 5–6
compound, honeybee, **176,** 309–311
development, in *Drosophila melanogaster,* **163,** 127–128
Drosophilia
ommatidia and, **139,** 143–146
R7 cell fate and, **139,** 146–148
human, transplantation, techniques, **138,** 39–41
melanocytes and, **146,** 192, 199
tissues, growth factor-induced retinal regeneration and, **146,** 57, 59, 66
Eye disease, RPE applications
in animal models
photoreceptor cell rescue
control for RPE cell transplantation studies, **138,** 25–26
first RPE cell transplantation results, **138,** 13–17
mechanism of action of photoreceptor cell rescue, **138,** 26–33

Eye disease (*continued*)
optimal conditions for RPE cell transplantation results, **138,** 24–25
structural and functional characteristics, **138,** 17–24
prevention of neovascularization in retinas, **138,** 33–34
RPE cell isolation and culture, **138,** 10–13
support of photoreceptor cell survival, **138,** 34–39
in human
first human RPE transplantation studies, **138,** 41–42
techniques for possible human eye transplantation, **138,** 39–41
various diseases possibly treated by human RPE transplants, **138,** 41
Eyespot
apparatus in flagellate algae, **148,** 232–235
components, **148,** 271–272
dinoflagellate, **148,** 247–251
type A, **148,** 235–240
type B, **148,** 240–243
type C, **148,** 243–247
reflection in algal phototaxis, **148,** 257–269
Ezrin, regulation of MT–microfilament interactions, **173,** 229–230

F

Face, in human development, **158,** 218–219
Facial nerve, amphibian development, **169,** 99
FACS sorting, **166,** 151–153
FAK, *see* Focal adhesion kinase; $Pp125^{FAK}$
FAK-related non-kinase
description, **167,** 166
in regulating FAK activity, **167,** 170–171
Far-red light, environmental stimuli in plant cells and
chloroplasts, **145,** 272, 277, 280–281
nucleus, **145,** 288
photodinesis, **145,** 254, 258–259, 266
Fascia adherens, proteins in striated muscle and, **142,** 114, 126
Fascin, **156,** 239, 255
Fas ligand, role in apoptosis, **157,** 204
Fast low-angle shot sequences, in NMR, **145,** 29
Fast-twitch glycolytic fibers, proteins in striated muscle, **142,** 120, 127
Fast-twitch oxidative-glycolytic fibers, proteins in striated muscle, **142,** 120, 127–128
Fate map
brain, **175,** 264–266
early neural primordium at neural stage
forebrain territories, **175,** 257–260
postepiphysis cephalic neural fold, **175,** 260–261
encephalic vesicles, **175,** 246, 261–262
plasticity of NC cells, **138,** 271–273
Fatty acids
in fungal signal transduction cascade model, **172,** 82–84
in plant defense transmembrane signaling, **172,** 74
polysaturated, effects on volume regulation, **161,** 224–225
polysaturated fatty acids, **161,** 224–225
synthesis, higher plant plastids and, **145,** 150, 171, 182–190, 203
Fatty acid synthetase, higher plant plastids and, **145,** 182–187
FBPase, *see* Fructose 1,6-bisphosphatase
Fc receptors, macrophage, for IgG, **161,** 291–292
FcR-like proteins, role in sperm binding, **163,** 269–272
FdUMP, *see* 5-Fluorouridine-5′-monophosphate
Feedback
in embryonic tissue
intercalation role, **150,** 27–28
requirements, **150,** 1–2
stress-force
cell columnarization models, **150,** 23–24
definition, **150,** 2
hormonal control of amphibian metamorphosis and, **145,** 117–119
sensory, in control of rhythmic behaviors, **147,** 17–18
Feedback inhibitors
cell proliferation and
assays, **167,** 196–198
cell cycle regulation, **167,** 192–194

microenvironmental influences and self-renewal, **167,** 194–196
clinical perspectives
bone marrow protection, **167,** 239–242
hemopoeitic progenitor cell mobilization by MIP-1α, **167,** 242–243
overview, **167,** 238–239
conclusions, **167,** 243–244
in feedback inhibition
hemopoietic structure and, **167,** 189–192
principles of, **167,** 189
of hemopoeitic stem cell proliferation
hemoregulatory pentapeptide, **167,** 216–220
hemoregulatory tetrapeptide, **167,** 221–224
MIP-1, **167,** 198–208
other inhibitors, **167,** 224–232
TGF-β, **167,** 208–216
overview, **167,** 185–189
tumor tissues and
growth inhibitory factors, **167,** 234–235
growth modulation by proliferation inhibitors, **167,** 235–237
malignant transformation process, **167,** 232–234
tumor resistance mechanisms, **167,** 237–238
Feedback loops
biological clocks in retina and, **146,** 99–100, 115–116, 129, 134
in homeostasis, **167,** 185–186
interferons in autocrine, **167,** 231
Feedback regulators, tumor tissue
growth inhibitory factors, **167,** 234–235
growth modulation by proliferation inhibitors, **167,** 235–237
malignant transformation, **167,** 232–233
resistance mechanisms, **167,** 237–238
Female germ unit
conclusions, **140,** 287
cytoskeletal organization in megagametophyte, **140,** 284–287
development of megagametophyte, **140,** 234–247
cellularization of embryo sac, **140,** 242–247
megagametogenesis and cytoplasmic maturation, **140,** 240–242
megasporocyte, **140,** 234–238
meiotic products and functional megaspore determination, **140,** 238–240
history, **140,** 233–234
isolation and characterization of isolated megagametophytes
embryo sacs and component cells, **140,** 269–280
technique and protocols, **140,** 270–271
viability and quality, **140,** 271–280
megagametophytes, isolation and characterization, embryo sacs and component cells, **140,** 281–284
organization
central cell, **140,** 264–269
concept of female germ unit, **140,** 247–249
egg cell, **140,** 260–264
organization and ultrastructure of female germ unit, **140,** 249–269
synergid, **140,** 250–260
cell wall and filiform apparatus, **140,** 250–256
cytoplasmic features, **140,** 256–258
histochemistry and cytochemistry of, **140,** 258–260
prospects, **140,** 287–288
Female–male interactions
and advances in compatibility interactions, **140,** 567–581
and self-incompatibility in plant reproductive biology, **140,** 581–586
Female quality, control in *in vivo* reproductive process, **140,** 396–397
Female sexual organs, transformation *in situ* of, **140,** 441–442
Fermentation, vertebrate cell culture technology and
bioreactors, **142,** 166, 169–170, 179
optimization, **142,** 193, 196, 198, 201
Ferns, effects of microgravity, **171,** 31
Ferredoxin, higher plant plastids and, **145,** 192–194, 200–201

Ferritin, *see also* Isoferritins
binding to M cells and enterocytes, **167,** 123–124
colchicine inhibition of M cell endocytotic uptake of, **167,** 129–130
as contaminant in isolation of coated vesicles, **167,** 9
vesicles participating in endocytosis of exogenous
in amphibian spinal ganglia, **167,** 1
in plants, **167,** 4–5
Ferrodoxin-dependent glutamate synthase, flowering plants, **143,** 304–305
Ferromagnetic substances, nuclear magnetic resonance and, **145,** 30
Fertilin, role in sperm binding, **163,** 273
Fertilization
calcium channel role, **149,** 31–32
calcium regulation, **149,** 9–11, 15–17
decapacitation factor, **149,** 6
diversity in, **163,** 255
double, *see* Double fertilization
experimental
attempts to obtain artificial zygotes, **140,** 579
plant transformation using sexual route, **140,** 579–581
pollen selection, **140,** 578–579
genetic approach to, **163,** 286–287
ionic changes, **149,** 38
and mammalian egg activation by calcium signal, *see* Calcium signal
and membrane fusion, models, **140,** 372–377
mitosis, **144,** 241
potassium effects, **149,** 20
preferential fertilization, **140,** 377–378
sodium requirements, **149,** 20–22, 23–24
sperm capacitation, **149,** 1–2, 4–5
in vitro
chromosomal abnormalities, **164,** 142–143
oocyte, **164,** 143–144
spermatozoa, **164,** 144–145
conceptus developmental potential
cell cycle regulation, **164,** 161–164
cell number/integrity, **164,** 148–155
culture conditions effects, **164,** 165–166
culture media, **164,** 168–170
cytoskeleton role in multinuclear blastomeres, **164,** 160–161
DNA content, **164,** 155–160
endometrial synchronization, **164,** 166–168
nucleocytoplasmic ratio effects, **164,** 172
cytoplasmic maturation effects, **164,** 145–146
genomic activation, **164,** 172–173
oocyte maturation effects, **164,** 140–142
preimplantation diagnosis and
embryo quality, **151,** 18
ethical issues, **151,** 20
Ferulate-5-hydroxylase, in phenylpropanoid pathway, **172,** 255; **176,** 102–105
Ferulic acid, extracellular matrix and, **146,** 271, 279, 283
Fetal alcohol syndrome
animal models, **158,** 261
clinical features, **158,** 260, 262
incidence, **158,** 260
pathogenesis, **158,** 260–261
Fetus
ethics of prenatal disease diagnosis, **158,** 264
head development, **158,** 216–222
intrauterine surgery, **158,** 263
parathyroid hormone-related protein, **166,** 253–255
FF, *see* Formaldehyde fluorescamine
FGF, *see* Fibroblast growth factor
FHsub2, *see* 7,8-Dihydrofolate
FHsub4, *see* 5,6,7,8-Tetrahydrofolate
Fiber-bed reactors, vertebrate cell culture technology and, **142,** 185–187
Fiber cells, lens
elongating, **163,** 310–312, 337–338, 338
end curvatures, **163,** 316
epithelial–fiber cell interface, **163,** 343
fiber compaction phenomenon, **163,** 333
fusion zones, **163,** 313–316
mature, **163,** 312
polarization, **163,** 334
primary, formation, **163,** 309
radial cell columns, **163,** 306, 313–314
secondary, formation, **163,** 309–310
sutures, *see* Sutures
transitional zone, **163,** 310

Fibers
DNA, radioautography analysis
basic principles, **154,** 265–269
critical data, **154,** 269–274
double-pulse labeling prior to, **154,** 274–280
method, **154,** 263
replicon model, **154,** 296
skeletal muscle
fiber type transitions
mechanism, **170,** 188–192
related to energy metabolism, **170,** 152–153
related to myosin isoform, **170,** 153–155
fiber type transitions, factors
aging, **170,** 186–188
denervation, **170,** 176–177
development, **170,** 158–162
electrostimulation-induced fast-to-slow transitions, **170,** 163–170
electrostimulation-induced slow-to-fast transitions, **170,** 177–178
exercise training, **170,** 171–172
hindlimb suspension, **170,** 181–182
hormone action, **170,** 184–186
immobilization in shortened position, **170,** 180–181
innervation, **170,** 162–163
mechanical overloading, **170,** 172–175
microgravity, **170,** 182–183
myotonia, **170,** 170–171
spinal cord transection and isolation, **170,** 177
thyroid hormone action, **170,** 183–184
metabolic enzyme-based types, **170,** 148–150
myosin ATPase-based types, **170,** 144–148
myosin heavy chain isoforms, **170,** 150–152
stress, *see* Stress Fibers
typing, proteins in striated muscle and, **142,** 67–69, 71
Fibrillarin, immunological characterization, **162B,** 119–121
Fibrillar networks
basal body-associated
basal foot in ciliated epithelium, **173,** 208–209
ciliated protozoa
centrosomes, **173,** 214
ciliary motility, **173,** 213
MT–microfilament complex, **173,** 211–213
proximal structures, **173,** 211
neurosensory epithelium
cochlea hair cells, **173,** 210
MT–microfilament interactions, **173,** 210
photoreceptor cells, **173,** 209–210
nucleolar centers
anchoring sites for rDNA loops, **162B,** 126
and fibrillarin distribution, **162B,** 121
storage of transcription proteins, **162A,** 138–139
surrounded by dense fibrillar components, **162B,** 305–306
Fibrils
anchoring, role in collagen fibril anchoring, **173,** 136–137
beaded, role in collagen fibril anchoring, **173,** 136
chloroplasts and, **144,** 291
cholesteric ordering, **166,** 86, 87, 90
collagen, *see* Collagen fibrils
collagen I assembly into, **173,** 83–85
cytoplasm
contractility, **156,** 203, 205
differentiation, **156,** 206–207
plasmodium, **156,** 203–207
spatial localization, **156,** 205
Drosophila muscles and
contractile protein genes, **143,** 112, 120, 122, 125
development, **143,** 72, 77, 86
fibrillin, role in connective tissue, **173,** 134–135
fish chromatophores and, **143,** 194–195
karyosphere and
oocytes of insects, **144,** 11, 14–15, 17, 21
vertebrates, **144,** 25, 30, 32, 37–39
microfibrils, *see* Microfibrils
myofibrils, *see* Myofibrils
neurofibril, microtubule-associated protein, **165,** 249–256
oligotene fibril, polytene chromosome, **165,** 93–94, 95–96

Fibrils (*continued*)
regenerative cementogenesis, **142,** 51
established root surfaces *in vitro,* **142,** 31–33, 35–37
established root surfaces *in vivo,* **142,** 46–47, 51
growing root surfaces, **142,** 9, 11–13, 15, 19, 21–23, 27–28
role in collagen fibril anchoring
anchoring fibrils, **173,** 136–137
beaded fibrils, **173,** 136
role in connective tissue
fibrillin fibrils, **173,** 134–135
fibronectin fibrils, **173,** 135–136
Fibrin
and cytokine-mediated tissue factor induction, **152,** 59
and endothelial cell tissue factor induction, **152,** 71–72
in tumor cells, **152,** 81
Fibrinogen, host defense function, **170,** 239
Fibroblast growth factor receptor, and retinal regeneration, **146,** 57, 67–69
Fibroblast growth factors
acidic
binding to extracellular matrix, **161,** 3
migration induction by, **155,** 70–71
neurite initiation and elongation role, **153,** 253
actin
effects on
retinal precursor cells, **146,** 176, 178
retinal regeneration, **146,** 56–58, 64, 66–69
expression in
melanocytes, **146,** 240
retinal pigmented epithelium, **146,** 77, 79
antibodies, for vascular disease control, **161,** 30
association with gap junctions, **157,** 327, 335
avian retina and, **146,** 29
basic
binding to extracellular matrix, **161,** 3
effect on arterial smooth muscle cell culture proliferation, **169,** 210–211
effect on ascidian notochord, **176,** 275–276
melanocytes and, **146,** 240–241
migration induction by, **155,** 71–73
neural induction in amphibians, **169,** 123
neurite initiation and elongation role, **153,** 253
retinal pigmented epithelium and, **146,** 77–79
retinal precursor cells and, **146,** 176–178
retinal regeneration and, **146,** 67–69
induction, **146,** 61–62, 64, 67
ocular tissues, **146,** 56–59
synergism with vascular endothelial growth factor, **161,** 3
synthesis in arterial smooth muscle cell culture, **169,** 221–222
effect on cell migration, **159,** 136
effect on pituitary cell proliferation, **157,** 51–52
family members, **159,** 310
fgf-2, role in apototic processes, **163,** 118
fusion proteins, therapeutic use in vascular disease, **161,** 31–32
heparan sulfate proteoglycan interaction, **161,** 6
melanocytes and, **146,** 240–241
migration induction by, **155,** 70–73
acidic fibroblast growth factors, FGF-1, **155,** 70–71
basic fibroblast growth factors, FGF-2, **155,** 71–73
neurite initiation and elongation role, **153,** 253–254
receptor proteins, **161,** 6
retinal pigmented epithelium and, **146,** 77–79
and retinal pigment epithelial cells
fate regulation, **150,** 127
transdifferentiation, **150,** 130–131
retinal precursor cells and, **146,** 173, 177–178, 182
retinal regeneration and, **146,** 49, 56–59, 67–69
induction, **146,** 61–62, 64, 66–67
ocular tissues, **146,** 56–59
role
angiogenesis, **159,** 129, 135–136
chondrocyte differentiation, **159,** 308–312
secretion, **159,** 135
and Sertoli-germ cell communication network, **147,** 49, 78

stimulation of vascular cells, **161,** 5–7
tissue distribution, **159,** 135
Fibroblasts
actin cytoskeleton, **148,** 56
actin retraction in, **148,** 63
apoptosis, **169,** 165
cadherin, expression effects, **165,** 172–174
cytoskeletal dynamics, **144,** 88, 96, 100, 103, 106, 115
in dome epithelium, **167,** 110
Drosophila muscles and, **143,** 114
growth factor-induced retinal regeneration and, **146,** 57
growth factor synthesis, **169,** 161–164
melanocytes and
genetics of melanoma, **146,** 236, 245–246, 248
pigment-cell genetics, **146,** 208, 213–214
migration, **169,** 163
motor activity, **148,** 41–42
myofibrillogenesis and, **143,** 168, 175, 179
proteins in striated muscle and, **142,** 97, 113
regenerative cementogenesis and, **142,** 4
established root surfaces *in vitro,* **142,** 28–29, 31–33, 35, 37–42
established root surfaces *in vivo,* **142,** 45–46, 50
growing root surfaces, **142,** 7, 9, 11–13, 22, 27
retinal pigmented epithelium and, **146,** 76
retinal precursor cells and, **146,** 153, 157
Sertoli-germ cell communication network, **147,** 34
tissue factor on, **152,** 73–74
transcriptional regulation and, **143,** 16, 28, 46
transdifferentiation in medusae and, **142,** 218
tropomyosin
exon-specific functions, **170,** 23–24
generation, **170,** 3
role in cytokinesis, **170,** 22
vertebrate cell culture technology and, 151
bioreactors, **142,** 174, 176, 189
three-dimensional cultures, **142,** 156, 158–160
wound healing role, **169,** 160–165
Fibrogenesis, role of collagen III, **173,** 87–88
Fibronectin
complexity of protein-binding interactions, **168,** 82
effect on arterial smooth muscle cell culture phenotype, **169,** 199, 201
fibrils, role in connective tissue, **173,** 135–136
MAP kinase stimulation by cell adhesion to, **167,** 174–175
neurite initiation and elongation role, **153,** 243–244
plant analog, **174,** 271–272
proteins in striated muscle and, **142,** 99
regenerative cementogenesis and
established root surfaces in uitro, **142,** 32–33, 36, 39, 41
growing root surfaces, **142,** 13, 25
related proteins, **176,** 152–154
role in cartilage formation, **159,** 273, 293
synthesis in arterial smooth muscle cell culture, **169,** 226
transdifferentiation in medusae and, **142,** 222
vertebrate cell culture technology and, **142,** 188
Fick's law, **162B,** 186–189
Ficoll/D_2O gradient technique, **167,** 8–9
Field vole, trophoblast
cell morphology, **165,** 63–65
secondary trophoblast giant cells, **165,** 63
trophoblast supergiant cells, **165,** 63–65
Filament protein
intermediate
M cell, vimentin expression in rabbit, **167,** 104, 129
M cell transport function and structure, **167,** 128–130
Tetrahymena
amino acid sequence, **164,** 21
citrate synthase
identification, **164,** 13–16
modulation, **164,** 18–20
gene analysis, **164,** 9–13
intramitochondrial filamentous inclusions, **164,** 16–18
and molecular evolution, **164,** 29
in oral morphogenesis during cell cycle, **164,** 4–6

Filament protein (*continued*)
in oral replacement during conception, **164,** 6
polymers, comparison with actin and tubulin, **164,** 4
in pronuclear behavior during conjugation, **164,** 7–9
properties, **164,** 2–4, 20–22
Filaments
actin
biochemical characteristics, **176,** 141–143
in pollen tube cytoskeleton, **176,** 146–149
structure, **140,** 86–87
bacterial flagellar motor, **147,** 97–100, 112, 117, 121, 135
core
nuclear matrix, **162A,** 91–97
NuMA localization, **162B,** 17–18
NuMA as part, **162A,** 164–165
cytoplasmic, and NPCs, **162B,** 232–233
intermediate
antibodies, **162B,** 100–102
complex with nuclear matrix, **162A,** 87–90
cultured cells, **162A,** 492
family of proteins, **162B,** 129
plant cytoplasmic, **162B,** 90–95
relationship to lamins, **162B,** 143–145
matrix protein, poorly staining, **162A,** 18–23
Filamin
cytoskeletal dynamics, **144,** 96, 106
myofibrillogenesis, **143,** 182
in striated muscle, **142,** 90–91, 108–109
Filopodia
growth cone, **165,** 256–258, 266–267
in lamellipodia
definition, **150,** 39–40
rearward traction forces, **150,** 41–42
in retrograde surface transport, **150,** 48, 53
Filter membrane assay, PMN random motility, **139,** 180–182
Fimbrin, **156,** 239; **175,** 55
Finite element methods, for developmental biology
application, **150,** 100–101
approach, **150,** 98–99
definition, **150,** 95
formulataion
assembly, **150,** 112–114
discretization, **150,** 107–108
element geometries, **150,** 103–105
equations, **150,** 102–103
external driving forces, **150,** 112–113
physical properties, **150,** 109–112
quantities, **150,** 103–104
shape functions, **150,** 104–106
solution techniques, **150,** 114
hourglassing, **150,** 110–111
modeling
evolution, **150,** 100, 102
role, **150,** 95–98
practical considerations, **150,** 116
recent advances, **150,** 116
time-dependent phenomena, **150,** 114–115
verification of results, **150,** 115
FISH, *see* Fluorescence *in situ* hybridization
Fish, *see also specific fish*
accessory respiratory organs, **157,** 303–304
avian salt glands and, **144,** 129
chromatophores, cytophysiology, **143,** 191–192, 245–250
cytoskeleton, **143,** 220–224
dermal chromatophores, **143,** 193–194
epidermal chromatophores, **143,** 239–242
erythrophores, **143,** 197–199
iridophores
motile, **143,** 210–220
optics, **143,** 206–208
platelets, **143,** 206
static, **143,** 209–210
leucophores, **143,** 201–205
melanophores, **143,** 195–197
motility control
calcium, **143,** 242–245
cyclic AMP, **143,** 239–242
hormonal signals, **143,** 234–237
membrane potential, **143,** 238–239
modulating agencies, **143,** 237–238
nervous signals, **143,** 225–234
signals, **143,** 224–225
varieties, **143,** 192–195
xanthophores, **143,** 197–200
diffuse neuroendocrine system
gills, **157,** 296–303
lungs, **157,** 292–293, 296

egg surface sperm-binding proteins, **163,** 260, 263–264
enameloids, crystal structure, **172,** 145
endocrine glands, **159,** 7
growth factor-induced retinal regeneration, **146,** 54
karyosphere and, **144,** 25
localization of neurosecretory cells, **159,** 2, 7
lung structure of air-breathing fish, **157,** 278–279
melanocytes, **146,** 232–234
microgravity experiments, **156,** 335
parathyroid hormone-related protein, **166,** 264–266
retinal precursor cells, **146,** 157, 163, 174
scale, extracellular matrix, **166,** 77, 78
vitellogenin receptor, **166,** 129–130
Fission zone, mitosis and, **144,** 228
FK506, effect on T cell receptor activation, **157,** 218–219
Flagella
axonemal variations, **173,** 5–9
basic axenome, **173,** 2–5
beat cycle
effect of nucleotides, **173,** 30–31
mechanics, **173,** 31–34
minimal requirements, **173,** 29–31
role of Ca^{2+}, **173,** 30–31
beating regulation, **164,** 285–289
chloroplasts and, **144,** 277–278, 280, 284, 289
compound, natural development, **173,** 9–10
doublet sliding, **155,** 142–145
dyneins, *see* Dyneins, flagellar
effect of nickel ion, **173,** 21
endosymbiont hypothesis, **141,** 330–331
kinesin-like proteins in, **164,** 305–306
mitosis and, **144,** 224
motility, **173,** 213
motility, regulation
effect of vanadate, **173,** 27–28
role in living cell, **173,** 23–25
role of Ca^{2+}, **173,** 25–29
role of cAMP, **173,** 25–29
signal pathways, **173,** 21–23
physical model, **173,** 39–46
physical parameters of movement, **173,** 34–39
sliding converted to bending, **155,** 145–146
structure, **155,** 142–145; **173,** 1–2
transdifferentiation in medusae, **142,** 252
initiation, **142,** 238, 244
potential, **142,** 223–224, 227, 229, 232
Flagellar motor, bacterial, *see* Bacterial flagellar motor
Flavanoid glycoside, chalcone synthase and, **147,** 271
Flavanone 3-hydroxylase, chalcone synthase and, **147,** 252, 260–261
Flavanones, chalcone synthase and, **147,** 248–249, 252
Flavin, environmental stimuli in plant cells and, **145,** 253, 271
Flavinoids, extracellular matrix and, **146,** 278–281, 301
Flavins, as photoreceptors in algae, **148,** 287
Flavolan, **156,** 39
Flavones, chalcone synthase and, **147,** 246–247, 249, 252
Flavonoids
biosynthesis genes, **156,** 38
chalcone synthase, **147,** 233, 277
control of activity, **147,** 252–253, 256, 258, 260–261
control of gene, **147,** 267, 271
functions of products, **147,** 246–252
nodulation, **156,** 1, 2, 7, 8–9, 14–16, 38, 46, 49, 51
properties, **140,** 43–45
Flavonolglycosides
chalcone synthase, **147,** 246–247, 261
isolated from pollen, **140,** 44–45, 49
Flavonols
chalcone synthase, **147,** 246
effect on pollen tube growth, **176,** 180
FLiG, in bacterial flagellar motor, **147,** 109–110, 145
FLiM, in bacterial flagellar motor, **147,** 109–110, 145
FLiN, in bacterial flagellar motor, **147,** 109–110, 145
Floor plate
formation, **175,** 251
Hensen's node origin, **175,** 255
Flow, cortical, *see* Cortical flow
Flowering plants
amino acid pathways, **143,** 318, 323–324
aspartate, **143,** 318–321
proline, **143,** 321–323

Flowering plants (*continued*)
ammonium, **143,** 303
asparagine synthetase, **143,** 306–307
genes, **143,** 308–309
glutamine synthetase, **143,** 305–308
photorespiration, **143,** 303–305
aromatic amino acid biosynthesis, **143,** 309–310
auxin, **143,** 317–318
shikimate pathway, **143,** 310–313
tryptophan, **143,** 313–318
list of anther-and pollenspecific cDNA clones isolated from, **140,** 21
nitrate, **143,** 299
future directions, **143,** 302–303
nitrate reductase, **143,** 299–302
nitrite reductase, **143,** 302
reproductive barriers, **140,** 450–452
definition, **140,** 450–452
interspecific, **140,** 450, 451
intraspecific, **140,** 450, 451
types of systems, **140,** 452
sexual reproduction in, **140,** 297–304
extracellular matrix in plants, **140,** 300–304
cell wall, **140,** 300–301
stylar wall, **140,** 301–304
extracellular matrix in animal systems, structure function, **140,** 299–300
Flow imaging, nuclear magnetic resonance and, **145,** 25–26
FLP recognition target site, tagged circular construct, **162A,** 438–440
Flt-3/flk-2, role in B cell development, **157,** 152
Fludarabine phosphate, inhibition of DNA primase, **162A,** 552–554
Fluidized-bed reactors, vertebrate cell culture technology and, **142,** 178–180
Fluid movement, microgravity environment, **156,** 345
Fluorescein, proteins in striated muscle and, **142,** 65, 117
Fluorescein isothiocyanate
environmental stimuli in plant cells and, **145,** 261, 278
in FISH diagnosis of genetic diseases, **153,** 11–12
Fluorescence
biological clocks in retina, **146,** 122, 139
calcium channels and, **147,** 218
chloroplasts and, **144,** 280
control of rhythmic behaviors and, **147,** 4
cryptochromes as photoreceptors in algae, **148,** 287–289
cytoskeletal dynamics and, **144,** 99, 101
and electron microscope evidence of actin in GSP, **140,** 161–162
environmental stimuli in plant cells and
light, **145,** 263, 277–278, 280–282
low temperature, **145,** 290
wounding, **145,** 296
EXGT, **173,** 177–179
growth factor-induced retinal regeneration, **146,** 52
hormonal control of amphibian metamorphosis and, **145,** 122
intracellular calcium messenger system and, **142,** 315, 327
karyosphere and, **144,** 37
Physarum microfilament system analysis, **149,** 186–190, 193
in plant mutants, flowering plants and, **143,** 315–317
proteins in striated muscle and
cardiac muscle, **142,** 110, 114
sarcolemma, **142,** 124
skeletal muscle, **142,** 63, 80, 87, 95, 99
retinal precursor cells, **146,** 150, 160
transdifferentiation in medusae and, **142,** 226
Fluorescence *in situ* hybridization
for diagnosis of genetic diseases
applications, **153,** 6–14
autosomal imbalances, **153,** 25–33
chromosomes
anomalies, **153,** 28, 30–31
painting, **153,** 16–17, 28, 30–31
Philadelphia, **153,** 30
specific repeat probes, **153,** 25–27
trisomy detection, **153,** 27–29
comparative genome hybridization, **153,** 21
DNA
deletions, **153,** 30–33
duplications, **153,** 33
halo preparations, **153,** 21
isotopic, **153,** 7
multicolor FISH, **153,** 19–20

polymerase chain reaction technology, **153,** 20
procedure, **153,** 1–2
samples
postnatal, **153,** 3
preimplantation embryo, **153,** 4–6
prenatal, **153,** 3–4
sexing of human cells
clinical application, **153,** 24–25
early studies, **153,** 22–23
patient strategies, **153,** 25
postnatal samples, **153,** 21
preimplantation embryos, **153,** 22–25
prenatal samples, **153,** 22
research needs, **153,** 22
single-color FISH, **153,** 24
studies, **153,** 23–24
single locus detection, **153,** 17–19
tandem repetitive probes, **153,** 14–16
technological advances, **153,** 33–34
three-dimensional, **153,** 20–21
total genomic probes, **153,** 18
types, **153,** 6–14
embroyo cells at interphase, assessing all, **167,** 267–268
impact on cytogenic research, **167,** 264
in preimplantation diagnosis of genetic defects, **151,** 13–15
sperm assays
centromere size in nondisjunction, **167,** 285–286
disomy in men exposed to mutagens, **167,** 283
genetic factors in nondisjunction, **167,** 283–286
versus humster method, **167,** 281
yeast artificial chromosome clones, **158,** 233
Fluorescence microscopy
cytoskeletal dynamics and, **144,** 113
proteins in striated muscle and, **142,** 62
in whole-chromosome hybridization, **153,** 183–187
Fluorescent dyes, avian retina and, **146,** 4, 10, 16
Fluoride, incorporation in apatite, **172,** 155–157
Fluorine, nuclear magnetic resonance and, **145,** 20, 32, 41
Fluoroapatite, in fish enameloids, **172,** 145
Fluorocarbons transdifferentiation in medusae and, **142,** 244–245
Fluorochromes, in whole-chromosome hybridization
chromosomal aberration analysis, **153,** 206
techniques, **153,** 184–188
5-Fluoroindole
flowering plants, **143,** 315
resistance, flowering plants, **143,** 315
Fluorosis, dental enamel and, **172,** 151–152
5-Fluorouracil, **156,** 116–120
5-Fluorouridine-5′-monophosphate, **156,** 117–118
Flux, membrane, cationic drugs, **171,** 139–141
Fly
karyosphere, **144,** 9, 43
Na^+/K^+-ATPase localization, **176,** 329–332
photoreceptors, Na^+/K^+-ATPase maintenance and distribution, **176,** 337–338
FMRF-amide
biological clocks in retina and, **146,** 108, 111–112
related peptides, in control of rhythmic behaviors, **147,** 12, 14–15
transdifferentiation in medusae and, **142,** 224, 230
Focal adhesion
organization, **167,** 162–164
overview, **167,** 161–162
$pp125^{FAK}$ in
conclusions, **167,** 178–179
discovery, **167,** 164–165
downstream effects of, **167,** 172–178
regulation of activity of, **167,** 168–172
structure of, **167,** 165–168
targeting sequence, FAK variants with mutations of, **167,** 166
Focal adhesion kinase
conclusions, **167,** 178–179
discovery, **167,** 164–165
downstream effects
substrates and binding partners, **167,** 172–175
in vivo effects, **167,** 175–178
integrin-mediated signal transduction
autophosphorylation, **168,** 90, 93
calcium role, **168,** 87–88

Focal adhesion kinase (*continued*)
inhibitors, **168,** 88–89
protein kinase C role, **168,** 87–88
regulation of gene expression, **168,** 94–95
Rho role, **168,** 88
Src complex
cell migration role, **168,** 105, 107–108
cell proliferation role, **168,** 103
formation, **168,** 89–91
Ras-MAP kinase pathway activation, **168,** 93–95
substrates, **168,** 91–93
tyrosine phosphorylation, **168,** 84, 86, 90–91, 93–94
integrin receptor signal transduction, **169,** 199, 201
organization of focal adhesion, **167,** 162–164
phosphorylation via crosstalking of other signal transduction pathways, **168,** 99–102
regulation
cytoplasmic proteins in, **167,** 170–171
integrins in, **167,** 168–170
nonintegrin receptors and signaling crosstalk, **167,** 171–172
sequence homology between species, **168,** 84
structure, **167,** 165–168; **168,** 84–86
subcellular localization, **168,** 86–87
Focal contact, *see* Focal adhesion
Fodrin, **144,** 108; **156,** 241; **173,** 229–230
Folate
antagonists, **156,** 108–116
flowering plants and, **143,** 311
metabolism, methotrexate, **156,** 110–112
Folic acid
cytoskeletal dynamics and, **144,** 110–111
transport, **156,** 109
Follicles
lymphoid
dome area association with, **167,** 99
forming Peyer's patches, **167,** 94
melanocytes and
biology, **146,** 192, 194–195
pigment-cell genetics, **146,** 203, 209–211, 220
ovarian in maternal age effect hypothesis for aneuploidy, **167,** 278
Follicle-stimulating hormone
control of amphibian metamorphosis end, **145,** 113, 115
control of release, **159,** 68
effects on testicular macrophages, **149,** 109, 118–119, 129
feedback mechanisms, **159,** 79–80
GnRH-encoding gene in teleosts and, **147,** 165, 185
regulation of activity, **159,** 69, 72
Sertoli cell effects
metabolism, **160,** 194
phagocytic activity, **160,** 197
spermatid binding, **160,** 192, 203
steroid hormone synthesis, **160,** 198–199
Sertoli-germ cell communication network and, **147,** 35, 80–81
function, **147,** 39, 41, 44–45, 58, 61
germ cell depletion, **147,** 62, 64–66
germ cell in vitro, **147,** 67–68, 70–71
mechanisms, **147,** 78–79
sex differences in expression, **159,** 73
synthesis, **159,** 68–69
Follicle-stimulating hormone receptor, **155,** 27; **167,** 56
Follicular dendritic cells
albumin gradients, **166,** 149
antibody response, **166,** 146–147
antigen-antibody-C3 complex function, **166,** 145–148
antigen-dependent immune response initiation, **166,** 163
antigen retention mechanisms, **166,** 145
B cell
CD40 ligation, **166,** 157–158
cytokines, **166,** 157–158
proliferation, **166,** 156–157
survival, **166,** 156–157
in vitro study, **166,** 156–158
centroblast in dark zone, **166,** 163–165
detailed phenotypic analysis, **166,** 149–153
extrafollicular responses, **166,** 163
FACS sorting, **166,** 151–153
follicular lymphoma, **166,** 158–160
functions, **166,** 144–148
germinal center
B cell proliferation, **166,** 145–146
HIV, **166,** 160–161

HIV, **166,** 160–163
B cell repertoire restriction, **166,** 161
infectiousness, **166,** 161–162
lymphoid follicles, **166,** 162–163
identification, **166,** 144
isolation, **166,** 149–153
MACS sorting, **166,** 149
memory B cell clone maintenance, **166,** 146–147
memory T cell maintenance, **166,** 147–148
memory T cell selection, **166,** 147–148
monoclonal antibody
cell lines, **166,** 153–155
generation, **166,** 153–155
mouse
iccosomes, **166,** 148–149, 150
isolation, **166,** 148–149
origin, **166,** 155–156
Percoll gradients, **166,** 149
Follicular lymphoma, follicular dendritic cell, **166,** 158–160
Folliculi lymphatici aggregati, *see* Peyer's patches
Folliculogenesis, **166,** 106–107
Folylpolyglutamate synthetase, **156,** 111, 114–115
Foot-and-mouth disease, vertebrate cell culture technology, **142,** 146, 151, 153, 176
Foot protein, *see* Ryanodine receptor
Forebrain
cell migration in, **175,** 274–277
distribution in early neural plate, **175,** 257–260
Formaldehyde fluorescamine methods
development and early consideration, **138,** 52–53
identification of FF-reactive compound, **138,** 53–58
5-Formyltetrahydrofolate, *see* Leucoverin
Forskolin
avian salt glands, **144,** 200–201
biological clocks in retina and, **146,** 110
calcium channels, **147,** 201
melanocytes and, **146,** 210
Founder cells, *Drosophila* muscles and, **143,** 71, 83–84
Fourier transform, in NMR, **145,** 4, 6, 29
Fox, trophoblast, cell morphology, **165,** 65–66
FPGS, *see* Folylpolyglutamate synthetase
Fractionation, *see also* Subfractionation
cellular
induced insolubility, **162A,** 528–531
in situ, in preparation of nuclear matrix, **162A,** 492–495
chromatin domains, **162A,** 221–226
for estimating total quantities, **161,** 128–129
nucleoli, **162B,** 115–116
plant lamina, **162B,** 97–98
in situ cell, **162A,** 492–495
Fragmin, **144,** 107; **149,** 156, 161–162; **156,** 205, 214, 227, 230–232, 243, 254
Free fall vehicles, **156,** 349–351, 355, 357
Free induction decay, in NMR, **145,** 4
Free-running rhythms, biological clocks in retina
Bulla, **146,** 137–138
concepts, **146,** 96–98
pacemaker mechanisms, **146,** 115–116, 118, 122
Freeze–fracture studies
avian salt glands and, **144,** 138
chloroplasts and, **144,** 292
male reproductive cells
cell division and differentiation, **140,** 187–188
conclusions, **140,** 202–203
generative cells, **140,** 193–201
generative nucleus, **140,** 194, 195, 197
interface between vegetative and generative cells, **140,** 193–197
organelles, **140,** 197–201
prospects, **140,** 203
sperm cells, **140,** 188–189
organelles, **140,** 202
plasma membrane, **140,** 201–202
sperm nucleus, **140,** 202
usefulness, **140,** 189–190
vegetative cells in pollen and pollen tubes
cell walls, **140,** 192, 193, 194, 196, 198–200
external plasma membranes, **140,** 190–191
organelles, **140,** 193
vegetative nucleus, **140,** 191–193

FRNK, *see* FAK-related non-kinase
Frog
microgravity experiments, **156,** 334
vitellogenin receptor, **166,** 129–130
Frontal lobe, degeneration with pathological Tau triplet, **171,** 199
Fructose bisphosphatase
degradation, **139,** 77–79
higher plant plastids, **145,** 177, 180
Fruit fly, microgravity experiments, **156,** 335, 355, 356
FSH, *see* Follicle-stimulating hormone
Fucose, M cells and
bacterial adhesins specific for, **167,** 124–126
lectin histochemistry in studying, **167,** 103–104
Fucosylation, in xyloglucan–cellulose microfibril interaction, **173,** 165–166
Fucosyltransferase, Sertoli-germ cell communication network and, **147,** 33
Fucoxanthin, chloroplasts and, **144,** 278, 282, 288–290, 294
Fucus
instability of homogeneous state
localized activity, **154,** 344–345
metabolic patterns, **154,** 359
sperm-binding proteins, **163,** 270, 278
Fucus serratus, chloroplasts and, **144,** 281
Fumagellin, inhibition of angiogenesis, **159,** 139
Funaria, environmental stimuli in, **145,** 268, 271, 288
Funaria hygrometrica
cell wall
changes under microgravity
role of Ca^{2+}, **170,** 65–66
role of enzymes, **170,** 62–64
parenchyma, weightlessness effect on structure, **170,** 49–50
polysaccharide content, effect of clinorotation, **170,** 57–60
effect of microgravity, **171,** 24, 30–31
intracellular calcium messenger system and, **142,** 309
Fundulus, fish chromatophores
cytoskeleton, **143,** 221–222
iridophores, **143,** 219
leucophores, **143,** 201, 203, 205
motility control, **143,** 228, 239
Fundulus heteroclitus, microgravity experiments, **156,** 335
Fungi
animal mtDNA and, **141,** 192
Arabidopsis thaliana and, **144,** 61, 74, 76
pathogens, **144,** 56, 58–59
phenotypes, **144,** 64–66
chloroplasts and, **144,** 277
endosymbiont hypothesis and
gene expression, **141,** 298, 300–302, 306
gene structure, **141,** 313–314
genome organization, **141,** 290, 293, 295
mitochondrial 282-283, **141,** 286, 320–321, 327
nuclear genome, **141,** 250
extracellular matrix, **146,** 309, 311
flowering plants, **143,** 311–312
infection, chromoplast-like plastids in infected tissue, **163,** 185–186
microgravity experiments, **156,** 352
mitochondrial genomes, **141,** 89–90, 119–121
DNA diversity, **141,** 90–91
ascomycetous yeasts, **141,** 97–103
basidiomycetes, **141,** 103–106
filamentous ascomycetes, **141,** 93–97
hypochytridiomycetes, **141,** 91–93
oomycetes, **141,** 91–93
generation of mtDNA diversity, **141,** 106
codon changes, **141,** 117–119
length mutations, **141,** 106–113
rearrangements, **141,** 113–117
mitosis and, **144,** 221–222, 224
Funnel-web spider toxin, calcium channels and, **147,** 212–213
Fura-2, calcium quantitation, **158,** 296–298
Furosemide, avian salt glands and, **144,** 154, 162, 173, 201
Fusicoccin, intracellular calcium messenger system and, **142,** 308
Fusion
effect in pattern of cytoplasmic inheritance, **140,** 379–383
paternal cytoplasmic inheritance mechanism, **140,** 382–383
uniparental, maternal cytoplasmic inheritance mechanism, **140,** 379–382
fiber cells, **163,** 313–316

gamete, *see* Gamete fusion
mitochondrial, *see* Mitochondrial fusion
technique for hamster, **167,** 269, 270
technique for human-hamster fusion, **167,** 269, 270
Fusion proteins
mitoxins, therapeutic use in vascular disease, **161,** 31
PML–RARα, **162A,** 80, 169–170
Fusosyl transferase, extracellular matrix and, **146,** 272, 275, 277
FXPRLa peptides, characterization, **164,** 226–227

G

GABA, *see* γ-Aminobutyric acid
Gadolinium
calcium channel blocking, **168,** 168
inflow, nuclear magnetic resonance and, **145,** 29
Gadus morhua, animal mtDNA and, **141,** 202–204
GAGA factor, chromatin modifications by, **154,** 36
GAL, *see* Galanin
Galactans, extracellular matrix and, **146,** 264
Galactose
hepatic galactose receptor, **175,** 145–146
in M cells, **167,** 125
recognition, **175,** 151–153
residue masking in sialic acid, **175,** 144–145
vertebrate cell culture technology and, **142,** 195
Galactose receptor
hepatic, in recognition site masking, **175,** 145–146
Kupffer cell, **175,** 147
peritoneal macrophage
morphological studies, **175,** 148
Sias masking effect, **175,** 148–150
Galactosyltransferase, Sertoli-germ cell communication network and, **147,** 33
Galacturan 1,4-α-galacturonidase, role in cell wall changes, **170,** 63
Galacturonides, in plant defense response mediation, **148,** 13
Galanin
effect on pituitary cell proliferation, **157,** 48–49
effects, **138,** 169
immunohistochemical localization of, **138,** 166–169
Gallus domesticus, animal mtDNA
protein genes, **141,** 183, 185, 188
sequences, **141,** 203–205
GALT, *see* Gut-associated lymphoid tissue
Gamete fusion
in animal systems, **140,** 372–375
fertilization in animals, **140,** 372–373
mechanism of membrane fusion, **140,** 373–375
in plants and animals, **140,** 372–378
gametic fusion, **140,** 377
membrane fusion in angiosperms, **140,** 375–377
models of fertilization and membrane fusion, **140,** 372–377
fertilization in animals, **140,** 372–373
gametic fusion in animal systems, **140,** 372–375
mechanism of membrane fusion, **140,** 373–375
preferential fertilization, **140,** 377–378
and recognition in animals and plants, **140,** 408–414
Gametes, *see also* Oocytes; Sperm
autosomal trisomy incidence in female *versus* male, **167,** 272–273
intrafallopian transfer treatment, **164,** 146, 167
karyosphere, **144,** 2–7, 15, 40
male
changes, **140,** 410–411
delivery of in embryo sac and their condition prior to fusion, **140,** 370–372
and generative cells, *see* Generative cells, and male gametes
nullisomic, monosomy arising from, **167,** 264–265
recognition
history of research, **163,** 254–256
sperm-binding proteins of egg plasma membrane, **163,** 283–284
sperm receptors of egg outer coat, **163,** 280–282

Gametes (*continued*)
surface components, **163,** 253
surface proteins, **163,** 254–256
single, micromanipulation of, **140,** 414–419
and syngamic system, data from, **140,** 586–587
in vitro fusion, and zygote production
conclusions, **140,** 419–420
egg cells and zygotes, culture of, **140,** 418–419
future directions, **140,** 420–421
micromanipulation of single gametes, synergids, and central cells, **140,** 414–419
electrofusion, **140,** 416–418
isolation and selection, **140,** 414–416
recognition and gamete fusion in animals and plants, **140,** 408–414
intergametic recognition during fertilization process, **140,** 413–414
IVF for higher plants by mixing male and female gametes without chemical or physical devices, **140,** 411–413
sperm capacitation in animals and plants, **140,** 409–411
male gamete changes, **140,** 410–411
pollen tube changes, **140,** 410
in vitro fertilization in animals, **140,** 408–409
studies using light and electron microscopy, **140,** 407–408

Gametogenesis
Chlamydomonas
activation by staurosporine, **168,** 292–293
assay systems, **168,** 262–263, 282
blue light signaling
action spectrum, **168,** 273
agglutinin complex activation, **168,** 295
assay, **168,** 272–273
continuous signal input, **168,** 274–276
gene induction, **168,** 273, 276
photoreceptor system, **168,** 273–274
rationale, **168,** 271–272
signal transduction pathway, **168,** 277, 296
steps, **168,** 276–277
cell culture, **168,** 259–260
Chlamydomonas eugametos
cell cycle regulation, **168,** 294
light activation, **168,** 294–295
nitrogen deprivation signaling, **168,** 293
Chlamydomonas reinhardtii
cell cycle regulation, **168,** 278–281
kinetics, **168,** 272, 274, 280
mating reaction, steps, **168,** 261–262
gamete division and growth, **168,** 281
gene characterization
gas genes, **168,** 286–287
lrg genes, **168,** 286, 288–290, 293
genomic imprinting, **160,** 55–57
inhibitors
cyclic AMP, **168,** 292
genistein, **168,** 290, 292
SC-10, **168,** 292
karyosphere and, **144,** 7–8, 41
model, **168,** 264–265
mutants
agglutinin-deficient mutants, **168,** 284
C4 mutants, **168,** 283–284
cell fusion mutants, **168,** 285
classification, **168,** 282–283
gametic lytic enzyme mutants, **168,** 285
nitrogen deprivation signaling
gene induction, **168,** 266, 268–269
nitrogen sources, **168,** 267–268
nucleic acid metabolism effects, **168,** 269–270
protein metabolism effects, **168,** 269–271
rationale, **168,** 265–266
sensing, **168,** 266–267
sexual life cycle, **168,** 260–261
synchronization of cell cycle, **168,** 278
ultrastructural changes during gametogenesis, **168,** 263–264

Gametophytes
development, generative cell and male gamete formation, **140,** 205–206
male, hypotheses on functions, **140,** 74–76

Gametophytic self-incompatibility
biochemical basis
identification of S allele-associated proteins, **140,** 459–461

temporal and tissue-specific expression of S alleles, **140,** 461–463
in vitro assays of S protein function, **140,** 463–465
cellular and physiological aspects, **140,** 457–459
genetics, **140,** 452–457
control, **140,** 453
genetic basis, **140,** 452–454
genetic breakdown, **140,** 454–457
self-incompatibility due to mutation outside S locus, **140,** 457
S locus mutants, **140,** 454–456
conclusions, **140,** 479–481
evolutionary aspects, **140,** 475–477
gene cloning, **143,** 284–287
genetic analysis, **143,** 280–281
glycoproteins, **143,** 283–284
historical background, **140,** 449–450
homomorphic
gene cloning, **143,** 284–287
genetic analysis, **143,** 280–281
glycoproteins, **143,** 283–284
physiology, **143,** 281–283
ribonuclease, **143,** 287–288
SLG, **143,** 288–289
mechanism of S protein function, **140,** 472–475
biological relevance of ribonuclease activity of S proteins, **140,** 472–475
ribonucleolytic properties of S proteins, **140,** 472
molecular characteristics of S gene, **140,** 465–471
cloning of DNA encoding S proteins, **140,** 465–466
molecular organization of S locus, **140,** 469–471
primary structure feature of S proteins, **140,** 466–469
conserved residues, **140,** 467
extreme sequence diversity, **140,** 466–467
variable and hyperactive residues, **140,** 467
in *Nicotiana tabacum,* **143,** 278
physiology, **143,** 281–283
plant breeding, potential applications, **140,** 477–479
reproductive barriers in flowering plants, **140,** 450–452
definition, **140,** 450–452
interspecific, **140,** 450, 451
intraspecific, **140,** 450, 451
types of systems, **140,** 452
ribonuclease, **143,** 287–288
SLG, **143,** 288–289
in various plant families, **143,** 258–259
$\gamma 5$
gene deletion in transgenic mice, **157,** 192
marker of B cell differentiation, **157,** 140–141
role in B cell development, **157,** 157
Ganglion cells
avian retina and, **146,** 1–5, 12, 15–16, 21
avian salt glands and, **144,** 155, 157–159, 163, 193, 196
biological clocks in retina and, **146,** 108–109, 114
cnidarian
anatomical locations, **167,** 40
lacking cilium in *Calliactis parasitica,* **167,** 42
control of rhythmic behaviors and, **147,** 3
chemical modulation, **147,** 19
ion channels, **147,** 5, 8
neuromodulation, **147,** 13–14, 16
growth factor-induced retinal regeneration and, **146,** 52–53, 57, 62–63
retinal
growth factor-induced retinal regeneration and, **146,** 62–64
retinal precursor cells and, **146,** 148, 151, 171
retinal precursor cells and, **146,** 148, 172, 179
Ganglionic synapses
axotomy effects, **159,** 195
effect of animal age, **159,** 213–214
electrophysiology
frogs, **159,** 216, 219–220
mammals, **159,** 210–211, 221
morphology
frogs, **159,** 216–220
mammals, **159,** 211, 213
deafferentation effect on ultrastructure, **159,** 195
morphology
amphibians, **159,** 205–206
mammals, **159,** 202–205

Ganglionic synapses (*continued*)
morphology, total deafferentation effect
frogs, **159,** 230–232
mammals, **159,** 224–227
neurotransmitters, **159,** 199
physiology of transmission
amphibians, **159,** 201–202
mammals, **159,** 199–201
quantitation, **159,** 206–207
synaptic delay, **159,** 199–200
types, **159,** 202–203
Gangliosides
avian retina and, **146,** 24
melanocytes and, **146,** 235
microtubule-associated protein response, **151,** 125
GAP, *see* Gonadotropin-releasing hormone-associated peptide
Gap junction
adhesion molecule role in communication, **157,** 336–337
astrocytes, **157,** 342–347
cellular diversity, **157,** 319–320
channel diameter, **157,** 317
components, **157,** 318
degradation, **157,** 339–340
discovery, **157,** 316
freeze-fracturing, **157,** 322, 325, 327, 350–354
in glucose transfer
blood-tissue barriers, **172,** 8
ciliary body epithelium, **172,** 21–22
rat placenta, **172,** 31
lens
in cortical and nuclear fiber cells, **163,** 338–343
in elongating fibers, **163,** 339
in epithelial cells, **163,** 337
species differences, **163,** 344–345
neurons, **157,** 347–348
phosphorylation, **157,** 333–335
protein turnover, **157,** 320, 332–333
retina, **157,** 348–355
role
disease, **157,** 340
embryonic development, **157,** 336–337
oncogenesis, **157,** 337–339
Sertoli-germ cell communication network and, **147,** 28–29, 32, 39
tight junction association, **157,** 330–332
transfer of fluorescence dyes, **157,** 317
Garden cress, microgravity experiments, **156,** 321, 323, 328, 330
Gas exchange, eggshells and biomineralization and, **145,** 230, 235
Gastric glands, urease in plant cells and, **145,** 83
Gastric receptor, anterior, control of rhythmic behaviors, **147,** 17
Gastrin-related peptide, immunohistochemical localization, **138,** 153–154
Gastrin-releasing peptide, production by neuroendocrine cells, **157,** 283, 289
Gastropyloric receptor, control of rhythmic behaviors and, **147,** 17–18
Gastrulation
in ascidian embryogenesis, **176,** 249–250
cell death during, **163,** 134–135
cytoplasmic determinants for movements, **176,** 267
in embryonic induction, definition, **150,** 373, 397–404
and Hensen's node, **175,** 255
platelet-derived growth factors in
activity, **172,** 111–112
distribution, **172,** 104
G^1 cyclins, regulation of cell cycle, **157,** 249
GDP, *see* Guanosine 5′-diphosphate
Geese, avian salt glands and, **144,** 131, 153, 155, 158–159, 162
Gel, collagen IV, **173,** 94
Gelation, cytoskeletal dynamics and, **144,** 86, 88, 105
Gelation factor, in *Dictyostelium discoideum,* **175,** 51–55
Gel electrophoresis
denaturing gradient, detection of gene mutations, **158,** 235–236
extracellular matrix-associated proteins, **146,** 275, 277, 299
polyacrylamide, 2D, nuclear matrix proteins, **162A,** 24–29
proteins related to biological clocks in retina, **146,** 107, 110, 139
pulsed-field
generation of long-range restriction map, **158,** 231–232
resolution, **158,** 231
SDS–PAGE
intracellular calcium messenger system and, **142,** 310–311

in study of Sertoli-germ cell communication network and, **147,** 66, 69, 72, 76

Gel formation, in protein fiber networks, **150,** 140

Gelsolin
- cytoskeletal dynamics and, **144,** 107
- G-actins, **156,** 233–235
- myofibrillogenesis and, **143,** 172–174
- proteins in striated muscle and, **142,** 109–110
- regulation by membrane phospholipids, **139,** 292–299
- solation of actin gel, **156,** 242–243

Gene expression
- animal mtDNA and, **141,** 187, 217, 226–229
- *Arabidopsis thaliana* and, **144,** 74
- carotenogenic gene, during chromoplast development, **163,** 214–216
- *Drosophila* muscles and, **143,** 64
  - contractile protein genes, **143,** 117, 126
  - development, **143,** 78–82
- endosymbiont hypothesis and
  - mitochondrial 280-281, **141,** 298–308
  - plastids, **141,** 266–267
- flowering plants and, **143,** 298
  - amino acids, **143,** 316, 324
  - ammonium, **143,** 305, 308–309
  - nitrate, **143,** 301–302
- during formation of generative cells and male gametes, **140,** 211–213
- higher plant mitochondrial genomes and, **141,** 145, 158
- higher plant plastids and, **145,** 159–162, 164
- hormonal control of amphibian metamorphosis and, **145,** 105, 121
- kinetoplastid mtDNA and, **141,** 84
- mitochondrial genomes in fungi and, **141,** 100, 112, 115
- patterns
  - class I interferons, **162A,** 390–393
  - histone, cell cycle-specific, **162A,** 263–264
  - in interphase nucleus, **150,** 150, 167–168
  - lamin multigene family, **162B,** 146–148
  - in melanocytes, **146,** 222, 224
  - osteocalcin, during osteoblast differentiation, **162A,** 265–269
  - phenotype-specific, **162A,** 105–106
  - steroid-regulated, **162B,** 338–344
  - two groups during pollen development, **140,** 20–22
- patterns in RPE
  - *Chx-10* homeobox genes, **171,** 229–230
  - *Mi* microphthalmia gene, **171,** 231
  - Msh-like homeobox genes, **171,** 228–229
  - *Otx-1* and *Otx-2* homeobox genes, **171,** 229
  - *Pax-2* paired-box-containing gene, **171,** 229
  - *pax-6* paired-box-containing gene, **171,** 229
  - *Rx* homeodomain gene, **171,** 230
- self-incompatibility and, **143,** 277
- transcriptional regulation and, **143,** 1, 31
  - function, **143,** 2–3, 5–6
  - genes, **143,** 11–12, 18
- transdifferentiation in medusae and, **142,** 213, 215, 237, 249, 256
- vertebrate cell culture technology and, **142,** 173

Gene linkage
- computer analysis, **158,** 228–229
- distance between flanking markers, **158,** 230–231
- familial studies, **158,** 227–230

Generative cells
- cycle
  - heterochrony, **140,** 180
  - MT dynamics, **140,** 179–180
  - prophase, **140,** 179
- division, **140,** 170–180
  - conventional spindles and phragmoplast cell plates, **140,** 170–172
  - further considerations
    - mitotic apparatus, architecture, **140,** 175–177
    - PPB, absence of, **140,** 177–178
    - vegetative cell interactions, **140,** 178–179
  - generative cell cycle
    - heterochrony, **140,** 180
    - MT dynamics, **140,** 179–180
    - prophase, **140,** 179
  - GSP locomotion
    - background, **140,** 168–169
    - not self-propelled, **140,** 169–170

Generative cells (*continued*)
GSP MTs and cell shape
evidence in favor, **140,** 165–167
MTs, sufficiency of to maintain cell shape, **140,** 167–168
history, **140,** 170, 172–175
generative nucleus, **140,** 194–195, 197
interface between vegetative and generative cells, **140,** 193–197
and male gametes
characterization of in situ, **140,** 207–213
gene expression, **140,** 211–213
structural characteristics, **140,** 207–211
conclusions, **140,** 228–229
formation during gametophyte development, **140,** 205–206
isolation, **140,** 213–218
purification of released cells, **140,** 216–218
release, **140,** 215–216
physiology and biochemistry, **140,** 218–228
biochemical characteristics, **140,** 226–228
immunological characterization, **140,** 226–227
polypeptide-protein characterization, **140,** 227–228
functional characteristics, **140,** 224–226
structural characteristics, **140,** 218–221
viability and maintenance, **140,** 221–224
functional characteristics
metabolic activity, **140,** 224–225
transcriptional activity, **140,** 225
translational activity, **140,** 225–226
structural characteristics
generative cells, **140,** 218–219
male gametes, **140,** 219–221
viability and maintenance, **140,** 221–224
organelles, **140,** 197–201
and sperm cytoskeleton
actin in, **140,** 160–165
correlative studies, **140,** 162–163
fluorescence and electron microscope evidence, **140,** 161–162
implications, **140,** 163–165
aspects of GSP, **140,** 150–151
conclusions, **140,** 180
division, **140,** 149–150
functions
evolutionary considerations, **140,** 165
generative cell division, **140,** 170–180
conventional spindles and phragmoplast-cell plates, **140,** 170–172
further considerations, **140,** 175–179
mitotic apparatus, architecture, **140,** 175–177
PPB, absence of, **140,** 177–178
vegetative cell, interactions with, **140,** 178–179
generative cell cycle, **140,** 179–180
heterochrony, **140,** 180
MT dynamics, **140,** 179–180
prophase, **140,** 179
history, **140,** 170
unconventional arrangement of division apparatus as typified by generative cell of *Tradescantia,* **140,** 172–175
anaphase-telophase, **140,** 174–175
background, **140,** 172
mitotic apparatus, establishment, **140,** 172–174
GSP locomotion
background, **140,** 168–169
not self-propelled, **140,** 169–170
GSP MTs and cell shape
evidence in favor, **140,** 165–167
MTs, sufficiency of to maintain cell shape, **140,** 167–168
microtubules, **140,** 151–160
considerations, **140,** 159–160
structural organization, **140,** 151–158
comparison with MTs, **140,** 155–158

general characteristics, **140,** 151–155
possible variations, **140,** 155
Gene-related peptide, calcitonin, **138,** 164–166
Genes, *see also* Alleles; Oncogenes; Promoters; Protooncogenes; Pseudogenes
2BE2121, **162A,** 469–476
accessory, urease in plant cells and, **145,** 86, 88, 92, 98–99
acetylcholinesterase, in avian neurogenesis, **151,** 140, 169
Adhl, **140,** 4
ALD2, **165,** 29
Ameiotic I, role in cell commitment to meiosis, **158,** 93–94
γ-aminobutyric acid metabolism and, **147,** 313
amplification
Drosophila muscles and, **143,** 64
drug resistance, **156,** 78, 113–114
and karyosphere, **144,** 27
anther- and pollen-specific, function, **140,** 29–32
antisense, chalcone synthase and, **147,** 257–258
armadillo, **139,** 151
atp6, higher plant mitochondrial genomes and, **141,** 133, 142, 148, 160
atp9
endosymbiont hypothesis and, **141,** 309
higher plant mitochondrial genomes and, **141,** 158, 160, 162–163
structure, **141,** 141, 145, 147–148, 151
atpA, higher plant mitochondrial genomes and, **141,** 133
recombination, **141,** 150–153
structure, **141,** 136, 142, 145, 148
auxin-regulated
ACC synthase, **152,** 132–133
arcA, cDNA clone, **152,** 129
biochemical analysis, **152,** 110
cloning, **152,** 111
dbp in *Arabidopsis,* **152,** 133
downregulation, **152,** 135
function, **152,** 136
GH3, expression, **152,** 115
GUS, activity, **152,** 126, 128, 134
parA
characterization, **152,** 121
responsiveness, **152,** 126
parB, in tobacco mesophyll protoplasts, **152,** 123–124
prokaryotic, **152,** 133–134
rolB, activation, **152,** 134
SAURs, characterization, **152,** 115
in strawberry fruits, **152,** 133
targeting, **152,** 136
avirulence
as hypersensitive response elicitors, **148,** 7
in pathogen-plant interactions, **148,** 29
avr, Arabidopsis thaliana and
biochemistry, **144,** 73, 75
genetic model, **144,** 61
models, **144,** 76–77
pathogens, **144,** 56–57
phenotypes, **144,** 67, 69
avr9, encoded hypersensitive response elicitor, **148,** 7
bacterial flagellar motor and, **147,** 98, 100, 102, 104–106, 108–110
Bcpl transcript, size, **140,** 23
β-type adaptin, **167,** 18
Bp4, **140,** 20, 31
Bp19, **140,** 9, 14, 20, 31
mRNA, **140,** 6, 14, 20, 31
Brachyury, role in mesoderm formation, **176,** 276–278
butylcholinesterase, in avian neurogenesis, **151,** 140
cAMP-responsive, **139,** 318–319
cdc2, **156,** 37
ced, in *Caenorhabditis elegans,* **163,** 123–126
cell death-related, **163,** 116–117
in *Caenorhabditis elegans,* **163,** 123–126
in *Drosophila melanogaster,* **163,** 127–129
cell-division cycle, **164,** 162–163
cellular transfer DNA, in habituated cells
expression, **152,** 289–290
homology with transfer DNA of Ri plasmid, **152,** 287–289
role, **152,** 287, 293
chalcone synthase, **147,** 233, 277; **148,** 27–28

Genes (*continued*)
chaperone
organization, **174,** 163–166
stress response, **174,** 161
chicken α-globin, **162A,** 332–333
chimeric, higher plant mitochondrial genomes and, **141,** 147–148, 153
chloroplasts and, **144,** 263, 281
classic improvement, **165,** 2
c-Myb, chalcone synthase and, **147,** 272
c-*myc*
amplified gene locus, **162B,** 430–434
chalcone synthase and, **147,** 272
change of localization, **162B,** 395
gene, **167,** 194
mRNA levels, **162B,** 342
promoter region, **162B,** 355–356
regulation of cell cycle, **157,** 249–250, 253–254
role in apoptosis, **157,** 253–254
role in chondrocyte differentiation, **159,** 321
cob, endosymbiont hypothesis, **141,** 308–309, 325, 328
collagen I, **173,** 81–82
collagen III, **173,** 87–88
collagen VII, **173,** 102–103
collagen XVII, **173,** 104–105
conserved
AT-rich sequences, **148,** 89–91
time of expression, **148,** 83–86, 112
control, **147,** 254, 256–268
control of rhythmic behaviors and, **147,** 4
4-coumarate–CoA ligase, in plant defense, **148,** 27–28
cox1
endosymbiont hypothesis and, **141,** 290, 293, 308–309, 325, 328
higher plant mitochondrial genomes and, **141,** 133, 148, 160
cox2
endosymbiont hypothesis and, **141,** 324–325
higher plant mitochondrial genomes and, **141,** 133, 147–148, 150, 160–162
cox3 endosymbiont hypothesis and, **141,** 325
c-*rel,* transcriptional regulation and, **143,** 19, 29, 34, 37, 43
C-rich template, kinetoplastid mtDNA, **141,** 72, 78, 84
cyanelle genome organization
algal plastid DNA, **151,** 38
biosynthetic pathway proteins, **151,** 49–50
photosynthetic apparatus, **151,** 45–49
proteins of translation apparatus, **151,** 42–45
ribosomal RNA, **151,** 39–41
transfer RNA, **151,** 41–42
cytoskeletal dynamics and, **144,** 94, 104, 109–110, 119–120
DDR2, **165,** 28–29
DDR48, **165,** 29
Deg-1, role in neurodegeneration, **163,** 119
degenerins, **163,** 119
Dekkera mitochondrial, in fungi, **141,** 101–103
delila, chalcone synthase and, **147,** 275–276
det1, chalcone synthase and, **147,** 277
Dfd, **174,** 48–52
DHFR, **162A,** 470–476
distribution in chromosomal DNA loops, **162B,** 416–425
Drosophila, **139,** 126–134
Drosophila head, expression pattern, **174,** 2–3
Drosophila muscles, *see Drosophila* muscles, genetic analysis
early nodulin, **156,** 33–37
echinus, in *Drosophila melanogaster,* **163,** 129
ein, Arabidopsis thaliana, **144,** 75–76
en, **139,** 128–134
ENA1
calcineurin, **165,** 33
HAL3 protein, **165,** 33
endothelin, expression studies, **171,** 269
endothelin receptor subtypes, expression studies, **171,** 270–271
engrailed, **174,** 11–13
enhancers, **147,** 269–271
Enod, **156,** 32–37

env, transcriptional regulation and, **143,** 29
enzymatic activity, **147,** 235–236, 241
evolution rate, and gene product temporal expression in development, **148,** 84
eye-absent, in *Drosophila melanogaster,* **163,** 127–128
F2S transcripts, **140,** 7
flowering plants and, *see* Flowering plants
functions
  mitosis and, **144,** 252
  product functions, **147,** 246–252
G
  evolution, **158,** 8
  matching for incompatibility, **158,** 8
GACI, **165,** 29
gap, **139,** 124–125, 125; **174,** 23–26
globin, expression in temporal sequence, **148,** 84–85
gls, germ/soma differentiation in *Volvox carteri* embryos, **139,** 206–207
glutamate metabolism and, **147,** 306
P-glycoprotein, **171,** 124–125
GmEnod, **156,** 35–37
gonadotropin releasing hormone, in teleosts, **147,** 165–166, 186–187
GPDI, **165,** 10, 29
GSI, breakdown, **140,** 454–457
  self-incompatibility due to mutation outside S locus, **140,** 457
  S locus mutants, **140,** 454–456
    tripartite mode of, **140,** 456
GUS
  Arabidopsis thaliana, **144,** 71, 73
  self-incompatibility and, **143,** 277–278
H19
  disease role, **160,** 86–87
  methylation, **160,** 67–68, 72–73
  regulation by imprinting, **160,** 75–78
  replication timing, **160,** 70, 73
HAL, **165,** 15, 16
HAL1, **165,** 15, 29
HAL2, **165,** 15, 16
HAL3, **165,** 15
HAL3 and *ENAI,* **165,** 33
heterozygosity loss, melanocytes and, **146,** 247
hierarchical expression, **148,** 109–104
higher plant plastids and, **145,** 204
  evolution, **145,** 165–166
  metabolism, **145,** 173, 186, 194
  structure, **145,** 153–164
histone
  codon bias in, **139,** 251–254
  H1, **139,** 227, 233–234
  H1 as, **139,** 227, 233–234
  H5, **162A,** 231–232
  nucleosomal, **139,** 226, 230–233
  and osteocalcin, **162A,** 253–263
homeobox
  and development of neural crest, **138,** 274–283
  Drosophila muscles and, **143,** 79–82, 84
  E2A-PBX1, tumorigenesis in transgenic mice bearing, **164,** 52–54
  role in cell death, **163,** 145
homeotic, role in differentiation, **148,** 103–104
Hox
  mutations on development of branchial arch derivatives, **138,** 275–278
  nomenclature, **159,** 276
  regional specification, **175,** 271–272
  retinoic acid effect on expression, **159,** 277
  role in cartilage formation, **159,** 275–277
Hox 2, in components of branchial arches, expression, **138,** 278–279
Hox 7-msh family and branchial arch development, **138,** 279–283
hrp, Arabidopsis thaliana, **144,** 57
HSP12, **165,** 29
HSP26, **165,** 29
HSP104, **165,** 29
in identification of MAR sequences, **162A,** 289–293
Igf2
  expression in cancer, **160,** 77–78, 86–87
  methylation, **160,** 64–68, 72–73, 88
  regulation by imprinting, **160,** 75–78
  replication timing, **160,** 70, 73
Igf2R
  imprinting, **160,** 55–57, 62
  methylation, **160,** 64–65, 67–68, 72–73

Genes (*continued*)
regulation by imprinting, **160,** 75–78
replication timing, **160,** 70, 73
immediate early I, transcriptional regulation and, **143,** 44
incolorata, chalcone synthase and, **147,** 261
interferon inhibition of expression of growth-inducing, **167,** 231
IRA, in adenylate cyclase regulation, **139,** 313–314
kit, melanocytes and, **146,** 214, 247
Knal, salt tolerance, **165,** 37
lab, **174,** 48–52
LAT52, **140,** 8
promoter, **140,** 15, 16
LAT56 promoter, **140,** 16
LatS9, **140,** 8
promoter, **140,** 16
LAT series, **140,** 27, 31–32
Lolpl, **140,** 29
LolplB, **140,** 25, 29, 31
luxury, melanocytes and, **146,** 198
major resistance, *Arabidopsis thaliana,* **144,** 56, 58, 60, 66, 68, 76
master, retinal precursor cells and, **146,** 166
mcd, role in muscle cell death, **163,** 130
mdr, **156,** 79, 81–82, 97
MDR1, **162A,** 563–564
mdrl, **156,** 79, 81, 97
mec-4 degenerin, role in neurodegeneration, **163,** 119
melanocytes and, **146,** 242–246
melanoma-associated expression, **146,** 228, 237, 239, 241–242, 250
MON, role in meiosis, **158,** 105
mot, bacterial flagellar motor and, **147,** 105–108
Mpsl, spindle pole body functions, **169,** 67
MsEnod, **156,** 34, 36
mSOS1, role in lymphocyte signal transduction, **157,** 224–226
Msx-1, role in cartilage formation, **159,** 277–278
mul, **139,** 206–207
in multigene families, differential expression, **148,** 107–108
muscle fiber type-specific programs of expression, **170,** 155–158
myb, chalcone synthase and, **147,** 268, 273–274
myc, chalcone synthase and, **147,** 268
nad, endosymbiont hypothesis, **141,** 282, 290, 293, 295
nadl
endosymbiont hypothesis and, **141,** 304
higher plant mitochondrial genomes and, **141,** 157–159, 162
neurogenic, **139,** 140–143
neurogenic gene, **139,** 140–143
NIC96p, **162B,** 286–287
nitrate reductase, **139,** 227, 238–239
nivea, chalcone synthase and, **147,** 259–261
nod, **156,** 1–3, 5–7
activators, **156,** 31–32
auxiliary functions, **156,** 27–29
chalcone synthase and, **147,** 248–250
inducers, **156,** 8–9, 14–15, 49–51
nonflavonoids, **156,** 9, 49
Nod factors and, **156,** 23–29
regulation, **156,** 7, 8
soybean mutants, **156,** 41–43
transfer, **156,** 30
Nod^{DTW}, mutation, **167,** 278–279
nodABC, **156,** 5, 24, 27
nod box, **156,** 7, 9–10, 12, 14–16
nodD, **156,** 12–13, 19
regulation of expression, **156,** 13–14, 16, 17
nodDEF, **156,** 5
nodE, **156,** 24, 30
nodEF, **156,** 5, 24, 27
mutants, **156,** 25, 26
nodF, **156,** 24
nodFEGHPQ, **156,** 30
nodG, **156,** 25
nodH, **156,** 26
nodI, mutants, **156,** 28
nodIJ, **156,** 27
nodL, **156,** 25
mutation, **156,** 31
nodM, **156,** 24–25
mutation, **156,** 31
nodN, **156,** 28
nodO, **156,** 28
nodP, **156,** 26
nodQ, **156,** 26
$nodQ_1$, mutation, **156,** 26, 31
nodS, **156,** 25
mutation, **156,** 31
nodT, **156,** 28
nodVW, **156,** 17

nodX, **156,** 26
nodZ, **156,** 26, 27
mutation, **156,** 31
nol, **156,** 5
nolA, **156,** 18
nol B, **156,** 45
nolFGHI, **156,** 28
nolJ, **156,** 16
nolK, **156,** 26
nolO, **156,** 26
nolR, **156,** 17
NSP1p, **162B,** 280–281
nts, mutation, **156,** 42, 43
nuclear, role in microtubule assembly, **175,** 118
NUP1p, **162B,** 279–280
nur 77, role in apoptosis, **157,** 251, 255–256
ontogenic open system, **155,** 36
ORF400, and ciliates, **141,** 36–38
overlap and pollen gene expression, **140,** 567–568
p35 tumor suppressor, role in apoptosis, **163,** 115
P53
as inhibitory regulator of cell proliferation, **167,** 233
phosphoprotein, TGF-β and, **167,** 216
PA13, **165,** 29
PAL, Arabidopsis thaliana, **144,** 72–73, 77
pathogenesis-related protein, expression, **148,** 26
pcf, higher plant mitochondrial genomes and, **141,** 132, 147, 149
pgp1, **156,** 81
phenylalanine ammonia-lyase, activation in plants, **148,** 27–28
photosynthetic, transcription, **163,** 195–197
plant defense
activation, **148,** 26, 30
promoters, **148,** 28
POM152, **162B,** 290
PoPo, petunia plants, **140,** 13
in preimplantation diagnosis of genetic defects
myelin basic protein, **151,** 18
proteolipid protein, **151,** 18
proopiomelanocortin, **170,** 80–85
psbA, higher plant plastids and, **145,** 162–163
psbB, higher plant plastids and, **145,** 163–164
PsEnod, **156,** 16, 31, 33, 36
ptc, **139,** 128–134
RAS, in adenylate cyclase regulation, **139,** 311–313
ras, intracellular calcium messenger system and, **142,** 311
rbcL, higher plant plastids, **145,** 163–164
reaper, in *Drosophila melanogaster,* **163,** 127
recombinant aneuploidy and aberrant genes, **167,** 282–283
regulation
Drosophila muscles and, **143,** 64–65, 82
transcription and, **143,** 1, 11–12, 25, 47–48
inflammation, **143,** 17–19
T lymphocytes, **143,** 12–17
regulatory
Drosophila body plan, **174,** 30–33
intermediate filaments, *see* Intermediate filament, gene-regulatory function
in patterning insect body plan, **174,** 45–48
Dll and *cnc* expression, **174,** 52–53
initiation and segment-specific stages, **174,** 48–52
significance, **174,** 53–54
rel, transcriptional regulation and, **143,** 47
biochemistry, **143,** 24
molecular cloning, **143,** 25–26, 28, 33–34, 37–39
ribosomal
protein, in cyanelle DNA, **151,** 44
in *Volvox carteri,* **139,** 224–225
rjl, soybean mutant, **156,** 41
rolA, **156,** 45
rolABC, **156,** 45
role in *Drosophila* development, **174,** 40–45
role in *Tribolium* larval development, **174,** 43–45
roughest, in *Drosophila melanogaster,* **163,** 129
rox-8, in *Drosophila melanogaster,* **163,** 127
RPMI, Arabidopsis thaliana and, **144,** 67–68, 74
rps4, higher plant mitochondrial genomes and, **141,** 149

Genes (*continued*)
rrn26, higher plant mitochondrial genomes and, **141,** 142, 150–151
rrnl8, higher plant mitochondrial genomes and, **141,** 142, 157
S, *see* S genes
saa, **156,** 26
Saccharomyces cerevisiae, see Saccharomyces cerevisiae
Scr, **174,** 48–52
self-incompatibility, **143,** 258, 290
gametophytes, **143,** 280–286, 289
sporophytes, **143,** 258–261, 264, 273, 275, 278
Sertoli-germ cell communication network and, **147,** 37, 57, 60, 69, 74, 77
sev, **139,** 148–149
sex-determining, *see* Sex-determining genes
sex-inducer, codon bias and, **139,** 254
SIP18, **165,** 29
SLG
expression in *Brassica* stigmas, **140,** 512–514
homologous sequences from anthers and leaves, **140,** 519–521
homologous transcripts in *Brassica* anthers, **140,** 517–521
and promoter constructs in transgenic plants, **140,** 514–517
SLG and *SRK,* isolation, **140,** 487–493
SLR sequences, isolation, **140,** 494–496
SLR1 subfamily, **140,** 503–506
SLR2 subfamily, **140,** 506–507
SNFI, stress tolerance, **165,** 33
Snrpn, methylation, **160,** 67
Sos, in *Drosophilia,* **139,** 149
SRA, self-incompatibility and, **143,** 272–276
SRB, self-incompatibility and, **143,** 272–276
SrEnod2, **156,** 35
SRP1p, **162B,** 287–289
SRV2(CAPI), in adenylate cyclase regulation, **139,** 313–314
SRY
characterization, **154,** 233–237, 247
cloning, **154,** 233–237
demonstration of sex determination, **154,** 236, 238
evolution of function, **154,** 238–241
evolution of organization, **154,** 241–245
identification, **154,** 246
mechanism, **154,** 238–241
sym pea mutants, **156,** 40, 41
syrM, **156,** 12, 14, 15, 19
T*me*, imprinting, **160,** 55–56
TA29, **140,** 27, 31
TA36, **140,** 27, 31
TA56, **140,** 27, 31–32
Tau proteins, organization, **171,** 170
testis-determining factor, **154,** 229–233
tissue-specific
coordinate transcription, **148,** 86–88
quantitative regulation, **148,** 88–89
TM 8, role in meiosis, **158,** 105
TPK1, protein kinase A, **165,** 32–33
TPK2, protein kinase A, **165,** 32–33
TPK3, protein kinase A, **165,** 32–33
Tpr/p265, sensitive to proteolysis, **162B,** 273–274
TPS1, **165,** 9–10, 29
TPS2, **165,** 9–10, 29
trans-acting factors, **147,** 271–277
transcription, **147,** 268–269
MAR sequence role, **162B,** 356–359
transcriptional regulation, **143,** 7, 11, 24, 47
molecular cloning, **143,** 25–26, 29–30, 34–35
proteins, **143,** 45–47
viruses, **143,** 40, 43–44
tropomyosin, **170,** 5–7
tubulin, **139,** 225–226, 228–230
codon bias in, **139,** 251–253
tumor cell control by recessive tumor suppressor, **167,** 233
tumor-specific, and habituation, **152,** 290–291
tyrosinase, transcription, regulation, **171,** 233–235
UB14, **165,** 29
ubiquitin, **139,** 227, 234–237
upl, chalcone synthase and, **147,** 274–276
urease
products urease in plant cells and, **145,** 88–92
structural genes, in plant cells, **145,** 84–90
urfl3, higher plant mitochondrial genomes and, **141,** 131–132, 148, 150, 154, 160

in vacuolar biogenesis, **139,** 102–103
mutants as, **139,** 93–97
VPS15 gene as, **139,** 103–106
VPSI gene as, **139,** 106–107
V-ATPase, molecular cloning
16- and 17-kDa subunit gene, **174,** 97–98
55- and 62-kDa subunit gene, **174,** 97
67- to 73-kDa subunit gene, **174,** 96–97
95- to 116-kDa subunit gene, **174,** 98–99
for V-H^+-ATPase subunits, **167,** 24
viral, *see* Viral genes
V-myc, role in chondrocyte differentiation, **159,** 322
volvocine, **139,** 204–218, 224–246
Volvox carteri, **139,** 224–246
VPS15, in vacuolar biogenesis, **139,** 103–106
VPSI, in vacuolar biogenesis, **139,** 106–107
wg, **139,** 128–134
whp, chalcone synthase and, **147,** 254, 256, 268
Wnt-1, mutation, **175,** 269
Wnt/wingless signaling pathway, β-catenin, **165,** 187–191
Xist
expression in embryo, **160,** 81–82
methylation, **160,** 67, 80–82
X chromosome regulation, **160,** 80–82
xyloglucan-related proteins, **173,** 182–184
by auxin, **173,** 188–189
by brassinosteroids, **173,** 189–191
environmental signals, **173,** 191–192
by other hormones, **173,** 191
spatial and temporal regulation, **173,** 186–188
ypt
codon bias, **139,** 254
introns, **139,** 250–251
Z, role in self-incompatibility, **158,** 48
zeste, product of *Drosophila*
function, **154,** 65
matrix element, **154,** 24–26
zfh-l, Drosophila muscles and, **143,** 81
Gene therapy, modulation of arterial smooth muscle cell behavior, **169,** 229
Genetic control
Drosophila larval epidermis
antennal cuticle, **174,** 37
clypeolabrum, **174,** 35–36
dorsal ridge, **174,** 40
intercalary cuticle, **174,** 37–38
labial cuticle, **174,** 39–40
mandibular cuticle, **174,** 38–39
maxillary cuticle, **174,** 39
ocular cuticle, **174,** 36–37
required genes, **174,** 33–35
evolution, **174,** 54–56
dorsal ridge evolution, **174,** 60–62
insects as arthropods, **174,** 63–65
mouthpart evolution, **174,** 56–60
segmentation
clypeolabral region, **174,** 26–27
dorsal ridge, **174,** 27
head gap genes, **174,** 23–26
labial segments, **174,** 20–23
maternal components, **174,** 19–20
maternal polarity systems, **174,** 18
maxillary segments, **174,** 20–23
segment-polarity, **174,** 18–19
Genetic defects
diagnosis, *see* Fluorescence *in situ* hybridization
preimplantation diagnosis
animal models of human genetic disease, **151,** 1–3
biopsy techniques
blastocyst, **151,** 9–12
culture of biopsied embryo, **151,** 13
early cleavage, **151,** 7–9
polar body, **151,** 6–7
proliferation of cells, **151,** 12
CA repeat polymorphisms, **151,** 19
chromosome *c* banding technique, **151,** 15
cystic fibrosis, **151,** 5
cytogenetics as chromosome analysis method, **151,** 15–16
DNA analysis, **151,** 17–19
ethical issues, **151,** 20
human embryo quality, **151,** 16–17
hybridization, *in situ,* fluorescent techniques, **151,** 13–15
hypoxanthine phosphoribosyl transferase activity as indicator, **151,** 2
legal issues, **151,** 20
lipid storage diseases, **151,** 14–15
myelin basic protein gene, **151,** 2–3
oligonucleotide ligation assay, **151,** 18

Genetic defects (*continued*)
polymerase chain reaction techniques, **151,** 2–5, 17–19
prevention role, **151,** 1
primer extension amplification reaction, **151,** 19
proteolipid protein, **151,** 2
proteolipid protein gene, **151,** 18
sickle cell anemia, **151,** 3
in situ hybridization, fluorescent techniques, **151,** 13–15
technical innovations, **151,** 17–19
Genetic engineering
higher plant mitochondrial genomes and, **141,** 165
vertebrate cell culture technology and, **142,** 146, 178, 191, 194
Genetic factors
in habituated tumorous plant cells
induction, **152,** 271–272
molecular aspects, **152,** 277–279
in nondisjunction of paternal origin, **167,** 283–286
Genetic imprinting
egg cytoplasm, genome modifications, **160,** 57–61
evidence in mammals, **160,** 54
gametogenesis, **160,** 55–57
gene methylation, **160,** 57, 62–63, 67–69
allelic inactivation, mechanisms
chromatin role, **160,** 71–72
imprinted gene function, **160,** 72–73
replication timing, **160,** 70
continuity between gametes and embryonic cells, **160,** 63–65
transgenes, **160,** 65–66
gene regulation, **160,** 75–78
human disease, role, **160,** 84–87
mating-type switches in yeast, **160,** 61–62
RNA modulation in DDK genome, **160,** 58–59
transgenic mice, **160,** 56, 59, 65–66
X chromosome inactivation, **160,** 54, 63
Genetic studies
Arabidopsis thaliana and
biochemistry, **144,** 72–74
genetic model, **144,** 60–64
models, **144,** 76–77
pathogens, **144,** 55–60
phenotypes, **144,** 66, 68, 70
avian retina and, **146,** 26, 35
biological clocks in retina and, **146,** 83–84, 116
coat-color, melanocytes and, **146,** 208, 214, 251
extracellular matrix and, **146,** 261–262, 313–314
characterization, **146,** 262, 268–269, 280, 282
modification, **146,** 287, 289, 291–292, 295, 299–301
stress, **146,** 303–304, 307, 310–313
GSI
control, **140,** 453
genetic basis, **140,** 452–454
genetic breakdown, **140,** 454–457
self-incompatibility due to mutation outside S locus, **140,** 457
S locus mutants, **140,** 454–456
instability of homogeneous state, **154,** 362–367
melanocytes and, **146,** 191, 227, 248–249
biology, **146,** 193–195
development, **146,** 214–224
function, **146,** 202–214
mouse coat-color loci, **146,** 197–202
melanoma, **146,** 224–225, 250–251
biology, **146,** 235–237
clinical evidence, **146,** 225–232
fish, **146,** 232–234
growth factors, **146,** 237–241
oncogenes, **146,** 241–247
tumor suppressor genes, **146,** 247–250
mitosis and, **144,** 226–233, 252
relationship to embryonic induction, **150,** 375–380
retinal pigmented epithelium and, **146,** 77–78
retinal precursor cells and, **146,** 166, 174, 178–179, 184
reverse, melanocytes and, **146,** 234
urease in plant cells and, **145,** 65, 79, 98–99
accessory genes, **145,** 86, 88
gene products, **145,** 88–92
metabolic origins, **145,** 75, 78
structural genes, **145,** 84–87
whole-chromosome hybridization in, **153,** 214–219
Genetic tumors, and habituation
auxin effects, **152,** 265–266, 285–286, 292
cDNA clones, specificity, **152,** 291

cellular transfer DNA genes, **152,** 287–290, 293
characterization, **152,** 279–280
cytokinin effects, **152,** 265–266, 286–287, 292
development, **152,** 266–267
hormonal aspects, **152,** 265–266, 285–287, 292
induction
in vitro, **152,** 282–285
in vivo, **152,** 282
interconversion
molecular basis, **152,** 287–291, 293
in vitro system, **152,** 283–285
in *Nicotiana,* **152,** 280–282
protoplast fusion, **152,** 282–283
stress-induced formation, **152,** 290
tumor-specific genes, **152,** 290–291
Gene transfer
endosymbiont hypothesis and, **141,** 237
mitochondrial 320, **141,** 324–325
nuclear genome, **141,** 249–250
plastids, **141,** 277–278
flowering plants, **143,** 317
in microspores and microspore-derived embryos
Agrobacterium-mediated gene transfer, **140,** 427–428
direct DNA delivery to microsporederived embryos, **140,** 428–431
biolistic DNA delivery, **140,** 429–431
microinjection, **140,** 428–429
Genistein
effect on pituitary cell proliferation, **157,** 54
inhibition of angiogenesis, **159,** 145
inhibition of *Chlamydomonas* gametogenesis, **168,** 290, 292
tyrosine kinase inhibition, **157,** 53–54
Genital abnormalities
antihormones and, **139,** 13–26
sex hormones and, **139,** 3–13
Genium, **139,** 197–198
Genomes
chloroplast
algae containing chorophyll *b,* **151,** 36
higher plants, **151,** 35–37
cyanelle, *see* Cyanelle genome
eukaryotic, functional organization, **162A,** 456–463
function, nuclear matrix as structural milieu, **162A,** 1–54
hybridization, comparative, **153,** 192–193, 218
mitochondrial
ciliates, *see* Ciliates, mitochondrial genomes
fungi, *see* Fungi, mitochondrial genomes
higher plant, **141,** 129, 164–165
abnormal phenotypes, **141,** 129–130
cytoplasmic male sterility, **141,** 130–134
NCS mutants, **141,** 134–135
gene location, **141,** 157
protein-coding genes, **141,** 157–159
tRNA, **141,** 159
intron splicing, **141,** 161–162
post-transcriptional regulation, **141,** 161
protein synthesis, **141,** 163–164
RNA editing, **141,** 162–163
structure
physical characterization, **141,** 135–137
plasmids, **141,** 137
recombination events, **141,** 147–157
replication, **141,** 146–147
restriction mapping, **141,** 137–146
transcription, **141,** 160–161
nuclear, endosymbiont hypothesis and, **141,** 236
origin, **141,** 247–253
plastids, **141,** 277–278
organization, endosymbiont hypothesis and, **141,** 288–298
organization and evolution, **162B,** 435–438
SacII sites, in fungi, **141,** 110
Genotype
Arabidopsis thaliana, **144,** 64, 74
pathogens, **144,** 55, 57–58
phenotypes, **144,** 65–70, 72
higher plant mitochondrial genomes, **141,** 134, 149
melanocytes, **146,** 208–209, 220–221, 233
self-incompatibility, **143,** 272–273, 277, 280, 284, 286
urease in plant cells, **145,** 93, 97
vertebrate cell culture technology, **142,** 192

Gentamicin, **156,** 290
Geodomes, in lens cytoskeleton, **163,** 326
Geometric clutch model
and experimental data, **173,** 50–54
flagellum, **173,** 39–46
oscillations in, **173,** 48–50
role of t-force, **173,** 53
as rudimentary model, **173,** 54–55
t-force in, **173,** 46–47
Geometry, *see* Cholesteric geometry
Gephyrin
associated with glycine receptor, purification, **176,** 205–207
diversity of GlyR, **176,** 223
in glycine receptor
GlyR assembly, **176,** 220–223
GlyR-rich postsynaptic microdomains, **176,** 219–220
at nonglycinergic synapses, **176,** 228–230
Geranium, plastids and, **145,** 155
Gerbera jamesonii sperm cell, **140,** 197, 198, 199, 201–202
Germ cells
endocytotic activity, **160,** 187
evolution, **148,** 83
junctions with Sertoli cells, **160,** 184, 186–188, 208
male
endocytosis, **154,** 128–130
secretory functions, **154,** 126, 128
primordial, Sertoli-germ cell communication network and, **147,** 34, 38, 62
regulation of secretion of Sertoli cell proteins, **160,** 203
-Sertoli cell adhesion molecules, structure, **147,** 36
-Sertoli cell communication network, *see* Sertoli-germ cell communication network
–soma differentiation, in asexual *Volvox carteri* embryos
asymmetric division and cell size in, **139,** 208–211
cell type-specific genes in absence of asymmetric division in, **139,** 214–215
cytological control of division symmetry in, **139,** 216–218
differential gene expression in, **139,** 211–214
model for genetic control of, **139,** 206–208
mutations as disruption in, **139,** 204–206
translocation from basal to adluminal compartment, **160,** 165, 191–193
Germinal center
affinity maturation model, **166,** 165
antigen-dependent immune response initiation, **166,** 163
centroblast in dark zone, **166,** 163–165
dendritic cell, **166,** 158
extrafollicular responses, **166,** 163
follicular dendritic cell, HIV, **166,** 160–161
hypothetical model for induction, **166,** 163, 164
lymphocyte production
antigen-driven, **166,** 140–141
memory B, **166,** 141
memory B cell generation
mechanisms, **166,** 142–144
steps, **166,** 142–144
selection of centrocytes in light zone, **166,** 165–167
antigen receptors, **166,** 166–167
structural features, **166,** 141–142
synchronized formation kinetics, **166,** 141–142
Germinal vesicles
breakdown, mitosis and, **144,** 247
karyosphere and, **144,** 44, 46–47
Germination
angiosperm seed, effect of microgravity, **171,** 32
higher plant plastids and, **145,** 177
pollen, **176,** 135–137
spore, mitochondrial fusion during, **161,** 75–76
urease in plant cells and
elimination, **145,** 80–83
metabolic origins, **145,** 71–72
nickel metabolism, **145,** 92, 94–95
zone in lens, **163,** 309, 312–313, 326
Germ unit, *see* Female germ unit; Male germ unit
Gerontoplasts, categorization, **163,** 185
Gerstmann–Straussler–Scheinker disease, pathological Tau proteins, **171,** 202–203
GH, *see* Growth hormone

GHRH, *see* Growth hormone-releasing hormone
Giant cells
in bone graft interface, **172,** 168
commissural, *see* Commissural giant cells
Giant nucleus
fragmentation, **165,** 91–92
mitosis, **144,** 230–231
Gibberellic acid, intracellular calcium messenger system, **142,** 315, 326
Gibberellins
chromoplast reversion, **163,** 202
cytoskeleton, **166,** 39
extracellular matrix and, **146,** 294
plant hormone, **166,** 39
Gingiva, regenerative cementogenesis and
established root surfaces in vitro, **142,** 29, 31–32, 37–39, 41–42
established root surfaces in vivo 44- 47, **142,** 50
growing root surfaces, **142,** 11, 17, 22
Gingko biloba, potential double fertilization in, **140,** 327–328
basic aspects of reproduction, **140,** 327–328
potential for intra-archegonial double fertilization, **140,** 328
β-Glucanase, *Arabidopsis thaliana* and, **144,** 73, 77
Glandular hypertrophy, avian salt glands and, **144,** 140–141
Glans penis, perinatal exposure to antihormones and, **139,** 25–26
Glass-bead reactors, vertebrate cell culture technology and, **142,** 189
Glaucocystis nostochinearum, cyanelle biochemical analysis in, **151,** 33
Glaucophytes, chloroplasts and, **144,** 264
Glenodinium foliaceum
eyespot reflection in, **148,** 259, 267–268
photoreceptive apparatuses, **148,** 249–250
Glial cells
γ-aminobutyric acid metabolism and, **147,** 311, 318–319
avian retina and, **146,** 40, 42
cellular processes, **146,** 25, 31–34
normal development, **146,** 3, 5
retinogenesis *in vitro,* **146,** 16–17, 19
avian salt glands and, **144,** 182
differentiation in peripheral nervous system, **138,** 294–304
4B3 antigen, **138,** 295–297
environmental control of glial cell differentiation, **138,** 300–304
glial percursors, identification of, **138,** 299–300
phenotypic diversity of peripheral glial cells, **138,** 294–295
Schwann cell myelin protein, **138,** 297–299
glutamate metabolism and, **147,** 286, 305, 308, 322
aminotransferases, **147,** 289–291
glutamate dehydrogenase, **147,** 295, 298–299
growth factor-induced retinal regeneration and, **146,** 52
interaction with neurons, **149,** 274–276
migration in amphibian brain development, **169,** 131–132
pathological Tau protein isoforms in, **171,** 207
percursors, identification, **138,** 299–300
retinal precursor cells and, **146,** 148, 153, 177, 178, 181–182
Glide symmetry model, of ion release and binding, **161,** 191–192
Gliogenesis, platelet-derived growth factors in, **172,** 114
Gliosis, nuclear magnetic resonance and, **145,** 50
Globin, genetic evolution, **140,** 499–502
Globulin, urease in plant cells and, **145,** 68
Gloeochaete wittrockiana, cyanelles, **151,** 33
Glomerular capillaries
length, **161,** 154–155
loop, schematic, **166,** 182
mean length, **161,** 158
number, **161,** 156–157
surface area, **161,** 155–156, 158
volume density of capillary lumina, **161,** 158–159
Glomerular epithelial cells
average volumes and surface areas, **161,** 151–154
quantity estimation in average glomeruli, **161,** 148–151
Glomeruli, kidney, *see* Renal glomeruli
Glorin, cytoskeletal dynamics and, **144,** 102
GLP-1, *see* Glucagon-like peptide 1
GLP-2, *see* Glucagon-like peptide 2
Glucagon, fish
amino acid sequence, **168,** 194–195

Glucagon (*continued*)
clearance
kidney, **168,** 222
liver, **168,** 219, 222
discovery, **168,** 190–191
enzyme regulation, **168,** 227–228
functions
fish, **168,** 225, 227–234
mammals, **168,** 223, 236–237
nutritional status effects, **168,** 228, 230
hepatocyte response, **168,** 229–232
induction, **168,** 214–216
localization in pancreas, **168,** 210, 212–214
plasma level, **168,** 216–218
precursor, *see* Proglucagon
processing, **168,** 192, 203, 213
sequence homology between species, **168,** 200–203, 240
signal transduction
adenylyl cyclase activation, essential residues, **168,** 201
hepatocyte-binding affinity, **168,** 235–236
pathway, **168,** 234–235, 237
receptors, **168,** 240
superfamily members, **168,** 188
synthesis sites, **168,** 187, 190, 198, 200
Glucagon-related peptide 1, fish
amino acid sequence, **168,** 196–197
clearance
kidney, **168,** 222
liver, **168,** 219, 222
functions
fish, **168,** 225–226, 228–229, 231–234, 241
mammals
glucagonostatic activity, **168,** 224
insulinotropic activity, **168,** 223–224, 242–243
nutritional status effects, **168,** 228, 230
hepatocyte response, **168,** 229–232
induction, **168,** 216
localization in pancreas, **168,** 210, 212–214
nomenclature, **168,** 203–204
plasma level, **168,** 216, 218
precursor, *see* Proglucagon
processing, **168,** 192, 203–204, 206, 208
sequence homology between species, **168,** 204–206
signal transduction
adenylyl cyclase activation, essential residues, **168,** 205
pathway, **168,** 237–238
receptors, **168,** 238–240
synthesis sites, **168,** 187, 198, 200, 241, 243
Glucagon-related peptide 2, fish
amino acid sequence, **168,** 208–209
discovery, **168,** 198
functions
fish, **168,** 234
mammals, **168,** 225
precursor, *see* Proglucagon
synthesis sites, **168,** 187, 200, 208, 243
Glucan, in plant cell wall, effect of clinorotation, **170,** 59
Glucan 1,4-β-glucosidase, role in cell wall changes under microgravity, **170,** 63
β-1,3-Glucanase, role in cell wall changes under microgravity, **170,** 61
Glucanases, extracellular matrix and, **146,** 300, 304, 310–311
Glucan-binding proteins, *see* Proteins, glucan-binding
Glucans
binding sites in soybeans, **148,** 16–17
extracellular matrix
characterization, **146,** 264
modification, **146,** 293–294, 300–301, 307
synthesis, **146,** 273–274
β-1,3-Glucan synthase, synthesis and activity, effect of clinorotation, **170,** 60
Glucocorticoid receptor
accessory factors, **162B,** 361–362
clustered distribution, **162A,** 156
Harderian gland, **168,** 42
Glucocorticoid responsive elements, osteocalcin gene, **162A,** 261–263
Glucocorticoids
control of amphibian metamorphosis and, **145,** 122
in cortical thymocytolysis, **156,** 183
glutamate metabolism, **147,** 302
role in thymocyte apoptosis, **163,** 117–118
secretion, role of endothelin, **171,** 277–278, 295
thymocyte sensitivity and resistance to, **156,** 168–170
transcriptional regulation, **143,** 38

Glucomannans, extracellular matrix and, **146,** 264, 295
Glucosamine synthase, **156,** 23
Glucose
avian salt glands, **144,** 178, 180, 192
consumption, microgravity experiments, **156,** 319
dietary absorption, **172,** 5–6
effect of on radioactivity, **138,** 81, 82
and extracellular matrix, **146,** 273, 275, 293
fetal/maternal concentrations, **172,** 28–29
higher plant plastids and, **145,** 177, 179–181, 201
isoosmotic, avian salt glands, **144,** 178, 180
level in aqueous humor, **172,** 17
nuclear magnetic resonance and, **145,** 37–38
plant
effect of clinorotation, **170,** 58
effect of microgravity, **170,** 56
regulation of GLUT1 expression, **172,** 35
vertebrate cell culture technology, **142,** 164–165, 195, 200, 204
Glucose-1-phosphate, higher plant plastids and, **145,** 169, 179–180
Glucose-6-phosphate, higher plant plastids and, **145,** 169, 180, 193–194, 201, 203
Glucose-6-phosphate dehydrogenase, avian salt glands and, **144,** 145, 147
Glucose pyrophosphorylase, higher plant plastids and, **145,** 181–182
Glucose transporters
in blood-tissue barriers
aqueous humor, **172,** 19–22
brain, **172,** 10–13
cerebrospinal fluid, **172,** 13–14
human placenta, **172,** 26–29
inner ear, **172,** 23–24
intestines, **172,** 5–6
iris, **172,** 22–23
peripheral nerves, **172,** 24–25
rat placenta, **172,** 29–31
retina, **172,** 15–17
testis, **172,** 31–32
thymus, **172,** 32–33
facilitated-diffusion family, forms and distribution, **172,** 2–5
GLUT1
in astrocytes, **172,** 11
in blood-brain barrier, **172,** 10–12
in choroid plexus, **172,** 11–12, 13–14
in ciliary body epithelium, **172,** 19–21
in endonuerial blood vessels, **172,** 25
in human placenta, **172,** 27–28
in inner ear, **172,** 23–24
in iris, **172,** 22–23
in perineurium, **172,** 24–25
in rat placenta, **172,** 29–31
regulation
developmental, **172,** 33–34
effects of degenerative disease on, **172,** 36
effects of ischemia and hypoxia on, **172,** 35–36
effects of tumors on, **172,** 36–37
by glucose, **172,** 35
in hypoglycorrhachia, **172,** 37
induction in blood vessels, **172,** 34
in retinal blood vessels, **172,** 15
in retinal pigment epithelium, **172,** 15–16
in testis, **172,** 32
GLUT2
in dietary absorption of glucose, **172,** 6
in retina, **172,** 16–17
GLUT3
in blood-brain barrier, **172,** 12
in human placenta, **172,** 27–28
in neurons, **172,** 12
in rat placenta, **172,** 29, 31
regulation
developmental, **172,** 33
effects of ischemia on, **172,** 35
effects of tumors on, **172,** 36–37
in retina, **172,** 17
GLUT4
in human placenta, **172,** 27–28
in insulin-stimulated glucose transport, **172,** 4
GLUT5, in blood-brain barrier, **172,** 12
regulation
developmental, **172,** 33–34
effects of degenerative disease on, **172,** 36
effects of ischemia and hypoxia on, **172,** 35–36
effects of tumors on, **172,** 36–37
by glucose, **172,** 35
in hypoglycorrhachia, **172,** 37
induced expression in blood vessels, **172,** 34

Glucose transporters (*continued*)
sodium-dependent family, *see also* SGLT **172,** 1
in blood-brain barrier, **172,** 13
forms and activity, **172,** 5
Glucose triphosphate, **167,** 21
α-Glucosidase, **167,** 131
Glucosteroid, **166,** 25
Glucosyl transferase, extracellular matrix and, **146,** 274–275
β-Glucuronidase
chalcone synthase and, **147,** 255–256, 266–267, 276
flowering plants and, **143,** 306, 325
reporter gene, **140,** 13
Glu-Gly-Arg-Phe-NH_2, *see* Antho-RFamide
Glu-Gly-Leu-Arg-Trp-NH_2, *see* Antho-RWamide neuropeptides
Glu-Ser-Leu-Arg-Trp-NH_2, *see* Antho-RWamide neuropeptides
Glutamate
aminotransferases, **147,** 288–294
control of rhythmic behaviors and, **147,** 14
flowering plants and, **143,** 302–303, 321
glutamate dehydrogenase, **147,** 294–301
glutaminase, **147,** 306–311
glutamine synthetase, **147,** 301–306
higher plant plastids and, **145,** 169, 171
in long-term potentiation of dendritic spines, **139,** 294
metabolism, **147,** 285–287, 322–323
neurotransmission, **149,** 256, 259–260, 263
synthesis in *Escherichia coli,* **165,** 7–8
urease in plant cells and, **145,** 69
vertebrate cell culture technology and, **142,** 204
Glutamate decarboxylase
neurotransmission end, **147,** 311–318, 321, 323
urease in plant cells and, **145,** 69
Glutamate dehydrogenase, glutamate metabolism and, **147,** 287, 303–304, 319, 322
Glutamate receptor
metabotropic, **164,** 212–213
neurotransmission, **149,** 260–262
L-Glutamate receptor
characterization, **164,** 192
molecular cloning, **164,** 192–193
Glutamate synthase–glutamine synthetase, **143,** 303–304
Glutamic acid decarboxylase, neurotransmission, **149,** 262–263
Glutaminase
glutamate metabolism and, **147,** 306–311, 322
phosphate-activated, neurotransmission, **147,** 287, 291, 306–311, 322
Glutamine
flowering plants and, **143,** 301–303, 306, 308, 322
higher plant plastids and, **145,** 169, 199–200
urease in plant cells and, **145,** 74, 78
vertebrate cell culture technology and, **142,** 165, 195, 204
Glutamine synthetase
avian retina and, **146,** 10, 16, 32
flowering plants and, **143,** 302–309
glutamate metabolism and, **147,** 287, 300–306, 309, 319, 322
–glutamate synthase, flowering plants and, **143,** 303–304
higher plant plastids and, **145,** 197–201
γ-Glutamylcysteinglycine, **156,** 134
γ-Glutamyl transpeptidase, Sertoli-germ cell communication network and, **147,** 52, 68, 76
Glutathione, **156,** 134; **167,** 26
Glutathione reductase, salt tolerance, **165,** 15
Glutathione S-transferase
and auxin-regulated genes
expression, **152,** 124
functions, **152,** 128
neoplastic transformation, **152,** 126
parB homology to, **152,** 123
role in alkylating anthracycline activity, **156,** 132
role in alkylation resistance, **156,** 134–137
Glycans
cell surface, resialylation, **175,** 150–151
extracellular matrix and, **146,** 277–278
Glycation, role in Tau protein aggregation, **171,** 194
Glycerol
intracellular concentrations, **165,** 10
metabolic cycle, **165,** 10
Glycine, as inhibitory neurotransmitter, **149,** 256

Glycine decarboxylase complex, higher plant plastids and, **145,** 196, 198
Glycine max, nodule development, **156,** 3, 31
Glycine receptor
channel properties, **176,** 212–215
gephyrin diversity, **176,** 223
GlyR assembly, **176,** 220–223
GlyR cluster formation, **176,** 223–228
GlyR-rich postsynaptic microdomains, **176,** 219–220
identification, **176,** 203–204
isoform and physiology relationships, **176,** 210–212
ligand binding, **176,** 212–215
as member of ligand-gated ionotropic receptor family, **176,** 207–208
neurotransmission, **149,** 256–257
purification, **176,** 205–207
spatial and temporal mapping
immunohistochemical studies, **176,** 219
in situ hybridization studies, **176,** 215–219
subunit isoform diversity, **176,** 208–210
Glycine-rich proteins, effects of overexpression, **152,** 279
Glycocalyx
dome *versus* villus enterocyte terminal saccharide composition, **167,** 131
implantation-associated changes, **160,** 35–36
M cell
versus enterocytes, **167,** 123
site-specific variations in, **167,** 115
structure of, **167,** 112
Glycoconjugates
in bacterial adherence to M cells, **167,** 124–126
secretion, cyclic AMP and CFTR regulation, **172,** 225
Glycogen, regenerative cementogenesis and, **142,** 40
Glycolipids
phenolic, bacterial, **161,** 288, 290
specificity of S-fimbriae, **175,** 177
Glycolysis
glutamate metabolism and, **147,** 286, 288
higher plant plastids and, **145,** 150, 171, 189–190, 201–203
nuclear magnetic resonance and, **145,** 37, 45
proteins in striated muscle and, **142,** 67, 76, 88, 120, 127–128
Sertoli-germ cell communication network and, **147,** 73
vertebrate cell culture technology and, **142,** 204
Glycoprotein hormone receptor, **167,** 56
Glycoproteins
acute phase proteins as, **170,** 264–265
algal cell wall, antibodies to, **171,** 106–107
avian retina and, **146,** 23
binding to TGF-β, **167,** 214–215
cell wall-bound proteins, **176,** 168–169
cytoskeletal dynamics and, **144,** 99, 107
definition, **174,** 202–203
eggshells and biomineralization and, **145,** 221, 231, 235
extensin-like, as cell wall-bound protein, **176,** 169–170
extracellular matrix and, **146,** 268–269, 278, 285, 305, 311
female, homology modeling, **140,** 544–551
gp1, sulfated
Sertoli-germ cell communication network and, **147,** 52, 61, 76
sperm maturation role
endocytosis, **154,** 134–135, 139, 143–145
epididymis, synthesis by, **154,** 167–169
secretion by Sertoli cells, **154,** 123–124
synthesis in secondary lysosomes of Sertoli cells, **154,** 118–120
gp2, sulfated
Sertoli-germ cell communication network and, **147,** 52, 76–78
in sperm maturation
endocytosis, **154,** 134–135, 139, 142
epididymis, **154,** 155, 157, 164, 166
secretion by Sertoli cells, **154,** 124–127
gp62, in messenger transport, **162B,** 203
gp140, inhibition of endothelial cell growth, **161,** 18–19
gp210
role as NPC anchor, **162B,** 272–273
role in fusion event, **162B,** 291

Glycoproteins (*continued*)
homologous with S locusspecific glycoproteins, other families of, **140,** 533–543
hormonal control of amphibian metamorphosis and, **145,** 113, 132
hydroxyproline-rich, extracellular matrix
characterization, **146,** 267–270
immunology, **146,** 284
modification, **146,** 295, 297, 303–304
pathogenesis, **146,** 309, 311
synthesis, **146,** 277, 283
intracellular calcium messenger system and, **142,** 307, 320
karyosphere and, **144,** 41
lactoferrin as hemopoietic progenitor cell inhibitor, **167,** 228
M cell apical membrane, site-specific variations in, **167,** 115
melanocytes and, **146,** 206, 235
myelin-associated gp, **175,** 200–202
O-linked, **162B,** 238–241, 265–272
proteins in striated muscle and, **142,** 99, 121–122, 124, 126
regenerative cementogenesis and, **142,** 13
retinal precursor cells and, **146,** 178–179
self-incompatibility and
gametophytes, **143,** 282–289
sporophytes, **143,** 263–279
sperm maturation role, in epididymal transit, **154,** 179–180
SSG 185, **139,** 245–246
transdifferentiation in medusae and, **142,** 213, 236
urease in plant cells and, **145,** 91
Glycosaminoglycans
eggshells and biomineralization and, **145,** 231, 237
regenerative cementogenesis and, **142,** 38
Glycosidases, extracellular matrix and, **146,** 267, 270, 293, 300
Glycosides, extracellular matrix and, **146,** 310
Glycosylation
acute phase proteins
alterations of APP, regulation
animal models, **170,** 268–269
in nontransformed cells, **170,** 269–270
in transfected transformed cells, **170,** 274
in transfected and transgenic nontransformed cells, **170,** 270–271
in transformed cells, **170,** 271–274
associated site in APP
effect of changes on function, **170,** 267–268
major microheterogeneity, **170,** 266
microheterogeneity, **170,** 265–266
minor microheterogeneity, **170,** 267
effect of cytokines
nontransformed cell-secreted APP, **170,** 275–276
transformed cell-secreted APP, **170,** 274–275
network of factors affecting, **170,** 276–279
advanced, endproducts, induced endothelial cell procoagulants, **152,** 70
avian retina and, **146,** 24
calcium channels and, **147,** 207
in CFTR mutants, **172,** 201
deglycosylation, monolignol glycosides, **172,** 270–271
dome epithelial cells, **167,** 103–104
ependymin, **165,** 125–126
in epididymal transit, **154,** 180–184
extracellular matrix and, **146,** 272, 276–278, 285
intermediate filament protein function modulated by, **154,** 42–43
melanocytes and, **146,** 204, 206
monolignols, **172,** 270
platelet-derived growth factors, **172,** 97
self-incompatibility and, **143,** 269, 283–284, 290
urease in plant cells and, **145,** 91
N-Glycosylation
at asparaginyl residues among *SLGs,* sequences from *Brassica* species, **140,** 536
Brassica species, potential sites, **140,** 534–535
O-Glycosylation, Tau proteins, **171,** 174
Glycosyl transferase
in epididymal transit, **154,** 180–181, 183
extracellular matrix and, **146,** 272, 274–275, 293
Glyoxylate, urease in plant cells and, **145,** 75–76

Glyoxysomes, urease in plant cells and, **145,** 76
GM-CFC, *see* Granulocyte–macrophage colony-forming cells
GM-CSF, *see* Granulocyte–macrophage colony-stimulating factor
GNEF, *see* Guanine nucleotide exchange factor
Gnetum, potential double fertilization in, **140,** 342–348
GnRH, *see* Gonadotropin-releasing hormone
Goblet cells
 dome epithelium, **167,** 123, 130
 negativity for alkaline phosphatase, **167,** 102
 suppression of differentiation of, **167,** 110
Golgi apparatus
 avian salt glands and, **144,** 135, 140, 143–144, 147, 171
 centrosomal localization, **169,** 35
 diatom morphogenesis role, **150,** 264–270
 –endosome communication, role of Rab9, **176,** 32–33
 epididymis
  caput epithelium, **154,** 159–161, 163
  initial segment, **154,** 154
 –ER interface trafficking, role of Rab, **176,** 30–31
 extracellular matrix and, **146,** 265–267, 272, 274, 285
 function in plant cell, effect of microgravity, **171,** 50
 higher plant plastids and, **145,** 149
 hormonal control of amphibian metamorphosis and, **145,** 116
 mitosis and, **144,** 219–220, 234
 plant cell coated vesicles
  plasma membrane recycling and, **167,** 5–6
  protein transport and, **167,** 7
  receptors in protein synthesis and transport to, **167,** 19–20
 in pollen tube, **176,** 139–140
 post-transport vesicles, role of Rab, **176,** 21–22
 regenerative cementogenesis and, **142,** 9, 11, 19, 21
 relationship to microtubule system in animal cells, **156,** 347
 role of Rab1, **176,** 20–21
 in *Saccharomyces cereuisiae,* **139,** 76–77
 self-incompatibility, **143,** 261, 270
 SNARE complex formation, role of Ypt1p, **176,** 60–62
trans Golgi network
 acidification, **172,** 226, 228
 clathrin-coated vesicles and
  partially coated reticulum as possible plant equivalent of, **167,** 5
  in protein transport and deposition into protein bodies, **167,** 8
  unresolved questions on assembly of, **167,** 27
 in membrane trafficking, **172,** 225–226
 polarity, **172,** 228–229
 sialylation of glycoproteins and, **172,** 227–228
 transport of CFTR and, **172,** 215
 tubule coat structures, **172,** 229
Golgi transport, intra-Golgi, role of Rab6, **176,** 31–32
Gomphonema parvulum, colloidal silica formation, **150,** 324
Gonadal differentiation
 developmental steps, **154,** 247
 higher vertebrates, **154,** 220–221
 hormone function, **154,** 216–218
 marsupials, **154,** 219–220
 testis determination, mammalian, **154,** 221–222
Gonadotropic hormone
 characteristics of cells, **159,** 68–69
 control of release, **159,** 43–45, 69, 72, 80, 83–84
 effect of hypothalamic lesions, **159,** 49
Gonadotropic hormone receptor, homology of related receptors, **155,** 27
Gonadotropin, **145,** 113; **149,** 121; **166,** 26
Gonadotropin-releasing hormone
 associated peptide
  modulation of release, **159,** 44–45
  neuron classification, **159,** 40–41
  precursor localization, **159,** 41–42, 44
  sequence homology between species, **159,** 42
  in teleosts, **147,** 166, 173, 176, 181, 184–185, 187
 effect on pituitary cell proliferation, **157,** 48

Gonadotropin-releasing hormone (*continued*)
encoding gene, in teleosts, **147,** 165–166, 186–187
antisense RNA, **147,** 185–186
estrogen, **147,** 175–177
evolution, **147,** 181–184
function, **147,** 168–170
GAP, **147,** 185
GnRH family, **147,** 166–168
molecular architecture, **147,** 171–174
promoters, **147,** 174–175
radiation, **147,** 176, 178–181
variants, **147,** 170–171
immunohistochemical localization of, **138,** 92–99
neuroendocrine and behavioral effects of, **138,** 99
Gonionemus, transdifferentiation in medusae and, **142,** 249
Gonosomes
in aneuploidy in live births, **167,** 265
in disomies in sperm, **167,** 269–270
as particularly prone to nondisjunction, **167,** 270
paternal age and incidence of, **167,** 281–282
G protein-coupled receptor
cloning, **167,** 55–56
computer models, **164,** 207
coupling specificity, **164,** 211–212
desensitization, **164,** 212
general properties, **164,** 205–213
identification, cultured insect cells for, **164,** 246–249
insect
dopamine, **164,** 218–219
histamine, **164,** 223–224
5-hydroxytryptamine, **164,** 219–220
muscarinic aceytlcholine receptor, **164,** 213–214
phenolamines, **164,** 222–223
ligand binding, **164,** 209–210
mammalian, expressed and characterized in insect cells, **164,** 248
MIP-1α binding and, **167,** 207
molecular cloning
classical approach, **164,** 232–233
expression in transfected cells, **164,** 233
hybridization under reduced stringency, **164,** 234
polymerase chain reaction, **164,** 234
in *Xenopus* oocytes, **164,** 233
molecular evolution, **164,** 243–246
signal transduction, **164,** 210–211
signal transduction crosstalk with integrin pathways, **168,** 101–102
topographical structure, **164,** 205–206
transmembrane domains, **164,** 205–207
G proteins
ATPase activity, **159,** 21
calcium channels and, **147,** 194
control of rhythmic behaviors and, **147,** 2, 17
environmental stimuli in plant cells and, **145,** 299
fish chromatophores and, **143,** 239, 246, 250
intracellular calcium messenger system and, **142,** 308–311, 338
regulation of protein synthesis, **159,** 21–23
role in angiogenesis, **159,** 125
role in nematocyst discharge, **156,** 285
signal transduction of receptors, **159,** 11
structure, **159,** 21
subunit dissociation, **159,** 22
types, **159,** 22
Gradient, Percoll, **166,** 149
Grafts
molecular, *see* Molecular grafting
selective, effect on epilepsy, **175,** 294–297
testis as site, **149,** 123–124
Gramicidin
effect on RVI, **161,** 185–186
osmotic conductance (P_{os}), **163,** 3–5
osmotic flow, bimodal analysis, **163,** 30–34
Gramineae, extracellular matrix and, **146,** 305
Gram-negative bacteria
cell membrane
lipopeptides, **161,** 281–284
lipopolysaccharides, **161,** 279–281
lipoproteins, **161,** 281–284
model of, **161,** 278
outer membrane proteins, **161,** 278
structure, **161,** 276–278
encapsulated, **161,** 284–285
soluble components, **161,** 286–287
Gram-positive bacteria
cell membrane
lipoteichoic acids, **161,** 270–272

peptidoglycans, **161,** 267–268
surface proteins, **161,** 272–274
teichoic acids, **161,** 270–272
encapsulated, **161,** 284–286
soluble components
ADP-ribosylating toxins, **161,** 274–275
exotoxins-275, **161,** 274
leucocidins, **161,** 274
Granules
intracellular movement, role of nonmuscle tropomyosin, **170,** 18–19
at preprophase band, **175,** 108
Granulocyte–macrophage colony-forming cells
accelerated chemotherapy delivery and, **167,** 240
accessory cell contamination of IFN-γ-induced inhibition of, **167,** 231–232
isoferritins as inhibitors, **167,** 227–228
LAI as inhibitor of, **167,** 234–235
reversible inhibition of by pEEDCK, **167,** 188
TNF-α augmentation of proliferation of, **167,** 230
Granulocyte–macrophage colony-stimulating factor
migration modulation by, **155,** 80
modulation of migration, **155,** 80
transcriptional regulation, **143,** 17
transcriptional regulation and, **143,** 14, 43, 47
Granulocytes
endometrial, pregnant uterus
immune-effector function, **153,** 123–125
nonimmunological function, **153,** 125–126
properties, **153,** 117, 119–121
progenitors, **167,** 191
Granulopoiesis inhibitors, **167,** 191
Granulopoietic cells, TGF-β as inhibitor of precursor, **167,** 211
Grasshopper nervous system, programmed cell death in, **163,** 113
Gravitropism, plants
agravitropism
effects
Ca^{2+} distribution, **168,** 142, 144, 146, 148, 150, 156–157, 164, 167
Ca^{2+}-induced cell death, **168,** 144
columnella cell ultrastructure, **168,** 137, 141–142, 148, 150–152, 155, 163
electron-dense inclusion induction, **168,** 137, 141, 144, 146, 158, 163–164
statocyte ultrastructure, **168,** 137, 141–142, 150, 161–162
induction
A23187, **168,** 154–157
calcium channel blockers, **168,** 164, 166–168
chlorprozamine, **168,** 157–158, 160–161
clinorotation, **168,** 137, 148
ethylene glycol-bis-(β-aminoethyl ether)-*N,N'*-tetraacetic acid, **168,** 153–154
lanthanium, **168,** 162–164
lithium, **168,** 168, 170
ruthenium red, **168,** 161–162
space flight, **168,** 124, 137
calcium, detection of transmembrane transfer, **168,** 129, 130–133
models of sensing, **168,** 170–172
role
calcium ATPase, **168,** 135, 150–151
calcium channels, **168,** 166–168
calmodulin, **168,** 134–135, 158
root cap involvement, **168,** 133
Gravity
altered, **156,** 302–357, 349–358
cytoplasmic streaming affected by, **155,** 127–129
manipulation, **156,** 349–358
clinorotation, **156,** 357–358
orbital spaceflight, **156,** 353–357
parabolic flight, **156,** 349–352
suborbital flight, **156,** 352–353
microgravity, *see* Microgravity
role in cell wall formation, Protodyn experiment, **170,** 42
GRB2 adaptor protein
FAK coimmunoprecipitation with, **167,** 175
role in lymphocyte signal transduction, **157,** 224–226
Green algae
chloroplasts, **144,** 264–265, 276, 293, 295
ancestry, **144,** 270–272
biochemistry, **144,** 272
cryptomonads, **144,** 276
envelope membranes, **144,** 266, 268–270

Green algae (*continued*)
euglenoids, **144,** 282–285
heterokont algae, **144,** 281
structure, **144,** 266–267
eyespot reflection, **148,** 258–259, 265–267
phototactic signal generation, **148,** 252
phototactic signal transduction, **148,** 290–299
phototaxis, calcium ion requirements, **148,** 290
retinal pigments, **148,** 277–284
Greig cephalopolysyndactyly syndrome
candidate genes, **158,** 248–249
chromosomal abnormalities, **158,** 248
clinical features, **158,** 248
mode of inheritance, **158,** 248
mouse model, **158,** 257–258
nomenclature, **158,** 247
GRF, *see* Growth hormone-releasing factor
Griffithsia pacifica, chloroplasts and, **144,** 263, 281
Growing pollen tubes, **140,** 78–80
Growth
acid, symplast and, **142,** 273–275
actin filament bundle formation, model, **175,** 64–65
adrenal cortex, effects of endothelin
cell hypertrophy and steroidogenic capacity, **171,** 286–288
cell proliferation, **171,** 288–289
mechanism, **171,** 289–290
role of ET receptor subtypes, **171,** 290–291
anchorage-dependent, cell, and FAK, **167,** 178
angiosperm seedling, effect of microgravity, **171,** 32
cellular, *see also* Proliferation, cellular
anchorage-dependent
FAK and, **167,** 178
TGF-β inhibition of tumor cell, **167,** 237
regulation of
template and antitemplate model, **167,** 186–187
tumor resistance mechanisms, **167,** 237–238
TGF-β effects, *versus* cell type and environment, **167,** 209
collagen fibrils *in vitro*
role of collagen V, **173,** 116–120
role of pNcollagen III, **173,** 114–116
neuronal, effect of MAG, **175,** 201
plant cell wall
cleavage of load-bearing crosslinks, **173,** 192–194
EXGT-mediated molecular grafting, **173,** 194–195
xyloglucan–cellulose microfibril interaction, **173,** 195–196
polarized, pollen tubes, affecting factors
fluctuated tip growth, **176,** 174–175
boron, **176,** 179–180
Ca^{2+} as secondary messenger, **176,** 175–177
cell wall properties, **176,** 177–178
flavonols, **176,** 180
H^+-ATPase, **176,** 179
K^+, **176,** 179
oligosaccharides as signals, **176,** 180–181
pH, **176,** 179
turgor pressure, **176,** 178–179
oscillatory and pulsatory growth, **176,** 174
pollen tube, **176,** 135–137
role of hormonal control of amphibian metamorphosis, **145,** 128–130, 132, 137
Growth-associated proteins, neurite initiation and elongation role, **153,** 262–263
Growth cone
active, tropomyosin function in, **170,** 12–13
ameboid movement, **165,** 241
anterograde movement of membrane antigens, **165,** 259
axonal elongation, **165,** 256–268
filopodia, **165,** 256–258, 266–267
motility, **165,** 256–259
motility relationship, **165,** 260–266
neurite initiation and elongation role, **153,** 258–262
new neurite shaft, **165,** 262, 263
retrograde actin escalator, **165,** 258–259
role in cell division and development, **173,** 221–222
steering, **165,** 266–268
Growth deformations, root hairs, rhizobia-induced, **156,** 2, 18
Growth factor receptor
encoding gene, tumorigenesis in transgenic mice bearing, **164,** 49–52

signal transduction crosstalk with integrin pathways, **168,** 99–101
Growth factors, *see also specific growth factors*
activation in microgravity, **156,** 317–318
cytokine, tissue factor as superfamily member, **152,** 51–53
effects on nuclear transport, **151,** 214–215, 218
encoding gene, tumorigenesis in transgenic mice bearing, **164,** 49–52
extracellular matrix and, **146,** 313
in FAK activation, **167,** 171
induced retinal regeneration, **146,** 49–74
induction of mRNA accumulation, **148,** 98
insulin-like, astrocyte reaction to, **148,** 144–146
insulin-like growth factor, **156,** 181
internalized, signaling mechanisms, **173,** 247–248
melanocytes and, **146,** 218, 222
melanoma and, **146,** 234, 236–243, 246–247
in migration, *see* Migration
nuclear magnetic resonance and, **145,** 17
regulation of apoptotic processes, **163,** 117–119
retinal pigment epithelium regeneration affected by, **152,** 253–254
retinal precursor cells and, **146,** 176–178
role in muscle fiber type pattern formation, **170,** 192
role in palatal cell death, **163,** 147–148
seminiferous, Sertoli-germ cell communication network and, **147,** 47
Growth hormone
aging effects on secretion, **157,** 70–71
biological activity, **157,** 34
cellular synthesis, *see* Mammosomatotroph; Somatotroph
control of amphibian metamorphosis, **145,** 113, 129, 131, 137–140
effect on endocrine organs, **156,** 180
effect on muscle fiber-type transitions, **170,** 185
evolution, **159,** 61, 67
gene structure, **159,** 61
inverse relationship to prolactin, **159,** 73–74, 91–92
producing cells
differentiation, gonadal steroid effects, **159,** 65–67
and prolactin cells, inverse relationship, **159,** 73–74
receptors, **159,** 58
regulation of insulin-like growth factor, **159,** 313–314
release, **159,** 38, 40, 58
role in cartilage formation, **159,** 312–314
role in immune response, **159,** 78
secretion in pituitary cells, **159,** 4, 58–59
size, **157,** 34
somatolactin homology, **169,** 1–2, 4–5
storage in granules, **159,** 66–67
stress response, **159,** 58, 68
Growth hormone-releasing hormone
analogs, **159,** 38
immunohistochemical localization, **138,** 119
mechanism of action, **159,** 38
neuroendocrine effects, **138,** 119–120
Growth inhibitory factor, in feedback regulation of tumor tissue, **167,** 234–235
Growth plates, epiphyseal, histological regions, **159,** 279
GRP, *see* Gastrin-releasing peptide
Gryllus firmis, animal mtDNA and, **141,** 203–204
GSH, *see* Glutathione
GSI, *see* Gametophytic self-incompatibility
GSP
GSP MTs and cell shape, **140,** 165–168
evidence in favor, **140,** 165–167
MTs, sufficiency of to maintain cell shape, **140,** 167–168
locomotion, **140,** 168–170
background, **140,** 168–169
not self-propelled, **140,** 169–170
GST, *see* Glutathione S-transferase
GTP, *see* Guanosine 5′-triphosphate
GTPase
cycle of Rab
role of GDP dissociation inhibitors, **176,** 49–51
role of GTPase-activating proteins, **176,** 52–53
role of guanine nucleotide exchange factors, **176,** 51–52

GTPase (*continued*)
intracellular calcium messenger system and, **142,** 309
Ran/TC4, role in nuclear import of proteins, **162B,** 246–248
GTPase-activating proteins, regulation of Rab function, **176,** 52–53
GTP-binding protein
fish chromatophores and, **143,** 227, 240
pEEDCK-like sequences in, **167,** 219–220
in plant defense transmembrane signaling, **172,** 73–74
β-Guanidinopropionate, urease in plant cells and, **145,** 72, 78
Guanine, fish chromatophores and, **143,** 206
Guanine nucleotide exchange factor
activation of Ras, **157,** 224
regulation of Rab function, **176,** 51–52
role in lymphocyte signal transduction, **157,** 223–225
types, **157,** 224–225
Guanosine 5′-diphosphate, intracellular calcium messenger system and, **142,** 308–311
Guanosine 5′-diphosphate dissociation inhibitor, regulation of Rab function, **176,** 49–51
Guanosine 5′-triphosphate
calcium channels and, **147,** 194
environmental stimuli in plant cells and, **145,** 299
glutamate metabolism and, **147,** 295
intracellular calcium messenger system and, **142,** 308–312, 323, 326
role in microtubule dynamics, **157,** 3, 22
Guide RNA
endosymbiont hypothesis and, **141,** 294, 305
kinetoplastid mtDNA and, **141,** 83–84
developmental control, **141,** 81–82
maxicircle genes, **141,** 72–74
maxicircle transcripts, **141,** 77–80
Guinea pig
genital abnormalities in, **139,** 3
nongenital abnormalities in, **139,** 28
perinatal exposure to antihormones and, **139,** 13–14
Gustatory signaling, insect, **164,** 231
Gymnodinium
chloroplasts, **144,** 288–290
photoreceptive apparatuses, **148,** 249–250
Gymnosperms
and angiosperms, differences, **176,** 91–93
cinnamoyl-CoA reductase substrate specificity in, **172,** 263
cinnamyl alcohol dehydrogenase in, **172,** 265
4-coumarate-CoA ligase activity in, **172,** 259–260
effects of microgravity, **171,** 31–32
extracellular matrix and, **146,** 271, 277
higher plant plastids and, **145,** 154
lignin types in, **172,** 245
monolignol gylcoside accumulation in, **172,** 269
pollen, isolation of exines from hydrated and autoclaved, **140,** 54–55
Gyrodan lividus mitochondrial genomes in fungi and, **141,** 103
Gyrodinium, chloroplasts and, **144,** 289

H

H-8, biological clocks in retina and, **146,** 108
HA4 protein
localization in hepatocytes, **158,** 148
phosphorylation, **158,** 148
sequence homology with ecto-ATPases, **158,** 148
Habituation, *see* Plant cells, tumorous, habituation in
Haematococcus, phototaxis
calcium ion requirements, **148,** 291, 294
eyespot reflection, **148,** 259, 261, 265
photosynthetic accessory pigments, **148,** 282–283, 286
signal generation, **148,** 256
Haemonoea, biological clocks in retina and, **146,** 86
Haemophilus influenza, binding to sialic acids, **175,** 181
Hafniomonas reticulata, eyespot apparatus, **148,** 238
Hair, melanocytes and, **146,** 211–212, 220
Hair bundles, anemones, **156,** 287–289
Hair cells, cochlea, in neurosensory epithelium, **173,** 210
β-Hairpins, structures, **140,** 547
Haliaectus leucogaster, avian salt glands and, **144,** 147

Halides, conductance by CFTR, **172,** 199
Halocynthia roretzi
cleavage pattern generation, **176,** 269–272
cytoplasmic determinants for gastrulation movements, **176,** 267
eggs, anterior–posterior specification, **176,** 289–291
embryo
development, **176,** 282–284
distribution of competence, **176,** 286
embryogenesis, **176,** 249–250
endoderm cells, **176,** 263–264
larval structure, **176,** 247–249
muscle development, **176,** 257–263
ooplasmic segregation, **176,** 253–255
sensory pigment cell induction, **176,** 279–281
trunk lateral cell induction, **176,** 281
Halophytic species
archaebacteria, salt tolerance, **165,** 14
seawater tolerance, **165,** 2
Halteria, replication band, **153,** 139
Hamster, *mdr* gene, **156,** 81
Hanks solution, **139,** 161, 177–178
Hansenula, mitochondrial genomes in fungi and, **141,** 101, 117–118
Hansenula mrakii, mitochondrial genomes of, **141,** 3
Hantzschia, valve formation, **150,** 268, 270
Hantzschia amphioxys, sintering, **150,** 306
Haploid transcription, evidence for, **140,** 3–5
Haplopappus gracilis
effect of microgravity, **171,** 42–44
microgravity experiments, **156,** 322, 323, 325, 327
Haptoglobin
angiogenic activity, **159,** 138
antioxidant activity, **170,** 242–243
Harderian gland
δ-aminolevulinate synthase expression, **168,** 44
comparative anatomy
amphibians, **168,** 2
birds, **168,** 4
reptiles, **168,** 3
rodents, **168,** 4
development
amphibians, **168,** 4–5
birds, **168,** 6–7
reptiles, **168,** 5–6
rodents, **168,** 7–8
factors affecting secretion
androgens, **168,** 52–53
estrogen, **168,** 53–54
glucocorticoids, **168,** 54
melatonin effects, **168,** 51
osmotic stress, **168,** 47–49
peptide hormones, **168,** 54–55
photoperiod, **168,** 46–47
pituitary hormones, **168,** 49–51
progesterone, **168,** 54
temperature, **168,** 46–47
thyroid hormones, **168,** 50–52
functions, **168,** 1, 56, 62–63
amphibians, **168,** 56–57
birds, **168,** 58–59
mammals, **168,** 59–61
reptiles, **168,** 57–58
glandular cell ultrastructure
amphibians, **168,** 13–14
armadillo, **168,** 25–26
birds, **168,** 19–20
degu, **168,** 25
desert rodents, **168,** 24
hamster, **168,** 22
mammals, **168,** 21–22, 24–26
mouse, **168,** 25
opossum, **168,** 26
rat, **168,** 24–25
reptiles, **168,** 16, 19
innervation, **168,** 45
interstitium ultrastructure
amphibians, **168,** 14–15
birds, **168,** 20–21
mammals
immunocompetent cells, **168,** 30
macrophages, **168,** 28
mast cells, **168,** 26, 28
melanocytes, **168,** 29–30
reptiles, **168,** 19
receptors
androgen receptor, **168,** 41, 43
estrogen receptor, **168,** 41–43
glucocorticoid receptor, **168,** 42
insulin receptor, **168,** 43
melatonin receptor, **168,** 43
progesterone receptor, **168,** 42
somatostatin receptor, **168,** 42–43
table, **168,** 40
vitamin D receptor, **168,** 42

Harderian gland (*continued*)
secretory products
amphibians
histochemistry, **168,** 30–32
immunohistochemistry, **168,** 32
birds
histochemistry, **168,** 33–34
immunohistochemistry, **168,** 34
mammals
histochemistry, **168,** 34–35
immunohistochemistry, **168,** 35
indole synthesis, **168,** 38–40
lipid synthesis, **168,** 37–38
porphyrin synthesis, **168,** 35–37
reptiles
histochemistry, **168,** 32–33
immunohistochemistry, **168,** 33
somatostatin expression, **168,** 45
structure
amphibians, **168,** 8–9
birds, **168,** 10–11
mammals, **168,** 11–13
reptiles, **168,** 9–10
tumor pathology, **168,** 61–62
Hatching, blastocyst, *in vitro* culture conditions effects, **164,** 165–166
Hawley's hypotheses, for maternal age effect in aneuploidy
first hypothesis, **167,** 278–279
second hypothesis, **167,** 279
HCG, *see* Human chorionic gonadotropin
Head
development in human fetus, **158,** 216–217
brain, **158,** 221–222
branchial arches, **158,** 217–218
face, **158,** 218–219
palate, **158,** 220
teeth, **158,** 221
tongue, **158,** 219
vertebrate, formation, **175,** 260–261
Heart
ecto-ATPase levels, **158,** 164
hormonal control of amphibian metamorphosis and, **145,** 107, 127
hypertrophy
decreased mechanical load effects, **150,** 81
myosin heavy chain isoenzyme transition during, **150,** 80
proto-oncogene expression in early stages, **150,** 79–80
RNA accumulation in early stages, **150,** 78
myofibrils, role of mechanical forces
in formation, **150,** 75–76
in maturation promotion, **150,** 74–75
nuclear magnetic resonance and, **145,** 43–44
platelet-derived growth factors
activity, **172,** 114–115
distribution, **172,** 109, 110
programmed cell death in, **163,** 139–140
tissues, response to mechanical loading, **150,** 72
Heart disease, spaceflight experiments, **156,** 320
Heat, discharge in angiosperms, effect of microgravity, **171,** 40
Heat shock
cell response, **148,** 171–172, 214–216
acquired tolerance, **148,** 190
heat shock protein induction, **148,** 176–177
primary thermostability increase, **148,** 182, 184–186
protein thermostability increase, **148,** 179–182
recovery rate of cells, **148,** 186–189
repair stimulation, mechanisms, **148,** 205–210
stability
cell functions, mechanisms, **148,** 203–205
increase, in tolerant zone, **148,** 210–214
stress-induced, **148,** 182–183, 190, 203–210
stress proteins
induction, **148,** 178–179
role in increased stability, **148,** 190–202
thermostability increase in, **148,** 172–175
response in reproductive tissues, **140,** 400–402
role of mitochondrial chaperones, **174,** 160
transcriptional regulation and, **143,** 29, 38
Heat shock cognate proteins, *Drosophila* muscles and, **143,** 114

Heat shock element, **148,** 176–177; **165,** 27–28
Heat shock factor, activation, **148,** 176–177
Heat shock proteins
DnaJ homologs, **174,** 139–140, 147–148
DnaK, DnaJ, and GrpE, **174,** 131–132
Drosophila muscles and, **143,** 113–114
HSP70, **148,** 176–177, 208; **167,** 22; **172,** 205, 207
induction, **148,** 176, 178
synthesis during cardiac hypertrophy, **150,** 79–80
in tolerant zone, **148,** 213–214
Heat shock transcription factor, **165,** 27–28
Heat tolerance, yeast cell, **165,** 28
Heavy meromyosin
on actin filaments, **156,** 249–250
in cytoplasmic fibrils, **156,** 203
cytoskeletal dynamics, **144,** 94
in F-acto-heavy meromyosin, **156,** 247
fish chromatophores and, **143,** 221
in microfilaments, **156,** 211
myofibrillogenesis and, **143,** 169, 172
in striated muscle, **142,** 63–64
Helianthus annuus
cell wall
changes under microgravity, role of enzymes, **170,** 61
polysaccharide content, effect of clinorotation, **170,** 57
higher plant mitochondrial genomes and, **141,** 133
microgravity experiments, **156,** 322
Helicobacter pylori
binding to sialic acids, **175,** 179–181
urease in plant cells and, **145,** 83
Helicoid, twisted liquid crystalline supramolecular arrangement, **166,** 95
Heliobacterium chlorum, endosymbiont hypothesis and, **141,** 255, 271
Helisoma
control of rhythmic behaviors and, **147,** 14–15
neuronal modulation
growth cone activity, **153,** 259–261
neurotransmitter effects, **153,** 257
synaptic connections, **153,** 269–270
Helix-hoop-helix, *Drosophila* muscles and, **143,** 82–84, 137
Helodea, environmental stimuli in, **145,** 257, 276
Hemangiomatosis, therapeutic strategies, **161,** 26, 32
Hematopoietin receptors, hepatocyte
post-activation intracellular events, **170,** 260–264
subunits, effect on APP synthesis, **170,** 257–259
Hematoxylin, vertebrate cell culture technology and, **142,** 172
Heme, ciliates and, **141,** 41
Heme-regulated eIF-2 kinase, transcriptional regulation and, **143,** 24
Hemerocallis, microgravity experiments, **156,** 322
Hemerocallis fulva, **140,** 55
Hemicellulose
antibodies to, generation, **171,** 92–93
in cell wall
effect of clinorotation, **170,** 58
production, effect of microgravity, **170,** 55–56
protoplast, effect of microgravity, **170,** 43
extracellular matrix and
characterization, **146,** 263–264, 266
modification, **146,** 293–295
synthesis, **146,** 274–275
Hemichannel
formation of intercellular channels, **157,** 322–325
freeze-fracturing, **157,** 322
intermolecular binding interactions, **157,** 322–325
Hemidesmosomes, intermediate filament associations, **164,** 105–107
Hemocytes
motility, effect of cytokines, **170,** 105–106
as nitric oxide source, **170,** 115–116
POMC products in, **170,** 89–93
Hemopexin, antioxidant activity, **170,** 242
Hemopoiesis
feedback inhibitors and structure of, **167,** 189–192
maintained in absence of exogenous growth factors, **167,** 194–195
melanocytes and, **146,** 214, 218
neglect of inhibitory factors in research on, **167,** 186
physiological inhibitors of, proposed, **167,** 225*t*

Hemopoietic chalones, inhibiting myelocyte proliferation, **167,** 188
Hemopoietic pentapeptide
in cell cycle, **167,** 193–194
description, **167,** 216–219
effects on other tissues, **167,** 220
initial synthesis of, **167,** 188
neoplastic cells and, **167,** 236
structure, **167,** 219–220
Hemopoietic stem cells
assays for inhibitory regulators of, **167,** 197
proliferation regulation of, **167,** 194–196
search for MIP-1α receptors on, **167,** 206–207
in structure of developing cell populations, **167,** 190–192
TGF-β receptors in, **167,** 214–215
Hemoregulatory pentapeptide
description, **167,** 216–220
neoplastic cells and, **167,** 236
Hemoregulatory tetrapeptide
in cell cycle, **167,** 193
clinical trial of, **167,** 244
description, **167,** 221–224
Hemorrhage, avian salt glands and, **144,** 185
Hensen's node
dorsal echoderm origin, **175,** 251–253
dorsal endoderm origin, **175,** 255
floor plate origin, **175,** 255
and gastrulation, **175,** 255
notochord origin, **175,** 255
Heparanitase, growth factor-induced retinal regeneration and, **146,** 56–57
Heparan sulfate
growth factor-induced retinal regeneration and, **146,** 56–57
inhibition of smooth muscle cell growth, **161,** 23–24
interaction with fibroblast growth factors, **161,** 6
retinal precursor cells and, **146,** 154
Heparin
growth factor-induced retinal regeneration and, **146,** 56
inhibition of angiogenesis, **159,** 141
inhibition of smooth muscle cell growth, **161,** 23–24
modulation of arterial smooth muscle cell behavior, **169,** 227
retinal pigmented epithelium and, **146,** 77
retinal precursor cells and, **146,** 182
Hepatic histidase, perinatal exposure to sex hormones and antihormones and, **139,** 28–29
Hepatitis B virus
human, tumorigenesis in transgenic mice bearing gene, **164,** 61–62
transcriptional regulation, **143,** 39, 44
Hepatocyte growth factor
–scatter factor
biological activity, **160,** 238–239
embryogenesis role, **160,** 240–241
hepatic development role, **160,** 241
migration induction, **155,** 63–64, 78–79
mutagenesis, **160,** 236
pathology role, **160,** 241–242
posttranslational processing, **160,** 235
structure, **160,** 234–236
tissue distribution, **160,** 239–240
variants, **160,** 236
stimulation of vascular cells, **161,** 13
Hepatocyte growth factor receptor–scatter factor, *see* c-Met receptor
Hepatocytes
AcSDKP conflicting effects on proliferation of, **167,** 222
for acute phase protein studies
normal hepatocytes, **170,** 250–251
secreted APP, glycosylation, effect of cytokines, **170,** 275–276
serum APP production, **170,** 243–244
peroxisome, *see* Peroxisome, hepatocyte
transcriptional regulation and, **143,** 18
transdifferentiation in medusae and, **142,** 217
Hepatoma, ecto-ATPase levels, **158,** 172–174
Hepatoma cells, for acute phase protein studies
APP studies, **170,** 251–252
regulation of APP glycosylation alterations, **170,** 271–274
secreted APP, glycosylation, effect of cytokines, **170,** 274–275
Hepoxilin A_3, role in volume regulation, **161,** 222–223
Heraeleum, environmental stimuli in, **145,** 261
Herbicides
cytoskeleton, **166,** 15

flowering plants and, **143,** 312, 318, 323–324
Herbimycin A, potential inhibition of FAK by, **167,** 176
Heritability
Arabidopsis thaliana and, **144,** 56
estimation, **158,** 227
formula, **158,** 227
multifactorial, **158,** 226–227
Herpes simplex virus, immediate early product ICP0, **162A,** 80–81
Herpesvirus, transcriptional regulation and, **143,** 39–40, 44
Heterochromatin
interphase nucleus marker, **165,** 73–75
mitosis and, **144,** 222, 230
role in plant chromosome breakage, **158,** 112–117
role in somoclonal variation, **158,** 112–113
Heterochrony, **140,** 180; **148,** 106
Heterodera schachtii, plant-pathogen interactions in, **144,** 72
Heteroduplex analysis, detection of gene mutations, **158,** 235–236
Heterogeneity
early, migrating neural crest cells, **138,** 284–285
microheterogeneity, acute phase proteins
major microheterogeneity, **170,** 266
minor microheterogeneity, **170,** 267
types, **170,** 265–266
Heterokont algae, chloroplasts
biochemistry, **144,** 281
dinoflagellates, **144,** 288, 290, 293
euglenoids, **144,** 283–286
eyespot, **144,** 278, 280–281
relationship, **144,** 282
structure, **144,** 277–279
Hexacorallia, *see* Sea anemones
Hexamethonium, avian salt glands and, **144,** 157
Hexokinase, avian salt glands and, **144,** 145
Hexose, higher plant plastids and, **145,** 179–180, 189
Hexose monophosphate, avian salt glands and, **144,** 145
High endothelial venules, in Peyer's patches, **167,** 95
Higher plants
mitochondrial genomes, *see* Mitochondrial genomes, higher plant
occurrence of sporopollenin, **140,** 52–53
tapetal and peritapetal cell walls, **140,** 52–53
Ubisch bodies (orbicules), **140,** 52
viscin threads, **140,** 53
plastids, *see* Plastids, higher plant
High-mobility group proteins
animal mtDNA, **141,** 220
with nuclear matrix activity, **162A,** 307–309
postsynthetic modifications, **162A,** 217–218
role in DNase I sensitivity, **162A,** 220–221
High-performance liquid chromatography
intracellular calcium messenger system, **142,** 314, 322
self-incompatibility, **143,** 265, 272
High-pressure liquid chromatography, in analysis of hormonal control, **145,** 115
High-voltage electron microscope, fish chromatophores, **143,** 221
Hindbrain
cell migration in, **175,** 280–285
regional specification, **175,** 271–272
Hippocampus
γ-aminobutyric acid metabolism and, **147,** 321
calcium channels and, **147,** 211
glutamate metabolism and
aminotransferases, **147,** 291, 294
glutamate dehydrogenase, **147,** 299, 300–301
glutaminase, **147,** 308
glutamine synthetase, **147,** 304–305
Hippurus vulgaris L.
cell wall changes under microgravity, role of enzymes, **170,** 62
lignin biosynthesis, effect of microgravity, **170,** 56
Histamine
control of rhythmic behaviors, **147,** 18
effect on prolactin release, **159,** 65
functions in insect CNS, **164,** 223–224
in neurotransmission, **149,** 251–254
Histaminergic receptors, **149,** 253
Histidine
environmental stimuli in plant cells and, **145,** 297–298
flowering plants, **143,** 324

Histiocytosis X, as dendritic cell malignancy, **153,** 86
Histiotypic structure, avian retina and, **146,** 8–9, 11, 27, 40
Histochemical markers
  in lignin detection, **176,** 95
  in M cell studies
    alkaline phosphatase, **167,** 102
    intermediate filament proteins, **167,** 104
    labeling pattern correlation, **167,** 104–107
    recognition as homogeneous population by, **167,** 117
    types, **167,** 101–102
Histochemistry
  avian salt glands, **144,** 159–161, 167, 171
  female germ unit, **140,** 258–260
  glutamate metabolism and, **147,** 287, 294, 298, 300–301
  M cell, **167,** 115–117
  melanocytes and, **146,** 195–196
  Sertoli-germ cell communication network and, **147,** 59
Histogenesis
  avian retina, *see* Retina, avian, histogenesis of
  growth factor-induced retinal regeneration and, **146,** 52
Histology
  control of rhythmic behaviors, **147,** 10, 14
  melanocytes and, **146,** 194, 211
  retinal precursor cells and, **146,** 151
Histolysis, *Drosophila* muscles and, **143,** 70–71
Histone acetyltransferase
  associated with nuclear matrix, **162A,** 208–210
  catalysis of histone acetylation, **162A,** 207–208
Histone deacetylase
  associated with nuclear matrix, **162A,** 208–210
  catalysis of histone acetylation, **162A,** 207–208
  as marker enzyme for internal nuclear matrix, **162A,** 210–211
Histones
  acetylation, reversible, **162A,** 207–208
  core
    acetylation, **162B,** 386–387
    modifications, **162A,** 203–214
    variants, **162A,** 202–203
  depleted chromosomes, **162B,** 410–411
  evolutionary aspects of DNA interactions with, **154,** 70
  genes, cell cycle-regulated, **162A,** 253–263
  gene variant H3-1, **156,** 37–38
  H1
    accumulation, effect of unbalanced growth, **148,** 97–99, 111–112
    in cell differentiation, **148,** 100–101, 104
    in chromatin loop, **162A,** 307
    in gene transcription, **148,** 112
    in genome expression, **148,** 101–103
    heterochrony effect, **148,** 106
    MARs as nucleation sites, **162A,** 459–460
    modifications, **162A,** 215–217
    with nuclear matrix activity, **162A,** 307–309
    role in chromatin structure, **162A,** 198–199
    subtypes, **162A,** 214–215
  H3, introns, **139,** 249–250
  H4, acetylated in cytoplasm, **162A,** 204–205
  intermediate filament protein interactions with, **154,** 7
  intracellular calcium messenger system and, **142,** 334
  nucleosomal or core, **162A,** 196–197
  nucleosomal genes, **139,** 226, 230–233
  stabilization of microtubules, **157,** 21
Histopathology, retinal pigmented epithelium and, **146,** 75
HIV, *see* Human immunodeficiency virus
HMM, *see* Heavy meromyosin
Hodgkin's disease
  alkylating agents for cure, **156,** 128–129
  dendritic cell association, **153,** 86
HOG pathway, **165,** 30–31
Hollow-fiber reactors, vertebrate cell culture technology, **142,** 180–183, 191
Holocentrus, fish chromatophores
  cytoskeleton, **143,** 221
  erythrophores, **143,** 248
  motility control, **143,** 228, 239, 242, 245
  pigment-containing vesicles, **143,** 198–199

Holocentrus ascensionis, fish
chromatophores, **143,** 242
Holotoxin, structure, **161,** 286–287
Homarus gammarus, control of rhythmic
behaviors and, **147,** 16
Homeobox genes
and development of neural crest, **138,** 274–283
Drosophila muscles and, **143,** 79–82, 84
E2A-PBX1, tumorigenesis in transgenic
mice bearing, **164,** 52–54
Homeodomain protein-binding elements, S/
MARs as, **162A,** 441
Homeodomain proteins, recognition and
binding sites, **162A,** 335–338
Homeostasis, feedback loops in, **167,** 185–186
Homogeneous state, instability
channels, localized activity, **154,** 343–346
Child's results, **154,** 346–347
colocalization, **154,** 313
cytoskeleton localization, **154,** 355–360
differentiation
metabolism, **154,** 360–362
Turing biochemical mechanism, **154,** 324–327
Drosophila
dorsoventral system, **154,** 335–339
terminal system, **154,** 335–339
electrophoresis, spontaneous endogenous, **154,** 339–342
epigenesis *versus* preformation, **154,** 313–322
lateral inhibition, **154,** 312
metabolic field, **154,** 355–360
morphogenesis
metabolism, **154,** 360–362
Turing biochemical mechanism, **154,** 324–325, 328–329, 334
preformation *versus* epigenesis, **154,** 313–322
principle, **154,** 309–311
proliferation metabolism, **154,** 360–362
pumps, localized activity, **154,** 343–346
reducibility of development to molecular
genetics, **154,** 362–367
reduction fields, **154,** 347–355
Turing biochemical mechanism, **154,** 322–335
Turing-Child metabolic system, **154,** 310
Homoiogenetic induction, embryonic,
definition, **150,** 408–409
Homology
animal mtDNA and, **141,** 174, 199, 209, 230–231
protein genes, **141,** 181, 184, 187
rRNA, **141,** 189–190
transcription, **141,** 219–220
Arabidopsis thaliana and, **144,** 68, 74, 76
avian salt glands and, **144,** 130, 176
chloroplasts and, **144,** 261, 263, 276, 283, 285, 292
ciliates and, **141,** 1, 55–56, 60
Paramecium aurelia, **141,** 36, 38, 40–41
rRNA genes, **141,** 48, 50, 52
structure, **141,** 8, 110
cytoskeletal dynamics and
current concepts, **144,** 97–98, 103
cytoskeletal components, **144,** 105–107, 109
theories, **144,** 94
Drosophila muscles and, **143,** 85, 108, 124–126
endosymbiont hypothesis and gene
expression, **141,** 300–304, 306–307
gene structure, **141,** 308–313, 315–317
gene transfer, **141,** 278
genome organization, **141,** 293–294
lineages of life, **141,** 238–239, 246
nuclear genome, **141,** 248–249
organelles, **141,** 330–331
Plasmodium, **141,** 326
plastids, **141,** 262–263, 265–268, 279
rRNA sequences, **141,** 319–320
sequences, **141,** 271
fish chromatophores and, **143,** 198
flowering plants and, **143,** 315
higher plant mitochondrial genomes and,
141, 137, 145, 147, 149, 158
higher plant plastids and, **145,** 155, 157, 168
hormonal control of amphibian
metamorphosis and, **145,** 115, 130, 133, 138
intracellular calcium messenger system
and, **142,** 311, 333
karyosphere and, **144,** 11
kinetoplastid mtDNA and, **141,** 72–76, 82
mitosis and, **144,** 228, 231, 250
myofibrillogenesis and
isoform change, **143,** 154–155

Homology (*continued*)
sarcomere formation, **143,** 183
thin filament proteins, **143,** 167, 172, 177–178
proteins in striated muscle and, **142,** 121–122, 125
self-incompatibility and
gametophytes, **143,** 284, 286–287
sporophytes, **143,** 269–270, 272–276
transcriptional regulation and, **143,** 1
genes, **143,** 19
molecular cloning, **143,** 29–30, 33–39
urease in plant cells and, **145,** 92
Homo-oligonucleotide repeats, short, in MAR/ORIs, **162A,** 343–344
Homo sapiens, animal mtDNA and, **141,** 183, 188, 196, 205
Honeybee
compound eye, ommatidium organization, **176,** 309–311
Na^+/K^+-ATPase localization, **176,** 325
photoreceptors, Na^+/K^+-ATPase maintenance and distribution, **176,** 334–337
Hook-basal body complex, bacterial flagellar motor and, **147,** 104–105
Hordeum vulgare
environmental stimuli, **145,** 253
flowering plants, **143,** 300
intracellular calcium messenger system, **142,** 315
microgravity experiments, **156,** 322, 325
Hormidium faccidum, environmental stimuli in, **145,** 270
Hormone receptor
characteristics, **155,** 37–38
multicellular organisms, **155,** 3
ontogeny
conclusions, **155,** 35–37
imprinting
cell-to-cell transmission, **155,** 34–35
in developing adult cells, **155,** 33–35
optimal time for, **155,** 29–30
pathological, **155,** 30–33
perinatal period, **155,** 24–26
steroid receptors, **155,** 32–33
maturation, **155,** 24–26
perinatal period
imprinting, **155,** 24–26
specificity, **155,** 27–29
phylogeny, lowest level, **155,** 1–3
unicellular organisms
imprinting mechanism, **155,** 19–24, 36
microbes, **155,** 3–7
protozoa, **155,** 3–7
second messengers, **155,** 8–11
signal molecules in receptor development, **155,** 11–18
specificity of hormone and receptor, **155,** 7–8
Hormones
aneuploidy and changes in ovarian, **167,** 278
Arabidopsis thaliana and, **144,** 72, 75
avian salt glands and, **144,** 130–131, 168, 188, 201
chalcone synthase and, **147,** 255
cnidarian neurohormones
G-protein-coupled receptor homology with glycoprotein family of, **167,** 56
locally acting as neurotransmitters, *see* Paracrine hormones
types possible, **167,** 77
control, **144,** 189–200
control of amphibian metamorphosis, **145,** 108–109, 140–141
adrenocortical hormones, **145,** 119–123
growth hormone, **145,** 137–140
prolactin
larval growth, **145,** 128–132
levels, **145,** 133–137
secretion, **145,** 123–128
thyroid hormone, **145,** 105–107, 110–112
feedback, **145,** 117–119
hypothalamic control, **145,** 115–117
pituitary control, **145,** 112–115
control of rhythmic behaviors and, **147,** 9–10, 19
cytoskeleton
adrenocorticotropic hormone, **166,** 24
androgen, **166,** 27
angiotensin, **166,** 25
behavior, **166,** 31, 34
bombesin, **166,** 20
bombesin-like peptide, **166,** 20
cellular level motility, **166,** 23, 30
chemotaxis, **166,** 23, 30
corticosteroid, **166,** 25
1,25-dihydroxyvitamin D_3, **166,** 25
direct effects, **166,** 16–31

endothelin, **166,** 25
estrogen, **166,** 27
follicle-stimulating hormone, **166,** 26
glucosteroid, **166,** 25
gonadotropin, **166,** 26
growth, **166,** 16–17
growth factors, **166,** 16–17
growth hormone, **166,** 18
hormone, **166,** 1–42
human chorionic gonadotropin, **166,** 26
indirect effects, **166,** 16–31
insulin, **166,** 16, 18–20
insulin-like growth factor, **166,** 18
invertebrates, **166,** 31–35, 36
luteinizing hormone, **166,** 26
mevalonate, **166,** 29
motility, **166,** 23–31
muscular level motility control, **166,** 23–31, 32–33
myotropin, **166,** 23–31, 32–33
oocyte maturation, **166,** 31, 35
organismal level movement, **166,** 31, 34
parathyroid functioning, **166,** 22–23, 28–29
parathyroid hormone, **166,** 29–30
progesterone, **166,** 27
steroidogenesis regulation, **166,** 22, 24–25, 26–27
thyroid, **166,** 22–23, 28–29
thyrotropin, **166,** 28
thyrotropin-releasing hormone, **166,** 28
thyroxine, **166,** 28–29
transepithelial transport, **166,** 17–21
vasopressin-mediated water, **166,** 17–21
effect on muscle fiber-type transitions, **170,** 184–186
environmental stimuli in plant cells and, **145,** 294, 297, 299–300
fish chromatophores and, **143,** 193, 246, 249–250
iridophores, **143,** 209
motility control, **143,** 225, 229, 234–238
flowering plants and, **143,** 297
and gonadal differentiation, **154,** 216–218
growth factor-induced retinal regeneration and, **146,** 56
internalized, signaling mechanisms, **173,** 247–248
nuclear magnetic resonance and, **145,** 37
plant, regulation of chromoplast development, **163,** 201–203
regulation of apoptotic processes, **163,** 117–119
regulation of XRP gene expression, **173,** 191
Sertoli-germ cell communication network and, **147,** 44, 70, 80
transcriptional regulation and, **143,** 18
vertebrate cell culture technology and, **142,** 147, 201–202
Horseradish peroxidase, seminal vesicle epithelium endocytosis, **160,** 279, 281, 292
Host–parasite specificity
determination, **172,** 63, 64–65
suppressors and, **172,** 84–85
plant resistance in, **172,** 55–56
role of cell wall in, **172,** 77–80
Host specificity, *Rhizobium*–plant interactions, **156,** 24–27, 29, 30–32, 49–51
Hot spots, genomic, **162B,** 435–438
H protein in striated muscle, **142,** 82
H-strand
promoters
animal mtDNA and, **141,** 218, 223, 226–227
endosymbiont hypothesis and, **141,** 299
synthesis, animal mtDNA and, **141,** 225, 229
5-HT, *see* 5-Hydroxytryptamine
HTLV-I, transcriptional regulation and, **143,** 6–7, 14, 39, 42–44
Human chorionic gonadotropin
effects on testicular macrophages, **149,** 119–120
perinatal exposure to, **139,** 5
tamoxffen and, **139,** 19
Human dendritic cells
adenoidal, **153,** 66–67
adhesion molecules
costimulation, **153,** 78–82
effects, **153,** 85–86
allergy, **153,** 90
autoimmune disease, **153,** 88–89
blood
adherence depletion, **153,** 49
contaminating cells, removal, **153,** 51–52
density-gradient separations, **153,** 49, 51
fresh cell isolation, **153,** 52–53

Human dendritic cells (*continued*)
immunophenotypic cell separations, **153,** 52
isolation protocols, **153,** 48–53
panning techniques, **153,** 51–52
phagocyte depletion, **153,** 52
properties, **153,** 53
bone marrow, **153,** 48, 69–70
characterization, **153,** 43, 45
clinical applications
immunization, **153,** 90
transplantation, **153,** 90–91
costimulator molecules, effects, **153,** 85–86
cytochemical characteristics, **153,** 46–47
cytokine receptors, costimulation, **153,** 82–83
cytokines, costimulation, **153,** 82–83
differentiation, **153,** 43–44
functional activities, **153,** 42–43
functional properties
allogeneic mixed leukocyte reaction, **153,** 76
antigen processing, **153,** 72, 74–75
antigen-specific T-lymphocyte sensitization, **153,** 77
autologous mixed leukocyte reaction, **153,** 76–77
cellular interactions, **153,** 83–84
costimulation, **153,** 78–86
cytokine regulation, **153,** 84–85
mitogenic assay stimulation, **153,** 77–78
regulation of activity, **153,** 84–86
hypersensitivity, **153,** 90
identification, **153,** 41–43, 45
immune response stimulation
primary, **153,** 76–77
secondary, **153,** 77–78
immunodeficiency, **153,** 89
infectious diseases
human immunodeficiency virus, **153,** 89
leprosy, **153,** 89–90
interstitial, **153,** 60–64
Langerhans cells
isolation, **153,** 55–57
properties, **153,** 54, 57–62
liver, **153,** 62–63
lung, **153,** 63
lymphatic, afferent, **153,** 65–66
lymphoid, **153,** 64–68
malignancies
histiocytosis X, **153,** 86
Hodgkin's disease, **153,** 86
monoclonal antibodies, **153,** 46
morphology, **153,** 46
mucosal, **153,** 63
ontogeny
blood cell precursors, **153,** 70–71
bone marrow, **153,** 69–70
migration, **153,** 71–72
myeloid lineages, relationship, **153,** 72–73
phenotypic comparisons of subtypes, **153,** 60–61
preparations, purity of, **153,** 45–47
psoriasis, **153,** 90
rheumatic disease, **153,** 88–89
splenic, **153,** 67
subtypes, **153,** 42
synovial fluid, **153,** 63–64
thymic, **153,** 68
tonsil, **153,** 66–67
transplantation, **153,** 87, 90–91
veiled, **153,** 65–66
Human immunodeficiency virus
dendritic cell role in, **153,** 89
follicular dendritic cell
B cell repertoire restriction, **166,** 161
germinal center, **166,** 160–161
infectiousness, **166,** 161–162
lymphoid follicles, **166,** 162–163
transcriptional regulation, **143,** 17, 33, 40–42, 44–45
tumorigenesis in transgenic mice bearing gene for, **164,** 63–64
type 1
mapping of NOS, **162B,** 308
preferential adherence to M cell apical membranes of, **167,** 123
Rev-responsive element, **162B,** 210–213
Humidity
eggshells and biomineralization and, **145,** 235
environmental stimuli in plant cells and, **145,** 251
higher plant plastids and, **145,** 176
Humoral factors, possible M cell interaction with, **167,** 143
Humoral hypercalcemia, malignancy, **166,** 231–232

Huntington's disease, glucose transporter expression in, **172,** 36
Huxley model, bacterial flagellar motor and, **147,** 156–157
Huxley-Simmons model, bacterial flagellar motor and, **147,** 156–158
Hyacinthus, environmental stimuli in, **145,** 263
Hyalodiscus, locomotion, **150,** 56, 61
Hyaluronic acid
  angiogenic activity of fragments, **159,** 138–139
  inhibition of angiogenesis, **159,** 142–143
Hybridization
  animal mtDNA and, **141,** 198
  Arabidopsis thaliana and, **144,** 63, 67, 76
  ciliates and, **141,** 10, 35–36, 47–48, 60
  comparative genomic
    clinical studies, **153,** 218
    technique, **153,** 21, 192–193
  Drosophila muscles and, **143,** 63, 101, 105, 137
  flowering plants and, **143,** 312, 315
  fluorescence *in situ, see* Fluorescence *in situ* hybridization
  higher plant mitochondrial genomes and, **141,** 133, 161
    recombination, **141,** 149, 151–152
    structure, **141,** 140–143, 146–147
  higher plant plastids and, **145,** 204
  hormonal control of amphibian metamorphosis and, **145,** 112, 116
  in situ
    γ-aminobutyric acid metabolism and, **147,** 308, 312, 315
    chalcone synthase and, **147,** 256
    Drosophila muscles and, **143,** 63
      actins, **143,** 111–112
      contractile protein genes, **143,** 100, 108, 119
      development, **143,** 80, 87
      myosin light chains, **143,** 105, 107
    extracellular matrix and, **146,** 287, 295
    generative cells and male gametes, **140,** 207–213
    glutamate metabolism and, **147,** 290–291, 294, 303, 308
    in glycine receptor mapping, **176,** 215–219
    GnRH-encoding gene in teleosts and, **147,** 169
    growth factor-induced retinal regeneration and, **146,** 57, 68
    localization of pollenspecific transcript by, **140,** 25–27
    myofibrillogenesis and, **143,** 179
    proteins in striated muscle and, **142,** 66
    self-incompatibility and, **143,** 285
    Sertoli-germ cell communication network and, **147,** 59, 76
  in situ end labeling, analysis of apoptotic cells, **163,** 121–123
  karyosphere and, **144,** 11
  kinetoplastid mtDNA and, **141,** 73–75
  melanocytes and, **146,** 233–234, 244, 248, 251
  mitochondrial genomes in fungi and, **141,** 91, 93, 98, 101, 105
  proteins in striated muscle and, **142,** 66, 76, 215
  under reduced stringency, cloning of G-protein-coupled receptors by, **164,** 234
  self-incompatibility and, **143,** 264, 269–270, 279, 284–285
  vertebrate cell culture technology and, **142,** 147
  whole-chromosome
    applications
      characteristics, **153,** 197–198
      chromosomal aberrations, **153,** 204–208
      clinical studies, **153,** 214–219
      evolutionary studies, **153,** 199–200
      genetic toxicology, **153,** 208–213
      mammals, **153,** 199–200
      nuclear architecture, **153,** 201–204
      plants, **153,** 200
      primates, **153,** 199–200
      reverse experiments, **153,** 217
      taxonomic studies, **153,** 199–200
    comparative genomic hybridization, **153,** 192–193
    development of techniques, **153,** 174–175
    experimental uses, **153,** 172–173
    in situ hybridization techniques
      detection methods, **153,** 181–184
      microscopy, **153,** 183–188
      nucleic acid, **153,** 175–178
      probe labeling methods, **153,** 178–181
      typical method, **153,** 188–190

Hybridization (*continued*)
methods, **153,** 171–173, 219
probes
cloned chromosome-specific DNA probe libraries, **153,** 193–195
genomic DNA, **153,** 191–193
labeling methods, **153,** 178–181
methods, **153,** 190–191
polymerase chain reaction methods, **153,** 195–198
yeast artificial chromosome probes, **153,** 195
Hybridoma cells, vertebrate cell culture technology, **142,** 146–147, 153, 177–178, 189, 194
Hydra, see also Cnidarians
instability of homogeneous state, reduction fields, **154,** 348
nervous system centralization and complexity, **167,** 44–46
reproduction, **167,** 38
RFamide neuropeptides
isolated from, **167,** 48
neuromuscular transmission and, **167,** 51
transdifferentiation in medusae, **142,** 216
Hydra attenuata, see Hydra vulgaris
Hydractinia echinata
density of neuronal plexus in body column, **167,** 44–45
neuropeptides and metamorphosis of, **167,** 54–55, 71–74
Hydra magnipillata, **167,** 75
Hydra oligactis, **167,** 44
Hydra-RFamide neuropeptides, biosynthesis, **167,** 74–77
Hydration, pollen, activation and germination, **140,** 77–78
Hydraulic conductivity, symplast
auxin, **142,** 278–279, 283
integration of activity, **142,** 287–288, 293, 297
Hydra vulgaris
agglomeration of sensory neurons, **167,** 44
animal mtDNA, **141,** 173, 193, 199
avian retina, **146,** 9
Hydrocarbons, aromatic, imprinting of steroid receptors, **155,** 32–33
Hydrogen
avian salt glands and, **144,** 173
bacterial flagellar motor and, **147,** 108
extracellular matrix and, **146,** 265–267
intracellular calcium messenger system and, **142,** 337
concentration, **142,** 312–318
phosphatidylinositol turnover, **142,** 324, 328
receptors, **142,** 307–308
symplast and, **142,** 267, 274–276
vertebrate cell culture technology and, **142,** 153, 171, 173
Hydrogen-bonding distances, **140,** 548–549
Hydrogenosomes, endosymbiont hypothesis and, **141,** 330
Hydrogen peroxidase, origin, **176,** 117
Hydrolagus colliei, **138,** 112
Hydrolase, extracellular matrix and, **146,** 304–305
Hydrolysis
cytoskeletal dynamics and, **144,** 88, 116, 118
eggshells and biomineralization and, **145,** 241
higher plant plastids and, **145,** 168, 187
Rab GTPase role, **176,** 5–14
urease in plant cells and
biochemical genetics, **145,** 89–90
elimination, **145,** 80, 82
metabolic origins, **145,** 71, 75, 77–78
Hydromedusae, transdifferentiation
DNA replication, **142,** 252–256
initiation, **142,** 236–251
isolation of tissues, **142,** 222–223
potential, **142,** 223–236
tissue organization, **142,** 218–222
Hydrophobicity, higher plant plastids and, **145,** 153, 168
Hydroxycinnamate CoA ligase, *see* 4-Coumarate–CoA ligase
Hydroxycinnamic acids, **156,** 9
Hydroxycinnamoyl–polysaccharides, in plant cell wall
biological roles, **151,** 257–259
biosynthesis, **151,** 252–253
oxidative coupling, **151,** 253–255
photodimerization, **151,** 256–257
Hydroxycinnamyl alcohols, *see* Monolignols
4-Hydroxycyclophosphamide, TNF-α and, **167,** 241

Hydroxylation
extracellular matrix and
characterization, **146,** 268–270
immunology, **146,** 285
synthesis, **146,** 276–277, 279, 284
melanocytes and, **146,** 193
3-α-Hydroxy-4-pregnen-20-one, Sertoli-germ cell communication network and, **147,** 49
Hydroxyproline
extracellular matrix and, **146,** 277–278
glycoproteins rich in, extracellular matrix and
characterization, **146,** 267–270
immunology, **146,** 284
modification, **146,** 295, 297, 303–304
pathogenesis, **146,** 309, 311
synthesis, **146,** 277, 283
3β-Hydroxysteroid dehydrogenase, control of amphibian metamorphosis and, **145,** 123–124
5-Hydroxytryptamine
effect on prolactin release, **159,** 65
functions in insect CNS, **164,** 219–220
localization in insect CNS, **164,** 217–218
target organs, **159,** 25
5-Hydroxytryptamine receptor, cloning, **164,** 236
Hydroxyurea, **143,** 72–73; **156,** 129–130; **167,** 204
Hydrozoa
centralization and complexity of nervous system, **167,** 44–46
life cycle of, **167,** 38
neuropeptide biosynthesis in, **167,** 74–77
Hyla regilla, **138,** 92, 97, 99
Hynobius nigrescens, hormonal control of metamorphosis and, **145,** 107
Hyoscyamus muticus, flowering plants and, **143,** 317
Hyperaldosteronism, idiopathic, role of endothelin, **171,** 294
Hyperammonemia, urease in plant cells and, **145,** 83
Hypercalcemia, transcriptional regulation and, **143,** 43
Hypercholesterolemia, familial
LDL receptor, molecular defects, **166,** 117–120
mammalian LDL receptor supergene family, **166,** 111, 112–120
cytoplasmic tail, **166,** 116–117
epidermal growth factor, **166,** 115–116
ligand binding domain, **166,** 115
low-density lipoprotein receptor biosynthesis, **166,** 113–117
low-density lipoprotein receptor structure, **166,** 113–117
membrane anchoring domain, **166,** 116
mutations, **166,** 112–120
O-linked sugar domain, **166,** 116
Hyperglycemia, avian salt glands and, **144,** 192
Hypergravity, **156,** 349
Hyperkalemia, avian salt glands and, **144,** 192
Hyperplasia
avian salt glands and, **144,** 141, 143
smooth muscle cells, **161,** 29–30
Hyperplexia, in human, **176,** 230–231
Hyperpolarization
bacterial flagellar motor and, **147,** 113
biological clocks in retina and, **146,** 91, 121, 128
Bulla, **146,** 129, 131–134, 136–138
entrainment pathway, **146,** 103, 107, 110–111
calcium channels and, **147,** 194
control of rhythmic behaviors and, **147,** 12–13
responses, in calcium-activated mammalian eggs, **152,** 185
symplast and, **142,** 274–276, 296
Hyperrestoration, in embryonic tissue
definition, **150,** 25
preexisting stress hypothesis, **150,** 24–25
stretching in cell intercalation, **150,** 27
Hypersecretion, hormonal control of amphibian metamorphosis and, **145,** 136
Hypersensitivity
Arabidopsis thaliana, **144,** 59–60, 75, 78
pathogens, **144,** 54, 57
phenotypes, **144,** 65, 67–69
delayed-type, and fibrin deposition, **152,** 59
dendritic cell role in, **153,** 90
plant elicitor recognition, **148,** 4, 6–7
protein stimulation, **148,** 11
Hypertension, therapeutic strategies, **161,** 29–30

Hypertonic stimulation, avian salt glands and, **144,** 182, 184–186, 198, 200
Hypertrophy
avian salt glands and, **144,** 140–144, 151, 158, 191–192
glandular, avian salt glands and, **144,** 140–141
proteins in striated muscle and, **142,** 118
Hyphochytridiomycetes, mitochondrial genomes in fungi and, **141,** 91–92
Hyphochytrium catenoides, mitochondrial genomes in fungi and, **141,** 91–92
Hypocotyl cells
epidermal wall, effect of weightlessness, **170,** 52
intracellular calcium messenger system and, **142,** 322, 324, 326, 328, 334
symplast and, **142,** 265
auxin, **142,** 280–281, 283
electrophysiological structure, **142,** 266–271, 273
integration of activity, **142,** 286–287, 291–292, 294–295, 298–299
proton pumps, **142,** 274–276
Hypoglycorrhachia, GLUT1 expression in, **172,** 37
Hypomethylation, DNA, induction by azacytidine, **158,** 116
Hypophyseal factors, hormonal control of amphibian metamorphosis, **145,** 123–126
Hypophysectomy
avian salt glands and, **144,** 192, 198
control of amphibian metamorphosis and
adrenocortical hormones, **145,** 122–125
growth hormone, **145,** 138–140
prolactin, **145,** 128, 131
thyroid hormone, **145,** 112, 118
effects on testicular macrophages, **149,** 121
Sertoli-germ cell communication network and, **147,** 44, 61, 73
Hypothalamic neurons
bidirectional communication, **159,** 23–24
differentiation in culture, **159,** 14, 17
lesion effect on hormone release, **159,** 48–49
maturation, **159,** 17, 20
migration during embryogenesis, **159,** 12–13, 23
organization, **159,** 23
origin, **159,** 12
proliferation rate, **159,** 12
role in sexuality, **159,** 43–44
species specificity
annelids, **159,** 7
fish, **159,** 7–8
synaptogenesis, **159,** 17, 20
termination sites, **159,** 5–6, 23
Hypothalamus
avian salt glands and
hormonal control, **144,** 197–198
receptive systems, **144,** 180, 182, 184, 188–189
control of amphibian metamorphosis and, **145,** 140–141
adrenocortical hormones, **145,** 122, 124
growth hormone, **145,** 140
prolactin, **145,** 128, 133, 136–137
thyroid hormone, **145,** 115–119
GnRH-encoding gene in teleosts and, **147,** 165, 173, 175
GAP, **147,** 185
GnRH decapeptide, **147,** 166, 168–169
hormonal control of amphibian metamorphosis and, **145,** 140–141
Hypotonic shock, Sertoli-germ cell communication network, **147,** 67–68, 71
Hypotrichs, replication band
characterization, **153,** 138–139
chromatin replication models, **153,** 162–165
early studies, **153,** 143
function, **153,** 148
Hypoxanthine-guaninephosphoribosyl-transferase, **156,** 121
Hypoxanthine-phosphoribosyltransferase, **151,** 2
Hypoxia
effect on connexon density, **157,** 327
effects on GLUT1 expression, **172,** 35–36
neuroendocrine cell response, **157,** 289, 302–303
NMR studies, **145,** 25, 42, 50
role in muscle fiber type transition, **170,** 191

I

I-band-associated protein, in striated muscle
calsequestrin distribution, **142,** 122

cardiac muscle, **142,** 109, 118
skeletal muscle, **142,** 84–85, 93–94, 101
Iberis amara, self-incompatibility, **143,** 259
Ibuprofen, effects in cystic fibrosis, **172,** 194–195
ICAM-1
intercellular adhesion molecule-1, **156,** 162
intracellular adhesion molecule-1, **167,** 110
Iccosome, **166,** 148–150
ICE, *see* Interleukin-1β-converting enzyme
ICP0, concentrated in promyelocytic bodies, **162A,** 80–81
IF, *see* Intermediate filament protein
IFN-γ, *see* Interferon-γ
IGF-I, *see* Insulin-like growth factor I
IL, *see* Interleukins
Ileum
Peyer's patches presence in, **167,** 94
reduction in M cell number in inflamed, **167,** 143
Image processing
cell motility
three-dimensional analysis, **163,** 70–73
two-dimensional analysis, **163,** 50–53
computer, multidimensional, **162A,** 50–53
intravoxel incoherent motion, **145,** 27, 29
in NMR
cardiovascular system, **145,** 31
diffusion, **145,** 26–29
magnetic angiography, **145,** 25–26
magnetization transfer, **145,** 29–30
metabolites, **145,** 30
microscopy, **145,** 31–32
perfusion, **145,** 29
receptor imaging, **145,** 30–31
Immobilin, sperm maturation role in epididymis
endocytosis, **154,** 171, 175–176
secretion, **154,** 157–158, 164, 166–168
Immortalization, melanocytes and, **146,** 203, 241–242, 250
Immune response
cell-mediated, and fibrin deposition, **152,** 59
to chaperones, **174,** 167–168
link to stress response, **170,** 98
M cells and
antigen interactions as initial step in, **167,** 122
induction in intestine, **167,** 142–143
mediating molecules, **170,** 121–122
MIP-1α in, **167,** 199
oral vaccine initiation of, **167,** 140
role of cytokines, **170,** 104–107
role of nitric oxide, **170,** 116–117
role of POMC products, **170,** 87–93
specific immunity, **167,** 142–143
suppression of, intraepithelial lymphocytes in, **167,** 97
transcriptional regulation and, **143,** 11, 16, 18–19
Immune system
cell activation in microgravity, **156,** 305–316
cell locations, **167,** 92–93
intestine and
as immunological barrier, **167,** 92
overview, **167,** 91–92
invertebrate
cytokines, **170,** 102–104
role of nitric oxide, **170,** 113–116
parathyroid hormonerelated protein, **166,** 252–253
vertebrate
cytokines, **170,** 102
POMC mRNA, **170,** 83–85
role of nitric oxide, **170,** 110–113
Immunization
Arabidopsis thaliana and, **144,** 77–78
dendritic cell role in, **153,** 90
Immunoblasts, **167,** 133
Immunochemistry, replication band of ciliated protozoa, **153,** 155–161
Immunocytochemistry
γ-aminobutyric acid metabolism and, **147,** 315, 321
glutamate metabolism and, **147,** 303, 309
GnRH-encoding gene in teleosts and, **147,** 169
localization of proteins in striated muscle, *see* Proteins in striated muscle
polyamine, **138,** 64–68
accessibility of polyamines, **138,** 66–68
polyamine antibodies, **138,** 64–65
tissue fixation, **138,** 65–66
retinal precursor cells and, **146,** 151, 154, 168–170, 173–175
Immunodeficiency, dendritic cell role in, **153,** 88–89
Immunoelectron microscopy, proteins in striated muscle
sarcolemma, **142,** 125

Immunoelectron microscopy (*continued*)
sarcoplasmic reticulum, **142,** 122
skeletal muscle, **142,** 73, 82, 88, 93–94, 97
Immunofluorescence
cytoskeletal dynamics and, **144,** 95–96, 110–111
environmental stimuli in plant cells and, **145,** 263, 285, 295–296
hormonal control of amphibian metamorphosis and, **145,** 110
indirect, environmental stimuli in plant cells, **145,** 263, 287
proteins in striated muscle and, **142,** 61
antibody use, **142,** 101
cardiac muscle, **142,** 107, 110–113, 118–119
sarcolemma, **142,** 127
sarcomere, **142,** 94, 96–97, 99
sarcoplasmic reticulum, **142,** 120, 122, 124
thick-filament proteins, **142,** 73, 82–83
thin-filament proteins, **142,** 87–88, 90–92
Sertoli-germ cell communication network and, **147,** 68
Immunoglobulin E receptor, signal transduction crosstalk, **168,** 102
Immunoglobulins, *see also* Antibodies
calcium channels and, **147,** 218, 220
expressed by lymphocytes in Peyer's patches
in dome epithelium, **167,** 133–134
in lamina propria, **167,** 98
in lymphoid follicles, **167,** 95
Fc receptor, **157,** 189
gene rearrangement, **157,** 135–136, 158, 186–188, 192
genetic evolution, **140,** 502–503
IgA, Peyer's patches
in lamina propria, **167,** 98
precursors in plasma cells secreting, **167,** 93
secretory, **167,** 126
IgB, lamina propria $CD4^+$ helper T cells and, **167,** 98
IgG
hormonal control of amphibian metamorphosis and, **145,** 113, 137
marker of B cell differentiation, **157,** 141
myofibrillogenesis and, **143,** 173
origin, **157,** 141
retinal pigmented epithelium and, **146,** 76–77
IgM, Peyer's patches, **167,** 98
marker of B cell differentiation, **157,** 135–136
melanocytes and, **146,** 228
myofibrillogenesis and, **143,** 178, 183
proteinsinstriatedmuscleand, **142,** 81, 110, 117
related superfamily, avian retina and, **146,** 23
role in B cell development, **157,** 156–158
Sertoli-germ cell communication network and, **147,** 33
structure, **157,** 184–188
transcriptional regulation and, **143,** 26, 40, 46
function, **143,** 2, 5–6
structure, **143,** 19
vertebrate cell culture technology and, **142,** 166
Immunohistochemistry
γ-aminobutyric acid metabolism and, **147,** 314–315
control of rhythmic behaviors and, **147,** 3–5, 19
endothelin, **171,** 269–270
endothelin receptor subtypes, **171,** 273
glutamate metabolism and, **147,** 294, 323
glutamate dehydrogenase, **147,** 297–298
glutaminase, **147,** 308–309
glutamine synthetase, **147,** 303–304
glycine receptor, **176,** 219
in localizations
of ANF, **138,** 133–135
of AVT and MT, **138,** 120–124
of bombesin-related peptides, **138,** 163
of carnosine, **138,** 171–172
of CRF, **138,** 106–110
of enkephalins, **138,** 147–148
extracellular matrix, **146,** 261, 275, 285, 292, 295
of galanin, **138,** 166–169
of gastrin- and CCK-related peptides, **138,** 153–154
of GNRH, **138,** 92–99
of GRF, **138,** 119
internal matrix proteins, **162B,** 105–111
of MCH, **138,** 149–151

nucleolar matrix proteins, **162B,** 119–123
of NYP, **138,** 127–131
of tachykinins, **138,** 156–160
Immunology
avian retina and, **146,** 10
avian salt glands and, **144,** 162–163, 196
in characterization of isolated cells, **140,** 226–227
in embryo sacs and component cells, **140,** 283–284
extracellular matrix and, **146,** 274, 278, 284–290
glutamate metabolism and, **147,** 306
growth factor-induced retinal regeneration and, **146,** 60
Sertoli-germ cell communication network and, **147,** 37, 44, 72
testicular macrophages, **149,** 122–124
vertebrate cell culture technology and, **142,** 147, 149
Immunoprecipitation, hormonal control of amphibian metamorphosis, **145,** 124
Immunoreactivity
biological clocks in retina and, **146,** 102
calcium channels and, **147,** 221, 223
control of amphibian metamorphosis
adrenocortical hormones, **145,** 122, 124, 126–127
growth hormone, **145,** 139–140
prolactin, **145,** 132, 137
thyroid hormone, **145,** 113
control of rhythmic behaviors and, **147,** 14, 19
glutamate metabolism and, **147,** 297, 303, 305, 308
GnRH-encoding gene in teleosts and, **147,** 165, 167–168, 176
growth factor-induced retinal regeneration and, **146,** 58
retinal precursor cells and, **146,** 160, 173, 176–183
Sertoli-germ cell communication network and, **147,** 39, 41, 67
transdifferentiation in medusae, **142,** 224, 229–230
Immunoregulators, and neurotransmission
function, **149,** 268
neuropeptides, **149,** 269–272
studies, **149,** 267
thiamine, **149,** 272–274
types, **149,** 268–269
Immunosuppression
melanocytes, **146,** 240
Sertoli-germ cell communication network, **147,** 44
transcriptional regulation, **143,** 16, 18
Immunotolerance, induction by antigens in gut epithelium, **167,** 142–143
Impatiens balsamina, cell wall, effect of weightlessness
epidermal cells, **170,** 51–52
meristem cells, **170,** 47
parenchyma cells, **170,** 48–49
Impedin, *see* Suppressors
Implantation, cell death during, **163,** 134
Importin
NLS receptor candidate, **162B,** 260
yeast homolog SRP1p, **162B,** 287–289
Import stimulatory factor, mitochondrial, as cytosolic chaperone, **174,** 141
Imprinting
genetic, *see* Genetic imprinting
hormone receptors
cell-to-cell transmission, **155,** 34–35
developing adult cells, **155,** 33–35
developmental phases as determining factor, **155,** 36–37
evolutionary studies, **155,** 37
induced by aromatic hydrocarbons, **155,** 32–33
learning process induction, **155,** 34
maturation affected by, **155,** 25
optimal time for, **155,** 29–30
pathology, **155,** 30–33
during perinatal period, **155,** 27–29
steroids, **155,** 32–33
unicellular organisms, **155,** 19–24, 36
Impulse activity, biological clocks in retina and, **146,** 93, 95, 133
Inachis io, male eupyrene meiosis, **152,** 20
Inclusion, cytoplasmic, cholesteric supramolecular order, **166,** 83–84
Independent effector hypothesis, **156,** 291
Indirect flight muscles, in *Drosophila*
actins, **143,** 111–113
contractile protein genes, **143,** 99–101, 103, 107–108
development, **143,** 66, 71, 75, 86–87
as dispensable muscles, **143,** 64, 138–139
interactions, **143,** 129

Indirect flight muscles (*continued*)
muscle protein genes, **143,** 126
tropomyosin, **143,** 116
troponin, **143,** 119–122
Z-disc proteins, **143,** 122–124
Indole-3-acetic acid
effect on nodulation, **156,** 39
flowering plants, **143,** 317–318
induced habituation in tumorous plant cells, **152,** 267, 276–277
production and uptake, **156,** 46
symplast, **142,** 274–277, 280, 282
Inducers
elicitors, *see* Elicitors
transcriptional regulation, **143,** 8–12, 24
Induction, *see* Embryonic induction
Infarct damage, proteins in striated muscle, **142,** 116
Infection
accumulation of SV40 virions during, **162A,** 522–523
Arabidopsis thaliana and, **144,** 55, 60, 64–66, 71–72
early and late phases, T-antigen activities, **162A,** 500–501
with influenza A virus, **175,** 163
production of SV40 progeny virus, **162A,** 489–490
role in immume response against chaperones, **174,** 167
root hairs by rhizobia, **156,** 2, 34
and testicular macrophages, **149,** 125–126
thread, **156,** 2–3, 39
via M cells
blocking receptors to prevent, **167,** 122
in dome epithelium as primary entry site for intestinal, **167,** 138–139
and virion formation, **162A,** 525–528
Infectious diseases, dendritic cell role
HIV, **153,** 89
leprosy, **153,** 89–90
Inflammation
CD11/CD18 antigen role, **152,** 75–76
monocyte/macrophage procoagulant activity stimulated by, **152,** 59
procoagulant expression suppression, **152,** 89
regenerative cementogenesis and, **142,** 2, 44
retinal pigmented epithelium and, **146,** 78
transcriptional regulation and, **143,** 14, 17–19
Inflammatory bowel disease, **167,** 143
Inflammatory response
mediating molecules, **170,** 121–122
role of macrophages, **170,** 118–122
Influenza A virus, binding to sialic acids, **175,** 162–169
Influenza A virus inhibitor, aglycon as, **175,** 165–166
Influenza B virus, binding to sialic acids, **175,** 162–169
Influenza C virus, receptor determinant, **175,** 169–170
Influenza virus NS1, **162B,** 210–213
Infrared absorption, in analysis of calcifications, **172,** 136–137
Inheritance
animal mtDNA, **141,** 175
cytoplasmic
higher plant plastids, **145,** 153
influence of fusion in pattern, **140,** 379–383
paternal cytoplasmic, mechanism, **140,** 382–383
uniparental, mitochondria, **161,** 63–64
Inhibin
secretion by Sertoli cells, **160,** 201
Sertoli-germ cell communication network and
function, **147,** 46, 57
germ cell depletion, **147,** 62, 64–66
germ cell *in vitro,* **147,** 67, 69–70
mechanisms, **147,** 72, 76, 79
Inhibition assay, for regulators of stem cell proliferation, **167,** 196–198
Inhibitors
animal mtDNA, **141,** 227
avian salt glands
hormonal control, **144,** 192–193, 195, 198–199
innervation, **144,** 157, 161–162, 165, 168
receptive systems, **144,** 178, 180, 182, 187–189
secretory mechanism, **144,** 169
stimulus-secretion coupling, **144,** 201, 204–205
clinical evaluation, challenges, **167,** 243–244
control of amphibian metamorphosis
adrenocortical hormones, **145,** 120, 127–128

growth hormone, **145,** 140
prolactin, **145,** 129, 133, 135–137
thyroid hormone, **145,** 110, 117
cytoskeletal dynamics, **144,** 94, 105, 107, 113
difficulty of determining cellular response, **167,** 243
Drosophila muscles, **143,** 72, 99, 117, 119, 136
effects on cytoskeleton, **140,** 112–113
eggshells and biomineralization, **145,** 226, 236, 240, 242, 244
endosymbiont hypothesis, **141,** 240, 307
environmental stimuli in plant cells, **145,** 300
chloroplasts, **145,** 271, 275–277, 279–281, 283–284, 286
photodinesis, **145,** 253–254, 260–261, 265–266
wounding, **145,** 291–296
as factors for cell entry into G_0 state, **167,** 193
feedback, *see* Feedback inhibitors
flowering plants, **143,** 297
amino acids, **143,** 309, 311, 321–322, 324
ammonium, **143,** 304
nitrate, **143,** 301–302
hemopoietic stem cell proliferation
AcSDKP, **167,** 221–224
G_1 state and, **167,** 193
in malignant transformation, **167,** 233–234
MIP-1α, **167,** 198–208
other inhibitors, **167,** 224–232
pEEDCK, **167,** 216–220
polyfunctional nature of, **167,** 239
as potential chemoprotective agents, **167,** 240
suicide assays in confirming effect of, **167,** 197–198
TGF-β, **167,** 208–216
types, **167,** 191–192
higher plant plastids
amino acid synthesis, **145,** 193, 196–197
metabolism, **145,** 168, 174–175, 188
intracellular calcium messenger system, **142,** 308
calmodulin, **142,** 330–331
concentration, **142,** 313, 315–317, 319–320
G proteins, **142,** 309, 311
phosphatidylinositol turnover, **142,** 322–324, 327
protein kineses, **142,** 333, 335–337
mitosis, **144,** 217, 252
chromosomes, **144,** 241, 243–246, 249
dissociation of events, **144,** 227–228, 234–237
mitotic events, **144,** 218
myofibrillogenesis, **143,** 158, 163–165, 168, 171
neoplastic cells, **167,** 236–237
overproduction in bone marrow suppression, **167,** 239
proteins in striated muscle
sarcoplasmic reticulum, **142,** 121, 123
skeletal muscle, **142,** 64–65, 74, 84, 100
regenerative cementogenesis, **142,** 38, 40
self-incompatibility, **143,** 261, 268, 282–283, 288
sporopollenin, application, **140,** 57–58
symplast, **142,** 270, 276
auxin, **142,** 281, 284–285
integration of activity, **142,** 286, 296, 299
transcriptional regulation
biochemistry, **143,** 23–25
function, **143,** 2, 7
genes, **143,** 13
inducible activation, **143,** 7, 11
molecular cloning, **143,** 25, 28–31, 33–34, 36–38
transdifferentiation in medusae, **142,** 235, 257
DNA replication, **142,** 252–253
initiation, **142,** 241, 243–245, 247, 249, 251
tumor growth modulation by proliferation, **167,** 235–237
tyrosine kinase, in studying FAK, **167,** 176
tyrosine phosphatase, in studying FAK function, **167,** 177
urease in plant cells, **145,** 95, 98
biochemical genetics, **145,** 90–92
elimination, **145,** 80, 82, 84
metabolic origins, **145,** 69, 71–72, 75, 77–79
vertebrate cell culture technology, **142,** 150
bioreactors, **142,** 164–165, 175, 182–183
optimization, **142,** 194–195

Inhibitory signals
in feedback regulation, **167,** 189
malignant transformation and loss of sensitivity to, **167,** 233, 235–236
Initiation
animal mtDNA and, **141,** 174, 191, 230–231
control region, **141,** 201–203
protein genes, **141,** 183, 187–188
regulation, **141,** 226
replication, **141,** 222–224
transcription, **141,** 218, 220–222
endosymbiont hypothesis
gene expression, **141,** 298–300, 303, 306–307
gene structure, **141,** 311, 315
genome organization, **141,** 293
higher plant mitochondrial genomes and, **141,** 160
kinetoplastid mtDNA and, **141,** 76, 82, 84
sites
ciliates, **141,** 9–11, 56
nascent strands, identification, **162A,** 466–473
Injury, neuronal, relationship to Tau protein, **171,** 203
Inner centromere protein
migration, **162B,** 48–54
mitosis and, **144,** 244–245
with MTOC properties, **175,** 115
Inner ear, blood barriers in, glucose transport, **172,** 23–24
Inner endomeningeal layer, **165,** 139–140
Inner plexiform layer
avian retina and, **146,** 3, 15–16, 19, 21, 34
retinal precursor cells, **146,** 149
Inner segments, avian retina, **146,** 17
Innervation
adrenergic, avian salt glands and, **144,** 167–168, 200
avian salt glands and, **144,** 131, 189, 195
adrenergic system, **144,** 167–168
cholinergic system, **144,** 159–162
nerve supply, **144,** 155–159
vasoactive intestinal polypeptide, **144,** 162–167
denervation
avian salt glands and, **144,** 158
effect on muscle fiber type transitions, **170,** 176–177
effect on muscle fiber type transitions, **170,** 162–163
Inoculation groups, *Rhizobium,* **156,** 1
Inorganic phosphate
eggshells and biomineralization and, **145,** 231, 234
higher plant plastids and, **145,** 168–169, 179, 190
nuclear magnetic resonance and, **145,** 2, 7, 19, 42–43
Inosine monophosphate dehydrogenase inhibitor, **162A,** 554–556
Inositol phosphate
avian salt glands, **144,** 200, 205
intracellular calcium messenger system, **142,** 309, 321–324
Inositol phospholipids, intracellular Ca^{2+} messenger system, **142,** 321–323, 328
Inositol 1,4,5-triphosphate
avian salt glands, **144,** 102, 200, 203, 205
diacylglycerol pathway, in Sertoli-germ cell communication network, **147,** 79, 81
elicitor-stimulated levels in plant cells, **148,** 23
fish chromatophores and, **143,** 227–228, 242–245, 250
intracellular calcium messenger system and, **142,** 311, 338
concentration, **142,** 318–320
turnover, **142,** 321–328
in plant defense transmembrane signaling, **172,** 71–73
role in calcium concentration regulation in plants, **148,** 22
role in calcium oscillation in mammalian egg, **152,** 191–194
Inositol 1,4,5-trisphosphate receptor, in Ca^{2+} oscillation, **152,** 197–199
Input pathway, biological clocks in retina, **146,** 98–99
Insects
as arthropods, **174,** 63–65
body plan, patterning, **174,** 45–54
ecto-ATPase
levels, **158,** 175–176
metal dependence, **158,** 176
head
clypeolabral region, **174,** 26–27
development, **174,** 1–2, 65–68
dorsal ridge, **174,** 27

form diversity, **174,** 4–8
gap genes, **174,** 23–26
segmental organization
dorsal ridge, **174,** 16–17
engrailed expression, **174,** 11–13
procephalon, **174,** 13–16
karyosphere, **144,** 25, 43
Diptera, **144,** 8–11
Neuroptera, **144,** 11–18
tenebrionid beetles, **144,** 19–25
microgravity experiments, **156,** 335
Na^+/K^+-ATPase
localization in photoreceptors, **176,** 323–324
molecular analysis, **176,** 320–323
photoreceptors, phototransduction, **176,** 314–316
phylogeny, **174,** 8–10
programmed cell death, **163,** 126–133
retina
ion concentration, **176,** 316–319
light-induced ion movements, **176,** 316–319
organization
ommatidium in honeybee compound eye, **176,** 309–311
photoreceptor cell plasma membrane, **176,** 311–313
protein mobility in photoreceptor plasma membrane, **176,** 313–314
Insolubility, induction by cell fractionation, **162A,** 528–531
Instrinsic curvature, DNA
in α-globin gene MAR, **162A,** 332–333
in MAR/ORIs, **162A,** 338–339
Insulin
astrocyte reaction to, **148,** 129, 144–146
binding capacity, **155,** 11, 13–15, 17, 26
effect on APP synthesis, **170,** 257
endocytosis, **173,** 249–250
glutamate metabolism and, **147,** 302
imprinting in *Tetrahymena,* **155,** 9–10
interactions
with cytoplasmic proteins, **173,** 261–265
with nuclear matrix, **173,** 265–266
internalization
effect of insulin receptor mutations, **173,** 250–252
mechanisms, **173,** 247–248
retinal regeneration and, **146,** 56, 58–59, 67
role in biological effects
cytoplasmic and nuclear insulin, **173,** 266–267
endosome-associated insulin, **173,** 253–258
sensitive phosphoprotein
pp120 as, **173,** 267–269
Shc as, **173,** 269–272
Sertoli-germ cell communication network and, **147,** 58
signal transduction network, **173,** 244–247
stimulation of FAK dephosphorylation by, **167,** 171–172
translocation to cytoplasm and nucleus, **173,** 258–260
Insulin-like growth factor
astrocyte reaction to, **148,** 129–130, 144–146
binding proteins, **159,** 312–313
control of amphibian metamorphosis and, **145,** 139–140
melanocytes and, **146,** 241
migration induction by, **155,** 77–78
regulation by growth hormone, **159,** 313–314
retinal regeneration and, **146,** 56, 58–59, 67
role in cartilage formation, **159,** 312–314
stimulation of vascular cells, **161,** 14
synthesis, **159,** 312
transcriptional regulation and, **143,** 19
Insulin-like growth factor-binding proteins, Sertoli-germ cell communication network and, **147,** 51
Insulin-like growth factor I
effect on arterial smooth muscle cell culture proliferation, **169,** 213
effect on pituitary cell proliferation, **157,** 50–51, 53
effects on nuclear transport, **151,** 214
as regulator of thymic epithelial cell function, **156,** 181
role in B cell development, **157,** 153–154
Sertoli-germ cell communication network and, **147,** 48
tamoxffen and, **139,** 42–43
Insulin receptor
endocytosis, **173,** 249–250

Insulin receptor (*continued*)
Harderian gland, **168,** 43
mutations, effects on insulin internalization, **173,** 250–252
Integration, retroviral, at multiple sites, **162A,** 436–437
Integration targets, S/MARs as, **162A,** 435–437
Integrin receptor, bacterial binding to, **161,** 293
Integrins
cellular responses to binding to ECM, **167,** 164
communications with nucleus, FAK and MAP kinase pathway in, **167,** 174–175
β_1 cytoplasmic tail of, FAK binding to, **167,** 167–168
in cytoskeleton–membrane connections, **176,** 151–152
Drosophila muscles, **143,** 85
in focal adhesion disassembly/reassembly during mitosis, **167,** 163
function, overview, **167,** 162
interaction with extracellular matrix, **156,** 347
intermediate plaque linkage in hemidesmosomes, **164,** 106–107
plant analog, **174,** 270–275
protein–protein interactions, **168,** 82–83
in protein–substratum linkage, **150,** 121–122
proteins binding to
FAK to cytoplasmic tail of β_1, **167,** 167–168
talin, **167,** 162
in regulation
anchorage-dependent cell growth, **167,** 178
FAK activity, **167,** 168–170
pp120 phosphorylation, **167,** 164–165
role in signaling, **163,** 285
Sertoli-germ cell communication network, **147,** 33
signaling role in transdifferentiation, **150,** 130–131
signal transduction
cellular functions regulated by signaling
apoptosis, **168,** 103–104
differentiation, **168,** 104
migration, **168,** 104–105, 107–108
proliferation, **168,** 102–103
crosstalk with other signaling pathways
G protein-coupled receptors, **168,** 101–102
growth factor receptors, **168,** 99–101
immunoglobulin E receptors, **168,** 102
cytosolic alkalinization, **168,** 96–97
intracellular calcium regulation, **168,** 95–96
phosphatidylinositol 3-kinase role, **168,** 98–99
phosphatidylinositol 4,5-bisphosphate regulation, **168,** 97–98
structure, **168,** 82
in suppressing apoptosis, **167,** 178
TGF-β activity and, **167,** 215
transduction in muscle cell differentiation, **150,** 72, 84–86
Integrity, cellular, during conceptus preimplantation development, **164,** 148–155
Integument, parathyroid hormone-related protein, **166,** 249–250
Intercalary cuticle, genetic control, **174,** 37–38
Intercalation movement, in embryonic tissue
capabilities, **150,** 27–28
capacities, lack of, **150,** 25–27
hyperrestoration, **150,** 25
relaxation of mechanical stresses, **150,** 9
tissue compression, **150,** 29
tissue stretching as promotion method, **150,** 22
Intercellular adhesion molecule-1, **156,** 162
Intercellular junctions, lens
in cortical fibers, **163,** 338–343
in elongating fibers, **163,** 337–338
in lens epithelial cells, **163,** 337
in nuclear fibers, **163,** 338–343
species variability, **163,** 342–343
Interchromatin granules
bright spots in speckled domain, **162B,** 311–312
clusters, **162A,** 16–23,160–164
internal matrix, **162B,** 111–112
Interchromatinic regions, initial observations, **162A,** 3–4

Interchromosomal space
centromeric protein role, **162B,** 40
extrusion of nascent RNA, **162A,** 174–175
Interference color, fish chromatophores and, **143,** 211
Interferon-α
inhibition of endothelial cell growth, **161,** 20
transcriptional regulation and, **143,** 17, 47
Interferon-β
chromatin structure, **162A,** 393–407
gene expression, **162A,** 390–393
inhibition of endothelial cell growth, **161,** 20
transcriptional regulation and, **143,** 17, 22, 43–45
Interferon-γ
endothelial cell tissue factor induction by, **152,** 69–70
intestinal lymphocytes producing, **167,** 98
microgravity, **156,** 309, 314
as procoagulant response regulator, **152,** 60–63
transcriptional regulation and, **143,** 5
Interferons
classes, **167,** 231
clinical potential of polyfunctional nature of, **167,** 239
diversity of effects of, **167,** 224
as effective antitumor agents, **167,** 237
hemopoietic progenitor cells and, **167,** 230–232
migration modulation, **155,** 81
TNF-α potential interactions with, **167,** 230
vertebrate cell culture technology and, **142,** 147, 151, 169, 174, 176, 186
Intergametic recognition, during fertilization, **140,** 413–414
Interleukin-1
dendritic cell interactions, **153,** 82–84
endothelial cell tissue factor induction by, **152,** 68–69, 71–72
Leydig cells influenced by, **149,** 113–115
macrophage production of, **161,** 310–311
melanocytes and, **146,** 240
and newurotransmission, **149,** 268
rat diencephalic neurons affected by, **153,** 275
role in immune response, **170,** 105–107
secretion by granulated metrial gland cells, **153,** 122
Sertoli-germ cell communication network and, **147,** 49
stimulation of vascular cells, **161,** 14
transcriptional regulation and, **143,** 24, 40
biochemistry, **143,** 16, 18
function, **143,** 2
inducible activation, **143,** 7, 10
Interleukin-1α
and Leydig cells, **149,** 114
Sertoli-germ cell communication network and, **147,** 71, 74
Interleukin-1β
and Leydig cells, **149,** 113–114
retinal pigmented epithelium and, **146,** 78
Interleukin-1β-converting enzyme, role in apoptosis, **157,** 251–252
Interleukin-1 receptor, and neurotransmission, **149,** 268
Interleukin-2
B cell activation, **157,** 206
effect on pituitary cell proliferation, **157,** 52
endometrial granulocyte-stimulated, **153,** 124
MIP-1α binding and, **167,** 207
and neurotransmission, **149,** 268
receptor, **157,** 155, 203, 213, 243–244
T cell growth factor, **157,** 202–203, 248, 253
transcriptional regulation and, **143,** 12–13
receptor, **143,** 14
Interleukin-4
B cell activation, **157,** 206, 248
procoagulant expression suppression by, **152,** 88–89
Interleukin-6
costimulation with dendritic cells, **153,** 83
effect on APP synthesis, **170,** 257–259
effect on pituitary cell proliferation, **157,** 52
macrophage production of, **161,** 311–312
and neurotransmission, **149,** 269
receptor, **157,** 155
regulation in transgenic mouse, **170,** 249
role in cytokine signal transduction pathways, **170,** 260–264
Sertoli-germ cell communication network and, **147,** 49, 71, 74
transcriptional regulation and, **143,** 14, 17–18

Interleukin-7
cloning, **157,** 148
effect on B cells
differentiation, **157,** 145–146, 150–151
lineage commitment, **157,** 162
lineage progression, **157,** 158–159
proliferation, **157,** 150
Interleukin-7 receptor
signal transduction, **157,** 150, 155
size, **157,** 149
types, **157,** 149–150
Interleukin-8
angiogenic activity, **159,** 137
transcriptional regulation and, **143,** 17
Interleukin-11
cloning, **157,** 152
induction in stromal cells, **157,** 152
receptor, **157,** 153, 155
role in B cell development, **157,** 153
Interleukins
intestinal lymphocytes producing
after stimulation with
lipopolysaccharide, **167,** 143
TH1 and TH2 cells, **167,** 98
microgravity, **156,** 306, 308, 310, 311, 313–315
role in apoptotic processes, **163,** 118
TNF-α potential interactions with, **167,** 230
transcriptional regulation and, **143,** 12
Intermediate filaments, *see also* Core filaments
antibodies, **162B,** 100–102
centrosome interactions, **164,** 112–114
classes, **164,** 93
complex with nuclear matrix, **162A,** 87–90
cultured cells, **162A,** 492
cytoplasmic
anchorage to plasma membrane, **164,** 102–109
cell nucleus interactions, **164,** 94–102
cytoplasmic organelle interactions, **164,** 109–116
cytoskeletal networks, **162B,** 167
desmosome associations, **164,** 102–105
erythrocyte membrane associations, **164,** 107–108
evolutionary aspects, **164,** 123
family of proteins, **162B,** 129
functions, **164,** 92–94
gene-regulatory function
characteristics, **154,** 1–2, 70–71
cytoskeletal functions, **154,** 2–4
differentiation systems
amphibian early development, **154,** 61–62
characterization, **154,** 48–49, 62–64
cultured cells, **154,** 58–59
cytokeratin filament role, **154,** 54–57
extension of cellular processes, **154,** 53–54
leukemia cell differentiation by vimentin, **154,** 52
mammalian early development, **154,** 60–61
muscle differentiation by desmin, **154,** 49–52
redundancy of function, **154,** 64–66
transgenic animals, **154,** 59
evolutionary aspects, **154,** 66–70
functions, **154,** 71–75
nuclear functions
analysis, **154,** 4
cytoskeleton role, **154,** 38–41
DNA loop formation, **154,** 23–24
double nuclear membrane barrier, **154,** 27–31
histones, interactions with, **154,** 5–7
matrix element, gene expression role, **154,** 24–26
nucleic acids, interactions with, **154,** 5–7
organization of nucleus, **154,** 7–13
physical associations, **154,** 26–37
post-translation modification, **154,** 41–48
repetitive DNA sequence interactions, **154,** 32–37
reverse transformation model, **154,** 38–48
transcription elongation, **154,** 20–23
and transcription factors, relationship, **154,** 13–18
transcription initiation, **154,** 18–20
nucleic acids, interactions with, **154,** 5–7
genes coding for, **164,** 93
hemidesmosome associations, **164,** 105–107
lens fiber cell plasma membrane associations, **164,** 108–109

lipid-containing organelle associations, **164,** 115–116
mechanical load transport, **150,** 176
mechanical stress response, **150,** 187
microtubule interactions, **164,** 112–114
mitochondrial associations, **164,** 114–115
modulation of nuclear architecture, **164,** 100–102
molecular properties, **164,** 92–94
morphogenesis regulation by, **150,** 173
mRNA bound, **162B,** 198–199
with MTOC properties, **175,** 112–114
myofibrillogenesis and, **143,** 14, 182
nuclear envelope associations, **164,** 95–99
nuclear lamin–inner nuclear membrane interactions, **164,** 116–118
overlapping with non-IF proteins, **164,** 122–123
phosphorylation, **154,** 357
plant cytoplasmic, **162B,** 90–95
relationship to lamins, **162B,** 143–145
role in neurogenesis, **159,** 20–21
skeletal muscle Z-disk associations, **164,** 109–112
smooth muscle dense body associations, **164,** 109–112
in striated muscle and, **142,** 96–98, 113–114
subunit heterogeneity, **164,** 94
as tensile stiffeners, **150,** 205–206
tension molding, **150,** 200
in vitro binding to nuclear envelope, **164,** 99–100
in vivo/de novo assembly, **164,** 119–120
Intermediate voltage electron microscopy, **140,** 264
Internal matrix
interchromatin granules, **162B,** 111–112
network domains, **162B,** 105–111
ultrastructure and stability, **162B,** 104–105
Interocular coupling, biological clocks in retina and, **146,** 112–114
Interphase, assessing embryo cells for aneuploidy at, **167,** 267
Interphase cells, lamin phosphorylation, **162B,** 156–158
Interphase nucleus
and chromosome territories, **162A,** 171–172
as dynamic structure
chromatin motion, **150,** 150–157, 167–168
chromosomes
functional state of cells, **150,** 160–164
patterns, **150,** 160–164
topology, **150,** 157–160
contractile proteins, **150,** 164–167
DNA sequence analysis, **150,** 167
gene expression, **150,** 167–168
intranuclear motor, **150,** 164–167
nuclear rotation, **150,** 150–157
spatial organization, **150,** 149–150
nascent pre-mRNA distribution, **162A,** 157–158
NuMA antibody pattern, **162B,** 8–11
NuMA structural role, **162B,** 17–19
Interphase nucleus marker
heterochromatin, **165,** 73–75
nucleolus, **165,** 73
sex chromatin, **165,** 70–72
Interphotoreceptor retinoid-binding protein, **146,** 173, 180
Interrenal cells, steroid secretion, effect of endothelin, **171,** 278–279
Interstitial fluid volume, avian salt glands, **144,** 185, 187–188
Interstitial retinal-binding protein, **138,** 42–43
Intestinal brush border, actin filament bundles
cross-linking proteins, **175,** 34–35
organization, **175,** 30–33
Intestinal peptide, vasoactive, **138,** 169–171
Intestine
glucose absorption, **172,** 5–6
immune system and
M cells in induction of immune response, **167,** 142–143
surveillance of, **167,** 91–92
Peyer's patches
locations of, **167,** 94
M cells as primary entry sites for pathogens, **167,** 138–139
morphology *versus* location in, **167,** 114–115
TGF-β antiproliferative effect on mucosa in, **167,** 212
Intine, cell walls, **140,** 193

Intra-archegonial double fertilization phenomena
in conifers, reports, **140,** 322–327
in cycads, potential for, **140,** 328
in *Ephedra,* early reports, **140,** 333–337
in *Gingko biloba,* potential for, **140,** 328
in nonflowering seed plants, **140,** 320–321
Intracellular adhesion molecule-1, **167,** 110
Intracellular dynamics, growth factor receptors, **138,** 221–222
Intron–exon structure, volvocalean genes
actin introns and, **139,** 248
histone H3 introns and, **139,** 249–250
nitrate reductase introns and, **139,** 248–249
α-tubulin introns and, **139,** 248
ypt genes and, **139,** 250–251
Introns
endosymbiont hypothesis and, **141,** 247, 328
gene expression, **141,** 301, 304
genome organization, **141,** 293, 296–297
mitochondrial genomes in fungi and, **141,** 106–109
splicing, higher plant mitochondrial genomes and, **141,** 161–162, 164
Invasion assay, as migration analysis method, **155,** 63–66
Invertebrates
integument, extracellular matrix, **166,** 71–74
microgravity experiments, **156,** 335
role of cytokines
in cells and organs, **170,** 102–104
in immune response, **170,** 105–107
in stress response, **170,** 107–108
role of nitric oxide, **170,** 113–116
role of POMC products
in chemotaxis, **170,** 88–93
in cytotoxicity, **170,** 95–97
in phagocytosis, **170,** 94–95
stress response evolution, **170,** 97–98
tropomyosin gene, **170,** 6
Inverted repeats
enriched MAR/ORIs, **162A,** 323–325
higher plant plastids, **145,** 155, 157, 160
In vitro fertilization
aneuploidy
enabling study of aneuploidy at conception, **167,** 267
oocytes, **167,** 269
in animals, **140,** 408–409
chromosomal abnormalities, **164,** 142–143
oocyte, **164,** 143–144
spermatozoa, **164,** 144–145
conceptus developmental potential
cell cycle regulation, **164,** 161–164
cell number/integrity, **164,** 148–155
culture conditions effects, **164,** 165–166
culture media, **164,** 168–170
cytoskeleton role in multinuclear blastomeres, **164,** 160–161
DNA content, **164,** 155–160
endometrial synchronization, **164,** 166–168
nucleocytoplasmic ratio effects, **164,** 172
cytoplasmic maturation effects, **164,** 145–146
genomic activation, **164,** 172–173
for higher plants, **140,** 411–413
oocyte maturation effects, **164,** 140–142
preimplantation diagnosis
embryo quality, **151,** 18
ethical issues, **151,** 20
Iodine
hormonal control of amphibian metamorphosis, **145,** 117, 131
proteins in striated muscle, **142,** 116–117, 119
Ion channels
biological clocks in retina and, **146,** 100
chloride, *see* Chloride channels
control of rhythmic behaviors and, **147,** 2
in drug transport, **171,** 129–130
K^+, neuroblastoma cells, tamaxifen and, **139,** 42
localized activity *versus* localized distribution, **154,** 343–346
Na^+
epithelial, possible CFTR regulation of, **172,** 231
modulation by CFTR, **172,** 213–214
in neurotransmission, **149,** 227–232
Ion exchangers
in drug transport, **171,** 129–130
Na^+–Ca^{2+}-
acrosomal exocytosis role, **149,** 29
sperm capacitation regulation, **149,** 14
sperm capacitation role, **149,** 26–27
Na^+–H^+-
acrosomal exocytosis role, **149,** 30–31, 38

protein phosphorylation/ dephosphorylation effects, **161,** 234–236
in RVI, **161,** 194–196
sperma capacitation role, **149,** 30–31
Ionic compounds, and intracellular pH in pollen tubes, **140,** 83
Ionotropic receptor, ligand-gated, glycine receptor as member, **176,** 207–208
Ion pumps
in drug transport, *see* Drug pumps
localized activity *versus* localized distribution, **154,** 343–346
Ions
cations
divalent, in adhesion, **139,** 170
uptake system, salt tolerance, **165,** 17–21
concentration in insect retina, **176,** 316–319
light-induced movements in insect retina, **176,** 316–319
Iopanoic acid, hormonal control of amphibian metamorphosis and, **145,** 110
Ipomaea, self-incompatibility, **143,** 262
Ipomaea batatas, self-incompatibility, **143,** 260
IRF-1, transcriptional regulation, **143,** 17
Iridial epithelium, GLUT1 in, **172,** 22–23
Iridial stroma, GLUT1 in, **172,** 22–23
Iridophores, cytophysiology
leucophores, **143,** 201, 203
motility, **143,** 210–220
motility control, **143,** 226, 235–236
optics, **143,** 206–208
platelets, **143,** 206
static iridophores, **143,** 209–210
Iris
glucose transport, **172,** 22–23
melanocytes, **146,** 192, 194
Iron binding factors, **167,** 227–228
Ischemia
effects on cardiac function, **150,** 82–83
effects on glucose transporter expression, **172,** 35
induced heart disease, spaceflight, experiments, **156,** 320
NMR analysis, **145,** 25, 29, 35, 42, 44, 50
proteins in striated muscle and, **142,** 115, 117
Islet-activating protein, fish chromatophores and, **143,** 227, 240
Isocitric dehydrogenase, avian salt glands and, **144,** 145, 147
Isodesmosine, eggshells and biomineralization and, **145,** 220–221
Isoelectric focusing, selfincompatibility and, **143,** 264–265, 277
Isoferritin, regulation of granulocyte–macrophage production, **167,** 227–228
Isoflavonoids, chalcone synthase and, **147,** 244, 247–250
Isoleucine, flowering plants and, **143,** 320, 323–324
Isomyosins, proteins in striated muscle and, **142,** 101, 103–104
Isopentenyl pyrophosphate, biosynthesis in chromoplasts, **163,** 211–219
Isoprenylation
in chromoplast biogenesis, **163,** 211–219
in cyanelle genome organization, **151,** 49–50
Rab, role of REP, **176,** 45–47
Rab C terminus, **176,** 44–47
role in lamin dynamics, **162B,** 158–160
Isoproterenol
avian salt glands, **144,** 167
thymocytes, **156,** 190
Isozymes, proteins in striated muscle and, **142,** 66–67, 70, 75, 77

J

Jasione montana
embryo sac in, **140,** 259
emergent FAs of, **140,** 251
Jasmonate, as endogenous signaling compound in plants, **148,** 25
Jasmonic acid
chalcone synthase, **147,** 248
as endogenous elicitor of plant defense, **148,** 14
systemic acquired resistance induction, **168,** 321–322
Jejunum
intraepithelial lymphocytes in human, **167,** 96
Peyer's patches in, **167,** 94

Jellyfish
centralization and complexity of nervous system, **167,** 47
life cycle of, **167,** 38
JNK kinase
activation, **157,** 253
role in apoptosis, **157,** 253
Jumping library, clone characterization, **158,** 231
Juvenile hormone, mechanism of action, **159,** 178–180
Juvenile hormone receptor, identification attempts, **164,** 204

K

K99-fimbriae, expression in *Escherichia coli,* **175,** 177–178
K-252, effect on endoreduplication, **158,** 75
Kainic acid, retinal precursor cells and, **146,** 154, 181
Kanamycin, spaceflight experiments, **156,** 340
Kar3, spindle pole body functions, **169,** 68, 71
Karyokinesis, mitosis and
dissociation of events, **144,** 228, 230, 232–235
mitotic events, **144,** 224–226
Karyology, tumor, in transgenic mice, **164,** 73
Karyolymph, historic designation, **162A,** 3–4
Karyosomes, vertebrates and, **144,** 25
Karyosphere
chromosome assembly, **144,** 42–43
function, **144,** 44
gametogenesis, **144,** 8
mammals, **144,** 38–39
nuclear matrix, **144,** 40–41
nuclei, **144,** 43–44, 46–47
oocytes of insects
Diptera, **144,** 8–11
Neuroptera, **144,** 11–18
tenebrionid beeties, **144,** 19–25
protein metabolism, **144,** 41–42
substructural elements, **144,** 44–46
vertebrates, **144,** 25–27
anurans, **144,** 27–36
birds, **144,** 37–38
reptiles, **144,** 36–37
Karyotypes, melanocytes and, **146,** 243
Katanin, microtubule-severing activity, **164,** 279–280
KBG7.2 clones, **140,** 9–10
Kearns-Sayre syndrome, animal mtDNA and, **141,** 207
Kentucky bluegrass clones, **140,** 10
Keoxifene, genital abnormalities in mice, **139,** 20, 22–23
Keratan sulfate, eggshells and biomineralization and, **145,** 234, 238
Keratin
eggshells and biomineralization and, **145,** 220–221, 237
hormonal control of amphibian metamorphosis and, **145,** 121
Keratin-like protein, with MTOC properties, **175,** 112–114
Keratinocyte growth factor
biological activity, **160,** 250–251
carcinoma role, **160,** 251
retinal regeneration and, **146,** 56
Keratinocyte growth factor receptor, epithelial cell expression, **160,** 224, 250
Keratinocytes
biology, **146,** 193–194
cadherin expression, **165,** 177–178
cytokine receptors, **169,** 166–168
epidermal pentapeptide inhibition of, **167,** 227
growth factor synthesis, **169,** 171–172
inhibition of EGF-stimulated, **167,** 193
inhibitory effect of MIP-1α on, **167,** 208
markers, **169,** 168
mechanical integrity, cytokeratin filament role, **154,** 54–57
metal ion regulation, **169,** 170
migration, **169,** 165–168, 170
pEEDCK-like peptide inhibition of, **167,** 220
wound healing role, **169,** 165–168, 170–172
Kernal, development, **140,** 399
Ketanserin, inhibition of smooth muscle cell growth, **161,** 25–26
225-Ketoacyl ACP-synthetase, higher plant plastids, **145,** 184

α-Ketoglutarate
  γ-aminobutyric acid metabolism and, **147,** 311, 317, 319
  glutamate metabolism, **147,** 286, 289, 322
    glutamate dehydrogenase, **147,** 294–295, 300
    glutamine synthetase, **147,** 302–304
Kevitone, chalcone synthase and, **147,** 252
Kex1p protein, turnover, **139,** 76–77
Kex2p protein, turnover, **139,** 76–77
Kidney
  avian salt glands and, **144,** 129–130, 176, 188, 199
  cancer, *mdr1* gene expression, **156,** 88
  cytoskeletal dynamics and, **144,** 106
  ecto-ATPase levels, **158,** 170–171
  embedded, volume estimation, **161,** 141
  glomerulus, *see* Renal glomerulus
  parathyroid hormone-related protein, neoplasm, **166,** 256
  platelet-derived growth factors in, distribution, **172,** 110
  programmed cell death, **163,** 140–141
Kinase, role in Tau proteins
  activity and specificity, **171,** 178–179
  Tau phosphorylation, **171,** 178
Kinectin, kinesin binding, **164,** 300–301
Kinesin
  axonal assembly, **165,** 270–271
  characteristics, **164,** 295–297
  cross-bridge cycle, **164,** 298
  cytoskeletal dynamics, **144,** 109
  fish chromatophores, **143,** 222–223, 242, 247
  function, **164,** 299–300
  and myosin, **140,** 93–94
    biochemical characteristics, **140,** 94–96
      kinesin, **140,** 95–96
      myosin, **140,** 94–95
      profilin, **140,** 96
    structure of molecules, **140,** 93–94
      kinesin, **140,** 94
      myosin, **140,** 93–94
  and myosin-like protein from pollen, **140,** 94–96
    myosin, **140,** 94–95
    profilin, **140,** 96
    structure of molecules, **140,** 93–94
      kinesin, **140,** 94
      myosin, **140,** 93–94
  regulation
    motor activity, **164,** 301
    organelle–kinesin interactions, **164,** 300
  step size, **164,** 298–299
  structure of molecules, **140,** 94
Kinesin-related proteins
  in cytoskeleton-based motor proteins, **176,** 157–159
  functions
    in flagella, **164,** 305–306
    in mitosis and meiosis, **164,** 304–305
    in organelle transport, **164,** 302–304
  K71S23, **140,** 111–112
  properties, **164,** 302
Kinetics
  Arabidopsis thaliana, **144,** 60, 70
  bacterial flagellar motor, **147,** 98, 111
    model comparisons, **147,** 146–150
    models, **147,** 133, 136, 139, 144–145
    motor energetics, **147,** 118, 120
  calcium channels and, **147,** 194
  in drug pump model, **171,** 134–135
  eggshells and biomineralization and, **145,** 239
  environmental stimuli in plant cells and, **145,** 279
  glutamate metabolism and, **147,** 290
  higher plant plastids and, **145,** 168–169
  histone acetylation and deacetylation, **162A,** 205–206
  nuclear magnetic resonance and, **145,** 33–34
  transcriptional regulation, **143,** 7
  translocation, **162B,** 203–204, 213
Kinetin, **166,** 39
Kinetochores
  –centromere region
    chromatin anchoring role, **162B,** 39
    and prekinetochore, characterization, **162B,** 36–38
    role in microtubule-based movement, **162B,** 45–46
  microtubules, lepidopteran
    meiosis
      female, **152,** 14
      male apyrene, **152,** 25
      male eupyrene, **152,** 22–23
    organization, **152,** 12
    post-translational modifications, **152,** 31
    structural features, **152,** 33
  mitosis, **144,** 217
    chromosomes, **144,** 239–241, 243–247

Kinetochores (*continued*)
dissociation of events, **144,** 227, 230, 237–238
mitotic events, **144,** 219–220, 222–223
plates, development, **162B,** 41–45
role in signal transduction, **162B,** 46–48
Kinetoplastids
chloroplasts, **144,** 264, 284–285, 294
mtDNA, **141,** 65, 84
developmental control, **141,** 80
alternate proteins, **141,** 82
editing, **141,** 81–82
mRNA abundance, **141,** 80–81
polyadenylation, **141,** 81
DNA loss, **141,** 68–69
maxicircle genes, **141,** 69–70
gene order, **141,** 74
guide RNA genes, **141,** 72–73
protein-coding genes, **141,** 70–72
variable region, **141,** 73–74
maxicircle and minicircle diversity, **141,** 66–68
maxicircle transcripts, **141,** 74–75
editing, **141,** 75–77, 79–80
guide RNA genes, **141,** 79
overlapping genes, **141,** 77–79
minicircles, **141,** 82–84
replication, **141,** 69
subcellular localization, **141,** 65–66
Klayveromyces lactis, mitochondrial genomes
DNA diversity, **141,** 98, 100–103
generation of mtDNA diversity, **141,** 108, 110, 115, 117–119
Klebsiella aerogenes, urease in plant cells, **145,** 89, 92
Kloeckera africana, mitochondrial genomes, **141,** 101, 103
Kluyveromyces, mitochondrial genomes, **141,** 101, 110, 118
Kupffer cells, galactose receptors, **175,** 147

L

Labeling
in drug pump model, **171,** 136
end
in situ, **163,** 121–123
TdT-mediated dUTP-biotin nick, **163,** 123
photoaffinity, glucan-binding proteins, **148,** 17–18
photolabeling, in drug pump model, **171,** 136
Labial cuticle, genetic control, **174,** 39–40
Labrum, insect, form diversity, **174,** 7
Labyrinthula, locomotion, **150,** 59–62
Laccase
in lignin biosynthesis, **176,** 111–114
in lignin polymerization, **172,** 273
Lacerta muralis, karyosphere and, **144,** 37
Lacrimal gland
development, **168,** 5–6, 8
similarity with Harderian gland, **168,** 3, 62–63
Lactase, expression by villus *versus* dome epithelial enterocytes, **167,** 131
Lactate
NMR analysis, **145,** 30, 43, 45–46
Sertoli-germ cell communication network, **147,** 55, 67
vertebrate cell culture technology, **142,** 195, 200, 204
Lactation, parathyroid hormone-related protein, **166,** 250
Lactic acid
symplast and, **142,** 275
vertebrate cell culture technology and, **142,** 164–165, 173, 195
Lactic dehydrogenase, avian salt glands and, **144,** 145
Lactoferrin, regulation of hemopoietic progenitor cells by, **167,** 228
Lactuca sativa, microgravity experiments, **156,** 321–323, 325–327
Lagging strand assay, **162A,** 469
LAI, *see* Leukemia inhibitory factor
Lambert-Eaton myasthenic syndrome, calcium channels and, **147,** 218, 220, 225
Lamella
chloroplasts and, **144,** 263, 266, 278, 282, 286, 290
cytoskeletal dynamics and, **144,** 94, 98, 101, 115
leading, definition, **150,** 38–39, 44
middle, in cell wall, **174,** 199–200
Lamellipodia
actin, role in movement, **150,** 64
definition, **150,** 38–40
rearward traction, **150,** 41
retrograde surface transport, **150,** 44, 46, 48–49, 53

Lamina-associated polypeptides, role in lamin linkage to nuclear envelope, **164,** 118
Lamina densa
collagen fibril anchoring to
anchoring fibrils, **173,** 136–137
beaded fibril role, **173,** 136
microfibril role, **173,** 134–136
microthreads, **173,** 137–138
role in organogenesis, **173,** 132–133
direct physical connection with collagen fibrils, **173,** 127–130
formation in absence of macromolecules, **173,** 123–125
–collagen fibril connection, light microscopy, **173,** 125–126
–collagen fibril interaction, related molecules
collagen I, **173,** 78–86
collagen III, **173,** 86–88
collagen IV, **173,** 88–94
collagen V, **173,** 95–101
collagen VII, **173,** 101–103
collagen XVII, **173,** 103–105
laminins, **173,** 105–108
nidogen, **173,** 108
osteonectin, **173,** 109
perlecan, **173,** 108–109
skeleton, and collagen IV, **173,** 120–123
structural components, **173,** 108–109
tissue location, **173,** 75–76
Lamina pores, karyosphere and, **144,** 30, 40
Lamina propria, intestinal lymphocytes in
definition, **167,** 94
description, **167,** 97–98
Laminaria angustata, intracellular calcium messenger system and, **142,** 336
Laminar organization, retinal precursor cells and, **146,** 148
Lamin-associated proteins, **162B,** 167–170,383
Lamination, avian retina and, **146,** 42
cellular processes, **146,** 36–40
normal development, **146,** 1–4, 6
retinogenesis *in vitro,* **146,** 10–13, 15
Lamin B receptor
interaction with other nuclear proteins, **162B,** 168–169
role in lamin linkage to nuclear envelope, **164,** 117–118
Laminin
basement lamina composition, **159,** 31–32
effect on arterial smooth muscle cell culture phenotype, **169,** 197–198
interactions with ECM molecules, **173,** 107–108
molecular structure, **173,** 106
neurite initiation and elongation role, **153,** 244–246
role in neurogenesis, **159,** 21
superfamily, roles *in vivo,* **173,** 106–107
transdifferentiation in medusae and, **142,** 222
Lamins
A
association with Rb protein, **162B,** 395–396
and C, appearance in differentiated cells, **162B,** 378–379
assembly and disassembly *in vivo,* **162B,** 151–160
B
association with lamin B receptor protein, **162B,** 379
and DNA replication, **162B,** 164–165
in perichromosome region, **162B,** 34–35
cell cycle dependence, **162B,** 382–384
expression changes, **162B,** 161–162
in formation of nuclear envelope, **162B,** 165–167
forming peripheral nuclear matrix, **162A,** 304
identification and primary structure, **162B,** 142–145
nuclear, inner nuclear membrane interactions, **164,** 116–118
nucleoplasmic foci, **162B,** 151–154
plant
immunological identification, **162B,** 100–102
nuclear localization, **162B,** 95–97
topological distribution, **162B,** 102–103
and unicellular organisms, **162B,** 148
proteins in striated muscle and, **142,** 122
as structural support for nuclear membrane, **162B,** 162–163
structure and physical properties, **162B,** 149–151
vertebrate and invertebrate, **162B,** 146–148
Langerhans cells
isolation, **153,** 55–57
MIP-1α transcripts in, **167,** 208
properties, **153,** 54, 57–62

Lanthanium
agravitropism induction in plants, **168,** 162–164
calcium substitution in plants, **168,** 162
Large-single copy sequence, higher plant plastids and, **145,** 155, 158
Larmor frequency, nuclear magnetic resonance and, **145,** 3–5
Larus, avian salt glands and, **144,** 133, 140–141
Larva
ascidian
epidermis development, **176,** 264–267
structure, **176,** 246–249
Drosophila, muscle
contractile protein genes, **143,** 101, 109, 111
development, **143,** 66–71, 73, 84–87
Drosophila epidermis, genetic control
antennal cuticle, **174,** 37
clypeolabrum, **174,** 35–36
dorsal ridge, **174,** 40
intercalary cuticle, **174,** 37–38
labial cuticle, **174,** 39–40
mandibular cuticle, **174,** 38–39
maxillary cuticle, **174,** 39
ocular cuticle, **174,** 36–37
required genes, **174,** 33–35
growth, hormonal control of amphibian metamorphosis, **145,** 128–132, 134, 137–139
planula
as cnidarian developmental stage, **167,** 38
neuropeptides and metamorphosis of *Hydractinia echinata,* **167,** 54–55
Tribolium castanaeum, genetics of development, **174,** 43–45
Laser scanning confocal microscopy, **162A,** 50–53
Lateralization, avian retina and, **146,** 42
Lathyrus saliva, urease in plant cells and, **145,** 67
Latissimus dorsi, anterior, proteins
antibody use, **142,** 100–101
thick-filament proteins, **142,** 68, 70, 77–78
thin-filament proteins, **142,** 90–92
Layer of Chievitz, avian retina and, **146,** 39, 42
LDL, *see* Low-density lipoproteins
Lead, in histochemical staining of ATPases, **158,** 143, 163
Leaf, photosynthetic cells, effect of microgravity, **171,** 36–38
Lebistes, fish chromatophores and, **143,** 201, 227
Lectins
accumulation, **156,** 48
avian retina and, **146,** 10, 16–17, 24
–concanavalin A, for analysis of APP microheterogeneity, **170,** 265–266
extracellular matrix and, **146,** 264, 277
histochemical studies
dome *versus* villus enterocyte composition, **167,** 131
M cells
dome epithelial cell ratio to, among species, **167,** 114
versus enterocytes, **167,** 115
M cell identification, **167,** 103–104
terminal saccharide presence in glycocalyx of, **167,** 124–126
undifferentiated crypt cells, **167,** 108–109
immunological target in multiple sclerosis, **159,** 23
peanut
avian retina and, **146,** 4, 16, 19
retinal precursor cells and, **146,** 153, 160
role in neurogenesis, **159,** 22–23
self-incompatibility and, **143,** 269
Sertoli-germ cell communication network and, **147,** 36
transcriptional regulation and, **143,** 14, 16
transdifferentiation in medusae and, **142,** 250–251
urease in plant cells and, **145,** 84, 98
LEF-1, *see* Lymphoid-specific enhancer factor
Legumes
higher plant plastids and, **145,** 163
mutations, **156,** 39–41
Rhizobium–plant interactions, **156,** 1–51
seeds, origin of protein bodies in, **167,** 6–7
urease in plant cells and, **145,** 74, 98
Leishmania, kinetoplastid mtDNA and, **141,** 66, 79
Leishmania tarentolae, kinetoplastid mtDNA and
maxicircle genes, **141,** 69–74

maxicircle transcripts, **141,** 74–76, 79–80
minicircles, **141,** 83
Lemna
environmental stimuli in, **145,** 268, 271
intracellular calcium messenger system and, **142,** 309, 336
Lemna paucicostata, flowering plants and, **143,** 320–321
Length, renal capillaries, **161,** 154–155
Length density, parameter specification, **161,** 119–121, 119–221
Lens
biological clocks in retina and, **146,** 90, 92, 101
cell populations, **163,** 312–313
embryological, **163,** 309
epithelial cells
central zone, **163,** 312–313, 323, 326
epithelial–fiber cell interface, **163,** 343
intercellular junctions, **163,** 337
polarization, **163,** 326
pregerminative zone, **163,** 312–313, 326
transitional zone, **163,** 312–313
ultrastructural characteristics, **163,** 323–326
fiber cells
elongating, **163,** 310–312, 337–338, 338
end curvatures, **163,** 316
epithelial–fiber cell interface, **163,** 343
fiber compaction phenomenon, **163,** 333
fusion zones, **163,** 313–316
mature, **163,** 312
plasma membrane, intermediate filament associations, **164,** 108–109
polarization, **163,** 334
primary, formation, **163,** 309
radial cell columns, **163,** 306, 313–314
secondary, formation, **163,** 309–310
sutures, *see* Sutures
transitional zone, **163,** 310
germinative zone, **163,** 309
gross structure, **163,** 306–307
growth, **163,** 309–313
growth factor-induced retinal regeneration and, **146,** 57–58
transdifferentiation in medusae and, **142,** 217–218
ultrastructure
elongating fibers, **163,** 326–330
epithelium, **163,** 323–326
mature fibers, **163,** 326–330
senescent fibers, **163,** 330–334
Lens culinaris, microgravity experiments, **156,** 323, 325–327, 329
Lentils, microgravity experiments, **156,** 323, 325–327, 329
Lepidium sativum
intracellular calcium messenger system and, **142,** 315
microgravity experiments, **156,** 321, 323, 328, 330
Lepidodinium viride, chloroplasts and, **144,** 288
Lepidoptera, spindle structure
characterization, **152,** 1–3
chromosomes
kinetic organization, **152,** 8–12
structure, **152,** 33–34
experimental systems, **152,** 32–35
meiosis
female, **152,** 13–15
male apyrene, **152,** 24–30, 35–36
male eupyrene, **152,** 16–24, 34–36
microtubules, post-translational modifications, **152,** 30–32
mitosis, **152,** 12–13
technical aspects of analysis, **152,** 3–8
infectious particles, **152,** 4
tissue preparation, **152,** 6
Leprosy
dendritic cell role in, **153,** 89–90
immune responses in, **161,** 289
Leptinotarsin, calcium channels and, **147,** 211
Leptomeninges, **165,** 129, 147–148
Lesions, autoimmune, chaperone levels, **174,** 169
Lethocerus, Drosophila muscles and, **143,** 121, 136
Lettuce, microgravity experiments, **156,** 321–323, 325
Leucine, flowering plants and, **143,** 323–324
Leucocidins, Gram-positive bacteria, **161,** 274
Leucophores, cytophysiology
comparison to other chromatophores, **143,** 194
hormonal signals, **143,** 234–237
motile iridophores, **143,** 210
nervous signals, **143,** 226, 228–229, 231

Leucophores (*continued*)
as reflecting chromatophore, **143,** 198, 201–205
Leucosomes, fish chromatophores
leucophores, **143,** 201, 203, 205
motility control, **143,** 228, 233, 237, 241
Leucoverin, **156,** 110–111, 116, 119
Leukemia
acute
cytosine arabinoside, **156,** 120–121
mdr1 gene expression, **156,** 88
MDR reversal, **156,** 91
acute promyelocytic, retinoic acid effect, **159,** 182–183
cell differentiation by vimentin, **154,** 52
chronic, **156,** 129
drug resistance, **156,** 127
mdr1 gene expression, **156,** 88
topoisomerase II, **156,** 107
chronic lymphatic
paraneoplastic effect, **167,** 235
progressive bone marrow failure in, **167,** 237–238
chronic myeloid
altered growth kinetics in, **167,** 186
IFN-α effects on, **167,** 232
inhibitor cooperation and, **167,** 203–204
microenvironment and stem cell inhibition in studies of, **167,** 195–196
MIP-1α in potential treatment of, **167,** 239
resistance to inhibitors, **167,** 222, 236
cytosine arabinoside, **156,** 120–121
drug resistance, **156,** 127
hemopoietic inhibitors and bone marrow failure in, **167,** 234–235
malignancy, ecto-ATPase as marker, **158,** 173
MDR, **156,** 100, 106
mdr1 gene expression, **156,** 88
MDR reversal, **156,** 91
melanocytes and, **146,** 228, 250
microgravity experiments, **156,** 313
and procoagulants, **152,** 83–85
topoisomerase II, **156,** 106, 107
transcriptional regulation and, **143,** 27, 37–38, 42–43
variations in response to growth factors and cytotoxic agents, **167,** 236–237
Leukemia inhibitory factor
effect on hemopoiesis, **167,** 234–235
neuropeptide Y expression affected by, **153,** 271
role in apoptotic processes, **163,** 118
secretion by granulated metrial gland cells, **153,** 122–123
Leukocytes
cytoskeletal dynamics and, **144,** 99, 102
–endothelial interaction, selectin role, **175,** 183–186
mixed reaction, dendritic cell role
allogeneic, **153,** 76
autologous, **153,** 76–77
mobilization by BB10010 and G-CSF, **167,** 242–243
polymorphonuclear, *see* Polymorphonuclear leukocytes
retinal pigmented epithelium and, **146,** 78
Leukoplasts, higher plant plastids and, **145,** 169–170, 179, 190
Leukotrienes, effects on arterial smooth muscle cell culture
phenotype, **169,** 206
proliferation, **169,** 217
Leydig cells
in autoimmune disease, **149,** 125
in cryptorchidism, **149,** 126
photoperiod effects, **149,** 121–122
Sertoli-germ cell communication network and, **147,** 60, 77
testicular macrophage interactions with, **149,** 101–102, 129
functional, **149,** 109–111
phagocytosis, **149,** 123
secretory products in, **149,** 112–115
structural, **149,** 105–109
LFA, *see* Lymphocyte function-associated antigens
LH, *see* Luteinizing hormone
Library, linking, clone generation, **158,** 231
Licmophora, morphogenesis
branching costae, **150,** 271
nucleating structure, **150,** 274
Lidhium, avian salt glands and, **144,** 184
LIF, *see* Leukemia inhibitory factor
Life span, M cell *versus* enterocyte, **167,** 110–111
Ligand receptor, acidification, **167,** 23–26
Ligands
avian salt glands and, **144,** 165, 168

bacterial flagellar motor and, **147,** 130–131, 136, 140, 142, 149
control of rhythmic behaviors and, **147,** 2, 17
C-terminal *versus* N-terminal targeting, **167,** 27
cytoskeletal dynamics and, **144,** 102
Drosophila muscles and, **143,** 85
endocytosis into clathrin-coated vesicles, receptors and, **167,** 18
Fas, role in apoptosis, **157,** 204
fish chromatophores and, **143,** 244
GnRH-encoding gene in teleosts and, **147,** 186
interactions with receptors in antigen binding to M cells, **167,** 124
for sea anemone glycoprotein hormone receptor, **167,** 56
Sertoli-germ cell communication network and, **147,** 36
transcriptional regulation and, **143,** 16, 22
transdifferentiation in medusae, **142,** 220, 249–250
Ligation assay, oligonucleotide in diagnosis of genetic defects, **151,** 18
Light
biological clocks in retina and
Bulla, **146,** 129–130
cycles, **146,** 86, 96–97, 100
entrainment pathway, **146,** 101–107, 109, 111–112, 114
opisthobranch retina, **146,** 90–93
pacemaker mechanisms, **146,** 116, 121
constant, biological clocks in retina and, **146,** 138
cytoplasmic streaming affected by, **155,** 125
environmental stimuli in plant cells and, **145,** 251, 299, 301
chemicals, **145,** 297–298
chloroplasts, **145,** 268–275
motile systems, **145,** 259–263
nucleus, **145,** 287–288
photodinesis, **145,** 252–259
signal transduction, **145,** 263–268
transduction chain, **145,** 275–287
extracellular matrix and, **146,** 301
far-red, environmental stimuli in plant cells
chloroplasts, **145,** 272, 277, 280–281
nucleus, **145,** 288
photodinesis, **145,** 254, 258–259, 266
inducibility by, chalcone synthase and, **147,** 263–265, 269, 273–274
intracellular calcium messenger system and, **142,** 305, 318, 327, 329–331, 333
neurons sensitive to, **167,** 39–40
regulation of chromoplast development, **163,** 199
retinal precursor cells and, **146,** 145, 148, 155, 162, 164–165
symplast and, **142,** 270
Light chains, proteins in striated muscle
cardiac muscle, **142,** 106
myosin, **142,** 64–65, 69, 71, 76–77, 79
Light meromyosin
myofibrillogenesis and, **143,** 177
in striated muscle, **142,** 63
Light microscopy
avian salt glands and, **144,** 133, 148, 167, 180, 182
cnidarian nervous system anatomy studies, **167,** 40
collagen fibrillamina densa connection, **173,** 125–126
control of rhythmic behaviors and, **147,** 10
glutamate metabolism and, **147,** 290
growth factor-induced retinal regeneration and, **146,** 54, 61–62
karyosphere and, **144,** 11, 18, 39, 45
melanocytes and, **146,** 196
mitosis and, **144,** 225, 230, 232
of Peyer's patches, M cell identification using, **167,** 99–100
proteins in striated muscle and, **142,** 66, 129
regenerative cementogenesis and, **142,** 32
retinal precursor cells and, **146,** 151, 157
studies using, **140,** 407–408
in study of microgravity effects on statocytes, **171,** 11
Lignan, production, **176,** 93–94
Lignification, cell wall, model systems, **176,** 98–101
Lignin
analysis, **176,** 94–95
antibodies to, generation, **171,** 94–95
Arabidopsis thaliana and, **144,** 60
biosynthesis, **176,** 101–102
associated enzymes, **176,** 107–108

Lignin (*continued*)
caffeoyl-CoA 3-*O*-methyltransferase in, **172,** 258
carbon flow regulation, **176,** 119–122
cell-to-cell cooperation during, **176,** 108–109
cinnamate-4-hydroxylase in, **172,** 253–254
cinnamoyl-CoA reductase in, **172,** 262–264
cinnamyl alcohol dehydrogenase in, **172,** 264–269
coumarate 3-hydroxylase in, **172,** 254–255
4-coumarate-CoA ligase in, **172,** 259–262
ferulate 5-hydroxylase in, **172,** 255
general phenylpropanoid pathway and, **172,** 246
lignin branch pathway and, **172,** 246–247
O-methyltransferase in, **172,** 256–258
monomeric composition regulation, **176,** 122
phenylalanine ammonia lyase in, **172,** 247–252
polymerization step, **176,** 109–111
role of laccase, **176,** 111–114
role of peroxidase, **176,** 114–117
shikimate pathway and, **172,** 246
tyrosine ammonia lyase in, **172,** 252
carbon flow regulation, **176,** 119–122
composition, **172,** 245; **176,** 89–91
content in mung bean, effect of microgravity, **171,** 39
cytochemical probes, **176,** 95
deposition in cell wall, **176,** 117–119
determination, **176,** 95–96
distribution, **176,** 89–91
distribution in plants, **172,** 245
extracellular matrix and
characterization, **146,** 271
modification, **146,** 293–295, 297, 300
pathogenesis, **146,** 305, 309–310
stress, **146,** 301, 304
synthesis, **146,** 278–279, 281–284
histochemical probes, **176,** 95
inhibitors, effects *in planta,* **172,** 251–252
manipulation
future directions, **172,** 278–279
problems in, **172,** 274–276
use of mutants in, **172,** 277–278
microgravity experiments, **156,** 332
monomer composition, **176,** 96–97
nature, **176,** 89–91
phenylpropanoid pathway products as precursors, **176,** 102–105
in plant cell wall, dynamic aspects
biological roles, **151,** 246–247
biosynthesis
control, **151,** 238–239
process, **151,** 233–238
composition, **151,** 231–233
detection *in situ,* **151,** 229–231
grass, structural features, **151,** 243, 245–246
heterogeneity, **151,** 231–233
occurrence, **151,** 229–231
polysaccharides, bonds with
formation, **151,** 240–245
linkage, **151,** 239–243
polymerization
laccases and, **172,** 273
peroxidase and, **172,** 271–273
phenoxyradicals and, **172,** 271, 273
production in plant, effect of microgravity, **170,** 55–56
related compounds, production, **176,** 93–94
residues, **172,** 245
roles in plants, **172,** 244, 274–275
structure probing *in situ,* **176,** 97–98
Lignin-specific pathway
associated enzymes, **176,** 107–108
in lignin biosynthesis, **172,** 246
cinnamoyl-CoA reductase, **172,** 262–264
cinnamyl alcohol dehydrogenase, **172,** 264–269
role of cinnamoyl-CoA reductase, **176,** 105–106
role of cinnamyl alcohol dehydrogenase, **176,** 105–106
role of cinnamyl alcohol storage forms, **176,** 106–107
Lilium
environmental stimuli in, **145,** 261
intracellular calcium messenger system and, **142,** 324
Lilium henryi, anthers of, **140,** 57
Lilium humboldtii, **140,** 190, 193

Lilium longiflorum
cytokinin-induced habituation, **152,** 272
pollen tubes, **140,** 190, 191, 195
self-incompatibility and, **143,** 282–283
Limb bud
development, programmed cell death in, **163,** 143–146
platelet-derived growth factor distribution in, **172,** 108–109
LIM domains, in zyxin and cCRP, **167,** 163
Limited pool hypothesis, for maternal age effect in aneuploidy, **167,** 277
Limonium, symplast and, **142,** 292
Limulus, Drosophila muscles and, **143,** 99
Limulus polyphemus, acrosome reaction, **175,** 37–40
Linkage
Arabidopsis thaliana and, **144,** 63
endosymbiont hypothesis and, **141,** 235, 293, 295, 303
mitochondrial genomes in fungi and, **141,** 97, 101–102, 120
Linking library, clone generation, **158,** 231
Linomide, inhibition of angiogenesis, **159,** 142
Lipid peroxidation, in angiosperm seedlings, effect of microgravity, **171,** 40–41
Lipids
actin network interaction with, **148,** 44
acyl, chromoplast
in *Capsicum* and *Narcissus,* **163,** 190
characterization, **163,** 188
metabolism, **163,** 219–220
avian salt glands and, **144,** 141, 144, 169, 204
bacterial flagellar motor and, **147,** 105, 137
calcium channels and, **147,** 209
chloroplasts and, **144,** 268
cytoskeletal dynamics and, **144,** 99
directed flow hypothesis, **148,** 45
endosymbiont hypothesis and, **141,** 240, 246
extracellular matrix and, **146,** 277, 301, 305, 311–312
fish chromatophores and, **143,** 198, 245
in fluid mosaic model of cell membrane, **148,** 43
Harderian gland secretion
biosynthesis, **168,** 38
functions, **168,** 60–61
structures in mammals, **168,** 37
higher plant plastids and, **145,** 204
hormonal control of amphibian metamorphosis and, **145,** 122–123
intracellular calcium messenger system and, **142,** 334, 338
karyosphere and, **144,** 28, 41
nuclear magnetic resonance and, **145,** 45
organelles containing, **164,** 115–116
reserve in plant cells, effect of microgravity, **171,** 51
retinal precursor cells and, **146,** 147, 153, 155–156
role in membrane flow, **148,** 46, 48–49
self-incompatibility and, **143,** 261
Sertoli-germ cell communication network and, **147,** 40, 73–74
storage diseases, preimplantation diagnosis, **151,** 14–15
symplast and, **142,** 289
Lipochitooligosaccharides, **156,** 19
Lipooligosaccharides, **156,** 19, 24, 30, 44
Lipopentasaccharides, **156,** 19, 21
Lipopeptides
Gram-negative bacteria, **161,** 281–284
uptake by macrophages, **161,** 294–295
Lipopolysaccharides
coagulation effects
endothelial cell tissue factor induction, **152,** 66–69, 71–72
monocytendash;macrophage tissue factor induction, **152,** 57–58
procoagulants
activity, **152,** 79, 81, 85
expresssion modulation, **152,** 86–88
tissue factor induction, **152,** 61–63
effect on testicular macrophages, **149,** 109
Gram-negative bacteria, **161,** 279–281
intracellular events induced by, **161,** 296–297
macrophage receptors, **161,** 292–293
ring, bacterial flagellar motor end, **147,** 100–102, 105
transcriptional regulation and, **143,** 2, 7, 9–10, 41
Lipoproteins
Gram-negative bacteria, **161,** 281–284
lipoprotein(a), stimulation of vascular cells, **161,** 14–15

Lipoproteins (*continued*)
low-density
effect on arterial smooth muscle cell proliferation, **169,** 218–219
endothelial cell tissue factor induction, **152,** 70–71
mechanism, **166,** 109–111
transport components, **166,** 109–111
Liposomes
in inducing immunity against streptococci, **167,** 141–142
intracellular calcium messenger system and, **142,** 310, 314, 317
V-ATPase reconstitution into, **174,** 105–113
Lipoteichoic acids, bacterial cell wall, **161,** 270–272
Lipotropin
evolution, **159,** 52, 56
precursors, **159,** 52, 54–55
Lipovitellin, **166,** 107
Lipoxygenase
pathway for arachidonic acid metabolism, **161,** 221
role in volume regulation, **161,** 222–223
Liquid crystal
arced patterns in transmission electron microscopy, **166,** 66–68
artifacts, **166,** 68–70
biological analogues, **166,** 63
cholesterics, **166,** 62–63
cylindrical twist, **166,** 64, 66
defects, **166,** 68–70
deviations, **166,** 68–70
intermediate states, **166,** 61–62
lyotropics, **166,** 62
mesophases, **166,** 62
nematics, **166,** 62–63
periodic extinctions in polarized light, **166,** 65–66, 67
planar twist, **166,** 64, 66
properties, **166,** 61–70
smectics, **166,** 62–63
thermotropics, **166,** 62
toroidal twist, **166,** 64, 66
Lithium
agravitropism induction in plants, **168,** 168, 170
biological clocks in retina and, **146,** 122
Lithium 3,5-diiodosalicylate
in nuclear isolation, **162A,** 12
preparation of scaffolds, **162B,** 422–424
Liveborns
aneuploidy in, incidence of, **167,** 265
mosaicism in, incidence of, **167,** 287
Liver
acute phase protein production, **170,** 243–244
dendritic cells, **153,** 62–63
ecto-ATPase
levels, **158,** 169–170
localization, **158,** 170
purification, **158,** 170
eggshells and biomineralization and, **145,** 234
hormonal control of amphibian metamorphosis and, **145,** 107, 131–132, 140
karyosphere and, **144,** 41
mitosis and, **144,** 225
nuclear magnetic resonance and, **145,** 31, 35, 43–44
TGF-βmRNA production by fetal cells, **167,** 210
transcriptional regulation and, **143,** 18, 36
urease in plant cells and, **145,** 83
Liverwort, plastids and, **145,** 153, 155, 157, 166–167
Lizards
avian salt glands and, **144,** 133, 147, 155
karyosphere and, **144,** 37
Lobopodia, tissue cell
definition, **150,** 40
rearward traction forces, **150,** 41
Lobsters, control of rhythmic behaviors, **147,** 11
Localization, cytoplasmic localization, **138,** 78–83
Lockhart equations, symplast and, **142,** 277–285
Locus control regions
association with MAR elements, **162A,** 233–234
copy number dependence, **162A,** 427–428
linked γ-globin gene, **162A,** 434
and S/MAR enhancers, general properties, **162A,** 417–421
transcription factors loading into, **162A,** 228–230
Locust
Na^+/K^+-ATPase localization, **176,** 325–327
vitellogenin receptor, **166,** 130–131
Lolium, extracellular matrix and, **146,** 269

Lolium perenne, **140,** 11, 24–25
Long terminal repeats, transcriptional regulation and, **143,** 33
Long-term potentiation
interphase chromatin reorganization by, **150,** 163–164
mechanism, **153,** 267
Loose coupling model, bacterial flagellar motor and, **147,** 134–135
Lotus corniculatus, flowering plants and, **143,** 308
Low-density lipoprotein receptor
biosynthesis, **166,** 113–117
human domain model, **166,** 114
pathway, **166,** 111–112
structure, **166,** 113–117
supergene family, mammalian, familial hypercholesterolemia
cytoplasmic tail, **166,** 116–117
epidermal growth factor, **166,** 115–116
ligand binding domain, **166,** 115
low-density lipoprotein receptor biosynthesis, **166,** 113–117
low-density lipoprotein receptor structure, **166,** 113–117
membrane anchoring domain, **166,** 116
mutations, **166,** 112–120
O-linked sugar domain, **166,** 116
Low-density lipoproteins
effect on arterial smooth muscle cell proliferation, **169,** 218–219
endothelial cell tissue factor induction, **152,** 70–71
Loxodes striatus, karyosphere and, **144,** 36
LRH, *see* Lutein-releasing hormone
L-strand
promoters
animal mtDNA and, **141,** 218, 223, 226–227
endosymbiont hypothesis and, **141,** 299–300
synthesis, animal mtDNA and, **141,** 225–226
LTBMC, *see* Long-term bone marrow culture
LTCIC, *see* Long-term culture-initiating cell
Lucifer yellow, avian retina and, **146,** 4, 16
Lüger model, bacterial flagellar motor
model comparisons, **147,** 146, 148, 150–151, 153–155
models of muscular contraction, **147,** 156–159
Lumicolchicine, cytoskeleton, **166,** 15
Lung
dendritic cells, **153,** 63
ecto-ATPase levels, **158,** 167
parathyroid hormone-related protein, neoplasm, **166,** 257
platelet-derived growth factors
activity, **172,** 115–116
distribution, **172,** 109
Lung cancer
MDR, **156,** 100
MDR reversal, **156,** 92–93
Lupinus texensis, urease in plant cells and, **145,** 82
Lupus nephritis model, **166,** 212–214
Luteinizing hormone
control of amphibian metamorphosis and, **145,** 113, 115–116, 119
effect on Harderian gland secretion, **168,** 51
β-endorphin effect on, **149,** 115
feedback mechanisms, **159,** 79–80
GnRH-encoding gene in teleosts, **147,** 165, 167, 185
regulation of activity, **159,** 69, 72
response to interleukin-1, **149,** 113–114
synthesis, **159,** 68–69
and testicular macrophages, **149,** 109, 119
Luteinizing hormone receptor
cloning, **167,** 56
homology with related receptors, **155,** 27
Luteinizing hormone-releasing hormone, control of metamorphosis, **145,** 116
Lutein-releasing hormone
analogs, **159,** 42
biological activity, **159,** 42–43
estrogen effect on release, **159,** 82–85
feedback mechanisms, **159,** 79–80
forms, **159,** 80
gene, **159,** 42
messenger RNA levels, **159,** 9
modulation of release, **159,** 44–45
neuron differentiation, **159,** 40, 43
precursor localization, **159,** 41
Lutein-releasing hormone receptor
signal transduction, **159,** 41, 43
structure, **159,** 41
Luteolin
chalcone synthase and, **147,** 247, 249
effect on *nod* box binding, **156,** 15
effect on *nodD* expression, **156,** 14
Ly6A, marker of B cell differentiation, **157,** 136–137, 142

LY117018, genital abnormalities in rats and, **139,** 16–17
Lycopersicon
environmental stimuli in, **145,** 261, 289
self-incompatibility, **143,** 280–281
Lycopersicon peruoianum, selfincompatibility and, **143,** 283–284
Lymnaea, control of rhythmic behaviors and, **147,** 12, 14
Lymnea stagnalis, **167,** 56
Lymphatic dendritic cells, in human, **153,** 64–68
Lymphoblasts
in Peyer's patches, migration, **167,** 93
vertebrate cell culture technology, **142,** 166, 170
Lymphocytes
associated with M cells in dome epithelium, distribution of, **167,** 133
B, *see* B cells; B lymphocytes
ecto-ATPase levels, **158,** 156–157
function-associated antigens, **156,** 162
granulated, pregnant uterus
characterization, **153,** 105, 126
decidual tissue-associated cells, **153,** 107–109
natural killer cells, **153,** 107–108
neutrophils, **153,** 108–109
suppressor cells, **153,** 109
endometrium, human, **153,** 117, 119–121
functions
cytokine secretion, **153,** 121–123
immunological, **153,** 123–125
nonimmunological, **153,** 125–126
metrial gland cells, rodent
characteristics, **153,** 110–111
morphology, **153,** 113–119
origin, **153,** 111–113
phenotypes, **153,** 113–119
in gut wall, *see* Lymphoid tissue, gut-associated
interaction with testicular macrophages, **149,** 108
intraepithelial
definition, **167,** 94
description, **167,** 96–97
lack of correlation with immature M cells, **167,** 110
as major structural characteristic of dome epithelium, **167,** 99
M cell number increase per dome with increased number of, **167,** 143
possible presentation of endocytosed antigens by M cells to, **167,** 128
site-specific variations in, **167,** 115
melanocytes and, **146,** 248
microgravity, **156,** 306
–peritoneal macrophage interaction, **175,** 150–151
in Peyer's patches
description, **167,** 95–96
role in dome epithelium, **167,** 133–135
production, germinal center
antigen-driven, **166,** 140–141
memory B, **166,** 141
production of ACTH and β-endorphin-like molecules, **170,** 83–84
T, *see* T cells
TNF-α production by, **167,** 229
transcriptional regulation, **143,** 28
transcriptional regulation and, **143,** 2, 10–17, 19, 22, 39–41
vertebrate cell culture technology and, **142,** 147, 150
Lymphoepithelium
in BALT, **167,** 136–137
in NALT, DALT, and CALT, **167,** 137–138
Lymphoid follicles
in Peyer's patches
location, **167,** 95
in M cell formation, **167,** 110
overview, **167,** 94
species variations in presence in BALT, **167,** 136–137
Lymphoid-specific enhancer factor, **157,** 162
Lymphoid tissue
bronchus-associated tissue, **167,** 136–137
conjunctiva-associated tissue, **167,** 138
duct-associated tissue, **167,** 137–138
gut-associated
definition, **167,** 92–94
dome *versus* villus enterocyte terminal saccharides in rabbit, **167,** 131
immunological functions, **167,** 92
localization of, **167,** 92–98
M cell presence in
overview, **167,** 92
species, **167,** 100
vimentin immunohistochemistry in detecting, **167,** 104

migration through high endothelial venues, **167,** 95
versus ordinary gut epithelium in antigen uptake, **167,** 139
locations, **167,** 135
in M cell formation, **167,** 110
mucosa-associated tissue
lymphocyte migration through mucosal system and, **167,** 93–94
M cell identification in epithelia of, **167,** 100
nasal-associated tissue, **167,** 137
parathyroid hormonerelated protein, neoplasm, **166,** 261–262
submucosal tissue, conjunctiva-associated lymphoid tissue as, **167,** 138
Lymphokine, microgravity, **156,** 309
Lymphoma
Burkitt lymphoma, **156,** 103
cell transcriptional regulation, **143,** 10, 17, 26–27, 37, 46
MDR, **156,** 103
MDR reversal, **156,** 91
whole-chromosome hybridization use in clinical studies, **153,** 216
Lymphotoxin
similarities to TNF-α of, **167,** 228–229
transcriptional regulation and, **143,** 43
Lymph sinus, in sheep and rabbit Peyer's patches, **167,** 95–96
Lysin
V-lysin, **139,** 196–197
in zona pellucida, **164,** 165–166
Lysine
eggshells and biomineralization and, **145,** 220–221
flowering plants and, **143,** 318–321
Lysophosphatidic acid, effect on arterial muscle cells, **169,** 219
Lysosomes
acidification, **172,** 226
mannose-6-phosphate receptors in protein transport to, **167,** 19
matrix substructure, **160,** 339
M cell, reduction in volume fraction of, **167,** 128
secondary, Sertoli cells, sulfated glycoprotein-1 synthesis, **154,** 118–120
Sertoli-germ cell communication network and, **147,** 74
transdifferentiation in medusae and, **142,** 232
transformation of vesicular to tubular form, **160,** 338–339
Lysozyme, bactericidal activity, **161,** 308
Lysyl oxidase, eggshells and biomineralization and, **145,** 221, 237
Lytechinus pictus, **138,** 246, 262

M

α_2-Macroglobulin
binding to influenza A virus, **175,** 166
inhibition of proteinase, **170,** 240–242
Macromolecules
absence in lamina densa formation, **173,** 123–125
translocation into cytoplasm, mechanisms, **173,** 260–261
Macrophage inflammatory protein-1α
CML progenitor cell resistances to inhibitory actions of, **167,** 195
effects on nonhemopoietic tissues, **167,** 208
events following inhibitor-receptor binding, **167,** 193–194
growth modulation of epithelial tissues, possible, **167,** 242
hemopoeitic progenitor cell cycle regulation by, **167,** 200–206
identification and characterization of, **167,** 198–199
increased expression of, in aplastic anemia and myelodysplasia, **167,** 239
mobilization of hemopoietic progenitor cells by, **167,** 242–243
neoplastic cell lines and, **167,** 236
nonhemopoietic tumor cell line lack of sensitivity to, **167,** 237
polymerization, **167,** 199–200
as potential chemoprotective agent, **167,** 240, 241
receptors for, **167,** 206–207
TGF-β functional overlap with, **167,** 208–209
Macrophage inflammatory protein-1β
blocking AcSDKP, **167,** 222
shared receptor with MIP-1α, **167,** 206
Macrophage procoagulant-inducing factor
production, **152,** 60
tissue factor expression, stimulation of, **152,** 63

Macrophages
activation, defined, **161,** 291
–bacterial interactions, *see* Bacteria, –macrophage interactions
chemotaxis, role of POMC products, **170,** 88–93
and coagulation
cellular activators, **152,** 78–79
cytokine-mediated induction of tissue factor, **152,** 59–64
pathway protein expression by, **152,** 77
procoagulant activity, **152,** 85
suppression of procoagulant expression, **152,** 86, 88–89
tissue factor induction, **152,** 81–82
in colonization of bone graft interface, **172,** 174–175
cytokine production, **169,** 156–157, 159
cytoskeletal dynamics and, **144,** 99
differentiation, **169,** 159
in dome epithelium
as major structural characteristic of, **167,** 99
role, **167,** 131–133
inhibition of CFUS-S from entry into DNA synthesis by marrow, **167,** 192
microgravity experiments, **156,** 312, 314
nitric oxide localization, **170,** 112–113
ovarian, comparisons with testicular macrophages, **149,** 116–117
peritoneal
galactose receptor, **175,** 148–150
interaction with lymphocytes, **175,** 150–151
phagocytosis, role of POMC products, **170,** 94–95
retinal pigmented epithelium and, **146,** 77–78
role in inflammatory response, **170,** 118–121
surface receptors, **161,** 291–293
testicular, *see* Testicular macrophages
TNF-α potential interactions with, **167,** 230
TNF-α production by, **167,** 229
transcriptional regulation and, **143,** 7, 14, 18, 41
wound healing role, **169,** 156–157, 159–160
Macrophage tissue factor, **152,** 57–59
Macroplasmodia, *Physarum,* **149,** 157, 172–175
MAD3 protein, transcriptional regulation, **143,** 36–37
MAG, *see* Myelin-associated glycoprotein
Magainins, **138,** 175–176
Magnesium
avian salt glands and, **144,** 171, 184
cytoplasmic streaming requirements, **155,** 114–115
cytoskeletal dynamics and, **144,** 88, 97, 103, 106, 108
environmental stimuli in plant cells and, **145,** 259, 261, 263, 265
higher plant plastids and, **145,** 172–174, 203
incorporation in apatite, **172,** 157–158, 165–166
intracellular calcium messenger system and, **142,** 312–316, 322
myofibrillogenesis and, **143,** 163
and nitrate reductase, **149,** 79–82
nuclear magnetic resonance and, **145,** 3, 19–20
proteins in striated muscle and, **142,** 121, 127
Magnetic field effects
biological effects, **158,** 286
on cells
acetylcholinesterase activity, **158,** 301
breast cancer proliferation, **158,** 319–320
calcium flux, **158,** 298–299
DNA synthesis, **158,** 305–307
membrane hydrophobicity, **158,** 301
signal transduction, **158,** 282
induction of electric field, **158,** 286, 291–292
pulsed devices, **158,** 292
Magnetic resonance imaging, **145,** 7, 10, 23, 31–32, 41–46, 50–51
Magnetization transfer contrast, and pathology, **145,** 30
Main intrinsic protein, cortical and nuclear fiber cells, **163,** 338–340
Maize, *see Zea mays*
Major histocompatibility complex
antigen presentation, **157,** 190
class I, encoding gene, transcriptional regulation, **143,** 6, 22, 44–45, 47

class I and II molecules, antigen processing, **161,** 299–302
class II molecules
dendritic antigen-presenting cell expression of, **167,** 132
M cells and
in basolateral and lysosomal membranes, **167,** 128
expression of, **167,** 143
intraepithelial lymphocyte cytotoxic function and, **167,** 97
protein classes, **157,** 190
in transcytosis of antigens by M cells, **167,** 128
Malate, higher plant plastids and, **145,** 170–171
Male–female interactions
and advances in compatibility interactions
experimental fertilization, **140,** 578–581
attempts to obtain artificial zygotes, **140,** 579
plant transformation using sexual route, **140,** 579–581
pollen selection, **140,** 578–579
pistil receptivity and embryo sac isolation, **140,** 571–573
embryo sac and gamete analyses, **140,** 573
pistil receptivity and *in vitro* pollination, **140,** 571–572
pollen, **140,** 567–571
cellular organization and quality, **140,** 569–570
pollen gene expression and genetic overlap, **140,** 567–568
pollen wall as living structure, **140,** 568–569
sperm cells, **140,** 570–571
progamic phase, **140,** 573–576
pollen tube chemotropism, **140,** 575–576
pollen tube growth, **140,** 574–575
pollen tube wall formation, **140,** 573–574
syngamic phase, **140,** 576–578
cytoplasmic DNA inheritance, **140,** 577–578
MGU, preferentiality and synchronization of double fertilization, **140,** 576–577
preprogrammed fertilization, **140,** 578
pollen selection, **140,** 578–579
and self-incompatibility in plant reproductive biology, **140,** 581–586
Male germ unit
concepts and misconceptions, **140,** 130–131
sperms, true cells or naked nuclei, **140,** 130
two sperms of pair, identity of, **140,** 131
conclusions and prospects, **140,** 145
functional significance of, **140,** 142–145
sperm dimorphism, **140,** 143–144
sperm-to-sperm connections, **140,** 142–143
vegetative nucleus, **140,** 144–145
pollen tube, evolution of, **140,** 129–130
preferentiality and synchronization of double fertilization, **140,** 576–577
results from quantitative
connection-associations between DNA containing components of male heredity, **140,** 131–132
exceptions to rule, **140,** 136–142
implications of male germ unit, **140,** 133–136
sperm dimorphism in *Plumbago,* **140,** 133
three-dimensional studies and concept of, **140,** 131–142
Male quality, control in *in vivo* reproductive process, **140,** 396–397
Malic enzyme, avian salt glands and, **144,** 145, 147
Malignancies, *see also* Cancer; Carcinoma; Tumors
hemopoietic, effects of IFN-α effects, **167,** 232
nuclear magnetic resonance and, **145,** 50
and nuclear matrix, **162A,** 106–110
and procoagulants, **152,** 82–86
transcriptional regulation and, **143,** 43
Malonyl coenzyme A, chalcone synthase and, **147,** 234–235, 244

MALT, *see* Mucosa-associated lymphoid tissues
Malus domestica, self-incompatibility and, **143,** 282
Mammalian cells
DNA replication initiation, **162A,** 463–476
duration of cell cycle stages, **167,** 192–193
nucleus, functional organization, **162B,** 303–327
permeabilized, **162A,** 42–49
polytene chromosome, **165,** 53–107
interphase nucleus markers, **165,** 70–75
nonelassic form characteristics, **165,** 75–81
sex chromatin, **165,** 70–72
TGF-β receptors, **167,** 209, 214
Mammals
DNA replication, seeDNA replication, in mammals
early development, differentiation systems, **154,** 60–61
microgravity experiments, **156,** 336
sex chromosomes, *see* Sex chromosomes, mammalian
sperm-binding proteins, **163,** 269–273
in whole-chromosome hybridization studies, **153,** 199–200
Mammary gland, parathyroid hormone-related protein, **166,** 244–246
Mammillary layer, eggshells and biomineralization and, **145,** 243–244
fabrication, **145,** 238–240
structure, **145,** 219–220, 224–231
Mammosomatotroph
development, **157,** 73–74, 76
differentiation
initiation by Pit-1 transcription factor, **157,** 38
milk-borne peptide factors, **157,** 39
regional differences in rats, **157,** 38–39
hormone synthesis, **157,** 34, 72–74
identification, **157,** 72
physiological significance, **157,** 74–77
species distribution, **157,** 72–73, 78
tissue distribution, **157,** 77
Mammotroph
dissociation of cells, **157,** 59
effects of age
DNA content, **157,** 68
hormone secretion, **157,** 70–71
morphology, **157,** 67–68
prolactin secretion rate, **157,** 67
prolactin synthesis, **157,** 69–70
proliferation, **157,** 69
functional heterogeneity
assay, **157,** 59–60, 78
dopamine receptor variants, **157,** 62
electrophysiological variants, **157,** 62–63
location dependence, **157,** 61
prolactin cell surface antigen, **157,** 60–61
role of extracellular matrix, **157,** 78
secretion of prolactin variants, **157,** 62–64
identification, **157,** 35, 44–45
immunoelectron microscopy, **157,** 40–41, 44–45
maturation
bromocriptine effects, **157,** 56
estrogen effects, **157,** 55
morphology, **157,** 35, 40, 67–68, 78
ontogenesis, **157,** 35–37, 43–44
prolactin synthesis, **157,** 33, 38
proliferation
age-related changes, **157,** 45
circadian variations in cell division, **157,** 44
factors affecting proliferation
cytokines, **157,** 52
dopamine, **157,** 47–48
endorphin, **157,** 48
estrogen, **157,** 46
galanin, **157,** 48–49
growth factors, **157,** 49–52
melanocyte-stimulating hormone, **157,** 48
identification of mitotic cells, **157,** 44
streptozotocin-induced diabetic mice, **157,** 53–54
sex differences in cell number, **157,** 52–53
subtypes
age-related changes, **157,** 54–55
effects of estrogen, **157,** 57–58
identification, **157,** 40–41
Mandibular cuticle, genetic control, **174,** 38–39

Mandibulofacial dystosis
candidate genes, **158,** 244–247
clinical features, **158,** 244, 262
incidence, **158,** 244
mode of inheritance, **158,** 243–244
Manduca sexta, programmed cell death in, **163,** 130–133
Manganese
flowering plants and, **143,** 311
intracellular calcium messenger system and, **142,** 316, 322
mitochondrial genomes in fungi and, **141,** 115
Mannitol, **144,** 183–184; **165,** 11–12
Mannose
activation of sucrose phosphate synthase, **149,** 70–71
regenerative cementogenesis and, **142,** 13, 25
self-incompatibility, **143,** 284
α-Mannose, **139,** 74–75
Mannose-binding protein, specificity, **175,** 187–189
Mannose-6-phosphate receptor, in clathrin-coated vesicle formation, **167,** 19
Mannosyl–fucosyl receptor, macrophage, **161,** 292
MAP, *see* Microtubule-associated proteins
MAP kinase, *see* Mitogen-activated protein kinase
Mapping
Arabidopsis thaliana and, **144,** 63–68
chloroplasts and, **144,** 265, 281
cytoskeletal dynamics and, **144,** 106, 109
Drosophila muscles and, **143,** 63–64, 139–140
contractile protein genes, **143,** 104, 108, 119–120
development, **143,** 65, 73, 81, 86–87
fate, *see* Fate map
flowering plants and, **143,** 316
higher plant plastids, **145,** 161
melanocytes
development, **146,** 214, 217–220, 222–223
genetics of melanoma, **146,** 227, 230–234, 245
pigment-cell genetics, **146,** 203, 205, 209
spatial and temporal, glycine receptor
immunohistochemical studies, **176,** 219
in situ hybridization, **176,** 215–219
MAR, *see* Matrix-attached regions
Marchantia
higher plant mitochondrial genomes and, **141,** 164
mitochondrial genomes of, **141,** 37
Marchantia polymorpha
higher plant mitochondrial genomes, **141,** 141, 157
plastids, **145,** 153, 155, 166–167
Markers
for conceptus developmental potential, **164,** 173–174
histochemical, *see* Histochemical markers
interphase nucleus, *see* Interphase nucleus marker
molecular, RPE during differentiation, **171,** 235–237
recognition, sialic acids as, **175,** 155–156
Marsilea, **138,** 250
Marsupials
gonadal differentiation, **154,** 219–220
sexual dimorphisms, **154,** 219–220
Maser, **156,** 352–353
Mast cell growth factor
role in B cell development, **157,** 151–152
soluble form, **157,** 151
Mast cell growth factor receptor, **157,** 151
Mast cells
degranulation, and chemotactic peptides, **138,** 174–175
melanocytes and, **146,** 215, 218, 247
Mastocytoma, **156,** 104
Masugi nephritis, **166,** 207–209
Maternal polarity, gene systems, **174,** 18
Matrices
pairwise amino acid identification among, **140,** 470
volume density, renal glomeruli, **161,** 160–161
zone, avian retina and, **146,** 27–28
Matrigel
angiogenesis assay, **159,** 123–129
batch variation, **159,** 128
density dependence of tube formation, **159,** 123, 125
preparation, **159,** 123
Matrin, nuclear, distinct from nuclear lamins, **162A,** 29–36
Matrin 3, **162A,** 32–33, 304; **166,** 6
Matrin 4, **166,** 6
Matrin 12, **166,** 6

Matrin 13, **166,** 6
Matrin cyclophilin, characterization, **162A,** 33–34
Matrin D, **166,** 6
Matrin G, **166,** 6
Matrin p50, and RNA polymerase II-LS, **162A,** 34–37
Matrix-attached regions, *see also* Scaffold-attached regions
  association with locus control region, **162A,** 233–234
  A–T-rich sequences, **162A,** 101–102,542–543
  binding proteins, **154,** 12
  chromosomal loop anchoring sites, **162B,** 416
  cohabitation
    with potential ORIs, **162A,** 344–345
    with transcriptional enhancers, **162A,** 345–346
  functional properties, **162A,** 544–545
  functional significance, **162A,** 362–366
  gene expression effects, **154,** 8
  gene expression regulation role, **150,** 88
  high- and low-salt extraction, **162A,** 199–201
  identification, **162A,** 281–288
  identification with genes, **162A,** 289–293
  as interface of matrix and DNA loop, **162A,** 458–460
  ORIs as, **162A,** 312–314
  –ORI sequences
    active, nucleosome-free, **162A,** 347–349
    class, haboring kinked DNA, **162A,** 340–341
    composition and genomic position, **162A,** 356–358
    containing A–T tracts, **162A,** 320–322
    direct repeats in, **162A,** 343–344
    DNase I-hypersensitive sites, **162A,** 346–347
    DNA unwinding elements, **162A,** 322–323
    enriched
      in ATTA and ATTTA motifs, **162A,** 335–338
      in inverted repeats, **162A,** 323–325
    flanked by left-handed and triple helical structures, **162A,** 339–340
    interaction with ssDNA-binding proteins, **162A,** 349–350
    intrinsically curved DNA, **162A,** 338–339
    narrow minor grooves, **162A,** 350–351
    polymorphic and conserved sequences, **162A,** 351–352
    prediction, rules diagnostic for, **162A,** 317–356
    prognosis perspectives, **162A,** 355–356
    replication initiator sites, **162A,** 333–335
    with transcription factor binding sites, **162A,** 325–333
    vulnerability to mutagens, **162A,** 352–354
  in plants, **162B,** 127–128
  prediction, and chromatin domains, **162A,** 279–366
  proteins binding, **162A,** 201–202, 543–544; **162B,** 26–29, 131
  random cloning, **162A,** 359–362
  rebinding to affinity sites, **162B,** 418
  rebinding to isolated nuclear matrix, **162B,** 420
  role in gene transcription, **162B,** 356–359
  as transcriptional enhancers, **162A,** 309–312
  transcription elongation role, **154,** 21
  transcription factor sites, **162A,** 329–332
Matrix chaperones
  Cpn10, bioassay with early pregnancy factor, **174,** 150–151
  Cpn10 and Cpn60, **174,** 149–150
  Hsp70, roles
    Brownian ratchet model, **174,** 146–147
    force-generated motor, **174,** 147
  role for matrix-located Hsp70
    Brownian ratchet model, **174,** 146–147
    force-generated motor, **174,** 147
  role in protein folding, **174,** 151–152
  sequential action in protein folding, **174,** 152
Matrix-driven translocation, *see* Translocation, matrix-driven
Matrix metalloproteinase
  in neovascularization, **161,** 4
  types, **169,** 164
  wound healing role, **169,** 152, 160–161, 164–165, 170–171
Matrix protein filaments, poorly staining, **162A,** 18–23

Matrix proteins
classical, **162A,** 303–309
in cytoskeleton–membrane connections, **176,** 150–154
extracellular
ependymin, **165,** 148–150
in phenotypic expression, **150,** 129
Matter, intermediate states, **166,** 61–62
Maturation, proteolytic, mitochondrial preproteins, **174,** 145–146
Maturation-promoting factor
in calcium activation of mammalian eggs, **152,** 210
in cell cycle, **164,** 162
mitosis and, **144,** 218, 229, 231, 242, 245, 247–250, 252
regulation of microtubule behavior, **164,** 279
Maxicircles
endosymbiont hypothesis and, **141,** 294, 305
genes, **141,** 69–74
kinetoplastid mtDNA and, **141,** 65–69, 84
transcripts, **141,** 74–79
Maxicircle unidentifred reading frame
endosymbiont hypothesis and, **141,** 294
kinetoplastid mtDNA and, **141,** 72, 75–76, 78, 80
Maxillary cuticle, genetic control, **174,** 39
M-band proteins
Mb-1, marker of B cell differentiation, **157,** 141
in striated muscle, **142,** 70
M cells, in intestinal Peyer's patches, *see also* Dome epithelium
characteristics
general, **167,** 111–112
variations among species and locations, **167,** 112–117
clinical aspects
intestinal immune response induction and, **167,** 142–143
as potential entry sites for oral vaccines, **167,** 139–142
as targets for enteropathogenic microorganisms, **167,** 138–139
functions
antigen and tracer transport, **167,** 117–122
antigen interaction with surface of, **167,** 122–126
cytoskeleton, **167,** 128–130
transcytosis of antigens, **167,** 126–128
identification, **167,** 100–107
alkaline phosphatase, **167,** 102
cross-correlation of labeling patterns, **167,** 104–107
cytoskeleton, **167,** 104
lectins, **167,** 103–104
monoclonal antibodies, **167,** 102
ultrastructure, **167,** 100–102
locations outside of gut
BALT, **167,** 136–137
NALT/DALT/CALT, **167,** 137–138
tonsils, **167,** 135–136
lymphoid cells in gut wall
immunological functions, **167,** 92
localization, **167,** 92–96
overview, **167,** 91–92
in patchwork arrangement with enterocytes, **167,** 130–131
MCH, *see* Melanin-concentrating hormone
MDR, *see* Multidrug resistance
MDR1 protein, *see* P-glycoprotein
ME2303, effect on murine MDR cells, **156,** 92
Mean glomerular volume, renal glomeruli, **161,** 141–143
Mean length, renal capillaries, **161,** 158
Media, culture, effects on conceptus development, **164,** 168–170
Median eminence
cellular composition, **159,** 29
neurohormone storage, **159,** 29–30
termination of hypothalamic neurons, **159,** 6, 23, 29–30
ultrastructure, **159,** 29–30
Medicago, **156,** 43–45
Medicago sativa, nodulation, **156,** 3, 9, 34, 38, 43, 44
Medicine, use of AGPs, **174,** 267–268
Medroxyprogesterone acetate, effects on mouse pelvis, **139,** 36
Medusae
as cnidarian developmental stage, **167,** 38
transdifferentiation, **142,** 213–214, 256–257
animals, **142,** 216–218
hydromedusae
DNA replication, **142,** 252–256
initiation, **142,** 236–251
isolation of tissues, **142,** 222–223

Medusae (*continued*)
potential, **142,** 223–236
tissue organization, **142,** 218–222
plants, **142,** 215–216
terminology, **142,** 214–215
Megagametophytes
characterization, embryo sacs and component cells
biochemical and immunological characterizations, **140,** 283–284
cytochemical characterization, **140,** 281–282
cytological characterization, **140,** 281
elemental characterization, **140,** 282–283
development
cellularization of embryo sac, **140,** 242–247
megagametogenesis and cytoplasmic maturation, **140,** 240–242
megasporocyte, **140,** 234–238
meiotic products and functional megaspore determination, **140,** 238–240
isolation, embryo sacs and component cells, **140,** 269–280
assessment of viability and quality, **140,** 271–280
technique and protocols, **140,** 270–271
Megakaryocytes
mitosis and, **144,** 225
progenitors, **167,** 191
Meiosis
GnRH-encoding gene in teleosts, **147,** 170
karyosphere and, **144,** 1, 7–8, 42, 44
oocytes of insects, **144,** 9, 11
vertebrates, **144,** 28, 36, 38–39
kinesin-like proteins function, **164,** 304–305
and lepidopteran spindle structure
female, **152,** 13–15
male apyrene, **152,** 24–30, 35–36
male eupyrene, **152,** 16–24, 34–36
nondisjunction occurrence
maternal age and, **167,** 276–277
in oocytes, **167,** 269
overview, **167,** 264
recombination errors and, **167,** 282–283
stages compared, **167,** 275*f*
plant cells
chromosome disjunction, **158,** 96–99
chromosome pairing, **158,** 95–96
commitment in carrot, **158,** 91–92, 94
cytological features, **158,** 123
embryo sac, **158,** 103–104
homeotic transformation, **158,** 92
homologous chromosome segregation, **158,** 87–93
hormonal triggering, **158,** 123
impairment of entry, **158,** 93–94
multipolar spindle, **158,** 100–102
nuclear DNA content, **158,** 105–108
pollen, **158,** 103
proembryogenic masses, **158,** 103–104
RAPD analysis, **158,** 109
RFLP analysis, **158,** 109
role of heat shock proteins, **158,** 123–125
somatic reduction, **158,** 86–91
synaptonemal complex, **158,** 95–96
variant phenotypes of carrot, **158,** 100
variations, **158,** 93–102
Sertoli-germ cell communication network and, **147,** 25, 80
function, **147,** 41, 46, 57, 70, 77
stage-dependent changes, **147,** 59, 61
in yeast cells, **139,** 322–323
Melanin
fish chromatophores and, **143,** 195
growth factor-induced retinal regeneration and, **146,** 55
synthesis in retinal pigment epithelium, **171,** 232–233
Melanin-concentrating hormone
fish, chromatophores and, **143,** 225, 235–236
immunohistochemical localization, **138,** 149–151
physiological effects, **138,** 151–152
Melanoblasts
biology, **146,** 194–195
genetics, **146,** 234, 247
pigment-cell genetics, **146,** 199, 214–218, 220–222, 224
Melanocortins, astrocyte reaction to, **148,** 140–141
Melanocytes
biology, **146,** 191–196
pigment-cell genetics, **146,** 196–202
development, **146,** 214–224
function, **146,** 202–214

premelanocytes
biology, **146,** 196
genetics, **146,** 235
Melanocyte-stimulating hormone
effect on pituitary cell proliferation, **157,** 48
evolution, **159,** 52, 56
α-MSH, central effects, **138,** 144–145
pigment-cell genetics and, **146,** 208, 210
as POMC component, **170,** 81–82
precursor, **159,** 46
secretion in pituitary cells, **159,** 4
Melanoma
biology, **146,** 191, 195–196
cutaneous malignant melanoma–dysplastic nevus syndrome, **146,** 230, 250
familial
–dysplastic nevus syndrome, **146,** 226–228
locus MLM, **146,** 230
tumor suppressor genes in, **146,** 247
genetics, **146,** 224–225, 250–251
biology, **146,** 235–237
clinical evidence, **146,** 225–232
fish, **146,** 232–234
growth factors, **146,** 237–241
oncogenes, **146,** 241–247
tumor suppressor genes, **146,** 247–250
pigment-cell genetics, **146,** 197, 199, 205–208, 210, 213–214
transcriptional regulation and, **143,** 18
Melanoma cells
NMR analysis, **145,** 39
vertebrate cell culture technology and, **142,** 175, 189
Melanoma-derived growth-stimulatory activity
genetics, **146,** 240–241
pigment-cell genetics, **146,** 219
Melanophores
cytophysiology, **143,** 192, 194–197, 201, 205, 217, 246
cytoskeleton, **143,** 221–223
motility, **143,** 226–231, 233–237
signal transduction, **143,** 238–243, 245
formation by neural crest cells, **148,** 82
hormonal control of amphibian metamorphosis and, **145,** 123
Melanophore-stimulating hormone, fish chromatophores, **143,** 234–235, 241
Melanosomes
biology, **146,** 193–195
eumelanosomes, biology, **146,** 193
fish chromatophores, **143,** 195, 197–198, 201, 223
motility control, **143,** 229, 231–232, 234–237
signal transduction, **143,** 240–242, 245
pigment-cell genetics, **146,** 197, 204, 206–207, 211
premelanosomes
biology, **146,** 195
fish chromatophores and, **143,** 195, 197
Melanostatin, regulation of pars intermedia cells, **159,** 53
Melatonin
biological clocks in retina and, **146,** 85
fish chromatophores, **143,** 236–237
Harderian gland secretion
amphibians, **168,** 39–40
biosynthesis, **168,** 39
mammals, **168,** 38–40
melatonin effects, **168,** 51
oncostatic activity, **158,** 319–320
retinal precursor cells and, **146,** 155, 164–165
synthesis, **158,** 319
Melatonin receptor, Harderian gland, **168,** 43
Meloidogyne incognita, plant-pathogen interactions in, **144,** 72
Meloidogyne jouanica, animal mtDNA and, **141,** 197–198, 203–204
Melonosis, genetics, **146,** 233–234
Melosira varians, cross-costae formation, **150,** 302
Melphalan, **156,** 81, 127, 129–130, 136
Membrana limitans externa, avian retina and, **146,** 3, 31, 34, 36
Membrana limitans interna, avian retina and, **146,** 3, 31, 36
Membranal ring, bacterial flagellar motor
models, **147,** 130–131, 134–136, 139–140, 142–143
structure, **147,** 100–101, 105
Membrane-binding proteins, cytoskeletal dynamics, **144,** 107–108
Membrane flow, in amoebae
actin polymerization, **148,** 68–69
capping, **148,** 43–44

Membrane flow (*continued*)
comparison with cultured tissue cells, **148,** 41–43
directed lipid flow hypothesis, **148,** 45
dynamics at leading margin of cells, regulation, **148,** 70–72
early theories, **148,** 38–40
membrane proteins
association with cytoskeleton, **148,** 50–53
hauling by cytoskeleton, **148,** 47–50
membrane-to-cytoskeleton association, **148,** 60–61, 72–73
primary role, **148,** 45–47
submembrane layer, lateral movement
cortical cytoskeleton, **148,** 53–55
cortical flow hypothesis, **148,** 55–59
detachment from leading edge, **148,** 59–64
Membrane potential effects
biological clocks in retina
Bulla, **146,** 129, 136
entrainment pathway, **146,** 102–103, 105–106, 114
opisthobranch retina, **146,** 91
output pathway, **146,** 123, 126
pacemaker mechanisms, **146,** 115–116
on cationic drug flux, **171,** 139–141
on drug accumulation rate, **171,** 144–149
on drug transport, **171,** 137–139, 153–155
fish chromatophores and, **143,** 238–239
intracellular calcium messenger system, **142,** 316, 327
on steady-state drug accumulation, **171,** 141–144
symplast and, **142,** 269–271
on weak base accumulation, **171,** 149–153
Membrane raking hypothesis, **150,** 48–51
Membrane ratchet model, ameboid movement, **139,** 166, 167–168
Membranes
addition, axonal assembly, **165,** 269–273
apical, *see* Apical membranes
associated proteins, retinal precursor cells and, **146,** 160
association with Rab GTPase, **176,** 44–47
basement, *see* Basement membranes
basolateral, M cell
connection to enterocytes by desmosomes, **167,** 131
description, **167,** 101
monoclonal antibody binding to epitopes on, **167,** 102
structure, **167,** 111
in transcytosis of antigens, **167,** 127, 128
biological clocks in retina and, **146,** 84, 134–137
bound organelles, movement
dynein-like proteins, **176,** 159–160
kinesin-like proteins, **176,** 157–159
motor protein interrelations, **176,** 160–163
myosin heavy chain polypeptides, **176,** 154–157
conductance, biological clocks in retina and, **146,** 116, 123, 127
cytoplasmic, effect of microgravity
on plant cells, **171,** 49
primary sites, **171,** 58–62
–cytoskeleton, cadherin, **165,** 192
–cytoskeleton connections
fibronectin-like proteins, **176,** 152–154
integrins, **176,** 151–152
spectrin, **176,** 150–151
vitronectin-like proteins, **176,** 152–154
erythrocyte, intermediate filaments associations, **164,** 107–108
fetal, human, GLUT1, **172,** 27
fluidity, **139,** 176
fusion and fertilization, models of, **140,** 372–377
fusion in angiosperms, **140,** 375–377
gametic fusion in animals, mechanism of, **140,** 373–375
ghosts
cell model systems, **175,** 6–9
as models of CMT organization, **175,** 18
homotypic fusion, role of Rab proteins, **176,** 41–42
inner limiting, avian retina and, **146,** 31–34, 41
mitochondrial
inner, translocation components, **174,** 144–145
outer, translocation components, **174,** 142–144
preprotein proteolytic maturation, **174,** 145–146
in mitotic apparatus, **138,** 240–241

nuclear, lamins as structural support system, **162B,** 162–163
outer limiting
avian retina and, **146,** 16–17, 25, 32, 34
retinal precursor cells and, **146,** 160
plasma, *see* Plasma membranes
Rab association
role of GDP dissociation inhibitors, **176,** 49–51
role of GTPase-activating proteins, **176,** 52–53
role of guanine nucleotide exchange factors, **176,** 51–52
reactors based on, and vertebrate cell culture technology, **142,** 183–185
recycling from M cell basolateral to apical membrane, **167,** 128
recycling model of ameboid movement, **139,** 164, 166–167
ruffings as common cellular response to growth factors, induction, **138,** 214–216
skeletal complex, *see* Cortex
sperm, modification during epididymal transit, **154,** 178–184
trafficking model, **156,** 93
transport machinery, linkage to Rab GTPases, **176,** 57–58
Membrane transport
methotrexate, **156,** 109–110, 113, 116
Pgp, **156,** 82
Memory B cells, generation, germinal center
mechanisms, **166,** 142–144
steps, **166,** 142–144
Mental retardation
FISH diagnosis, **153,** 26–29
trisomies as prevalent causes, **167,** 265
6-Mercaptopurine, **156,** 121–122
Mercury, inhibition of ecto-ATPases, **158,** 174, 178, 190
Meristem cells
apical, barley roots, **156,** 325–327
effect of microgravity
on angiosperm root, **171,** 32–34
shoot apical meristem, **171,** 35–36
wall regeneration, effect of microgravity, **170,** 47
Mesangial cells
average volumes and surface areas, **161,** 151–154
kidney glomerulus, **166,** 182
quantity estimation in average glomeruli, **161,** 148–151
Mesencephalon, structure specification, **175,** 266–271
Mesenchymal cells, stimulatory effect of TGF-β, **167,** 209
Mesenchyme, growth factor-induced retinal regeneration, **146,** 51, 67
Mesenteric window assay, angiogenesis quantitation, **159,** 118
Mesoderm
determination genes, *Drosophila* muscles and, **143,** 78–79
Drosophila muscles and, **143,** 65–67, 79–84, 87, 137
formation, **160,** 224
in *Halocynthia roretzi* larva, **176,** 248–249
myofibrillogenesis and, **143,** 157
role of *Brachyury* gene in formation, **176,** 276–278
Mesogloea, transdifferentiation in medusae and, **142,** 220–223, 232, 250
Mesotaenium, environmental stimuli in, **145,** 269, 272–273, 281, 299
Mesotocin
immunohistochemical localization of AVT and MT, **138,** 120–124
neuroendocrine and behavioral effects of AVT and MT, **138,** 124–126
Messenger RNA
accumulation, induction, **148,** 98–99
γ-aminobutyric acid metabolism and, **147,** 312, 315
animal mtDNA and, **141,** 190–191, 226
Arabidopsis thaliana and, **144,** 73–74
association with cytoskeleton, **162B,** 197–200
in auxin-regulated genes, **152,** 112–114, 116
in *B. campestris* with clone *BcpI,* **140,** 26
biological clocks in retina and, **146,** 84
calcium channels and, **147,** 220
chalcone synthase and, **147,** 247, 254–257, 261, 266
complex with P110, **162B,** 205–206
Drosophila muscles and
contractile protein genes, **143,** 94, 101, 103, 108–109
myosin light chains, **143,** 117, 119
troponin, **143,** 118, 120

Messenger RNA (*continued*)
endosymbiont hypothesis and
gene expression, **141,** 298, 304–305
gene transfer, **141,** 325
genome organization, **141,** 289, 294
mitochondria, **141,** 282
plastids, **141,** 267
expression, H1 histone role, **148,** 102
expression in mouse tumors, **152,** 57
extracellular matrix and
characterization, **146,** 267–268
modification, **146,** 291–292, 304–305, 309
flowering plants and, **143,** 322
amino acids, **143,** 311–312, 324
ammonium, **143,** 305–309
nitrate, **143,** 299, 301–312
glutamate metabolism and, **147,** 289, 291, 294, 308
GnRH-encoding gene in teleosts and, **147,** 169, 173
growth factor-induced retinal regeneration and, **146,** 57–58, 68
higher plant plastids and, **145,** 158–159, 192–193, 199
hormonal control of amphibian metamorphosis and, **145,** 111–112, 130, 133, 139, 141
in hybridization experiments of coordinate transcription, **148,** 86–87
and instability of homogeneous state
cytoskeletal localization, **154,** 355
metabolic field, **154,** 359–360
intracellular calcium messenger system and, **142,** 310, 328
karyosphere and, **144,** 42
kinetoplastid mtDNA and, **141,** 80–84
maxicircle genes, **141,** 70–73
maxicircle transcripts, **141,** 74, 76–80
melanocytes and, **146,** 206–207, 237
microfilament system, **149,** 147
mitochondrial genomes in fungi and, **141,** 100
mitotin, **162B,** 386
myofibrillogenesis and, **143,** 179
in pollens, **140,** 4–5
POMC, in immune system, **170,** 83–85, 87
pre-mRNA
association with nucleoskeleton, **162B,** 200–202
nascent, distribution, **162A,** 154–158
processing, **162A,** 104–105
splicing, **162A,** 159–162
prosystemin, accumulation in plant defense responses, **148,** 13–14
proteins in striated muscle and, **142,** 66, 107
retinal pigmented epithelium and, **146,** 77
SAUR
induction, **152,** 115
response to 2,4-dichlorophenoxyacetic acid, **152,** 116
self-incompatibility and, **143,** 270, 277, 285, 288
Sertoli-germ cell communication network and
function, **147,** 57, 60–61, 66, 69
mechanisms, **147,** 74, 76–78
SPG-1 encoded by, **154,** 167
spliced and unspliced, **162A,** 76,104
synthesis, induction, **148,** 100
transcriptional regulation and, **143,** 7, 43, 46
genes, **143,** 13, 16, 18
inducible activation, **143,** 11–12
molecular cloning, **143,** 28, 30–31, 33, 37
transport, **162B,** 184–186,193–197,438–440
tropomyosin, expression patterns, **170,** 7
and unbalanced growth, **148,** 99–100
Metabolism
activity of isolated cells, **140,** 224–225
arachidonic acid, stimulation by endothelin, **171,** 281–282
associated enzymes, muscle fiber types based on, **170,** 148–150
cAMP cascade in regulation of, **139,** 319–320
energy, related muscle fiber transitions, **170,** 152–153
field, and cytoskeletal localization, **154,** 355–360
in plant cells, role of V-ATPase, **174,** 89–92
sialic acid, **175,** 142–143
Turing–Child system
metabolic field, **154,** 356–359
metabolic pattern, **154,** 344
phosphorylation field, **154,** 345–346
principle, **154,** 310

Metabolites
antimetabolites
affecting DNA replication, **162A,** 549–556
cytosine arabinoside, **156,** 120–121
definition, **156,** 108
5-fluorouracil, **156,** 116–120
methotrexate, **156,** 108–116
purine antimetabolites, **156,** 121–122
NMR analysis, **145,** 30, 43–49
Metacholine, avian salt glands and, **144,** 161
Metallothionein, encoding genes, **156,** 79
Metallothionein-1, melanocytes and, **146,** 243
Metamorphosin, Antho-LWamide similarity to, **167,** 71–74
Metamorphosis
amphibian, *see* Hormones, control of amphibian metamorphosis
cell death during, **163,** 146
Hydractinia echinata, **167,** 54–55
neuropeptides in sea anemones inducing, biosynthesis of, **167,** 71–74
planula larva
as cnidarian developmental stage, **167,** 38
neuropeptides and *Hydractinia echinata,* **167,** 54–55
Metaphase
chromosome loss in, **167,** 264–265
sperm studies of aneuploidy and, **167,** 269–270
transition to anaphase, spindle midzone during, **162B,** 48–54
Metaphytes, endosymbiont hypothesis and, **141,** 272, 275–276, 324
Metastases, melanocytes and, **146,** 248–249
Metazoa, animal mtDNA
directly repeated sequences, **141,** 203–208
protein genes, **141,** 182, 184–185
rRNA, **141,** 189–190
structure, **141,** 173–177
tRNA, **141,** 192, 197, 199–201
Metencephalon
induction of *En2* production, **175,** 269
structure specification, **175,** 266–271
Methanogens, endosymbiont hypothesis and, **141,** 241, 243
Methionine
biosynthetic pathway, **165,** 17
flowering plants and, **143,** 318, 320–321
L-Methionine sulfoximine, glutamate metabolism and, **147,** 302, 305–306
Methotrexate
drug resistance, **156,** 78, 81, 113
folate metabolism, **156,** 110–112
membrane transport, **156,** 109–110, 113, 116
polyglutamylation, **156,** 111–112, 114–115
Methoxydiphenylfuranone, **138,** 53
3-Methyladenine, **156,** 127
3-Methyladenine DNA glcosylase, **156,** 127
5-Methylanthranilate, flowering plants and, **143,** 313, 316
N-Methyl-D-aspartate, microtubule-associated protein dephosphorylation response, **151,** 126–127
Methylation
chalcone synthase, **147,** 257–258
DNA, *see* DNA methylation
Drosophila muscles and, **143,** 112–113
higher plant mitochondrial genomes and, **141,** 159
higher plant plastids and, **145,** 162–163
histone, **162A,** 211
hypomethylation, DNA, induction by azacytidine, **158,** 116
transcriptional regulation and, **143,** 40
Methylene blue, staining of cnidarian neurons, **167,** 40
O^6-Methylguanine-DNA methyltransferase, **156,** 126–127
Methyl jasmonate
chalcone synthase and, **147,** 248
as endogenous elicitor of plant defense, **148,** 14
4-Methylmorpholine N-oxide, isolation of exines using, **140,** 54
Methylobacterium, urease in plant cells and, **145,** 93
N-Methylpurine-DNA glycosylase, **156,** 127
O-Methyltransferase
expression
antisense gene, **172,** 257–258
developmental, **172,** 257
pathogen stress, **172,** 257
forms, **172,** 258
localization, **172,** 256
in phenylpropanoid pathway, **172,** 255–256; **176,** 102–105
substrate specificity types, **172,** 256

Metrial gland cells, granulated, rodent
colony-stimulating factor-1 secretion by, **153,** 122
cytokine secretion, **153,** 122
immune-effector function, **153,** 123–125
morphology, **153,** 113–119
nonimmunological functions, **153,** 125–126
origins, **153,** 111–113
phenotype, **153,** 113–119
properties, **153,** 105
Metridium senile, animal mtDNA
genome content, **141,** 178–179, 181
noncoding nucleotide sequences, **141,** 208–209
nucleotide bias, **141,** 204–205
protein genes, **141,** 182–183, 187–189
rRNA, **141,** 190
tRNA, **141,** 193, 199
Mevalonate, **166,** 29
Mezerein, transdifferentiation in medusae and, **142,** 247–248, 257
MGF, *see* Mast cell growth factor
MGMT, *see* O^6-Methylguanine-DNA methyltransferase
MHC, *see* Major histocompatibility complex
Microbes
hormone receptors, **155,** 3–7
–plant cell wall interactions, **171,** 107–109
Microcarrier assay, as migration analysis method, **155,** 63
Microcarriers, vertebrate cell culture technology
bioreactors, **142,** 166, 171–177, 180
optimization, **142,** 196–199
Microdissection, transillumination-assisted, **147,** 59–60
Microencapsulated-based reactors, **142,** 177–178
Microfibrils
cellulose, **173,** 162
distribution and orientation, **176,** 165–167
interaction with xyloglucans, **173,** 195–196
fucosylation, **173,** 165–166
pectic polysaccharide network, **173,** 167
synthesis, **173,** 168–169
extracellular matrix, **146,** 263–264, 284
role in collagen fibril anchoring to lamina densa
fibrillin fibrils, **173,** 134–135
fibronectin fibrils, **173,** 135–136
Microfilaments
in actin cytoskeletal models, **150,** 192, 196
in diatom valve morphogenesis, **150,** 332–335
environmental stimuli in plant cells, **145,** 259–262, 276–277, 279, 296
fish chromatophores and, **143,** 199
karyosphere and, **144,** 41
M cell, **167,** 129
mechanical forces
stress response, **150,** 186
tension in tissue pattern establishment, **150,** 176
morphogenesis regulation by, **150,** 173
–MT complex, **173,** 211–213
–MT interaction, in cell division and development
centriole migration, **173,** 216–217
centrosome movement, **173,** 214–215
cortical movement, **173,** 215–216
cytokinesis, **173,** 217–218
cytoplasmic streaming in ring canals, **173,** 220–221
role of actin filament arrays, **173,** 218–219
role of growth cones, **173,** 221–222
role of nuclear migration, **173,** 219–220
spindle orientation, **173,** 216
spindle positioning, **173,** 217–218
–MT interaction, regulation
by multisubunit complexes, **173,** 232–233
role of different motors, **173,** 232
role for MT and actin filament dynamics, **173,** 231–232
role of posttranslational modifications, **173,** 230
by tubulin- and actin-binding proteins, **173,** 226–230
myofibrillogenesis and, **143,** 174–175, 177, 179
organelle transport, model systems, **173,** 222–226
Physarum as model
biochemical composition, **149,** 155–156
actin, **149,** 155, 157–158
actin-binding proteins, **149,** 160–164

actomyosin thread models, **149,** 164–167
myosin, **149,** 157, 159–169
contractile activities, **149,** 184, 186
axenically cultured microplasmodia, **149,** 191–195
caffeine droplets, **149,** 190–191
computer controlled data evaluation, **149,** 188–198
digital image processing, **149,** 188–197
regulation, **149,** 199–202
small sandwich plasmodia, **149,** 195–198
video-enhanced microscopy, **149,** 188–197
experimental research, **149,** 151–155, 202–203
organization, **149,** 167–168
experimental stages, **149,** 175–185
life cycle stages, **149,** 170–175
membrane skeleton, **149,** 168–169, 171
in pollen, methods to visualize actin filaments in, **140,** 88–90
regenerative cementogenesis and, **142,** 40
tension molding, **150,** 196–200
transdifferentiation in medusae and, **142,** 232
β_2-Microglobulin, transcriptional regulation and, **143,** 6
Microgravity effects
altered
animals, **156,** 305, 320, 334–339, 344–348, 352, 354–356
growth factor activation, **156,** 317–318
immune cells, activation, **156,** 305–316
nonactivated cells, **156,** 318–320
plants, **156,** 303, 305, 319–333, 344–348
unicellular organisms, **156,** 339–343
animals, **156,** 305, 320, 334–339, 344–348, 352, 354–356
biological effects, range and mechanism, **171,** 2
on cytoplasmic membrane, primary site, **171,** 58–62
gravitropism, *see* Gravitropism, plants
growth factor, activation, **156,** 317–318
immune cells, activation, **156,** 305–316
manipulation, **156,** 349, 358
clinorotation, **156,** 357–358
orbital spaceflight, **156,** 353–357
parabolic flight, **156,** 349–352
suborbital flight, **156,** 352–353
on muscle fiber type transitions, **170,** 182–183
nonactivated cells, **156,** 318–320
on plant cells, **171,** 3–4
adaptation strategy, **171,** 56–58
aging, **171,** 53–56
algae, **171,** 25–30
angiosperms, *see* Angiosperms
associated weightlessness, definition, **171,** 4
differentiation, **171,** 53–56
ferns, **171,** 31
gymnosperms, **171,** 31–32
moss, **171,** 30–31
organelle structural–functional rearrangements, **171,** 46–53
space vehicle cabin environment effects, **171,** 5
specialized graviperceptive cells, **171,** 10–11, 20–22
techniques and instruments for measurement, **171,** 5–7
tip-growing cells, **171,** 22–25
on plant cell walls
biological effects, **170,** 41–42
epidermal cells, **170,** 52–54
meristem cells, **170,** 47
parenchyma wall structure, **170,** 48–50
polysaccharide synthesis, **170,** 54–56
protoplasts, **170,** 42–47
role of calcium, **170,** 64–68
role of enzymes, **170,** 60–64
on plants, **156,** 303, 305, 319–333, 344–348
simulated, *see* Clinostating
as space flight factor, **170,** 40
Microheterogeneity, acute phase proteins
major microheterogeneity, **170,** 266
minor microheterogeneity, **170,** 267
types, **170,** 265–266
Microinjection, **140,** 428–429, 435
Micromanipulation, single gametes, synergids, and central cells
electrofusion, **140,** 416–418
isolation and selection, **140,** 414–416
Microorganisms
M cells and
preferential adherence to apical membranes, **167,** 123

Microorganisms (*continued*)
as targets for enteropathogenic, **167,** 138–139
microgravity experiments, **156,** 339–343
Peyer's patch dome epithelium uptake of, **167,** 177–122
sialic acids as recognition determinants
bacteria, **175,** 175
coronaviruses, **175,** 171–172
Escherichia coli, **175,** 175–178
Helicobacter pylori, **175,** 179–181
influenza A virus, **175,** 162–169
influenza B virus, **175,** 162–169
influenza C virus, **175,** 169–170
other bacteria, **175,** 181–182
paramyxoviruses, **175,** 170–171
polyomavirus, **175,** 172
protozoa, **175,** 182–183
reoviruses, **175,** 172–175
Streptococcus strains, **175,** 178–179
viruses, **175,** 157–162
Microparticles, polyacrylamide, as antigen-delivering carriers, **167,** 141
Microplasmodia, *Physarum,* axenically cultured, **149,** 178–185, 191–195
Microplicae, M cell
characterizing apical surface, **167,** 101
versus enterocyte, **167,** 111
variations among species, **167,** 114
Microsatellites
identification of disease genes, **158,** 230
mapping, **158,** 230
Microscopy
centrifuge, *see* Centrifuge microscopy
collagen fibrillamina densa connection, **173,** 125–126
confocal laser scanning
M cell identification using AP method, **167,** 102
retinal precursor cells and, **146,** 175
in whole-chromosome hybridization, **153,** 187
confocal laser scanning, in whole-chromosome hybridization, **153,** 187
dark-field, bacterial flagellar motor, **147,** 111, 113
direct, living polymorphonuclear leukocytes, **139,** 177–180
electron, *see* Electron microscopy
epifluorescence, **162A,** 49–50
in whole-chromosome hybridization, **153,** 184–187
in evaluation of nuclear matrix isolation procedures, **162A,** 84–85
fluorescence
cytoskeletal dynamics and, **144,** 113
proteins in striated muscle and, **142,** 62
in whole-chromosome hybridization, **153,** 183–187
immunoelectron, proteins in striated muscle
sarcolemma, **142,** 125
sarcoplasmic reticulum, **142,** 122
skeletal muscle, **142,** 73, 82, 88, 93–94, 97
intermediate voltage electron microscopy, **140,** 264
laser scanning confocal microscopy, **162A,** 50–53
light, *see* Light microscopy
phase, retinal precursor cells, **146,** 153, 168
phase-contrast, polytene chromosome, **165,** 66–68
plant cell wall architecture, **173,** 160–162
reflection confocal laser scanning, in eyespot reflection analysis, **148,** 261, 266–267, 269
scanning electron
in analysis of calcifications, **172,** 137–138
avian salt glands, **144,** 180, 182
eggshells and biomineralization, **145,** 219, 224, 239–240
STEM mass analysis, **162B,** 228–230
in study of microgravity effects on statocytes, **171,** 11
transmission electron, *see* Transmission electron microscopy
video-enhanced, in study of *Physarum* microfilament system, **149,** 188–197
in whole-chromosome hybridization, **153,** 183–188
Microsomes
avian salt glands, **144,** 205
intracellular calcium messenger system, **142,** 307, 330
concentration, **142,** 312, 319
G proteins, **142,** 309, 311
phosphatidylinositol turnover, **142,** 322–324, 326
Microspectrophotometry, in analysis of algal photoreception, **148,** 270, 284, 286

Microspores, gene transfer approaches
Agrobacterium -mediated gene transfer, **140,** 427–428
direct DNA delivery to microsporederived embryos, **140,** 428–431
biolistic DNA delivery, **140,** 429–431
microinjection, **140,** 428–429
Microthreads, role in collagen fibril anchors, **173,** 137–138
Microtrabecular lattice, fish chromatophores and, **143,** 221
Microtubule-associated protein kinase
conclusions, **138,** 234–235
cytoskeletal reorganization elicited by growth factors, **138,** 214–219
description, **138,** 211–213
function and properties
activation of MAP kinase during M phase, **138,** 227–228
possible functions of MAP kinase during cell cycle, **138,** 232–234
properties of MAP kinase, **138,** 228–232
mitogen-activated serine/threonine kinase, **138,** 223–226
mitogenic signal transduction by cytoplasmic MTs, **138,** 219–222
Microtubule-associated proteins
binding sites on tubulin, **157,** 7
classification, **157,** 7
distribution on microtubule, **157,** 3
effects
on cell architecture, **164,** 274–275
on microtubule–motor interaction, **164,** 277
on microtubule stability and dynamics, **164,** 275–277
environmental stimuli in plant cells and 285–287, **145,** 301
heat-labile MAP, **157,** 14
MAP1, **165,** 250
MAP1A, **165,** 250, 255
MAP1B, **165,** 250
MAP2, **139,** 280, 282–287; **165,** 250–255
bundling, **165,** 252–253
interaction, **165,** 252
structure, **165,** 252
MAP2B, **165,** 253, 254
MAP2C, **165,** 253, 254
MAP3, **165,** 250
in microtubule, **165,** 249–256
in microtubule cytoskeleton, **165,** 253–254
in nervous system
axonal compartments
220-kDa MAPs, **151,** 79–80
determination, **151,** 81–83
MAP 1A, **151,** 77–79
MAP 1B, **151,** 77–79
MAP 2, **151,** 70–74
tau protein, **151,** 74–77
axons
caliber, **151,** 87–99
cross-linkage, **151,** 95–99
density, **151,** 97–98
determination, **151,** 87
intertubular spacing, **151,** 89–94
oligomerization, **151,** 93, 95
projection domain, **151,** 88
shape, **151,** 83–84
size, **151,** 83–84
characteristics, **151,** 67–68
definition, **151,** 68–70
dendrites
branching, **151,** 84–85
caliber, **151,** 87–99
cross-linkage, **151,** 95–99
density, **151,** 97–98
determination, **151,** 87
intertubular spacing, **151,** 89–94
oligomerization, **151,** 93, 95
projection domain, **151,** 88
shape, **151,** 83–84
size, **151,** 83–84
spines, **151,** 85–87
developmental outgrowth
early MAPs, **151,** 111–118
MAP expression, **151,** 103–107
diversity, **151,** 128–129
function, **151,** 67–68, 127–128
high molecular mass, **151,** 68–69
intermediate molecular mass, **151,** 68–69
light molecular mass, **151,** 68–69
MAP 1A
distribution, **151,** 77–79
molecular structure, **151,** 113
somatodendritic compartments, **151,** 77–79

Microtubule-associated proteins (*continued*)
MAP 1B
distribution, **151,** 77–79
as lability protein, **151,** 113
phosphorylation, **151,** 114–115, 118
remodeling role, **151,** 121
somatodendritic compartments, **151,** 77–79
MAP 2
dendrite branching roll, **151,** 84
in dendritic spines, **151,** 85–87
dephosphorylation, **151,** 126–127
differential expression, **151,** 82–83
function, **151,** 129
intertubule spacing, **151,** 92–94
phosphorylation, **151,** 114, 116, 125–127
as rigidifying agent, **151,** 100
somatodendritic compartments, **151,** 70–74
MAP 3, distribution, **151,** 79–80
MAP 5, phosphorylation, **151,** 114
MAP U, distribution, **151,** 79–80
microtubule assembly
bundling, **151,** 109–111
neurite outgrowth, **151,** 107–108
neuritic lability, role in remodeling, **151,** 118–121
neuronal form regulation, **151,** 70
regulatory function, **151,** 126–127
somatodendritic compartments
220-kDa MAPs, **151,** 79–80
determination, **151,** 81–83
MAP 1A, **151,** 77–79
MAP 1B, **151,** 77–79
MAP 2, **151,** 70–74
tau protein, **151,** 74–77
stability
bundling, **151,** 109–111
early MAPs, **151,** 111–118
mechanisms, **151,** 108–109
neurite outgrowth, **151,** 107–108
remodeling, **151,** 118–121
stabilization, stimulation-dependent
early sensory experience, **151,** 125–126
estrogen, **151,** 125
excitatory amino acids, **151,** 126–127
gangliosides, **151,** 125
nerve growth factor, **151,** 122–125
neurotrophic factors, **151,** 122–125
process, **151,** 122
typology of cells, **151,** 99–103
neurofibril, **165,** 249–256
neurofilaments, **165,** 249–256
nucleotide sensitivity, **157,** 7
NuMA as, **162B,** 12–13
phosphorylation, **157,** 7
in pollen tube cytoskeleton, **176,** 143–145
properties, **164,** 273–274
regulation of MT–microfilament interactions, **173,** 226–228
role
cold adaptation of microtubules, **157,** 14–16
microtubule assembly, **157,** 12, 14, 16
microtubule bundling, **157,** 19–20
temperature effects on tubulin binding, **157,** 16
Microtubule motors
bidirectional transport, **164,** 309–310
dynein-based transport
axonemal, **164,** 283–289
characterization, **164,** 282–283
cytoplasmic, **164,** 289–295
kinesin-based transport
characterization, **164,** 295–297
conventional, **164,** 295–301
kinesin-like protein-driven, **164,** 302–309
Microtubule-organizing centers, *see also* Spindle pole body
in animal and plant cells, **175,** 94–97
antigens, **169,** 61–62
cell model systems, **175,** 13–14
cytoskeletal dynamics and, **144,** 110, 113
higher plants, **169,** 59–63
in lower eukaryotes, **175,** 93–94
MT–nucleus relationship, **175,** 119–121
nuclear proteins with associated properties, **175,** 111–115
in plant cells, model, **175,** 104–107
primary and secondary centers, **175,** 101–104
secondary, cytokinesis and segregation, **175,** 107–111
Microtubules
accessory MT, **173,** 10
animals *versus* plants, **169,** 60–61
assembly, **169,** 31, 69
assay, **157,** 8

critical concentration for assembly, **157,** 9, 12
mechanism, **157,** 3
posttranslational modification role, **157,** 6, 11–14, 18, 20–21, 23
temperature effects on assembly, **157,** 1–2, 4, 7–9, 13
assembly, role of Tau proteins
associated domain, **171,** 176
isoforms in neurons, **171,** 184–185
other interactions, **171,** 181–182
pathological Tau proteins, **171,** 204–206
projection domain, **171,** 175–176
proteins in nonneuronal cells, **171,** 185
Tau phosphorylation
associated kinases, **171,** 178–179
associated phosphatases, **171,** 179–180
regulations, **171,** 180–181
Tau sorting in nerve cells, **171,** 182–184
Tau–tubulin interactions, **171,** 176–177
associated mRNA, **162B,** 198–199
axon, **165,** 239–240, 246–247
based movement, centromere/kinetochore complex role, **162B,** 45–46
in basic axoneme, **173,** 2–3
behavior, regulation
by microtubule-associated proteins, **164,** 273–277
severing proteins, **164,** 279–282
by γ-tubulin, **164,** 271–272
bundling, **157,** 19–20, 23–24; **165,** 252–253
chloroplasts and, **144,** 284, 291
–chromosomal cycles, **175,** 115–119
cold adaptation
homeotherms, **157,** 16–22, 25
plants, **157,** 22–23
poikilotherms, **157,** 1–2, 9, 14
as compression-resistant struts in cellular tensegrity, **150,** 201–205
cortical
cell model systems
isolated CMTs, **175,** 9
membrane ghosts, **175,** 6–9
perfused cells, **175,** 6
membrane ghost models, **175,** 18
cytoplasmic, mitogenic signal transduction, **138,** 219–222
–cytoplasmic interactions, model, **175,** 104–107
cytoplasmic streaming and, **155,** 109–110
cytoskeleton
dynamics, **144,** 94, 100, 109–110, 113–114
in eukaryotic cells, **175,** 119–121
neuronal polarity, **165,** 248
dendrite, **165,** 239–240
in diatom valve morphogenesis, **150,** 329–331
disassembly by centrosome, **169,** 35–36
Drosophila muscles and, **143,** 67, 75, 84–85
dynamic instability, **157,** 3–4, 9
endoplasmic, in coenocytes, **175,** 104–107
environmental stimuli in plant cells and
light, **145,** 279, 284–287
wounding, **145,** 292–294, 296
in eukaryotic flagellum, **173,** 1–2
fish chromatophores and, **143,** 199, 242, 246
cytoskeleton, **143,** 222, 224
leucophores, **143,** 203, 205
function in cells, **157,** 2
generative and sperm cell, **140,** 151–160
biochemical and molecular considerations, **140,** 159–160
structural organization, **140,** 151–158
comparison with MTs in vegetative cytoplasm and somatic cells, **140,** 155–158
general characteristics, **140,** 151–155
possible variations, **140,** 155
GSP, molecular and biochemical considerations, **140,** 159–160
GTP capping, **157,** 3, 22
interactions with intermediate filaments, **164,** 112–114
lepidopteran
in experimental systems, **152,** 33–35
kinetic organization, **152,** 8, 12
kinetochore
meiosis, **152,** 14, 22–23, 25
organization, **152,** 12
post-translational modifications, **152,** 31
structure, **152,** 33

Microtubules (*continued*)
in meiosis
female, **152,** 14–15
male apyrene, **152,** 24–25, 27–28
male eupyrene, **152,** 16, 18–24
post-translational modifications, **152,** 30–32
structure, **152,** 1
lysosomal transformation, **160,** 338
maintainenance cell shape, **140,** 167–168
M cell, **167,** 129–130
mechanical forces
stress response, **150,** 187
in tissue pattern establishment, **150,** 176–177
–microfilament complex, **173,** 211–213
–microfilament interaction, in cell division and development
centriole migration, **173,** 216–217
centrosome movement, **173,** 214–215
cortical movement, **173,** 215–216
cytokinesis, **173,** 217–218
cytoplasmic streaming in ring canals, **173,** 220–221
role of actin filament arrays, **173,** 218–219
role of growth cones, **173,** 221–222
role of nuclear migration, **173,** 219–220
spindle orientation, **173,** 216
spindle positioning, **173,** 217–218
–microfilament interaction, regulation
by multisubunit complexes, **173,** 232–233
role of different motors, **173,** 232
role for MT and actin filament dynamics, **173,** 231–232
role of posttranslational modifications, **173,** 230
by tubulin- and actin-binding proteins, **173,** 226–230
microtubule-associated protein, **165,** 249–256
mitosis and, **144,** 217
chromosomes, **144,** 244–246, 251
dissociation of events, **144,** 232–233, 236–238
mitotic events, **144,** 219–224
in mitotic apparatus, **138,** 239–240
morphogenesis regulation by, **150,** 173
motive force generation mechanism, **155,** 109–110
MPF kinase role in nucleation, **169,** 33–34
myofibrillogenesis and, **143,** 177–178
neuronal
axon transport, **171,** 169
neuron morphology, **171,** 168–169
in nonneuronal cells, **171,** 169
nucleation, **169,** 26–28, 31–35, 44, 69
organelle transport, model systems, **173,** 222–226
organization by centrosome, **169,** 25–26, 31, 59–63
orientation of, **140,** 107–109
and other elements, intermicrotubular connection of association between, **140,** 109–111
overview, **140,** 102
parallel bundles, **162B,** 55
plant arrays, **169,** 59–61
pollen grains, **140,** 102–103
in pollen tube cytoskeleton, **176,** 146–149
pollen tubes
intermicrotubular connection, **140,** 109–111
localization, **140,** 103–107
orientation of microtubules, **140,** 107–109
proteins in striated muscle and, **142,** 118
in regulation of CFTR, **172,** 212
relationship to nucleus
in animal and plant cells, **175,** 94–97
in lower eukaryotes, **175,** 93–94
retinal precursor cells and, **146,** 155–157, 164
role in dynein–tubulin interactions, **173,** 18–19
role in neurogenesis, **159,** 20–21
Sertoli cell, role in spermatogenesis, **160,** 180–181, 205
shearing by centrosome, **169,** 36
spindle
disassembly, **162B,** 35–36
organization, NuMA role, **162B,** 12–15
stability in axons, **157,** 17–18
stabilization, **162B,** 53
stabilizers
histones, **157,** 21
taxol, **157,** 22
structure, **157,** 2–3; **159,** 21
tau
bundling, **165,** 252–253
interaction, **165,** 252

tension molding, **150,** 200
translocation *in vitro* by flagellar dyneins, **155,** 167–168
and tubulin, **140,** 90–93
biochemical characteristics of pollen microtubules, **140,** 91–92
methods to visualize microtubules in pollen, **140,** 92–93
structure of molecules and microtubules, **140,** 90–91
tubulin structure, **169,** 31

Microvilli
actin cross-linking proteins, **175,** 34–35
avian salt glands and, **144,** 138–139, 147, 172, 182
cytoskeletal dynamics and, **144,** 97, 106, 116
M cell
characterizing apical surface, **167,** 101
versus enterocyte, **167,** 111
length *versus* intestinal location, **167,** 114–115
terminal web development and, **167,** 129
in palantine tonsil crypt epithelium, **167,** 135–136
retinal pigment epithelium, function, **171,** 238

Microviscosity, in angiosperm seedlings, effect of microgravity, **171,** 40–41

Midbrain, induction of *En2* production, **175,** 269

Midkine, neurite initiation and elongation role, **153,** 255

mif[+] factor, in mitochondrial fusion, **161,** 81–84

Migration
avian retina and, **146,** 4
Boyden chamber assays, **155,** 53, 55–57
cell, *see* Cell migration
centrioles in cell division and development, **173,** 216–217
chemotaxis, **155,** 53, 55–57
coating assays, **155,** 60
definitions, **155,** 50
direct observation, **155,** 53–54
extracellular matrix-track assay, **155,** 58–60
growth factor-induced
acidic fibroblast growth factors, **155,** 70–71
autocrine motility factor, **155,** 79–80
basic fibroblast growth factors, **155,** 71–73
capacity, **155,** 52
characterization, **155,** 50
endothelial cell growth factor, **155,** 80
epidermal growth factor, **155,** 67–70
fibroblast growth factors, **155,** 70–73
hepatocyte growth factor/scattering factor, **155,** 63–64, 78–79
insulin-like growth factors, **155,** 77–78
migration-stimulating factor, **155,** 79–80
nerve growth factor, **155,** 80
platelet-derived growth factor, **155,** 75–77
transforming growth factor-α, **155,** 68–69
transforming growth factor-β, **155,** 74–75
inhibition, **155,** 51
initiation, **155,** 51
invasion assays, **155,** 63–66
melanocytes and, **146,** 194–195, 199, 216, 220–222
microcarrier assays, **155,** 63
migration track assays, **155,** 58–60
modulating factors
granulocyte-colony stimulating factor, **155,** 80
granulocyte macrophage-colony stimulating factor, **155,** 80
interferon, **155,** 81
tumor necrosis factor-α, **155,** 80
nuclear, role in cell division and development, **173,** 219–220
phagokinetic track assay, **155,** 58
phases, **155,** 51
reaction specificity, **155,** 51
retinal precursor cells and, **146,** 149, 168, 171
scatter assay, **155,** 63–64
spheroid assays, **155,** 62
studies, **155,** 49–50, 66, 81
time-lapse recording, **155,** 53
translocation, **155,** 51
"under agarose" assay, **155,** 57
in vitro wound assay, **155,** 61–62

Migration assay, asymmetric, PMN random motility, **139,** 180

Migration-stimulating factor, characterization, **155,** 79–80
Mildew, *Arabidopsis thaliana* and, **144,** 76
Mineralization
demineralization, regenerative cementogenesis and, **142,** 32–33, 35, 39, 42, 51
and eggshells, *see* Eggshells biomineralization and
extracellular matrix, **162A,** 256–263
regenerative cementogenesis and, **142,** 4, 45, 51
established root surfaces *in vitro,* **142,** 28–29, 42
growing root surfaces, **142,** 11–12, 15, 17, 21–22, 25, 27–28
Mineral nutrition, salt toxicity, **165,** 6
Mineralocorticoid receptor, clustered distribution, **162A,** 156
Mineralocorticoids, secretion, effect of endothelin, **171,** 275–277
Minichromosomes
attachment sites, **162A,** 140–141
SV40
DNA replication, **162A,** 495–498
subnuclear distribution, **162A,** 523–525
unpacked and encapsidated, **162A,** 520–521
virion-contained, **162A,** 525–528
Minicircles
endosymbiont hypothesis and, **141,** 294, 305
kinetoplastid mtDNA and, **141,** 65–69, 73, 79, 82–84
Minimyosin, **156,** 218
Miniparamyosin, *Drosophila* muscles and, **143,** 108–109
Miniprotoplasts
cytokinesis, **175,** 15–16
for cytoskeleton studies, **175,** 16–18
future studies, **175,** 19–20
Mink, trophoblast, cell morphology, **165,** 65–66
MIP-1α, *see* Macrophage inflammatory protein-1α
Mitochondiral enzyme, phosphorylation, **149,** 54–58, 92–93
Mitochondria
γ-aminobutyric acid metabolism and, **147,** 311, 318
associations with intermediate filaments, **164,** 114–115
avian retina and, **146,** 25
avian salt glands and
innervation, **144,** 159
secretory mechanism, **144,** 171
secretory tissue, **144,** 135, 139–141, 144, 147
chloroplasts and, **144,** 259, 288, 291
Drosophila muscles and, **143,** 66–67, 86
endosymbiont hypothesis and, **141,** 233–234, 237, 280–281, 283, 326–329
future prospects, **141,** 331–332
gene expression, **141,** 298–308
gene structure, **141,** 308–319
gene transfer, **141,** 324–325
genome organization, **141,** 288–298
molecular biology, **141,** 281–288
nuclear genome, **141,** 251–252
organelles, **141,** 329–330
Plasmodium, **141,** 325–326
plastids, **141,** 257–258, 271, 273, 278
rRNA sequences, **141,** 319–324
endosymbiotic hypothesis, **165,** 208
environmental stimuli in plant cells and, **145,** 279
fish chromatophores and, **143,** 230
flowering plants and, **143,** 303, 305
giant, morphology, **161,** 56–58
globoid, transformation to tubular type, **160,** 337–338
glutamate metabolism and, **147,** 286
aminotransferases, **147,** 289–291
glutamate dehydrogenase, **147,** 295, 297–298
glutaminase, **147,** 306, 308
higher plant plastids and
metabolism, **145,** 187–189, 196, 198
structure, **145,** 153–155, 158, 164–165
intracellular calcium messenger system and, **142,** 317–318, 330
intramitochondrial filamentous inclusions, **164,** 16–18
life cycle, **161,** 70–72
matrix substructure, **160,** 339
membranes, preprotein proteolytic maturation, **174,** 145–146
nucleic acid
escape, **165,** 207–228
escape consequenees, **165,** 226–227

migration, **165,** 207–228
migration consequenees, **165,** 226–227
plant, effect of microgravity, **171,** 46–48
preprotein import
chaperones at ribosomes, **174,** 136–138
and interaction with cytosolic chaperones, **174,** 138–139
posttranslational import, **174,** 135–136
protein degradation in, role of chaperones
stress proteins as proteases, **174,** 152–154
various studies, **174,** 154
proteins in striated muscle and, **142,** 67, 107, 120
recombination, *see* Recombination, mitochondrial
retinal precursor cells and, **146,** 147
sexuality
and bacterial sex, **161,** 98–99
and endosymbiotic theory, **161,** 100
evolutionary aspects, **161,** 97–98
translocation components
inner membrane, **174,** 144–145
outer membrane, **174,** 142–144
uniparental inheritance, **161,** 63
urease in plant cells and, **145,** 68–69, 71, 83
vegetative segregation, **161,** 63–64
Mitochondrial chaperones
assisted proteases
Clp homologs, **174,** 158–159
ClpP proteolytic subunit homologs, **174,** 158–159
FtsH homologs, **174,** 159–160
La protease homologs, **174,** 157–158
GrpE and DnaJ homologs, **174,** 147–148
role during heat shock, **174,** 160
Mitochondrial DNA
animal
control region, **141,** 201–203
genome content, **141,** 176–181
human diseases, **141,** 207
nucleotide bias, **141,** 204–206
protein genes
codes, **141,** 184–187
identification, **141,** 181–183
translation initiation, **141,** 187–188
translation termination, **141,** 189
regulation, **141,** 225–229
replication, **141,** 217, 222–225, 229–231
rRNA, **141,** 189–191
sequences, **141,** 203–204
structure, **141,** 173–176, 207–210
transcription, **141,** 217–222, 229–231
tRNA
anticodon-codon interactions, **141,** 191–193
gene evolution, **141,** 199–201
modified bases, **141,** 199
variant structural forms, **141,** 193–199
ciliates and, **141,** 3–4, 7, 35, 47
constitutive homologous recombination with mF plasmid, **161,** 91–95
endosymbiont hypothesis and, **141,** 280–282, 286, 288, 327–329
gene expression, **141,** 298–301, 303–305
gene structure, **141,** 308–311, 313–316
gene transfer, **141,** 324–325
genome organization, **141,** 289–298
Plasmodium, **141,** 325–326
rRNA sequences, **141,** 320, 324
genomes in fungi and, **141,** 89–91
higher plant mitochondrial genomes and, **141,** 161, 164–165
abnormal phenotypes, **141,** 130–134
gene location, **141,** 157–159
recombination, **141,** 148–149, 151–153, 156–157
structure, **141,** 135–137, 140, 142–147
kinetoplastids, *see* Kinetoplastid mtDNA
mitochondrial genomes in fungi, *see* Fungi, mitochondrial genomes
recombination, **161,** 64–65
transmission during plasmodium formation, **161,** 7881
Mitochondrial DNA polymerase, animal mtDNA and, **141,** 226
Mitochondrial fusion
characteristics, **161,** 58–60
mF plasmid, genetic organization, **161,** 86–87
mobile genetic elements, **161,** 69–70
nuclear, **161,** 60–63
Physarum polycephalum
defective strains, **161,** 76–78
differentiation during sclerotization and sporulation, **161,** 72–75
genetic element controlling, **161,** 81–84
isolation, **161,** 76–78

Mitochondrial fusion (*continued*)
mF plasmid-mediated constitutive recombination, **161,** 91–95
mtDNA transmission during plamodium formation, **161,** 7881
mt-nuclear fusion in zygotes, **161,** 96–97
nuclear alleles controlling, **161,** 84–86
during sporulation and spore germination, **161,** 75–76
Mitochondrial import stimulatory factor, as cytosolic chaperone, **174,** 141
Mitochondrial RNA
from animal, **141,** 174, 180, 189, 230
regulation, **141,** 226–228
replication, **141,** 223–224
transcription, **141,** 218–219, 222
genomes in fungi and, **141,** 89, 93, 107–108, 116
higher plant mitochondrial genomes and, **141,** 161
processing
animal mtDNA and, **141,** 224, 226, 228–230
endosymbiont hypothesis and, **141,** 300, 302
Mitochondrial RNA polymerase
animal mtDNA and
regulation, **141,** 226, 228–229
transcription, **141,** 220, 222
endosymbiont hypothesis and, **141,** 299, 301
Mitochondrial transport, in axons, **173,** 223–225
Mitogen-activated protein kinase
activation in lymphocyte signaling, **157,** 219–221
control by Ras, **157,** 222
FAK interaction with, **167,** 174–175
function and properties, **138,** 227–228
isoforms, **157,** 2213
MAP serine/threonine kinase, **138,** 223–226
phosphorylation, **157,** 221–222
regulation of transcrtiption factors, **157,** 245
role in mitosis, **158,** 73
species distribution, **158,** 73
translocation, **157,** 219–220, 245
types, **157,** 219
Mitogenesis, platelet-derived growth factors and, **172,** 95, 96, 116
Mitogenic assay, dendritic cell-stimulated, **153,** 77–78
Mitogens
avian retina and, **146,** 30
CD45RA$^+$ T cell proliferation response to, **167,** 98
as factors for cell entry into G_0 state, **167,** 193
growth factor-induced retinal regeneration and, **146,** 56
melanocytes and, **146,** 219, 236–237, 240, 242
retinal pigmented epithelium and, **146,** 76–78
signal transduction by cytoplasmic microtubules, **138,** 219–222
stimulation of MAP-like protein phosphorylation, **138,** 223–224
universal activation of MAP kinase by, **138,** 224–226
transcriptional regulation and, **143,** 13, 30, 33, 38, 46
Mitomyocin C, transdifferentiation in medusae and, **142,** 235, 252
Mitosis
associated NuMA functions, **162B,** 11–17
avian retina and, **146,** 4, 28–30, 39, 42
avian salt glands and, **144,** 135, 143
and calcium
apparatus related to Ca^{2+} regulation, structure and composition
calmodulin and calcium binding proteins, **138,** 241–244
membranes, **138,** 240–241
microtubules, **138,** 239–240
summary, **138,** 244
conclusions, **138,** 264
effect of extracellular calcium agonists/antagonists on cell division
modulation of extracellular calcium, **138,** 248–252
modulation of phosphatidylinositol cycle, **138,** 252–254
endogenous calcium transients during mitosis, **138,** 244–248
intracellular modulation of calcium and related metabolism, **138,** 254–261
regulation between calcium and cell-cycle regulatory proteins, **138,** 261–263

cytoskeletal dynamics, **144,** 91, 107, 113
dissociation of events, **144,** 217–218
chromosomal events, **144,** 238–252
cytokinesis, **144,** 233–235
experimental evidence, **144,** 235–238
genetic evidence, **144,** 226–233
Drosophila muscles and, **143,** 65, 71, 73
environmental stimuli in plant cells and, **145,** 292–294
and fate of lamin-associated proteins, **162B,** 169–170
fish chromatophores and, **143,** 222
growth factor-induced retinal regeneration and, **146,** 50, 53, 55
hormonal control of amphibian metamorphosis and, **145,** 140
integrins in regulating disassembly/reassembly of focal adhesions in, **167,** 163–164
interphase–metaphase transition, **162B,** 406
kinesin-like proteins function, **164,** 304–305
lamin phosphorylation during, **162B,** 154–156
and lepidopteran spindle structure, **152,** 12–13
mitotic events
higher eukaryotes, **144,** 224–226
history, **144,** 218–219
lower eukaryotes, **144,** 220–224
standard type, **144,** 219–220
NPCs during, **162B,** 293–294
plant cells
amitosis, **158,** 76–77
C-metaphase, **158,** 79–80, 83
cytokinesis failure, **158,** 84–86
endopolyploidy, **158,** 73–76
evolution, **158,** 125–126
indefinite growth, **158,** 78–79
nuclear envelope breakdown, failure, **158,** 72–79
phases, **158,** 72
prophase chromosome reduction, **158,** 77–78
reductional grouping, **158,** 80, 82–83
spindle assembly failure, **158,** 79–86
split spindle, **158,** 83–84
retinal pigmented epithelium and, **146,** 75
retinal precursor cells and, **146,** 149, 151, 167–170, 172
Sertoli-germ cell communication network and, **147,** 25, 61, 66, 73
spindle midzone, midbody, and poles during, **162B,** 48–59
as stage in cell cycle, **167,** 192
tissue-specific inhibitors of, *see* Chalones
transdifferentiation in medusae and, **142,** 216, 252
Mitotic apparatus
architecture, **140,** 175–177
PPB, absence of, **140,** 177–178
vegetative cell, interactions with, **140,** 178–179
establishment, **140,** 172–174
nuclear matrix protein targeting and relocation to, **162B,** 20–59
related to calcium regulation, structure and composition, **138,** 239–244
calmodulin and calcium binding proteins, **138,** 241–244
membranes, **138,** 240–241
microtubules, **138,** 239–240
summary, **138,** 244
Mitotic spindle, evolution, **158,** 126
Mitotin, mRNA analysis, **162B,** 386
Mitoxantrone, **156,** 80, 93, 97, 101, 103–104
Mitoxins, therapeutic use in vascular disease, **161,** 31–32
Mixing, cell, during neural crest formation, **175,** 253–255
Mizoribine, effect on primer RNA synthesis, **162A,** 555–556
MK196, effect on RVI, **161,** 186
MLR, *see* Leukocyte reaction, mixed
MMA, *see* Metamorphosin
MM-creatine kinase, in striated muscle, **142,** 81, 107
Mobile genetic elements
introns, mitochondrial, **161,** 65–67
and mitochondrial fusion, **161,** 69–70
Models
actin filament bundle formation *in vitro*
counterion condensation theory, **175,** 62–64
nucleation and growth mechanism, **175,** 64–65
ordered-phase model, **175,** 69–71
phase transition model, **175,** 65–68
thermodynamic–kinetic, **175,** 60–61
for acute phase protein studies
knockout mouse, **170,** 250

Models (*continued*)
laboratory animals, **170,** 246–248
regulation of APP glycosylation alterations, **170,** 268–269
transgenic mouse, **170,** 248–249
animal, *see* Animal models
architectural, plant cell wall, **173,** 160–162
assembly–disassembly model, **162B,** 206–207
assembly driven model, **162B,** 191
Berg-Khan, bacterial flagellar motor end, **147,** 146, 150–151, 153–159
Brownian ratchet, matrix-located Hsp70, **174,** 146–147
Bullough's ear model, **167,** 187–188
cascade, steroid hormone action, **162B,** 341–344
cell columnarization, mechanical stress, **150,** 23–24
cell systems, *see* Cell model systems
cell wall lignification, **176,** 98–101
classical, nondisjunction, **167,** 275
cleavage, for cell wall loosening, **173,** 169–171
clustersome model, **162A,** 41–48
diffusionist
generalizability, **162B,** 192–193
interpretation of pre-mRNA migration, **162B,** 214–215
intracellular transport, **162B,** 186–189
soluble RNA, **162B,** 184–186
direct and indirect, for MDR protein function, **171,** 127–128
direct transfer model, **162B,** 190–191, 205–207
Eisenberg-Hill, bacterial flagellar motor, **147,** 157–159
Escherichia coli for molecular chaperones
DnaK, DnaJ, and GrpE, **174,** 131–132
GroEL and GroES, **174,** 132–134
substrate recognition, **174,** 134–135
Fepi, avian reflex epilepsy, **175,** 293
Fepi/JA chimeras, **175,** 294
flagellum
physical model, **173,** 39–46
physical parameters of movement, **173,** 34–39
geometric clutch, *see* Geometric clutch model
glide symmetry, ion release and binding, **161,** 191–192
hindlimb suspension, for chronic mechanical unloading, **170,** 181–182
Huxley, bacterial flagellar motor, **147,** 156–157
loose coupling, bacterial flagellar motor, **147,** 134–135
Lüger, bacterial flagellar motor
model comparisons, **147,** 146, 148, 150–151, 153–155
models of muscular contraction, **147,** 156–159
lupus nephritis model, **166,** 212–214
membrane potential effects
on drug flux kinetics, **171,** 144–149
on steady-state drug accumulation, **171,** 141–144
membrane ratchet, ameboid movement, **139,** 166, 167–168
microtubule–cytoplasmic interactions, **175,** 104–107
molecular grafting, **173,** 171–172
motor driven model, **162B,** 191, 205–207
organelle transport, **173,** 222–223
for pollen tube extension via stylar matrix, **140,** 312–313
protoplast, for cell wall regeneration, **170,** 42–47
rolling balloon, ameboid movement, **139,** 163–164, 166
Mold prepattern hypothesis, of diatom shell morphogenesis, **150,** 255–258
Molecular chaperones
definition, **174,** 129–130
Escherichia coli as model
DnaK, DnaJ, and GrpE, **174,** 131–132
GroEL and GroES, **174,** 132–134
substrate recognition, **174,** 134–135
at ribosomes, **174,** 136–138
Molecular grafts
as cell wall model, **173,** 171–172
mediation by EXGT, **173,** 194–195
mediation by EXGT *in muro,* **173,** 179–180
Molecular markers, RPE during differentiation, **171,** 235–237
Molecular structure
collagen I, **173,** 80–81
collagen IV, **173,** 88–90

collagen V
COL1 domain, **173,** 97–99
COL2 domain, **173,** 96–97
C-terminal region, **173,** 99–100
NC2 domain, **173,** 96–97
NC3 domain, **173,** 96–97
collagen VII, **173,** 101–102
collagen XVII, **173,** 103–104
laminins, **173,** 106
V-ATPase, **174,** 92–95
Molecular switch, in Rab GTPase, **176,** 16–17
Molgula occulta, and *M. oculata,* evolution, comparison, **176,** 293–294
Molgula oculata, and *M. occulta,* evolution, comparison, **176,** 293–294
Mollusca, instability of homogeneous state, localized activity, **154,** 345
Molybdenum cofactor, flowering plants and, **143,** 300–301, 309
Monkeys, duct-associated lymphoid tissue in, **167,** 137–138
Monoamine oxidase
fish chromatophores, **143,** 230
perinatal exposure to sex hormones and antihormones, **139,** 28–29
Monoamines
fish chromatophores, **143,** 237
non-occurrence in cnidarian nervous systems, **167,** 44–45
Monoclonal antibodies
to algal cell wall components
alginates and fucans, **171,** 106
carrageenan, **171,** 105–106
glycoproteins, **171,** 106–107
to arabinogalactan-proteins, development, **171,** 101–102
B3, to nuclear matrix proteins, **162A,** 34–36
cytoskeletal dynamics, **144,** 109
dendritic cell-specific, **153,** 46
in dome epithelium macrophage identification, **167,** 132
Drosophila muscles and, **143,** 123–124
eggshells and biomineralization and, **145,** 237
environmental stimuli in plant cells and, **145,** 261, 263, 284
extracellular matrix and, **146,** 273, 275, 284
follicular dendritic cell
cell lines, **166,** 153–155
generation, **166,** 153–155
in hemopoietic stem cell purification, **167,** 191
in M cell identification, **167,** 102
melanocytes and, **146,** 195, 206
to melanoma, labeling of human RPE, **171,** 236
as molecular probes of plant cell surface, **171,** 83–85
MRK16, **156,** 91
myofibrillogenesis and, **143,** 166
PH30, **163,** 273
probing of plant cell wall polymers
enzymes, **171,** 100
extensin, **171,** 95–99
glycine-rich proteins, **171,** 99–100
lignin, **171,** 94–95
matrix polysaccharides, **171,** 92–94
pectin, **171,** 85–92
proline-rich proteins, **171,** 99
proteins in striated muscle and, **142,** 61, 128
antibody use, **142,** 100–101
cardiac muscle, **142,** 103–105, 110–111, 113, 116–117, 119
myosin, **142,** 64–71, 73–74, 76, 78
sarcolemma, **142,** 125–126
sarcomere, **142,** 93–97, 99
sarcoplasmic reticulum, **142,** 120–124
thin-filament proteins, **142,** 84–86, 91–92
Qe5, identification of NPC subcomplexes, **162B,** 242
recognition of plant plasma membranes, **171,** 105
retinal precursor cells and, **146,** 160, 172
Sertoli-germ cell communication network and, **147,** 72
in study of plant cell wall–microbe interactions, **171,** 107–109
transdifferentiation in medusae and, **142,** 217, 223, 227, 239, 249, 253
vertebrate cell culture technology and, **142,** 146–147, 203
bioreactors, **142,** 167, 177, 180, 188
traditional cultures, **142,** 148–149
Monocyte–macrophage tissue factor
adhesion receptors, induction by, **152,** 75–76
properties, **152,** 57–59

Monocytes
and coagulation
cellular activators, **152,** 78–79
cytokine-mediated induction of tissue factor, **152,** 59–64
pathway protein expression by, **152,** 77
prothrombinase activity, **152,** 80
tissue factor induction, **152,** 81
cold-stable microtubules, **157,** 18–19
effects on IFN-γ-induced inhibition of GM-CFC, **167,** 231–232
–macrophage procoagulant activity, *see* Procoagulants, monocyte–macrophage
progenitors, **167,** 191
role in wound healing, **169,** 157, 159
TGF-β production by, **167,** 210
TNF-α production by, **167,** 229
transcriptional regulation, **143,** 17, 22, 36, 39–42
Monolignol glycosides
accumulation in plant tissues, **172,** 269
deglycosylation, **172,** 270–271
transport, **172,** 269
Monolignols
glycosylation, **172,** 270
lignin residues and, **172,** 245
non-lignin fates, **172,** 276
polymerization into lignin, **172,** 271–273
stimulation of peroxidase and, **172,** 272
transport, **172,** 269
Monomer-binding proteins, cytoskeletal dynamics and, **144,** 108
Monomers
cholesteric ordering, acid pH, **166,** 88–90
lignin, composition, **176,** 96–97, 122
Monophyletic origins, endosymbiont hypothesis
acquisitions, **141,** 277
algae, **141,** 273
future prospects, **141,** 331–332
mitochondrial 281, **141,** 327
plastids, **141,** 253–254, 257, 261
sequences, **141,** 272–273
Monosodium glutamate, rat model of aging, **157,** 69
Monosomies
nullisomic gametes and, **167,** 264–265
in preimplantation embryos, **167,** 267–268
X as most common aneuploidy, **167,** 267
Morphogenesis
Acetabularia
calcium effects, **150,** 227
growth, **150,** 225–228
simulation, **150,** 232–237
cardiac, role of mechanical forces, **150,** 73
collagen III during, **173,** 86
cytoskeletal dynamics and, **144,** 112–113
cytoskeletal regulation
identification of molecules in, **150,** 173
mechanism, **150,** 174
tissue pattern establishment, **150,** 177
transmembrane mechanical coupling, **150,** 178
diatom, *see* Diatom morphogenesis, chemical basis
diatom shell, *see* Diatom shell morphogenesis, chemical basis
instability of homogeneous state as source, **154,** 309
karyosphere and, **144,** 7–8, 38, 45–47
metabolism, *versus* metabolism of proliferation, **154,** 360–362
movements
elementary movements, **175,** 251
in embryonic induction, **150,** 409–410
inverse movements, **175,** 255
in quail–chick chimera system, **175,** 249–251
neurulation types, **175,** 249
oral morphogenesis, **164,** 4–6
in protein fiber networks
liquid-like properties, **150,** 142–143
model tissue system, **150,** 147
rearrangements based on interfacial forces, **150,** 146
twisted liquid crystalline, supramolecular arrangements, **166,** 59–96
Morphogens, definition, **150,** 409–410
Morphology
Arabidopsis thaliana and, **144,** 72
avian salt glands and, **144,** 133, 155, 163, 167, 171, 182
cytoskeletal dynamics and, **144,** 113
developing conceptus, assessment, **164,** 147
Drosophila head, **174,** 2
karyosphere and, **144,** 1, 7, 40, 45
M cell
description, **167,** 100–102

histochemical markers and, **167,** 105–106
intestinal location *versus,* **167,** 114–115
maturation stages *versus,* **167,** 115–117
neuron, **171,** 168–169
peritoneal macrophage galatose receptor, **175,** 148
related properties, isolated nuclear matrix, **162A,** 15–23
Mosaic analysis, of neurogenic genes, **139,** 141
Mosaicism
chromosomal
incidence and origin, **167,** 287–288
mechanism of formation, **167,** 288–289
overview, **167,** 286–289
confined placental
incidence in spontaneously aborted material, **167,** 288
outcomes of pregnancies involving, **167,** 287
Mosquito
karyosphere and, **144,** 9–11, 18, 30, 44
vitellogenin receptor, **166,** 130–131
Moss
microgravity experiments, **156,** 333; **171,** 30–31
protonema apical cells, effect of microgravity, **171,** 23–25
Moth, Na^+/K^+-ATPase localization, **176,** 327–329
Motility
cell, *see* Cell motility
cilia, **173,** 213
definition, **155,** 50
in *Dictyostelium discoideum,* **175,** 50–56
flagellar, regulation
effect of vanadate, **173,** 27–28
role in living cell, **173,** 23–25
role of Ca^{2+}, **173,** 25–29
role of cAMP, **173,** 25–29
signal pathways, **173,** 21–23
growth eone, **165,** 256–259
related behavior under centrifugation
Paramecium, **157,** 116–117
Spirogyra organelles, **157,** 118
Tetrahymena, **157,** 116–117
Vallisneria chloroplasts, **157,** 117
Motion analysis systems
computer-assisted, **163,** 47–50
three-dimensional systems, **163,** 71–85
two-dimensional systems, **163,** 50–71
Motive force, measurement by centrifuge microscopy
cytoplasmic streaming, **157,** 100, 113–114
forcevelocity in actin-myosin system, **157,** 124
Paramecium propulsion, **157,** 123
Motor driven model, **162B,** 191, 205–207
Motor end plate, calcium channels
CGRP, **147,** 221, 223, 225, 227
electron microscopy, **147,** 214
inhibitors, **147,** 207, 211–212
postsynaptic cytoplasm, **147,** 218
Motorneurons, cnidarian, **167,** 40–42
Motor proteins, cytoskeleton-based, movement
dynein-like proteins, **176,** 159–160
kinesin-like proteins, **176,** 157–159
motor protein interrelations, **176,** 160–163
myosin heavy chain polypeptides, **176,** 154–157
Motors
bacterial flagellar, *see* Bacterial flagellar motor
dynein, *see* Dynein motor
microtubule, *see* Microtubule motors
role in MT–microfilament interactions, **173,** 232
system in amphibian development
motor neuron connection formation
interneurons, **169,** 106
Mauthner neurons, **169,** 105
reticular formation neurons, **169,** 105–106
Rohon Beard cells, **169,** 104
spinocerebellar projections, **169,** 106
motor neurons, **169,** 98–101
motor nuclei, morphological differentiation, **169,** 101–104
muscle innervation, **169,** 100–101
requirements of motor system, **169,** 97–98
types in organelle, **173,** 225–226
Mougeotia
environmental stimuli in, **145,** 269, 272, 277–279, 281–282, 299–300
intracellular calcium messenger system and, **142,** 308

Mouse
aortic smooth muscle cells, FAK and stress fiber assembly in, **167,** 176
FAK deficiencies in, **167,** 177–178
follicular dendritic cell
iccosomes, **166,** 148–149, 150
isolation, **166,** 148–149
M cells in, lectin histochemistry in studying, **167,** 104
mdr genes, **156,** 81
microgravity experiments, **156,** 352
neonatal treatment with sex hormones and, **139,** 33–36
parathyroid hormone-related protein, gene structure, **166,** 234–236
perinatal exposure to antihormones and, **139,** 17–21, 21–26
platelet-derived growth factors and
in embryogensis, **172,** 117–118
in *patch* mutation, **172,** 110–111
sexual dimorphism and, **139,** 31–33
transgenic, *see* Transgenic mouse
trophoblast, cell morphology, **165,** 58–61
Mouthparts
evolution, **174,** 56–60
insect, form diversity, **174,** 7–8
types, phylogenetic distribution, **174,** 10–11
Movement, *see* Intercalation movement
centripetal, cytoplasmic, **148,** 42
fluid, microgravity environment, **156,** 345
pregastrulation, in embryonic induction, definition, **150,** 391–397
6-MP, *see* 6-Mercaptopurine
MPA, *see* Medroxyprogesterone acetate
MPF kinase, role in microtubule nucleation, **169,** 33–34
M phase, activation of MAP kinase during, **138,** 227–228
M protein
gram-positive bacteria, **161,** 272–274
myofibrillogenesis and, **143,** 153, 176
in striated muscle, **142,** 80–81
MRA, *see* Marrow repopulating cells
MRP protein, **156,** 92
MSG, *see* Monosodium glutamate
MSH, *see* Melanocyte-stimulating hormone
MT, *see* Microtubules
MTOC, *see* Microtubule-organizing centers
MTX, *see* Methotrexate
Mucelage, microgravity experiments, **156,** 330–331
Mucin-gold, **156,** 284–285
Mucins
animal, comparison with plant AGPs, **174,** 268–270
binding by viruses, **175,** 157–162
eggshells and biomineralization and, **145,** 235
secretion in cystic fibrosis, **172,** 220–221
Mucin-type suppressors, interactions with ATPase, **172,** 69–70
Mucocytes, avian salt glands and, **144,** 147
Mucopolysaccharides, eggshells and biomineralization, **145,** 239
Mucosa
intestinal, functions of, **167,** 92
lymphocyte migration in, **167,** 93–94
M cell presence outside of gut, overview, **167,** 100
reducing damage to, in chemotherapy, **167,** 241–242
Mucosal dendritic cells, human, **153,** 63
Müller cells
avian retina and, **146,** 5, 40, 42
cellular processes, **146,** 31–34, 39
retinogenesis *in vitro,* **146,** 16–17, 19–20
cytokine production and, **146,** 79
GLUT2 expression in, **172,** 16–17
growth factor-induced retinal regeneration and, **146,** 52, 58
retinal precursor cells and, **146,** 148, 160, 182
Müllerian duct, programmed cell death in, **163,** 141
Müllerian inhibiting substance, gonadal dfferentiation role, **154,** 218
Multidrug resistance
associated protein, *see* P-glycoprotein
collateral sensitivity, **156,** 81
cross resistance, **156,** 81
definition, **171,** 122–123
mdr gene, **156,** 81–82, 87–88
NMR analysis, **145,** 50
non-glycoprotein mediated, **156,** 92–97
P-glycoprotein mediated, **156,** 79–92
reversal, **156,** 88–92
topoisomerase, **162A,** 563–564
topoisomerase II mediated, **156,** 100

Multivesicular bodies, sperm maturation role
endocytic apparatus, **154,** 145
structural properties, **154,** 109
Mung bean
effect of microgravity, **156,** 321–323, 325–327, 331–332
on lignin content, **171,** 39
on respiration–heat discharge, **171,** 40
nodulation, **156,** 17
Muramyl dipeptide
analogs, **161,** 269
antitumor effects, **161,** 269
macrophage stimulation, **161,** 268
pyrogenicity, **161,** 269
Murine spleen colony-forming unit assay, **167,** 190
Muscarinic acetylcholine receptor, insect
characterization, **164,** 213–214
cloned, expression, **164,** 234–235
Muscarinic agonists, avian salt glands, **144,** 200
Muscarinic receptor
avian salt glands
innervation, **144,** 161–162, 165, 167–168
secretory mechanism, **144,** 176
stimulus-secretion coupling, **144,** 203, 205
fish chromatophores, **143,** 229–230
Muscari racemosum
effect of altered gravity, **171,** 53–55
effect of microgravity, **171,** 42
Muscle
bacterial flagellar motor and, **147,** 111, 134–135
calcicum channels and, *see* Calcium channels, in neuromuscular junction
cardiac, *see* Cardiac muscle
contraction, bacterial flagellar motor and, **147,** 98–99, 135, 141, 156–159
control of rhythmic behaviors and, **147,** 5, 10
cytoskeletal dynamics, **144,** 96, 98, 105–106
differentiation, desmin role, **154,** 49–52
differentiation time, **148,** 82
in *Drosophila,* genetic analysis, *see* *Drosophila* muscles, genetic analysis
environmental stimuli in plant cells, **145,** 260–261, 263, 277
extraocular, proteins, **142,** 66, 68–69
for indirect flight, *see* Indirect flight muscles
insect, programmed cell death, **163,** 132
intermediate filament associations, **164,** 109–112
patagialis, proteins, **142,** 101
pathological Tau proteins in, **171,** 207
skeletal, *see* Skeletal muscle
smooth, *see* Smooth muscle
striated, *see* Striated muscle
Muscle cells
cnidarian neuropeptide effects on
Antho-RPamide I, **167,** 53–54, 68
Antho-RPamides II-IV, **167,** 53, 69
compared, **167,** 51–54, 77–78
development, **176,** 257–263
differentiation, mechanical stimulation of
cardiac myocytes
development, **150,** 73–75
physiology, **150,** 76–77
cellular response to loading, **150,** 72–73
future experiments, **150,** 89
integrins, role in signal transduction, **150,** 84–86
load changes
effects of decrease, **150,** 81–82
pathophysiological, **150,** 82–83
in vitro, **150,** 82
myofibril formation, **150,** 75–76
nuclear matrix, **150,** 88
process, **150,** 69–72, 88
receptors of mechanical force, **150,** 84
regulatory molecule response, **150,** 78–81
protein synthesis, **150,** 78–80
proto-oncogen expression, **150,** 78–80
RNA synthesis, **150,** 78
skeletal myocytes, **150,** 77–78
stretch-activated channels, **150,** 86–87
FAK and, MASMC studies, **167,** 176
sphincter, cnidarian
innervation in sea anemones, **167,** 51
neuropeptides and, **167,** 54
Muscle creatine kinase, in striated muscle, **142,** 81
Muscular dystrophy
prenatal screening with FISH, **153,** 30–32
proteins in striated muscle, **142,** 100–101

Mus musculus, animal mtDNA and, **141,** 183, 188, 190, 205
Mustard, chalcone synthase and, **147,** 236, 247
Mustard cress, microgravity experiments, **156,** 321, 323, 332
Mustard spinach, microgravity experiments, **156,** 329
Mutagenesis
  animal mtDNA and, **141,** 219
  Arabidopsis thaliana and
    biochemistry, **144,** 74
    genetic model, **144,** 62
    models, **144,** 76–77
    phenotypes, **144,** 66, 70, 72
  Drosophila muscles and, **143,** 113, 138
  flowering plants and, **143,** 305, 308, 315, 320
  higher plant plastids and, **145,** 196
Mutagens, MAR/ORI vulnerability, **162A,** 352–354
Mutants
  Arabidopsis thaliana and, **144,** 53, 57
    biochemistry, **144,** 74–75
    genetic model, **144,** 61–63
    models, **144,** 76–77
    phenotypes, **144,** 68–72
  chloroplasts and, **144,** 273
  cytoskeletal, motion analysis, computer-assisted
    ABP-120 mutants, **163,** 91–93
    myosin II heavy chain mutants, **163,** 86–89
    myosin I mutants, **163,** 89–91
    ponticulin mutants, **163,** 92–93
  cytoskeletal dynamics and, **144,** 94, 111–112, 119–120
  eye imaginal disc (*Drosophila melanogaster*), **163,** 127–128
  Hydra, **167,** 39
  karyosphere and, **144,** 9, 43
  lycopenic maize, chromoplast-like plastids, **163,** 187
  maize, chromoplast-like plastids, **163,** 187
  mitosis and
    chromosomes, **144,** 242, 245, 248–249, 251–252
    dissociation of events, **144,** 226–244
  non-nodulating, **156,** 40–42
  supernodulating, **156,** 42–43
Mutations
  affecting bristle development, **175,** 49–50
  affecting *Drosophila* oogenesis, **175,** 46–49
  animal mtDNA and, **141,** 207, 210
  bacterial flagellar motor and, **147,** 121, 140
    function, **147,** 111, 113–114
    structure, **147,** 106–109
  biological clocks in retina and, **146,** 84, 116
  cAMP cascade and, **139,** 310–311
  chalcone synthase and
    control of activity, **147,** 260
    control of gene, **147,** 264, 271–272, 275–276
    enzymatic activity, **147,** 237, 241, 243
    functions of products, **147,** 249–250
  ciliates and, **141,** 1, 10
  cytokine production and, **146,** 79
  deletion, *see* Deletions
  Drosophila muscles and, **143,** 63–64, 138–140
    actins, **143,** 112–114
    contractile protein genes, **143,** 103, 123, 134, 136–137
    development, **143,** 75, 78–84, 86–87
    interactions, **143,** 126–127, 131–132
    myosin light chains, **143,** 105, 107
    tropomyosin, **143,** 116–117
    troponin, **143,** 118–121
  endosymbiont hypothesis and, **141,** 310
  extracellular matrix and, **146,** 278, 280–281, 314
  flagellar dyneins, **155,** 171–173
  flowering plants and, **143,** 297–298, 325
    amino acid biosynthesis, **143,** 312–318
    amino acid pathways, **143,** 318–324
    ammonium, **143,** 303–305, 307–309
    nitrate, **143,** 299–302
  germ/soma differentiation in *Volvox carteri* embryos and, **139,** 204–206
  GnRH-encoding gene in teleosts and, **147,** 174, 182
  higher plant mitochondrial genomes and, **141,** 129, 158, 164
    abnormal phenotypes, **141,** 129–132, 134–135
    structure, **141,** 149–150, 154
  higher plant plastids and
    amino acid synthesis, **145,** 196–197, 199

metabolism, **145,** 170, 173, 176, 181–182
structure, **145,** 153, 160
insulin receptor, effects on insulin internalization, **173,** 250–252
kinetoplastid mtDNA and, **141,** 69
melanocytes and
clinical evidence, **146,** 225–228, 232
development, **146,** 214–222
genetics of melanoma, **146,** 233, 237, 251
oncogenes, **146,** 241, 244–245
pigment-cell genetics, **146,** 197–211, 213
tumor suppressor genes, **146,** 249–250
mitochondrial genomes in fungi and, **141,** 90, 119, 121
DNA diversity, **141,** 94, 101
generation of mtDNA diversity, **141,** 106–113, 116, 119
myofibrillogenesis and, **143,** 175, 183
proteins in striated muscle and, **142,** 118
retinal precursor cells and, **146,** 178
self-incompatibility and, **143,** 281, 289
septation, mitosis and, **144,** 229
transcription, **141,** 220
transcriptional regulation and, **143,** 13, 34
function, **143,** 2, 5
viruses, **143,** 41–42, 44
urease in plant cells and, **145,** 98–99
biochemical genetics, **145,** 85–86, 88, 90, 92
elimination, **145,** 80–81, 84
metabolic origins, **145,** 69, 71, 75, 78
nickel metabolism, **145,** 92–94
Wnt-1 gene, **175,** 269
Mutation–selection theory, **156,** 78
Myasthenic syndrome, calcium channels and, **147,** 218, 220, 225
Mycobacteria
infections, **161,** 288–289
killed, as immune system stimulant, **161,** 289–290
surface components, **161,** 287
Mycophenolic acid, effect on primer RNA synthesis, **162A,** 555–556
Mycoplasma, infection, **163,** 185–186
Mycoplasma pneumoniae, binding to sialic acids, **175,** 181–182
Mycoplasm capricolum, endosymbiont hypothesis and, **141,** 307, 315
Mycosphaerella
accessbility induction and, **172,** 62–65
activity of elicitors and suppressors in, **172,** 65–67
effects on host cell wall-bound ATPase, **172,** 77–80
effects on host membrane-bound protein phosphorylation, **172,** 75
effects on host peroxide generation, **172,** 80–81
suppressor production, **172,** 59–61
suppressor signal transduction cascade model, **172,** 82–84
Myelin, avian retina and, **146,** 23
Myelin-associated glycoprotein, **175,** 200–202
Myelin basic protein gene, in preimplantation diagnosis of genetic defects, **151,** 2–3
Myelocytes
granulocytic extract regulating proliferation of, **167,** 191
proliferation regulation by pEEDCK, **167,** 216–217
Myelodysplasia, increased MIP-1α mRNA in, **167,** 238–239
Myeloid cells, TGF-β stimulation of human progenitor, **167,** 212
Myeloma
IFN-α effects on multiple, **167,** 232
MDR reversal, **156,** 91
multiple, effect of IFN-α, **167,** 232
transcriptional regulation and, **143,** 5–6
Myelomonocytic antigen, macrophage characterization by expression of, **167,** 132
Myelosuppression, chemotherapy-induced, **167,** 239–240
Myoblasts
Drosophila muscles and, **143,** 67, 71, 73, 77, 79, 82–83, 137
myofibrillogenesis and, **143,** 165, 169, 174, 179, 183
proteins in striated muscle and
myosin, **142,** 71, 75
sarcolemma, **142,** 125
sarcomere, **142,** 95, 97–98
sarcoplasmic reticulum, **142,** 122
thin-filament proteins, **142,** 84, 87, 91
Myocardial infarction, nuclear magnetic resonance and, **145,** 43–44

Myocarditis, proteins in striated muscle and, **142,** 115–116, 119
Myocardium, proteins in, **142,** 103–105, 108, 114, 116–117
Myocytes
cardiac
development, **150,** 73–75
load change effects
pathophysiological, **150,** 82–83
in vitro, **150,** 82
physiology, **150,** 76–77
protein synthesis, **150,** 78
response to mechanical forces, **150,** 73–75, 78
proteins in striated muscle and, **142,** 106, 109, 111, 114
MyoD, role in muscle fiber type transition, **170,** 191–192
Myofibers
proteins in, **142,** 103–106
transdifferentiation in medusae and, **142,** 220, 224, 254, 256
Myofibrillogenesis, contractile proteins
differential assembly, **143,** 179–181
isoform change, **143,** 154–157
sarcomere formation, **143,** 181–184
thick filament proteins, **143,** 176–179
thin filament proteins
actin, **143,** 157–159
α-actinin, **143,** 174–175
actin isoforms, **143,** 159–160
actin polymerization, **143,** 160–172
capping proteins, **143,** 172–174
tropomyosin, **143,** 175–176
troponin, **143,** 175–176
Myofibrils
cardiac, role of mechanical forces
in formation, **150,** 75–76
in maturation promotion, **150,** 74–75
Drosophila muscles, **143,** 138
actins, **143,** 111–114
contractile protein genes, **143,** 94–99, 107–108, 134, 136
development, **143,** 66–67, 75–76, 84–85, 87
interactions, **143,** 127, 132
muscle protein genes, **143,** 125
tropomyosin, **143,** 116
troponin, **143,** 118–119
Z-disc proteins, **143,** 122–124
proteins in
antibody use, **142,** 101
cardiac muscle, **142,** 105, 107–115, 118–119
myosin, **142,** 65–67, 73, 78
sarcomere, **142,** 95–98
sarcoplasmic reticulum, **142,** 123–124
thick-filament proteins, **142,** 79, 82–83
thin-filament proteins, **142,** 84–85, 87–88, 91–92
Myofilaments
Drosophila muscles and, **143,** 67, 75–76, 86–87, 94, 119
myofibrillogenesis and, **143,** 175, 181, 183
transdifferentiation in medusae and, **142,** 224, 226, 229–230, 252
Myogenesis
Drosophila muscles and, **143,** 82–84
myofibrillogenesis and, **143,** 155, 157, 160, 169, 172–173
programmed cell death during, **163,** 138–139
Myogenin, role in muscle fiber type transition, **170,** 191–192
Myomesin, in striated muscle, **142,** 80–81, 107
Myosin, *see also* Actomyosin
Acanthamoeba, **156,** 221–223
–actin system, *see* Actin-myosin system
ATPase, muscle fiber types based on, **170,** 144–148
bacterial flagellar motor and, **147,** 98, 156–157
brush border myosin I, **149,** 149–150
cerebral myosin, **139,** 280–282
classification, **149,** 148–151
cytoplasmic streaming mechanism affected by
calcium sensitivity, **155,** 123–124
identification in plants, **155,** 103–108, 130
phosphorylation, **155,** 123
cytoskeletal dynamics and
current concepts, **144,** 95–98, 100, 103
cytoskeletal components, **144,** 103, 105, 108–109
Dictyostelium, **144,** 110–113
mechanism, **144,** 116–119
theories, **144,** 86–89, 94
Dictyostelium, **156,** 223–225

environmental stimuli in plant cells and, **145,** 289, 296, 301
chloroplasts, **145,** 279–282
photodinesis, **145,** 259, 261, 263, 265
fish chromatophores and, **143,** 248
heavy-chain isoforms
effect of electrostimulation, **170,** 164–167
as fiber type transition markers, **170,** 189–191
as muscle fiber type, **170,** 150–152
transitions, effect of development, **170,** 158–162
in interphase nucleus, **150,** 165–167
isoforms, related muscle fiber transitions, **170,** 153–155
and kinesin, **140,** 93–94, 111–112
biochemical characteristics of myosin and kinesin-like protein from pollen, **140,** 94–96
kinesin, **140,** 95–96
myosin, **140,** 94–95
profilin, **140,** 96
pollen tubes, **140,** 111–112
kinesin-like protein, **140,** 111–112
myosin, **140,** 111
structure of molecules, **140,** 93–94
kinesin, **140,** 94
myosin, **140,** 93–94
and kinesin-like protein from pollen, biochemical characteristics of, **140,** 94–96
kinesin, **140,** 95–96
myosin, **140,** 94–95
profilin, **140,** 96
structure of molecules, **140,** 93–94
kinesin, **140,** 94
myosin, **140,** 93–94
muscle fiber type-specific distribution, **170,** 157
myofibrillogenesis and, **143,** 153
assembly, **143,** 170–171, 175–178
differential assembly, **143,** 179–181
sarcomere formation, **143,** 181–183
Physarum, **149,** 156–157, 159–160; **156,** 219–221
in retraction of cytoskeleton, **148,** 64–67
in striated muscle, **142,** 62–79
antibody use, **142,** 100–102
cardiac muscle, **142,** 103–106, 115–117, 119
sarcomere, **142,** 97
structure of molecules, **140,** 93–94
transdifferentiation in medusae and, **142,** 227
Myosin heavy chains
Drosophila muscles and, **143,** 87, 138, 140
contractile protein genes, **143,** 98–104, 134–136
interactions, **143,** 126–127, 131
muscle protein genes, **143,** 126
tropomyosin, **143,** 115
troponin, **143,** 118, 121
in motor proteins, **176,** 154–157
myofibrillogenesis and, **143,** 154, 180
in striated muscle, **142,** 64, 66–71, 73–75, 78, 83, 128
antibody use, **142,** 102
cardiac muscle, **142,** 103–106, 117–119
sarcomere, **142,** 95–96, 99
sarcoplasmic reticulum, **142,** 120
thin-filament proteins, **142,** 91
transcriptional regulation and, **143,** 12
Myosin-I, **156,** 219, 221–222, 224; **163,** 89–91
Myosin-II, **156,** 218, 219, 222–225; **163,** 86–89
Myosin light chain kinase
Drosophila muscles and, **143,** 99
effect of electromagnetic field, **158,** 317
environmental stimuli in plant cells and, **145,** 281
Myosin light chains
Drosophila muscles and, **143,** 104–108, 127, 135
myofibrillogenesis and, **143,** 154–155, 180
Myosuppressins, insect, characterization, **164,** 227
Myotonia, effect on muscle fiber type transitions, **170,** 170–171
Myotonic dystrophy, with Tau 55, **171,** 201–202
Myotropin, **166,** 23–31, 32–33
Myotubes
Drosophila muscles and, **143,** 67, 77
myofibrillogenesis and
differential assembly, **143,** 179
isoform change, **143,** 155
thick filament proteins, **143,** 177–178
thin filament proteins, **143,** 165–166, 168–169, 172, 175
proteins in striated muscle and, **142,** 107, 125–128
antibody use, **142,** 100

Myotubes (*continued*)
sarcomere, **142,** 95, 98–99
thick-filament proteins, **142,** 71, 74, 76–78, 80–81
thin-filament proteins, **142,** 87–88
Myrothecium verrucaria, urease in plant cells and, **145,** 80
Mytilus, cytoskeletal dynamics and, **144,** 97
Mytilus edulis, animal mtDNA and, **141,** 180–181, 192
Myxamoebae, **149,** 171–172; **156,** 231–232
Myxine glutinosa, **138,** 112
Myxoflagellates, *Physarum,* **149,** 172
Myxoviruses, binding to mucins, **175,** 157–162
Myzocytosis, chloroplasts and, **144,** 292

N

NAD
biosynthetic pathway in cyanelle genes, **151,** 49
intracellular calcium messenger system and, **142,** 329–331, 337
NADH
ciliates and, **141,** 12, 37–38
higher plant plastids and, **145,** 200–201
NADH dehydrogenase
animal mtDNA and, **141,** 174, 180, 204, 206, 209
protein genes, **141,** 181–183, 185, 187
endosymbiont hypothesis and, **141,** 266, 282, 290, 309
higher plant mitochondrial genomes and, **141,** 157
kinetoplastid mtDNA and, **141,** 70, 75–76, 78, 81–82
mitochondrial genomes in fungi and, **141,** 95, 97–98, 105
NADH hydrogenase, higher plant plastids and, **145,** 158
NADH:nitrate reductase, phosphorylation, **149,** 79–82, 92
NAD kinase, intracellular calcium messenger system and, **142,** 329–332
NADP, intracellular calcium messenger system and, **142,** 329–332
NADPH
avian salt glands and, **144,** 147
chromoplast, **163,** 207–208
higher plant plastids and
fatty acid synthesis, **145,** 184–185, 189
metabolism, **145,** 193–194, 203
translocation, **145,** 170–171
Nafenoprin
effects on apoptosis, **160,** 327–328
hepatocarcinogenicity, **160,** 328
Nafoxidine, *see also* Antihormones; Sex hormones
genital abnormalities in mice and, **139,** 22–23, 220
genital abnormalities in rats and, **139,** 15–16
history of, **139,** 1
Nannostomus beckfordi, fish chromatophores and, **143,** 236
N-(1-Naphthyl)-phthalamic acid, **156,** 44
Naringenin, chalcone synthase and, **147,** 235, 249, 252–253
Naringenin-chalcone, **147,** 235, 244–246, 252–253
NASA, *see* National Aeronautics and Space Administration
Nascent strand start sites, identification, **162A,** 466–473
National Aeronautics and Space Administration, **156,** 350–352, 354
Natural killer cells
decidual, pregnant uterus, **153,** 107–108
TNF-α induced activation, **167,** 230
Navanax, biological clocks in retina and, **146,** 86
Navicula, morphogenesis
branching costae, **150,** 272
nucleating structure, **150,** 273–274
Navicula alba, vesicle formation, **150,** 265
Navicula pelliculosa, morphogenesis
colloidal silica, **150,** 324–325
nucleating structure, **150,** 273
stages, **150,** 254
vesicles, **150,** 270
NBME-IV hemopoietic regulator
action at G_0–S switch, **167,** 193
with cytosine arabinoside in inducing leukemia cell cytotoxity, **167,** 236
MIP-1α as active component, **167,** 198
switching cells from G_1 to G_0 state, **167,** 194
NC2 domain, collagen V, molecular structure, **173,** 96–97

NC3 domain, collagen V, molecular structure, **173,** 96–97
NCAM, *see* Neural cell adhesion molecule
Nebulin
  Drosophila muscles and, **143,** 124
  myofibrillogenesis and, **143,** 182–184
  in striated muscle, **142,** 62, 88–90, 101
Necrosis
  characteristics, **163,** 108
  shrinkage, **163,** 109
  zones, in developing limb bud, **163,** 144–146
Negative feedback
  regulation principles, **167,** 187*f,* 189
  templates and antitemplates in, **167,** 186–187
Negative regulatory protein, **167,** 225–226
Nematoblasts, transdifferentiation in medusae and, **142,** 230, 253
Nematocysts
  chemistry, **156,** 275–278
  contents, **156,** 278
  discharge
    mechanism, **156,** 278–283
    regulation, **156,** 283–295
  structural components, **156,** 275–278
Nematodes
  animal mtDNA and, **141,** 198, 201, 203–204, 206, 208
  Arabidopsis thaliana and, **144,** 71–72
Neoblasts, transdifferentiation in medusae and, **142,** 217
Neocallimastrix, mitochondrial genomes in fungi and, **141,** 116
Neoplasia, NMR analysis, **145,** 35, 42
Neoplasms, *see* Tumors
Neoplastic cell, parathyroid hormone related protein, **166,** 255–262
Neovascularization
  corneal, *see* Corneal neovascularization
  retinal pigment epithelium, **138,** 34
  in retinas of RCS dystrophic rats by RPE cell transplantation, **138,** 33–34
  serine protease role, **161,** 4
Nephrons, differentiation into adrenal cortical cells, **148,** 82
Nerve cells, Tau protein sorting in, **171,** 182–184
Nerve fibers
  autonomic fibers, **156,** 186
  catecholaminergic, thymic epithelial cells, **156,** 186–187
Nerve growth factor
  actin matrix of dendritic spines and, **139,** 287
  discovery, **155,** 49
  melanocytes and, **146,** 241
  microtubule-associated protein response, **151,** 122–125
  migration induction by, **155,** 80
  and neuronal plasticity
    neurite initiation and elongation, **153,** 247–252
    phenotypic expression, **153,** 273, 275–279
    studies, **153,** 235
  retinal precursor cells and, **146,** 178
  retinal regeneration and, **146,** 56, 58, 67
  role in apoptotic processes, **163,** 118
  role in neuron survival, **163,** 151
  Sertoli-germ cell communication network and, **147,** 78
Nerve growth factor receptor, Sertoli-germ cell communication network, **147,** 78
Nerve nets, cnidarian
  as basic organization of nervous system, **167,** 39–42
  bidirectional synapses in, **167,** 43
  centralization and complexity of, **167,** 44–47
Nerve plexus, density
  in *Hydractinia echinata,* **167,** 44–45
  in sea anemones, **167,** 47
Nerve rings
  in hydrozoan medusae, **167,** 39
  as possible syncytia, **167,** 43
Nerves
  cranial, avian salt glands and, **144,** 155, 158
  facial, amphibian development, **169,** 99
  optic, *see* Optic nerve
  peripheral, blood barrier
    glucose transport, **172,** 24–25
    structure, **172,** 24
  spinal, amphibian development, **169,** 99–100
Nervous system
  autonomic, *see* Autonomic nervous system
  avian retina and, **146,** 1–2, 6, 15
  biological clocks in retina and, **146,** 83, 85, 99, 139
  central, *see* Central nervous system

Nervous system (*continued*)
cnidarian
anatomy of
description, **167,** 39–42
simplicity of, **167,** 38
developing from stem cell implantation, **167,** 39
as earliest in evolution, **167,** 39, 77
construction, **175,** 242
developmental studies, **175,** 244
Drosophilia, **139,** 134–143
central, **139,** 134–137
neurogenic genes and, **139,** 140–143
peripheral, **139,** 134–140
ecto-ATPase levels in tissue, **158,** 165–167
FAK expression
rat, **167,** 175–176
in *Xenopus* embroyo, **167,** 168
melanocytes and, **146,** 211, 222
microtubule-associated proteins, *see* Microtubule-associated proteins
nitric oxide localization, **170,** 111–112
peripheral
growth factor-induced retinal regeneration and, **146,** 58
retinal precursor cells and, **146,** 171
programmed cell death in
Drosophila melanogaster, **163,** 129–130
grasshopper, **163,** 133
vertebrates, **163,** 148–153
signals, fish chromatophores and, **143,** 225–234, 237
Neural cell adhesion molecule
E-type ATPase activity, **158,** 150
isoforms, **158,** 150; **169,** 132
neural induction in amphibians, **169,** 123, 132–133
phosphorylation, **158,** 150
polysialylation, **169,** 132–133
role in chondrocyte aggregation, **159,** 294–295
Neural chimeras, quail–chick system
applications, **175,** 245–246
for cell migration during CNS development
in cerebellar cortex, **175,** 277–280
in forebrain, **175,** 274–277
in hindbrain, **175,** 280–285
in optic tectum, **175,** 280
techniques, **175,** 272–274
for cerebellum development, **175,** 262–263
construction, **175,** 245
initial development, **175,** 244–245
marker system, **175,** 245
morphogenetic movements, **175,** 249–251
neuronal connections in, **175,** 263–264
transplantation experiments
regional specifications, **175,** 271–272
tectal and cerebellar structures, **175,** 266–271
transplantation effects, **175,** 264–266
Neural circuits, in control of invertebrate rhythmic behavior, **147,** 1–20
Neural crest
cell transdifferentiation in medusae, **142,** 218
derivatives, development
abbreviations, **138,** 270
cell lineage segregation during ontogeny
clonal analysis, **138,** 288–293
early heterogeneity, **138,** 284–285
cephalic neural crest and patterning of vertebrate head, **138,** 273–274
conclusions, **138,** 310–311
environmental influences in patterning of peripheral nervous system ganglia, **138,** 304–310
fate map of and plasticity of cells, **138,** 271–273
formation, **138,** 269–270
glial cell differentiation in peripheral nervous system
4B3 antigen, **138,** 295–297
environmental control of glial cell differentiation, **138,** 300–304
glial percursors, identification of, **138,** 299–300
phenotypic diversity of peripheral glial cells, **138,** 294–295
Schwann cell myelin protein, **138,** 297–299
and homeobox-containing genes, **138,** 274–283
formation, cell mixing during, **175,** 253–255

melanocytes and, **146,** 194–196, 220–222, 224, 249
melanophore formation by cells, **148,** 82
platelet-derived growth factor distribution in, **172,** 105
programmed cell death in, **163,** 135–137
Neural fold, role in vertebrate head formation, **175,** 260–261
Neural induction, amphibian brain development
ectodermal dorsalization, **169,** 91
mesodermal invagination, **169,** 93–94
superimposed transformation, **169,** 91, 93
time-dependent changes in ectodermal competence, **169,** 93
Neural integration, avian salt glands and, **144,** 188–189, 193
Neural plate
amphibian development, **169,** 94–96
derivation from ectoderm, **175,** 255–257
early, forebrain territories distributions, **175,** 257–260
Neural primordium, early, fate map at neurula stage
forebrain territories, **175,** 257–260
postepiphysis cephalic neural fold, **175,** 260–261
Neural retina, growth factor-induced retinal regeneration and, **146,** 69
development, **146,** 50–52
fibroblast growth factors, **146,** 67–69
ocular tissues, **146,** 57–60, 63–64, 66
transdifferentiation, **146,** 52–55
Neural tubes
amphibian, relaxation of stress, **150,** 10–11
programmed cell death in, **163,** 135–137
Neuregulin–Neu differentiation factor
biological activity, **160,** 246–247
carcinoma role, **160,** 248–249
isoforms, **160,** 243–244
structure, **160,** 243–244
tissue distribution, **160,** 247–248
Neuregulin–Neu differentiation factor receptor, *see* c-ErbB receptor
Neurites
dystrophic, Tau proteins in, **171,** 187
initiation, **165,** 242–256
neuronal polarity development, **165,** 244–249
process outgrowth earliest events, **165,** 243–244
Neuroblastoma
ecto-ATPase levels, **158,** 174–175
effect of differentiating agents, **156,** 85
mdr1 mRNA levels, **156,** 88
melanocytes and, **146,** 233, 244
transcriptional regulation and, **143,** 47
Neuroblasts
avian retina and, **146,** 36–40, 42
differentiation, **159,** 13–14
growth factor-induced retinal regeneration and, **146,** 62
migration during embryogenesis, **159,** 12–13, 23
origin, **159,** 12
proliferation rate, **159,** 12
Neurodegenerative disorders
accumulation of Tau protein
dystrophic neurites, **171,** 187
neurofibrillary tangles, **171,** 185–187
Pick bodies, **171,** 187–188
without pathological Tau proteins, **171,** 203–204
Neuroectodermal cells, inhibitory effect of TGF-β, **167,** 209
Neuroendocrine cell, air tract
associated nerve profiles, **157,** 302–303
calcium-binding proteins, **157,** 300, 302
dense-cored vesicles, **157,** 282, 293, 296
development, **157,** 281
electron microscopy, **157,** 280, 282
evolution, **157,** 278, 289, 291–292, 305
function, **157,** 296, 300, 302–303
hypoxia response, **157,** 289, 302–303
markers, **157,** 286–287, 299–300, 305
morphology, **157,** 298, 305
role in cancer, **157,** 281
secretion
peptide hormones, **157,** 278, 280–281, 283–285, 298
serotonin, **157,** 278, 280, 282–283
structure, **157,** 280–282, 293
Neuroendocrine system
control of APP production, **170,** 244–246
and neurotransmitter functions
of ANF, **138,** 135–138
and behavioral effects of AVT and MT, **138,** 124–126
and behavioral effects of CRF, **138,** 110–112
and behavioral effects of GNRH, **138,** 99
of GRF, **138,** 119–120
of NYP, **138,** 131–133

Neuroendocrine system (*continued*)
of POMC-derived peptides, **138,** 139–144
of SRIF, **138,** 117
Neuroepithelial bodies
electron microscopy, **157,** 281, 287, 290–291
evolution, **157,** 291
secretory substances, **157,** 281, 287–289
serotonin release, **157,** 287–289
structure, **157,** 281–282
visualization, **157,** 281
Neuroepithelium
avian retina and, **146,** 41
cellular processes, **146,** 29, 32, 34, 39–40
normal development, **146,** 2, 4
growth factor-induced retinal regeneration and, **146,** 50, 54, 67
melanocytes and, **146,** 194
retinal precursor cells and, **146,** 173–174, 183
Neurofibril, microtubule-associated protein, **165,** 249–256
Neurofibromatosis, melanocytes and, **146,** 249
Neurofilament protein
ADP-ribosylation, **154,** 43
expression, subject to feedback control mechanism, **154,** 63
function, **165,** 240–241
microtubule-associated protein, **165,** 249–256
NF3, neural induction in amphibians, **169,** 123
Neurogenesis
morphogenetic changes during, **175,** 242
platelet-derived growth factors, **172,** 114
study perspectives, **165,** 236–237
Neurohormones, control of rhythmic behaviors and, **147,** 19
Neurohypophysis
neurohormone release, **159,** 29
steroidal effects, **159,** 29
termination of hypothalamic neurons, **159,** 6, 23, 27
Neurological diseases, animal mtDNA and, **141,** 207
Neuromediator receptor
control of blood-brain barrier function, **149,** 277
in non-dendrite sites, **149,** 222
in postsynaptic membrane, **149,** 231–233
Neuromediators
γ-aminobutyric acid interaction with, **149,** 265
blood-brain barrier function, **149,** 278
chemical, **149,** 222–227, 232–236
defined, **149,** 222
glial cells, **149,** 274–276
histaminergic, **149,** 253
ion channels, **149,** 227–232
release, synapse involvement, **149,** 220–222
storage, **149,** 226
transporters, **149,** 223, 225
Neuromedin, astrocyte reaction to, **148,** 129, 137–138
Neuromodulators
avian salt glands, **144,** 162, 165, 189
biological clocks in retina and, **146,** 114
retinal precursor cells and, **146,** 153–154, 164–165, 181–182
rhythmic behaviors, *see* Rhythmic behaviors, control
Neuromuscular activity, effect on muscle fiber type transitions
decreased activity
by denervation, **170,** 176–177
electrostimulation-induced slow-to-fast transitions, **170,** 177–178
by hindlimb suspension, **170,** 181–182
by immobilization in shortened position, **170,** 180–181
by microgravity, **170,** 182–183
by spinal cord transection and isolation, **170,** 177
increased activity
electrostimulation-induced fast-to-slow transitions, **170,** 163–170
by exercise training, **170,** 171–172
by mechanical overloading, **170,** 172–175
by myotonia, **170,** 170–171
Neuromuscular junction
calcium channels in, *see* Calcium channels, in neuromuscular junction
control of rhythmic behaviors and, **147,** 5
Neuromuscular transmission, neuropeptides involved in cnidarian, **167,** 51–54
Neuronal cells
ependymin, **165,** 144
transdifferentiation in medusae, **142,** 218

Neuronal dense-cored vesicles, cnidarian
  neurons containing, **167,** 40
  RFamide-like material in *Hydra,* **167,** 51
Neuronal plasticity
  cell cultures
    invertebrate neuronal cells, **153,** 236
    model systems, **153,** 235
    transformed cell lines, **153,** 235–236
    vertebrate neuronal cells, **153,** 236–237
  definition, **153,** 233
  future investigations, **153,** 280
  neurite initiation and elongation
    acetyl-L-carnitine, **153,** 255
    brain-derived neurotrophic factor, **153,** 247, 251–252
    cellular adhesion molecules, **153,** 246
    ciliary neurotrophic factor, **153,** 252–253
    concanavalin A, **153,** 247
    epidermal growth factor, **153,** 254–255
    fibroblast growth factors, **153,** 253–254
    fibronectin, **153,** 243–244
    growth-associated proteins, **153,** 262–263
    growth cones, **153,** 258–262
    laminin, **153,** 244–246
    mechanisms, **153,** 237–238
    midkine, **153,** 255
    nerve growth factor, **153,** 247–252
    neurite inhibitors, **153,** 256–258
    neurotrophins, **153,** 248–252
    patterned substrate role, **153,** 238–242
    retinoic acid, **153,** 255
    soluble neurite-promoting factor, **153,** 255
    substratum role, **153,** 238–247
    synthetic substrate role, **153,** 238–242
    transforming growth factors, **153,** 254–255
    trophic factors, **153,** 247–258
  phenotypic expression
    adult neurons, **153,** 275–279
    developing neurons, **153,** 271–275
    mechanism, **153,** 270–271
    neonatal neurons, **153,** 271–275
  studies, **153,** 234–235
  synaptic connections
    excitatory postsynaptic potentials, **153,** 265–267
    invertebrate neurons, **153,** 268–270
    mechanism of formation, **153,** 263–264
    vertebrate neurons, **153,** 264–268
  technological advances, **153,** 279–280
Neuronal polarity
  axon, **165,** 247
  dendrite, **165,** 247
  microtubule cytoskeleton, **165,** 248
Neurons
  avian retina and, **146,** 1–2, 42
    cellular processes, **146,** 31, 35, 39
    normal development, **146,** 2–4
    retinogenesis *in vitro,* **146,** 10, 19–20
  axonal growth, cytoplasmic mechanisms, **165,** 235–280
  basal retinal, biological clocks and
    entrainment pathway, **146,** 102–103, 111, 114
    opisthobranch retina, **146,** 89, 91–93, 95
    output pathway, **146,** 123
  biological clocks in retina and
    Bulla, **146,** 129–130, 132
    concepts, **146,** 99–100
    entrainment pathway, **146,** 101–105, 108, 114
    opisthobranch retina, **146,** 89, 91–93
    output pathway, **146,** 123, 127
    pacemaker mechanisms, **146,** 115
  cholinergic, avian retina and, **146,** 39
  circuits, role in avian reflex epilepsy, **175,** 291–293
  cnidarian
    description, **167,** 41–42
    light- and gravity-sensitive, **167,** 39–40
    multifunctional nature of, description, **167,** 41–42
    in nerve rings of hydrozoan medusae, **167,** 39
    transmitters and, **167,** 50–51
  conditional oscillators, *see* Conditional oscillators
  connections in chimeric brain, **175,** 263–264
  cytokine production and, **146,** 79
  cytoskeleton role in neurogenesis, **159,** 20–21
  dendritic growth, cytoplasmie mechanisms, **165,** 235–280
  development, role of tropomyosin, **170,** 12–13

Neurons (*continued*)
differentiation, **148,** 82, **159,** 4, 13–14
dorsal root ganglion, *see* Dorsal root ganglion neurons
early outgrowth, **165,** 248
growth, effect of MAG, **175,** 201
growth factor-induced retinal regeneration and, **146,** 56
hypothalamic, *see* Hypothalamic neuron; Hypothalamic neurons
injury, *see* Axotomy; Deafferentation
interneurons, in cnidarian nervous system anatomy, **167,** 40–41
ion channels, **159,** 11–12
large dense vesicle translocation, **159,** 20
maturation, **159,** 17, 20
melanocytes and, **146,** 246
morphology, **171,** 168–169
motility, **165,** 242
motorneurons, cnidarian, **167,** 40–42
neurite initiation, **165,** 242–256
nonubiquitous Rab expression, **176,** 18–20
pathological Tau protein variants
in glial cells, **171,** 207
in muscles, **171,** 207
Tau distribution, **171,** 206
peptidergic, cnidarian, **167,** 45
plasma membrane, **165,** 273
pyloric, control of rhythmic behaviors and, **147,** 12–19
receptor internalization, **159,** 20
retinal precursor cells and, **146,** 148–149, 174
chick embryo cells, **146,** 151, 153–154
experimental analysis, **146,** 164, 166–167
microenvironmental factors, **146,** 177–181
photoreceptor default hypothesis, **146,** 167–169
shape dependence, **165,** 237–238
slow oscillator, control of rhythmic behaviors and, **147,** 14
superior cervical ganglion, phenotypic plasticity, **153,** 271
survival, target-derived factors, **163,** 150–151
synaptogenesis, **159,** 17, 20
Tau protein isoforms in, **171,** 184–185
Neuron-specific enolase, production by neuroendocrine cells, **157,** 285
Neuropeptide receptor, astrocyte reaction to
angiotensin II, **148,** 127–128, 131
atrial natriuretic peptide, **148,** 127–128, 132–134
bombesin, **148,** 137–138
bradykinin, **148,** 127, 129, 137
brain natriuretic peptide, **148,** 127–128, 132
calcitonin, **148,** 146–147
corticotropin-releasing hormone, **148,** 129, 141–142
C-type natriuretic peptide, **148,** 128, 132–133
endothelins, **148,** 127–128, 133–134
insulin, **148,** 129, 144–146
insulin-like growth factors, **148,** 129–130, 144–146
neuromedin, **148,** 129, 137–138
opioids, **148,** 140–141
oxytocin, **148,** 129, 138–139
somatostatin, **148,** 129, 147–154
substance P, **148,** 128, 135–137
vasoactive intestinal peptide, **148,** 129, 142–144
vasopressin, **148,** 138–139
Neuropeptides
in amphibian brain
anatomical abbreviations, **138,** 179
anuran brain, diagram, **138,** 177–178
atrial natriuretic factor
immunohistochemical localization of ANF, **138,** 133–135
neuroendocrine and neurotransmitter functions of ANF, **138,** 135–138
bombesin-related peptides
effects of bombesin-related peptides, **138,** 163–164
immunohistochemical localization of bombesin-related peptides, **138,** 163
conclusions, **138,** 176–177
corticotropin-releasing factor
immunohistochemical localization of CRF, **138,** 106–100
neuroendocrine and behavioral effects of CRF, **138,** 110–112

gastrin- and cholecystokinin-related peptides
effects of CCK-related peptides, **138,** 154–155
immunohistochemical localization of gastrin- and CCK-related peptides, **138,** 153–154
gonadotropin-releasing hormones, **138,** 91–99
immunohistochemical localization of GnRH, **138,** 92–99
neuroendocrine and behavioral effects of GnRH, **138,** 99
growth hormone-releasing factor, **138,** 117120
immunohistochemical localization of GRF, **138,** 119
neuroendocrine effects of GRF, **138,** 119–120
melanin-concentrating hormone, **138,** 149–152
immunohistochemical localization of MCH, **138,** 149–151
physiological effects of MCH, **138,** 151–152
miscellaneous peptides, **138,** 164–176
calcitonin gene-related peptide, **138,** 164166
dipeptides, **138,** 171–174
galanin, **138,** 166–169
magainins, **138,** 175–176
mast cell degranulation and chemotactic peptides, **138,** 174–175
vasoactive intestinal peptide and related peptides, **138,** 169–171
neuropeptide findings, synoptic table, **138,** 180–187
neuropeptide tyrosine, **138,** 127–122
immunohistochemical localization of NPY, **138,** 127–131
neuroendocrine effects of NPY, **138,** 131–133
pro-enkephalin-related peptides, **138,** 145–149
immunchistochemical localization of enkephalins, **138,** 147–148
opiate peptides, effects of, **138,** 148–149
pro-opiomelanocortin-related peptides
central effects of alpha-MSH, **138,** 144–145
immunohistochemical localization of POMC-derived peptides, **138,** 139–144
somatotropin-release inhibiting factor, **138,** 112–117
study
amphibian brain, **138,** 90–91
limitations in methodology, **138,** 91
neuropeptides, **138,** 89–90
tachykinins
immunohistochemical localization of tachykinins, **138,** 156–160
substance P, effects of, **138,** 160–161
thyrotropin-releasing hormone, **138,** 99–106
immunohistochemical localization of TRH, **138,** 101–103
neuroendocrine effects of TRH, **138,** 103–106
vasotocin and mesotocin
immunohistochemical localization of AVT and MT, **138,** 120–124
neuroendocrine and behavioral effects of AVT and MT, **138,** 124–126
binding sites on astrocytomas, **148,** 153, 155–158
characterization, **159,** 3, 32
cnidarian
biosynthesis
Antho-RFamide in sea anemones, **167,** 58–64
Antho-RFamide in sea pansies, **167,** 64–67
Antho-RPamide I in sea anemones, **167,** 68–69
Antho-RPamides II-IV in sea anemones, **167,** 69–71
higher mammals *vs.,* **167,** 57–58
in Hydrozoans, **167,** 74–77
metamorphosis-inducing, in sea anemones, **167,** 71–74
cloning of receptors for, **167,** 55–56
isolation, **167,** 43–49
neuromuscular transmission, **167,** 51–54
neurons producing, **167,** 50–51
number isolated *versus* potential action of, **167,** 77–78

Neuropeptides (*continued*)
potential for isolating additional, **167,** 82
protective C- and N-terminal sequences, **167,** 49–50
reproduction, possible role in, **167,** 54–55
evolution, **159,** 8–10
evolutionary origin, **159,** 3–4
in FAK activation, **167,** 171
Gly-Arg-Phe-NH_2, *see* RFamide neuropeptides
hypothalamic regulation, **159,** 8
immunoregulation, **149,** 269–272
insect
adipokinetic hormones, **164,** 228–229
allatostatins, **164,** 227–228
characterization, **164,** 224–225
diuretic hormone, **164,** 230
eclosion hormone, **164,** 229
-FXPRLa peptides, **164,** 226–227
myosuppressins, **164,** 227
proctolin, **164,** 225–226
trypsin-modulating oostatic factors, **164,** 230
lifetime in coelenterates, **159,** 2
processing, **159,** 6, 9
research findings, synoptic table, **138,** 180–187
second messengers, **159,** 11
transcriptional regulation and, **143,** 16
type Y
phenotypic expression, **153,** 271
regulation of pars intermedia cells, **159,** 53
secretion by neuroendocrine cells, **157,** 298
tyrosine
immunohistochemical localization, **138,** 127–131
neuroendocrine effects, **138,** 131–133
Neurophysins
cellular localization, **159,** 34
messenger RNA levels, **159,** 9
precursors, **159,** 34–35
Neuroptera, karyosphere and, **144,** 8, 11–18
Neurosecretory cells
characterization, **159,** 1–2, 6, 89
environmental response
deer, **159,** 81–82
fish, **159,** 80–81, 89
experimental animals used, **159,** 5
localization
annelids, **159,** 2, 7
fish, **159,** 2, 7
mammals, **159,** 2
reptiles, **159,** 2
Neurospora
animal mtDNA and, **141,** 192
endosymbiont hypothesis and, **141,** 302, 306
hormone receptors, second messengers, **155,** 9
Neurospora crassa
endosymbiont hypothesis and, **141,** 298, 300, 311
higher plant mitochondrial genomes and, **141,** 158
mitochondrial genomes in fungi and
DNA diversity, **141,** 93–97, 100
generation of mtDNA diversity, **141,** 109–110, 113, 115, 117
mitochondrial genomes of, **141,** 12, 39–40, 50
Neurotransmission
action potentials, **149,** 227, 230
aminoacidergic, **149,** 255–266
blood-brain barrier, **149,** 277–278
catecholaminergic, **149,** 240–249
catecholaminergic neurotransmission, **149,** 240–249
chemical and electrical synapses in cnidarian, **167,** 42–43
chemical neuromediators, **149,** 222–227
cholinergic, **149,** 236–241
control of rhythmic behaviors and, **147,** 2, 15
endogenous anxiolytic molecules, **149,** 266–267
glutamate metabolism, *see* Glutamate metabolism
histaminergic, **149,** 251–254
and immunoregulators, **149,** 267–269
neuropeptides, **149,** 269–272
thiamine, **149,** 272–274
ion channels, **149,** 227–232
neuronal-glial interaction, **149,** 274–276
purinergic, **149,** 254–255
serotoninergic, **149,** 247–252
synapses, **149,** 219–222
Neurotransmitters
avian salt glands and, **144,** 131, 165

in cnidarians
lack of evidence of "classical," **167,** 43–44
nonsynaptic release and requirement for stability of, **167,** 50
peptides as earliest, **167,** 77
Drosophila muscles and, **143,** 137
fish chromatophores and, **143,** 247, 249
iridophores, **143,** 209
motility control, **143,** 226, 229–230, 238–239
nuclear magnetic resonance and, **145,** 45
retinal precursor cells and, **146,** 153–155, 181–182
role in nervous system function, **153,** 257
Neurotrophic factors
brain-derived, **153,** 247, 251–252
ciliary, **153,** 252–253, 274
effect on microtubule-associated proteins, **151,** 122–125
Neurotrophic theory, neuron survival, **163,** 151
Neurotrophins, neurite initiation and elongation, **153,** 248–252
Neurula, early neural primordium, fate map
forebrain territories, **175,** 257–260
postepiphysis cephalic neural fold, **175,** 260–261
Neurulation
in amniotes
floor plate formation during, **175,** 251
morphogenesis types, **175,** 249
notochord formation during, **175,** 251
studies, **175,** 248–249
associated movements, in embryonic induction, **150,** 405–406
growth factor-induced retinal regeneration, **146,** 50
Neutral–neutral 2-D gel assay, **162A,** 469–473
Neutral protease, bactericidal activity, **161,** 308
Neutrophils
decidual, of pregnant uterus, **153,** 108–109
lactoferrin in secondary granules of, **167,** 228
transcriptional regulation and, **143,** 17
Nevi, genetics, **146,** 226–229, 240, 244–245, 250
Nexin, role in t-force, **173,** 47
NHP-1, binding to estrogen receptor element, **162B,** 363–364
Nicardipine, effect on clofibrate-induced peroxisomes, **160,** 328
Nickel
effect on flagella, **173,** 21
urease in plant cells and, **145,** 65, 99
biochemical genetics, **145,** 88, 90–92
elimination, **145,** 80–81
metabolism, **145,** 92–97
Nick translation, *in situ,* **163,** 123
Nicotiana
environmental stimuli in, **145,** 263
habituation
cellular transfer DNA expression, **152,** 289–290
genetic tumors, **152,** 266, 280–282
tumor-specific genes, **152,** 290
self-incompatibility, **143,** 280, 282–284, 286–287
Nicotiana alata, **140,** 465–466; **143,** 282, 284, 286–288
Nicotiana bigelovii, cytokinin-induced habituation, **152,** 272
Nicotiana debneyi, genetic tumors, **152,** 281
Nicotiana glauca, habituation
cellular transfer DNA genes, **152,** 287, 289
DNA sequence analysis, **152,** 288–289
genetic tumors, **152,** 281–282
hormone concentration, **152,** 267
induction
by triiodobenzoic acid, **152,** 269
in vitro, **152,** 283
Nicotiana longiflorum, genetic tumors, **152,** 281–283
Nicotiana plumbaginifolia, flowering plants and, **143,** 300–302, 309, 321, 323–324
Nicotiana rustica, protoplast culture, effect of microgravity, **171,** 44–45
Nicotiana syluestris, hypersensitive response elicitors, **148,** 7
Nicotiana tabacum
anther-specific genes from, **140,** 23, 33
auxin-regulated genes in, **152,** 128
flowering plants and, **143,** 299
self-incompatibility and, **143,** 278, 287
habituation
cytokinin-induced, **152,** 277, 281
hormone concentration, **152,** 267, 269
temperature effects, **152,** 271

Nicotiana tabacum (*continued*)
intracellular calcium messenger system and, **142,** 322
plastids and, **145,** 153
protoplast culture, effect of microgravity, **171,** 44–45
Nicotinamide, systemic acquired resistance induction, **168,** 325
Nicotine, induction of chromoplast differentiation, **163,** 204
Nicotinic acetylcholine receptor
characterization, **164,** 191
molecular cloning, **164,** 191–192
Nidogen, as lamina densa component, **173,** 108
Niemann–Pick disease, type C, PHF-Tau protein in, **171,** 197
Nifedipine, **147,** 213; **156,** 289
Nigericin, role in potassium-regulated sperm function, **149,** 18
Nitella
cytoplasmic streaming
action potential as factor, **155,** 121
reconstruction experiments, **155,** 116
environmental stimuli in, **145,** 261
internodal cells, **156,** 201
intracellular calcium messenger system and, **142,** 325
symplast and, **142,** 280–281
Nitrate
flowering plants and, **143,** 298–303, 306
higher plant plastids, **145,** 200
inhibition, **156,** 38
as regulatory metabolite of plant enzymes, **149,** 86
Nitrate reductase
encoding gene, **139,** 227, 238–239
flowering plants and, **143,** 299–303, 308–309
introns, **139,** 248–249
phosphorylation, **149,** 86–87, 93
Nitric oxide
biological role in invertebrates, **170,** 113–116
effect on endothelin release, **171,** 292–293
interaction with POMC products and cytokines, **170,** 117–118
localization in vertebrates, **170,** 111–113
in long-term potentiation of dendritic spines, **139,** 293–295
relationship with cytokines, **170,** 107
role in immune response, **170,** 116–117, 121–122
role in inflammatory response, **170,** 118–122
role in macrophage antimicrobial activity, **161,** 309–310
role in physiological processes, **170,** 108–110
role in stress response, **170,** 116–117, 121–122
Nitric-oxide synthase
classes, **170,** 110
in invertebrates, **170,** 113–116
mapping in HIV-1 and HTLV-1, **162B,** 308–310
in vertebrates, **170,** 111–113
Nitrite
flowering plants and, **143,** 299, 303
higher plant plastids and, **145,** 179, 191, 193
Nitrite reductase
flowering plants and, **143,** 299, 303, 308–309
higher plant plastids and, **145,** 192–194
phosphorylation, **149,** 79
Nitrogen
chromplast metabolism, **163,** 220
combined, **156,** 38–39
control, nodulation and, **156,** 38–40
eggshells and biomineralization and, **145,** 239
flowering plants and, **143,** 297–298, 316, 325
ammonium, **143,** 303–304, 306–309
nitrate, **143,** 299, 302–303
glutamate metabolism and, **147,** 301, 303–304
higher plant plastids and
amino acid synthesis, **145,** 190–191, 193, 197, 199
metabolism, **145,** 167, 176, 203
nuclear magnetic resonance and, **145,** 3
urease in plant cells and, **145,** 97–99
biochemical genetics, **145,** 90
elimination, **145,** 80–83
metabolic origins, **145,** 66–68, 72–77, 79
nickel metabolism, **145,** 95–96

vertebrate cell culture technology and, **142,** 153, 171, 200
Nitrogen mustards
as alkylating agent, **156,** 122, 127–129
resistance to, **156,** 131
Nitrous oxide, inhibition of smooth muscle cell growth, **161,** 26–27
Nitzschia amphibia, morphogenesis, **150,** 255
NLS, *see* Nuclear localization sequences
NMR, *see* Nuclear magnetic resonance
NO, *see* Nitric oxide
Nocodazole
cytoskeleton, **166,** 15
mitosis and, **144,** 235, 246
NodA protein, **156,** 29
NodB protein, **156,** 29
NodC protein, **156,** 29
NodD protein
activator, **156,** 12–16
encoding genes, **156,** 5–8
flavonoids, interaction, **156,** 14–16
in *Rhizobium,* **156,** 1
NodE protein, **156,** 29
Nod factors
excretion, **156,** 27–28
nod genes and, **156,** 23–29
plant response to, **156,** 46–49
signaling, **156,** 30–32
structure, **156,** 18–22
subcellular location, **156,** 29–30
sulfation, **156,** 26
NodF protein, **156,** 29
NodG protein, **156,** 29
NodH protein, **156,** 29
NodI protein, **156,** 27, 29
NodJ protein, **156,** 27, 29
NodL protein, **156,** 29
NodM protein, **156,** 27, 29
NodO protein, **156,** 28, 32
NodP protein, **156,** 29
NodQ protein, **156,** 29
Nodulation
absence of *Rhizobium,* **156,** 43–44
autoregulation, **156,** 42, 43
genetic control, **156,** 38–40
mutants, **156,** 40–43
Nar phenotype, **156,** 43–44
plant hormones, **156,** 44–46
Rhizobium–plant interactions, **156,** 1–51
supernodulating mutants, **156,** 42–43
transcriptional regulation, **156,** 7–18
Nodules
determinate nodules, **156,** 3, 5
indeterminate nodules, **156,** 3
Nodulins
early, genes, **156,** 33–37
late nodulins, **156,** 33
NodV protein, **156,** 17
NodW protein, **156,** 17, 18
Noggin, neural induction in amphibians, **169,** 124
NolA protein, **156,** 18
NolR protein
binding, **156,** 11, 16–17
as repressor, **156,** 16–17
Nonchromosomal stripe mutants, higher plant mitochondrial genomes
abnormal phenotypes, **141,** 130, 134–135
structure, **141,** 149–150
Nondisjunction
aneuploidy resulting from, overview, **167,** 264
classical model, **167,** 275
confined mosaicism arising from mitotic, **167,** 288
etiology
mechanism, **167,** 275–276
as primarily *de novo* event, **167,** 274–275
gonosomes as particularly prone to, **167,** 270
models, pre-conception to livebirth, **167,** 268*f*
Nonflowering seed plants, double fertilization
cell biological features of, **140,** 348–349
in conifers, **140,** 321–327
basic aspects of reproduction, **140,** 321–322
reports of intra-archegonial double fertilization, **140,** 322–327
in *Ephedra*
basic features of reproduction, **140,** 328–333
early reports of intra-archegonial double fertilization in, **140,** 333–337
fate of second fertilization product in, **140,** 339–342

Nonflowering seed plants (*continued*)
recent studies offertilization in, **140,** 337–339
evolutionary considerations
among nonflowering seed plants, **140,** 349–350
double fertilization in *Ephedra* and angiosperms, **140,** 350–354
intra-archegonial double fertilization phenomena, **140,** 320–321
potential double fertilization in *Gingkobiloba* and cycads, **140,** 327–328
basic aspects of reproduction, **140,** 327–328
potential for intra-archegonial double fertilization, **140,** 328
potential double fertilization in *Gnetum* and *Welwitschia,* **140,** 342–348
Nonhistone proteins, mitosis and, **144,** 249
Non-Hodgkins lymphoma, *mdr1* gene expression, **156,** 88
Nonhypophyseal factors, hormonal control of amphibian metamorphosis, **145,** 126–128
Nonmammalian vertebrate, parathyroid hormone-related protein, **166,** 262–266
Nonneuronal cells
microtubules in, **171,** 169
Tau proteins in, **171,** 185
Nonphotosynthetic plastids, higher plant
amino acid synthesis, **145,** 199–201
carbohydrate oxidation, **145,** 201, 203
fatty acid synthesis, **145,** 182–183, 189–190
starch synthesis, **145,** 177, 180
structure, **145,** 150, 161, 163
translocation, **145,** 169
Nonrhythmic inputs, biological clocks in retina and, **146,** 100
Noradrenaline receptor, characteristics, **149,** 243–244
Noradrenergic system, drug interactions, **149,** 247–249
Norepinephrine
avian salt glands and, **144,** 167–168, 189, 200
fish chromatophores and, **143,** 227, 231–233, 242
retinal precursor cells and, **146,** 181–182
Northern blot
gene expression analysis, **158,** 237
RNA extraction, **158,** 237
Notochord
formation, **175,** 251
Hensen's node origin, **175,** 255
induction in ascidian tissue formation, **176,** 272–278
Notophthalmus viridescens
growth factor-induced retinal regeneration and, **146,** 53
hormonal control of amphibian metamorphosis and, **145,** 131
Novobiocin, **156,** 131
NPA, **156,** 44
NPCs, *see* Nuclear pore complexes
NPY, *see* Neuropeptides
NRP, *see* Negative regulatory protein
NS1 protein, inhibition of poly(A)$^+$ mRNA transport, **162B,** 210–213
NSP1p protein, characterization, **162B,** 280–281
Nuclear bodies, PML-containing, **162A,** 169–170
Nuclear–chromatin binding, role of receptor accessory factor, **162B,** 359–364
Nuclear envelope
adsorption to electron microscopic grid, **162B,** 226–228
biochemically distinct compartments, **162B,** 258–260
breakdown, mitosis and, **144,** 249, 251
centrosome-like properties, **175,** 98
characteristics, **151,** 183
cytoplasmic intermediate filament associations, **164,** 95–99
dynamic microtubules, **175,** 118
formation, lamin role, **162B,** 165–167,171–173
interconnection with nuclear lamina, **162B,** 378–379
lattice in oocytes, **162B,** 149
macromolecular transport across
energy requirements, **151,** 189–191
evidence, **151,** 189–191
pathways, **151,** 189–191
signals, **151,** 191–196
mitosis and, **144,** 217
chromosomes, **144,** 240–243, 245–247

dissociation of events, **144,** 230, 232–233, 235, 238
mitotic events, **144,** 218–225
nuclear localization signal receptors, **151,** 196–199
protein import signals, **151,** 191–193
RNA efflux signals, **151,** 193–195
U snRNA import, **151,** 195–196
Nuclear factors
activated T cells
biological activity, **157,** 247–248
regulation, **157,** 247–248
subcellular localization, **157,** 219
transcriptional regulation and, **143,** 12–13
NF-κB
biological activity, **157,** 246
regulation, **157,** 246
regulation of transcription, **143,** 1–2, 47–48
biochemistry, **143,** 22–25
function, **143,** 2–7
genes, **143,** 11–12
inflammation, **143,** 17–19
T Iymphocytes, **143,** 12–17
inducible activation, **143,** 7–11
molecular cloning, **143,** 25–27
ankyrin repeat proteins, **143,** 36–39
cDNA, **143,** 29–33
Drosophila, **143,** 35–36
modes of regulation, **143,** 34–35
p65 subunit, **143,** 33–34
Rel, **143,** 27–29
proteins, **143,** 45–47
structure of sites, **143,** 19–22
viruses, **143,** 39, 44
HIV, **143,** 39–42
HTLV-I, **143,** 42–44
structure, **157,** 246
transcriptional regulation, **143,** 11, 17, 46
biochemistry, **143,** 34–35
molecular cloning, **143,** 25, 29–30, 33, 35–39
Nuclear fragmentation, *see* Amitosis
Nuclear fusion, double fertilization, **140,** 383–385
Nuclear lamina
associated proteins, **162A,** 25–26
attachment of pores, **162A,** 89–90
complex with nuclear pores, **162A,** 194,209–210
interconnection with nuclear envelope, **162B,** 378–379
mitosis and, **144,** 245–246, 249
plant, **162B,** 93–103
characterization, **162B,** 95–97
fractionation, **162B,** 97–98
protein composition, **162B,** 98–100
ultrastructural organization, **162B,** 98
Nuclear localization sequences
identification in NPC, **162B,** 244–248
location plasticity, **162B,** 259–260
Nuclear localization signal receptor
macromolecular transport across nuclear envelope, **151,** 196–199
regulatory role, **151,** 212
Nuclear localization signals
function, **151,** 192–193
identification, **151,** 191–192
masking
by NRD transcription factors, **151,** 203–206
by steroid receptors, **151,** 203–206
number, effect, **151,** 201–202
transcriptional regulation and, **143,** 38
Nuclear magnetic resonance
in analysis of calcifications, **172,** 137
applications
ex vivo NMR spectroscopy, **145,** 34–40
imaging of metabolites, **145,** 43–49
in vivo NMR spectroscopy, **145,** 35, 39, 41–43, 50
basic concepts, **145,** 1–7
cell studies, **145,** 13–18
imaging, **145,** 41–42
cardiovascular system, **145,** 31
diffusion, **145,** 26–29
magnetic resonance angiography, **145,** 25–26
magnetization transfer, **145,** 29–30
metabolites, **145,** 30, 43–44
microscopy, **145,** 31–32
perfusion, **145,** 29
receptor imaging, **145,** 30–31
magnetization transfer, **145,** 29–30, 33–34, 44–45

Nuclear magnetic resonance (*continued*)
nuclear characteristics, **145,** 17, 19–20
observed parameters
chemical shift, **145,** 7–8
peak intensity, **145,** 11
peak shape, **145,** 12
relaxation times, **145,** 9–11
spin coupling, **145,** 8–9
spin echo, **145,** 11–12
pathology and, **145,** 2, 19, 43
proton density images, **145,** 25–26
spectroscopy, **145,** 35, 37–38
techniques, **145,** 26, 33
urease in plant cells and, **145,** 76
in vivo tissue studies, **145,** 20–24
Nuclear matrix
association
with DNA replication sites, **162A,** 549–556
with transcriptionally active chromatin, **162A,** 226–227,231–232
attachment site, **162A,** 264
bound enzymes, **162B,** 393–394
bound topoisomerase II, **162A,** 562–563
candidate acceptor protein, **162B,** 337–367
changes, and tumorigenesis, **162A,** 108–110
as channel system for mRNP transport, **162B,** 438–440
complex with intermediate filaments, **162A,** 87–90
constituents, cell cycle dependence, **162B,** 382–387
core filament network, **162A,** 194
corresponding to nuclear domains *in situ,* **162B,** 123–127
DNA
fragment size, **162B,** 410
sequence specificity, **162B,** 411–425
DNA topoisomerase II, **162A,** 557–559
functions, **162A,** 37–39
histone enzyme association, **162A,** 208–210
history and definition, **162A,** 81–86
interaction with insulin, **173,** 265–266
interaction with plant cytoskeleton, **162B,** 90–95
interface with DNA loop, **162A,** 458–460
isolated, morphological properties, **162A,** 15–23
isolation protocols, **162A,** 12–15, 83, 12–15,83–86,288–289; **162B,** 78–88
karyosphere and, **144,** 40–41
in muscle cell differentiation, **150,** 88
and nuclear domains, **162A,** 151–175
and ORIs, **162A,** 455–476
and phenotype-specific gene expression, **162A,** 105–106
preparation protocols, **162A,** 490–495
recognizing supercoiled DNA, **162A,** 349
repair occurrence, **162A,** 316–317
residual, prepared from *in situ* nuclei, **162A,** 514–516
RNA metabolism, architectural organization, **162A,** 99–105
role
chromatin organization, **162A,** 101,191–235,541–545
osteocalcin gene transcription, **162A,** 266–269
steroid hormone action, **162B,** 344–345
transcription, **162A,** 100–103,157–159
in situ
functional topography and dynamic properties, **162A,** 8–11
historic perspective, **162A,** 3–4
structural components defined, **162A,** 4–8
SV40 DNA replication at, **162A,** 518–519,525–528,531
and transcription factors, **162A,** 230–231,293–303
Nuclear matrix proteins
classification, **162A,** 293–309
general properties, **162A,** 23–29
malignancy-specific, **162A,** 107–108
in pre-mRNA processing, **162A,** 104–105
spindle-associated, **162B,** 52–54
synthesis, **162B,** 385–386
targeting and relocation to mitotic apparatus, **162B,** 20–59
Nuclear mitotic apparatus protein
coiled structure, **162A,** 28
distribution in proliferating cells, **162B,** 8–11
filament-forming capability, **162A,** 95–96
mitotic function, **162B,** 11–17
role in binding splicing machinery, **162A,** 164–165
at spindle poles, **162B,** 57–59

structural properties, **162B,** 4–8
structural role in interphase, **162B,** 17–19
Nuclear Overhauser effect, and NMR, **145,** 9, 44
Nuclear pore complexes, *see also* Pore–lamina complex
architecture, **162B,** 226–236
as channel, **162B,** 438–440
consensus model, **162B,** 236
dynamics, **162B,** 291–294
fibrillar system, **162B,** 205–206
gp62 nucleoporin, **162B,** 203
import of nuclear proteins, **162B,** 244–248
mRNA presentation to, **162B,** 204
passive diffusion through, **162B,** 243–244
peripheral components, **162B,** 231–233
structure and organization, **162B,** 260–264
subcomplexes, characterization, **162B,** 241–243
Nuclear pore complex proteins, **162B,** 238–241, 264–290
Nuclear proteins, with MTOC properties, **175,** 111–115
Nuclear tensegrity, **150,** 206–207
Nuclear trafficking, in regulation of cellular activity
macromolecular exchange
anchoring to cytoplasmic elements, **151,** 203
cell shape effect, **151,** 212–214
cytoplasmic factors, **151,** 211–212
diffusion, passive, **151,** 208
in dividing cells, **151,** 209–211
epidermal growth factor, **151,** 214
growth factors, **151,** 214–215
insulin-like growth factor-1, **151,** 214
mechanisms, **151,** 200–201, 217–218
NRD transcription factors, **151,** 203–206
platelet-derived growth factor, **151,** 214
pore complex, **151,** 211–212
post-translational modification, **151,** 207–208
quiescent cells, **151,** 209–211
signal accessibility, **151,** 203–207
signal composition, **151,** 201–203
steroid receptors, **151,** 206–207
SV40-transformed cells, **151,** 215–218
transport machinery regulation, **151,** 211–215, 218
transport machinery variation, **151,** 208–211
mechanisms, **151,** 183
nuclear localization signal receptors, **151,** 196–199
pore complex
composition, **151,** 185–189
structure, **151,** 184–186
Nucleases
activity, **140,** 438–439
endogenous and exogenous, excision by, **162B,** 432–433
sensitive chromatin domains, **162A,** 218–221
Nucleation, eggshells and biomineralization and, **145,** 242, 244
Nucleic acids
chloroplast
escape, **165,** 207–228
escape consequences, **165,** 226–227
migration, **165,** 207–228
migration consequences, **165,** 226–227
escape, organelle
limits to transfer of genetic information between compartments, **165,** 225–226
mechanisms, **165,** 218–226
transfer of escaped organellar DNA to nucleus, **165,** 221–222
intermediate filament protein interactions with, **154,** 5–7
mitochondria
escape, **165,** 207–228
escape consequences, **165,** 226–227
migration, **165,** 207–228
migration consequences, **165,** 226–227
nucleus
escape, **165,** 207–228
escape consequences, **165,** 226–227
migration, **165,** 207–228
migration consequences, **165,** 226–227
Nucleocytoplasmic ratio, effects on preimplantation development of human conceptuses, **164,** 172
Nucleocytoplasmic transport
import pathways, **162B,** 258–260
perspectives, models, and evidence, **162B,** 193–210

Nucleoids
chloroplasts and, **144,** 290
nicks, distance between, **162B,** 407–408
relaxed, DNA loop halo, **162B,** 408–410
Nucleolar factories, *see also* Replication factories
and mitosis, **162A,** 141
RNA polymerase I, **162A,** 137–139
Nucleolar matrix
characterization, **162B,** 112–114
nucleolar domain, **162B,** 114
preparation from isolated nucleoli, **162B,** 114–116
ultrastructure, **162B,** 116–117
Nucleolar matrix proteins, **162B,** 117–123
Nucleolar protein B-23, **166,** 6
Nucleolar targeting sequences, NOS, **162B,** 308–310
Nucleolin, immunological characterization, **162B,** 121–123
Nucleolus
interphase nucleus marker, **165,** 73
protein targeting to, **162B,** 305–310
Nucleolus-like bodies
karyosphere and, **144,** 46–47
oocytes of insects, **144,** 19, 21, 23, 25
vertebrates, **144,** 39
polytene chromosome, **165,** 81–87
Nucleomorphs
chloroplasts and, **144,** 275–276, 295
endosymbiont hypothesis and, **141,** 259, 276
Nucleoplasm
lamin foci, **162B,** 151–154
plant cell, **162B,** 124
T-antigen subclass distribution, **162A,** 502–508
Nucleoplasmic–cytoplasmic fraction, SV40 virions, **162A,** 525–528
Nucleoporins
biosynthesis and assembly, **162B,** 291–293
CAN/NUP214, **162B,** 269–270
composition, **151,** 185–188
gp62, **162B,** 203
gp210, **162B,** 272–273
interacting with XFXFG and GLFG family members, **162B,** 286–290
NUP107 and NUP155, **162B,** 273
NUP153, **162B,** 270–271
p62, **162B,** 268–269
POM121, **162B,** 271–272
Tpr/p265 and NUP180, **162B,** 273–274
vertebrate, **162B,** 264–274
yeast, **162B,** 275–290
GLFG family, **162B,** 282–286
XFXFG family, **162B,** 279–282
Nucleoside diphosphate kinase, melanocytes and, **146,** 249
Nucleosides, fish chromatophores and, **143,** 232–233
Nucleoside triphosphatase, composition, **151,** 188
Nucleoskeleton
association with pre-mRNA, **162B,** 200–202
and chromatin loops, **162A,** 129–131
core filaments, NuMA localization, **162B,** 17–18
interaction with plant cytoskeleton, **162B,** 91–93
NuMA isoforms, **162B,** 10
role in mRNA processing and transport, **162B,** 195–197,215
viral protein association, **162A,** 492–493
Nucleosomal positioning
determination with bendable DNA, **162A,** 348
S/MAR role, **162A,** 398–400
Nucleosomes, *see also* Polynucleosomes
active MAR/ORIs free of, **162A,** 347–349
arrays, folding, **162A,** 252–253
atypical, **162A,** 224–226
characterization, **162A,** 196
mapping, interferon-β, **162A,** 393–397
organization, **162A,** 263–269
reconstitution, **162A,** 398–400
structure alteration, **162A,** 206–207
transcription factor binding, **162A,** 229–230
Nucleotide excision repair, **156,** 124, 127–128
Nucleotides
animal mtDNA and, **141,** 174–177, 189, 201, 204–210
replication, **141,** 224
rRNA, **141,** 190–191
transcription, **141,** 218
tRNA, **141,** 192, 195–199
avian salt glands and, **144,** 201–205

chloroplasts and, **144,** 260, 272, 281, 284, 293
ciliates and, **141,** 49
cyclic, in plant defense transmembrane signaling, **172,** 71
Drosophila muscles and, **143,** 120, 126
endosymbiont hypothesis and
gene expression, **141,** 301
gene structure, **141,** 308–309
genome organization, **141,** 288, 293
lineages of life, **141,** 238
plastids, **141,** 268
rRNA sequences, **141,** 321
fish chromatophores and, **143,** 240, 243
glutamate metabolism and, **147,** 302
GnRH-encoding gene in teleosts and, **147,** 173, 181, 184
higher plant mitochondrial genomes and, **141,** 133, 137
higher plant plastids and, **145,** 153
intracellular calcium messenger system and, **142,** 309, 323, 328–329
kinetoplastid mtDNA and, **141,** 65, 69, 79, 83
mitochondrial genomes in fungi and, **141,** 100, 109
motifs in chalcone synthase gene control, **147,** 261, 263–267, 269–270
proteins in striated muscle and, **142,** 121
Rab GTPase role in binding, **176,** 5–14
self-incompatibility and, **143,** 284
transcriptional regulation and, **143,** 2, 19, 21, 39–40
urease in plant cells and, **145,** 78, 86, 91
Nucleotide sequence
affinity for nuclear matrix, **162A,** 199–201
Antho-RFamide precursor
from *Anthopleura,* **167,** 63*f*
from *Calliactis,* **167,** 59–61
from *Renilla,* **167,** 64–65*t*
Antho-RPamide I precursor, from *Anthopleura,* **167,** 68*t*
chalcone synthase, **147,** 237, 244
mammalian and plant clathrin, **167,** 13
MMA precursor, **167,** 71, 72*t*
Pol-RFamide precursor, **167,** 74–75
salmon gonadotropin-releasing hormone gene, **147,** 178–179
teleost gonadotropin-releasing hormone genes, **147,** 174
of TGF-β isoforms, **167,** 209
transcriptionally active, **162A,** 221–226
Nucleus
acceptor sites for steroid receptors, **162B,** 345–348
architecture
dynamic properties, **162A,** 10–11
link with gene expression, **162A,** 68
role in transcriptional control, **162A,** 251–270
changes in function, **162B,** 161–162
cholesteric supramolecular order, **166,** 81–83
condensed nucleus, **138,** 73–76
distribution of replication domains, **162A,** 165–166
division, in eupyrene meiosis of Lepidoptera, **152,** 21
endosymbiont hypothesis and, **141,** 233–235, 238, 250
future prospects, **141,** 332
gene structure, **141,** 310, 312–314
mitochondrial 301, **141,** 320, 324, 326–328
organelles, **141,** 329
plastids, **141,** 269, 276–279
environmental stimuli in plant cells and, **145,** 252, 300–301
light, **145,** 287–288
wounding, **145,** 290–296
in eukaryotic cells, **175,** 119–121
FAK potential involvement in communication between integrins and, **167,** 174–175
functional domains, **159,** 162–163; **162A,** 69–81,151–175
high salt-extracted, **162B,** 422–424
inner layer
avian retina and, **146,** 42
cellular processes, **146,** 28, 37
normal development, **146,** 2–5
retinogenesis *in vitro,* **146,** 15–16, 19
growth factor-induced retinal regeneration and, **146,** 57
retinal precursor cells and, **146,** 148–149, 151
insulin
role in biological effects, **173,** 266–267
translocation, **173,** 258–260
intact, fibrogranular RNP network, **162A,** 81–82

Nucleus (*continued*)
interphase, *see* Interphase nucleus
MAR/ORI sequences, estimates, **162A,** 358–359
migration in, role in cell division and development, **173,** 219–220
molecular evidence for genetic material transfer, **165,** 209–213
nuclear bodies, **159,** 164–165
nucleic acid
escape, **165,** 207–228
escape consequences, **165,** 226–227
migration, **165,** 207–228
migration consequences, **165,** 226–227
organizing entity, **162A,** 456–458
pollen grains and tubes, movements, **140,** 84–86
polytene, *see* Polytene nucleus
postmitotic reassembly, NuMA role, **162B,** 15–17
relationship to microtubules
in animal and plant cells, **175,** 94–97
in lower eukaryotes, **175,** 93–94
remodeling in response to steroids
cancer, **159,** 180–183
insects, **159,** 178–180
morphological changes, **159,** 183–184
reproductive tissues, **159,** 178–179
rooster liver, **159,** 174–178
structure, *see also* Subnuclear structures
alterations in malignancy, **162A,** 106–108
compartmentalization, **159,** 163–164
and histone gene expression, **162A,** 263–264
isolated, association of SV40 DNA and T-antigen, **162A,** 517–518
and osteocalcin gene expression, **162A,** 265–269
required for DNA replication, **162A,** 460–462
RNA role, **162A,** 97–99
SV40 DNA, *in situ* subfractionation, **162A,** 512–516
transcript distribution, **162A,** 157
vegetative cells, **140,** 191–193
Nucleus tractus solitarii, avian salt glands and, **144,** 188–189, 195
Nuf2 protein, spindle pole body functions, **169,** 68
Nullisomic gametes, monosomy arising from, **167,** 264–265
NuMA, *see* Nuclear mitotic apparatus protein
Numerical density
parameter specification, **161,** 121–122
renal glomeruli, **161,** 121–122
optical disector, **161,** 125
physical disector, **161,** 122–125
NUP1p, characterization, **162B,** 279–280
NUP2p, highly charged, **162B,** 281–282
NUP49p, NPC essential component, **162B,** 283
NUP57p, GLFG repeats, **162B,** 283
NUP100p, overlap with NUP116p, **162B,** 284
NUP107, containing leucine zipper, **162B,** 273
NUP116p, deletion effects, **162B,** 284–285
NUP133p, component of NPC, **162B,** 289–290
NUP145p, acidic residues, **162B,** 285–286
NUP153
characterization, **162B,** 270–271
3D localization, **162B,** 238–241
NUP155, NPC protein, **162B,** 273
NUP180, Tpr/p265 corresponding to, **162B,** 273–274
Nutrition
associated factors, in chromoplast development, **163,** 200–201
urease in plant cells and, **145,** 66, 75, 83
Nystatin
osmotic conductance (P_{os}), **163,** 3–5
osmotic flow, bimodal analysis, **163,** 30–34

O

Oats, urease in plant cells and, **145,** 67
Oat seedlings, microgravity experiments, **156,** 321, 324–326, 328, 331–332
Oceanodroma leucorrhoa, avian salt glands and, **144,** 133
Ocelli, of hydro-, cubo-, and scyphomedusae, description, **167,** 39–40
Ochromonas, phototactic signal generation, **148,** 255
Octocorallia, *see* Sea pansies

Octopamine, control of rhythmic behaviors and, **147,** 11–13
D-Octopamine
functions in insect CNS, **164,** 221–223
localization in insect CNS, **164,** 216–218
uptake systems, **164,** 216
D-Octopamine receptor
cloning, **164,** 236–239
subtypes, pharmacological profiles, **164,** 223
Octopus, cytoskeletal dynamics and, **144,** 107
Ocular cuticle, genetic control, **174,** 36–37
Odontoblasts, regenerative cementogenesis and, **142,** 11, 23
Odontobutis, fish chromatophores and, **143,** 218–219, 235
Oenothera
higher plant mitochondrial genomes, **141,** 161–162
gene location, **141,** 157–158
recombination, **141,** 149–151
structure, **141,** 136, 141–142, 145, 148
pollen, **140,** 6, 8
self-incompatibility, **143,** 285
Oenthera organensis, pollen-specific DNA from, **140,** 28
Okadaic acid
angiogenic activity, **159,** 137–138
mitosis and, **144,** 235, 245, 249, 251
Olfactory system
amphibian development
chemosensory epithelia, **169,** 107
olfaction requirements, **169,** 106–107
olfactory bulb, **169,** 107–108
olfactory projections, **169,** 108–109
vomeronasal projections, **169,** 108–110
signaling in insect, **164,** 231
Oligoasthenoteatozoospermia, nondisjunction and, **167,** 286
Oligodendrocytes
glutamate metabolism, **147,** 303
origin in neural epithelium, **175,** 285–291
precursors, platelet-derived growth factors in, **172,** 113
Oligogalacturonide, phytoalexin induction by, **148,** 12
Oligonucleotides
antisense, therapeutic use in vascular disease, **161,** 31
Arabidopsisthaliana, **144,** 64
flowering plants and, **143,** 313
intracellular calcium messenger system and, **142,** 333
kinetoplastid mtDNA and, **141,** 81
ligation assay, in preimplantation diagnosis of genetic defects, **151,** 18
MDR reversal, **156,** 91
mitochondrial genomes in fungi and, **141,** 111
transcriptional regulation and, **143,** 5
Oligopeptide inhibitors, **167,** 226–227
Oligosaccharides
effect on pollen tube growth, **176,** 180–181
extracellular matrix and, **146,** 269, 277, 285, 293, 311
Oligotene fibril, polytene chromosome, **165,** 93–94, 95–96
Oligouronides, as endogenous elicitors of plant defense, **148,** 13
Olisthodiscus luteus, chloroplasts and, **144,** 281
Ommatidium
Drosophilia, **139,** 143–146
organization in honeybee compound eye, **176,** 309–311
Oncogenes, *see also* Protooncogenes
extracellular matrix and, **146,** 301, 313
intracellular calcium messenger system and, **142,** 311
melanocytes and, **146,** 191
melanoma, **146,** 228, 234, 240–247, 249–250
pigment-cell genetics, **146,** 214, 218
regulation of cell cycle, **157,** 249–250, 253–254
role in RPE phenotype change, **171,** 255
transcription, electromagnetic field effects, **158,** 310–312
transcriptional regulation and, **143,** 40, 42, 47
Oncopeltus fasciatus
antennae, **174,** 6–7
complex eye, **174,** 5–6
epidermis, **174,** 8
labrum, **174,** 7
mouthparts, **174,** 7–8
Onion
lamina ultrastructure, **162B,** 98–100
nuclear matrix extraction, **162B,** 80–86

Onoclea, intracellular calcium messenger system and, **142,** 308
Ontogenesis, open system during, and hormonal imprinting, **155,** 36
Ontogeny, transdifferentiation in medusae and, **142,** 213, 230, 256–257
Oocytes, *see also* Egg
aneuploidy and
in hypotheses for maternal age effect, **167,** 277–279
incidence of, **167,** 269
autosomal trisomy originating in, **167,** 272
echinoderm
actin cross-linking proteins, **175,** 41–44
form and function, **175,** 40–41
growth, **166,** 104–107, 107–109
karyosphere and, **144,** 1, 43–44, 46–47
Diptera, **144,** 8–11
Neuroptera, **144,** 11–18
tenebrionid beetles, **144,** 19–25
vertebrates, **144,** 26–38
maturation, **166,** 106–107
maturation, an preimplantation failure, **164,** 140–142
mitosis and, **144,** 247
nuclear envelope lattice, **162B,** 149
ovarian hyperstimulation effects, **164,** 142
produced by Cnidarian medusa, **167,** 38
vitellogenin
genes, **166,** 108–109
precursor-product relationships, **166,** 107
receptor binding, **166,** 108–109
serum, **166,** 107
structure, **166,** 108–109
yolk, **166,** 107
vitellogenin receptor, future trends, **166,** 131–132
in vitro-fertilized, chromosomal abnormalities, **164,** 140–142
Oogenesis
Drosophila, mutations affecting, **175,** 46–49
karyosphere and, **144,** 1, 7, 41
anurans, **144,** 27–36
birds, **144,** 37–38
insects, **144,** 13, 17
mammals, **144,** 38–39
reptiles, **144,** 36–37
vertebrates, **144,** 25–27
nutrimental, karyosphere and, **144,** 13
sequence of events, **166,** 104–105
Oomycetes, mitochondrial genomes in fungi and, **141,** 89, 91–93, 103, 105
Ooplasm, segregation, in *Halocynthia roretzi* eggs, **176,** 253–255
Open reading frames
animal mtDNA and, **141,** 174, 177, 180–181
ciliates and, **141,** 12, 36–38
Drosophila muscles and, **143,** 128
endosymbiont hypothesis and, **141,** 263, 266, 292–293, 301, 306
higher plant mitochondrial genomes and, **141,** 132, 137, 148–149, 158
higher plant plastids and, **145,** 158
kinetoplastid mtDNA and, **141,** 72, 75–78, 83
mitochondrial genomes in fungi and, **141,** 120
DNA diversity, **141,** 96, 98, 100–101
generation of mtDNA diversity, **141,** 107–109, 117
myofibrillogenesis and, **143,** 166
urease in plant cells and, **145,** 91
Open sky technique, for transplantation of RPE cells, **138,** 9
Opioids, astrocyte reaction to, **148,** 140–141
Opisthobranch, biological clocks in retina
circadian rhythmicity, **146,** 115, 139
entrainment pathway, **146,** 100, 113
eye contents, **146,** 85
functional organization, **146,** 88–93
ocular rhythm, **146,** 86–88
ocular timing system, **146,** 128
pacemaker localization, **146,** 93–95
Opsin
biological clocks in retina and, **146,** 102
immunostained cells in rat retinal cell cultures, **138,** 37
insect
characterization, **164,** 232
cloning, **164,** 242–243
molecular evolution, **164,** 244
and Na^+/K^+-ATPase, **138,** 19–21
retinal precursor cells and, **146,** 148, 158, 173–177, 180, 183
Optical tweezers, bacterial flagellar motor, **147,** 117
Optic cup
biological clocks in retina and, **146,** 85
growth factor-induced retinal regeneration and, **146,** 50, 67–68
Optic fiber, avian retina, **146,** 2–3

Optic nerve
biological clocks in retina and, **146,** 118, 123, 133
entrainment pathway, **146,** 108–109, 114
opisthobranch retina, **146,** 88–89, 91–92
growth factor-induced retinal regeneration and, **146,** 53, 58, 62
platelet-derived growth factor activity in, **172,** 113
Optics, fish chromatophores and, **143,** 206–208, 215
Optic tectum, cell migration in, **175,** 280
Oral replacement, **164,** 6
ORCC, *see* Outwardly rectifying chloride channels
Orchiditis, experimental allergic, **149,** 124–125
Oreochromis niloticus, chromatophores, **143,** 222, 230–231, 243
Organelles
bidirectional transport regulation, **164,** 309–310
cytoplasmic, interactions with cytoplasmic intermediate filaments, **164,** 109–116
DNA escape detection, **165,** 213–218
endosymbiont hypothesis, **141,** 233, 235, 237, 329–332
environmental stimuli in plant cells and, **145,** 289, 295, 300
chloroplasts, **145,** 268, 270, 284–285
photodinesis, **145,** 252–253, 263
generative cells, **140,** 197–201
genetic information
loss, **165,** 209–211
sequence distribution in heterologous genomes, **165,** 211–213
transfer, **165,** 209–211
higher plant plastids and, **145,** 149–150, 204
metabolism, **145,** 189, 191, 196, 201, 203
starch synthesis, **145,** 177, 180
structure, **145,** 150–151, 160, 164–165
translocation, **145,** 167, 169, 171
kinesin–organelle interactions, **164,** 300
lipid-containing, intermediate filament associations, **164,** 115–116
membrane-bound, movement
dynein-like proteins, **176,** 159–160
kinesin-like proteins, **176,** 157–159
motor protein interrelations, **176,** 160–163
myosin heavy chain polypeptides, **176,** 154–157
microgravity environment, **156,** 345
molecular evidence for genetic material transfer, **165,** 209–213
nucleic acid escape
limits to transfer of genetic information between compartments, **165,** 225–226
mechanisms, **165,** 218–226
transfer of escaped organellar DNA to nucleus, **165,** 221–222
plant cell, effect of microgravity and clinostating, **171,** 46–53
pollen grains and tubes, movements, **140,** 84–86
specific stress signaling pathways, **174,** 161–163
sperm cells, **140,** 202
transport
chromatophores as model system, **164,** 309
kinesin-like proteins in, **164,** 302–304
urease in plant cells and, **145,** 90
vegetative cell walls, **140,** 193
Organelle transport
different motors, **173,** 225–226
model systems, **173,** 222–223
Organic acids, higher plant plastids and, **145,** 169–171
Organogenesis, role of connective tissue interactions, **173,** 132–133
Organoid culture assay, human carcinoma experiments, **155,** 248–252
Organs
generative, in plant cells, effect of microgravity, **171,** 41–42
invertebrate
cytokines, **170,** 102–104
proopiomelanocortin-derived peptides, **170,** 86–87
role of nitric oxide, **170,** 113–116
neurohemal, control of rhythmic behaviors, **147,** 9
potato, culture, effect of microgravity, **171,** 42
vertebrate
cytokines, **170,** 102
proopiomelanocortin gene, **170,** 82–85
role of nitric oxide, **170,** 110–113

Organum vasculosum laminae terminalis
avian salt glands and, **144,** 182, 197
innervation, **159,** 27
location, **159,** 27
neurohormone release, **159,** 27
Orgyia antiqua, male eupyrene meiosis, **152,** 20
Orgyia thyellina, male apyrene meiosis, **152,** 24–25
ORI, *see* Origins of replication
Origins of replication
–MAR sequences, **162A,** 312–314
active, nucleosome-free, **162A,** 347–349
class, haboring kinked DNA, **162A,** 340–341
composition and genomic position, **162A,** 356–358
containing A–T tracts, **162A,** 320–322
direct repeats in, **162A,** 343–344
DNase I-hypersensitive sites, **162A,** 346–347
DNA unwinding elements, **162A,** 322–323
enriched
in ATTA and ATTTA motifs, **162A,** 335–338
in inverted repeats, **162A,** 323–325
flanked by left-handed and triple helical structures, **162A,** 339–340
interaction with ssDNA-binding proteins, **162A,** 349–350
intrinsically curved DNA, **162A,** 338–339
narrow minor grooves, **162A,** 350–351
polymorphic and conserved sequences, **162A,** 351–352
prediction, rules diagnostic for, **162A,** 317–356
prognosis perspectives, **162A,** 355–356
replication initiator sites, **162A,** 333–335
with transcription factor binding sites, **162A,** 325–333
vulnerability to mutagens, **162A,** 352–354
and nuclear matrix, **162A,** 455–476
potential, cohabitation with MARs, **162A,** 344–345
SV40, **162A,** 498–501
Ornithine
flowering plants and, **143,** 321–322
urease in plant cells and, **145,** 68–69, 72
Ornithine aminotransferase, urease in plant cells and, **145,** 69
Ornithine decarboxylase, **138,** 72, 81; **158,** 314–315
Ornithine transcarbamylase, urease in plant cells and, **145,** 69
Orotate phosphoribosyl transferase, **156,** 118
Orthopedic implants, calcium phosphate coatings and, **172,** 165
Orthophosphate, eggshells and biomineralization and, **145,** 240
Orthovanadate
induction of phytoalexins by, **172,** 68–69
inhibition of ATPase activity, **172,** 67–69
inhibition of peroxide generation, **172,** 81
Oryctolagus cuniculus, animal mtDNA and, **141,** 203–204
Oryza saliva, plastids and, **145,** 153
Oryzias, fish chromatophores
cytoskeleton, **143,** 221
leucophores, **143,** 201, 203, 205
motility control, **143,** 226, 228, 231, 233, 235, 237
signal transduction, **143,** 241
Oryzias latipes, fish chromatophores
cytoskeleton, **143,** 222
leucophores, **143,** 205
melanophores, **143,** 195, 197
motility control, **143,** 228, 233, 242
Oscillations, in geometric clutch model, **173,** 48–50
Oscillators
biological clocks in retina, **146,** 83–84, 100, 123, 139
Bulla, **146,** 129, 133
concepts, **146,** 96, 98–100
pacemaker mechanisms, **146,** 115–116
conditional, control of rhythmic behaviors, **147,** 12
Osmolytes
membrane stabilization, **165,** 6–7
organic
induction, **161,** 238–239
taurine
efflux, comparison with Cl^-, **161,** 202
regulation, **161,** 198–199
stretch-activated channel, **161,** 199–201
synthesis
physiological role, **165,** 12–13

salt tolerance, tobacco mutants, **165,** 13
salt toxicity
Escherichia coli, **165,** 7–8
higher plants, **165,** 11–13
Saccharomyces cerevisiae, **165,** 8–11
transport, salt toxicity
Escherichia coli, **165,** 7–8
higher plants, **165,** 11–13
Saccharomyces cerevisiae, **165,** 8–11
Osmoprotectant, **165,** 8
Osmoregulation, avian salt glands and, **144,** 189, 197–198
Osmosensitivity, avian salt glands and, **144,** 183–184, 188
Osmosis
avian salt glands, **144,** 129, 151, 155
hormonal control, **144,** 192, 196
receptive systems, **144,** 177–178, 180, 184, 187
secretory tissue, **144,** 131, 139, 141, 144
bimodal theory
leaky pores, **163,** 21–26
semipermeable pores, **163,** 19–21
diffusion kinetics, **160,** 116–118
hydrodynamic theory
leaky pores, **163,** 8–10
semipermeable pores, **163,** 6–8
models in pores
bimodal theory, **160,** 121–124
impermeable solute, **160,** 118–119
permeable solute, **160,** 119–121
Poiseuille flow, **160,** 121–122
Vegard-Mauro model, **160,** 118–119
passive water transport, **160,** 100, 105–106, 128–129
single-file theory
in leaky pores, **163,** 16–18
in semipermeable pores, **163,** 11–16
symplast and, **142,** 263
auxin, **142,** 278–281
integration of activity, **142,** 287–290, 292, 297–299
Osmotic conductance
antibiotic pores, **163,** 3–5
bimodal theory, **163,** 24–26, 30–34
cellular aquapores, **163,** 5–6
Osmotic hypothesis, nematocyst discharge, **156,** 278, 281–282
Osmotic pressure, cytoskeletal dynamics and, **144,** 89
Osmotic shock, intracellular calcium messenger system and, **142,** 326
Osmotic stress, activation of sucrose phosphate synthase, **149,** 71–72
Osmotic toxicity, salt tolerance, **165,** 4–6
Os penis, perinatal exposure to antihormones, **139,** 23–25
Osteoblasts
differentiation, gene control and expression during, **162A,** 253–269
postproliferative, **162A,** 256–263
proliferating, histone gene expression, **162A,** 263–265
regenerative cementogenesis, **142,** 11, 25, 35, 40
Osteocalcin
marker of chondrocyte differentiation, **159,** 299
during osteoblast differentiation
gene control, **162A,** 253–263
gene expression, **162A,** 265–269
Osteoclasts
in bone graft interface, **172,** 168
transcriptional regulation, **143,** 43
Osteocytes, regenerative cementogenesis and, **142,** 23–25
Osteogenesis
in bone graft interface, **172,** 168
embryo, **159,** 256, 283–284
Osteonectin
expression in response to heat shock, **159,** 299
as lamina densa component, **173,** 109
Osteopontin, marker of chondrocyte differentiation, **159,** 298–299
Osteoporosis, crystal characteristics in, **172,** 152, 154
Otala, eggshells and biomineralization and, **145,** 221
Ouabain
acrosomal exocytosis, **149,** 28–29
avian salt glands
secretory mechanism, **144,** 169, 171, 173, 175–176
stimulus-secretion coupling, **144,** 201, 203–204
potassium-regulated sperm function, **149,** 18
sensitive Na^+/K^+-ATPase, stimulation by ET, **171,** 281
sodium regulation during sperm capacitation, **149,** 25–26

Outer dense fibers, structure and function, **173,** 10–13
Outer membrane, Gram-negative bacteria
host–microbe interactions, **161,** 284
proteins, **161,** 278, 284
structure, **161,** 276–278
Outer nuclear layer
avian retina and, **146,** 17
normal development, **146,** 2, 4
retinogenesis *in vitro,* **146,** 10, 12, 15–16, 19
growth factor-induced retinal regeneration and, **146,** 54
retinal precursor cells and, **146,** 148, 174
Outer plexiform layer
avian retina and, **146,** 2, 16–17, 19
retinal precursor cells and, **146,** 148–149
Outer segments
avian retina and, **146,** 17
growth factor-induced retinal regeneration and, **146,** 50, 52, 54
retinal pigmented epithelium and, **146,** 80
retinal precursor cells and, **146,** 180
Output pathway, biological clocks in retina and, **146,** 97–99, 123, 126–128
Outwardly rectifying chloride channels, CFTR modulation of, **172,** 214–215
Ovarian cancer, **156,** 130–131, 134
Ovaries
macrophages, comparisons with testicular macrophages, **149,** 116–117
parathyroid hormone-related protein, **166,** 250
polyovular follicles in, sex hormones, **139,** 8–10
self-incompatibility, **143,** 281–283
Oviduct
avian
estrogen RBF-1, **162B,** 354
progesterone RBF, **162B,** 348–353
ecto-ATPase levels, **158,** 167–168
Oviposition, eggshells and biomineralization pH and, **145,** 236, 240
OVLT, *see* Organum vasculosum laminae terminalis
Ovocalcin, eggshells and biomineralization and, **145,** 234
Ovotransferrin, role in chondrocyte differentiation, **159,** 314–315
Oxacillin, spaceflight experiments, **156,** 341
Oxaloacetate, higher plant plastids and, **145,** 170
Oxidation
Arabidopsis thaliana, **144,** 60
endosymbiont hypothesis and, **141,** 308–309, 330
environmental stimuli in plant cells and, **145,** 264, 276
higher plant plastids and, **145,** 193–194, 201–203
kinetoplastid mtDNA and, **141,** 70, 80
role in Tau protein aggregation, **171,** 194–195
Oxidative stress
in *Drosophila melanogaster,* **165,** 37
effects on conceptus development, **164,** 170–171
OXT, *see* Oxytocin
Oxygen
Arabidopsis thaliana and, **144,** 59
avian salt glands and, **144,** 151, 154, 161, 176, 201
bactericial products of respiratory burst, **161,** 306
ciliates and, **141,** 37
consumption rate, vertebrate cell culture technology and, **142,** 191
endosymbiont hypothesis and, **141,** 331
higher plant plastids and, **145,** 196–197
nuclear magnetic resonance and, **145,** 3, 14, 29, 32, 39, 41
symplast and, **142,** 264
transcriptional regulation, **143,** 17
vertebrate cell culture technology
bioreactors, **142,** 164, 171, 173, 175, 177–179, 184
optimization, **142,** 197–201
packed-bed reactors, **142,** 189, 191
traditional cultures, **142,** 149, 153
Oxygenation, vertebrate cell culture technology, **142,** 166, 184, 186, 189
Oxytocin
astrocyte reaction to, **148,** 138–139
cellular localization, **159,** 34
gene, **159,** 33
messenger RNA levels, **159,** 9
precursor for smaller hormones, **159,** 33–34
release, **159,** 36
Oxytricha, replication band
chromatin replication models, **153,** 165

cytochemical studies, **153,** 154
ultrastructure, **153,** 150, 152

P

P5CR, flowering plants and, **143,** 322
p34-cdc2–lamin B complex, **162B,** 381
p35, baculoviral, effects on cell death, **163,** 129
p49/p50B, transcriptional regulation, **143,** 29–34, 39
p50, transcriptional regulation
comparison to H2-TF1, **143,** 46
molecular cloning, **143,** 25, 27, 29–34, 36–37, 39
in NF-κB, **143,** 23, 42
structure, **143,** 21–22
p58, association with lamin B, **162B,** 383
p62
characterization, **162B,** 268–269
nonglycosylated, **162B,** 292–293
role in lymphocyte signal transduction, **157,** 223
3D localization, **162B,** 238–241
p65, transcriptional regulation
binding sites, **143,** 19
interaction with DNA, **143,** 23
molecular cloning, **143,** 25, 31, 33–34, 39
structure, **143,** 21–22
P110
complex with mRNA, **162B,** 206
phosphorylation, **162B,** 202–204
recognized by Rev, **162B,** 212
$p110^{RB}$, capable of binding lamins A and C, **162A,** 109–110
P120 protein, $pp125^{FAK}$ as, **167,** 164–165
p190, role in lymphocyte signal transduction, **157,** 223
PABA, flowering plants and, **143,** 311
Pacemakers, mechanisms, **146,** 115
Paired helical filaments–pathological Tau protein
associated disorders
Down's syndrome, **171,** 195
Parkinsonism with dementia, **171,** 195–196
phosphorylation sites, **171,** 189
Palate
cell death during development, **163,** 146–148
human development, **158,** 220, 253
Palinurus vulgaris, control of rhythmic behaviors and, **147,** 13, 15
Palisades, eggshells and biomineralization and, **145,** 219, 229–234, 238–240
L-PAM, **156,** 130
Pancreas
ecto-ATPase levels, **158,** 171–172
fish
Brockmann body, **168,** 189, 191
suitability for hormone studies, **168,** 189
insufficiency, in cystic fibrosis, **172,** 211
Pandorina, **139,** 197–198, *see also* Algae, volvocine
Panicum miliaceum, plastids and, **145,** 169
Panulirus interruptus, control of rhythmic behaviors, **147,** 15, 19
Papaver, self-incompatibility and, **143,** 282–284, 288
Papaver dubium, **140,** 195
Papilla, self-incompatibility, **143,** 260–262, 270, 278
Papillomavirus, tumorigenesis in transgenic mice bearing gene, **164,** 62–63
Parabolic flight, **156,** 349–352
Paracanthurus hepatus, chromatophores and, **143,** 217
Paracellular transport, antigens and tracers in gut epithelium, **167,** 117–118
Paracentrotus lividus
animal mtDNA, **141,** 180, 190, 196
nucleotide bias, **141,** 204–206
protein genes, **141,** 182–183
avian retina, **146,** 9
Paracheirodon innesi, chromatophores, **143,** 217
Paracoccus, endosymbiont hypothesis and, **141,** 310
Paracoccus denitrificans
endosymbiont hypothesis and, **141,** 308–309
mitochondrial genomes in fungi and, **141,** 120
Paracrine factors
control of rhythmic behaviors and, **147,** 2, 9
Sertoli-germ cell communication network and, **147,** 40, 77
Paracrine glands, growth factor-induced retinal regeneration, **146,** 56

Paracrine hormones, as neurotransmitters in cnidarians
evidence of presence, **167,** 42–43
unknown nature of, **167,** 43–44
Paracrine signaling, in platelet-derived growth factors, **172,** 104–105
Paraflagellar swelling, role in eyespot organization, **148,** 246
Paramagnetic substances, and NMR, **145,** 30–31
Paramecium
dynein, outer row composition, **155,** 147
mitochondrial genomes, **141,** 1–4, 58–60
DNA sequences, **141,** 35, 37, 41, 43
genetic code, **141,** 53, 55–56
rRNA genes, **141,** 46, 48
structure, **141,** 6–10
Paramecium aurelia
endosymbiont hypothesis and, **141,** 295, 314
genetic code, **141,** 52–54
mitochondrial genomes, **141,** 1, 3
mitochondrial genomes, DNA sequences, **141,** 12
comparisons, **141,** 41–42
divergence, **141,** 38–41
organization, **141,** 12–37
phylogenetic tree, **141,** 42–45
rRNA genes, **141,** 48–52
structure, **141,** 4–7, 10–11
Paramecium primaurelia, mitochondrial genomes, **141,** 6, 36, 43, 59
Paramecium tetraurelia
microgravity experiments, **156,** 342, 343, 356
mitochondrial genomes, **141,** 6, 59–60
DNA sequences, **141,** 13, 36, 42–43
rRNA genes, **141,** 46–47
Paramyosin
Drosophila muscles, **143,** 97, 108–109, 115, 139
myofibrillogenesis, **143,** 180
Paramyxoviruses, Sias as receptor determinant, **175,** 170–171
Paranemin, in striated muscle, **142,** 98–99
Parasites, *Arabidopsis thaliana,* **144,** 54–55, 71
Parathyroid gland
evolution, **166,** 236
functioning, **166,** 22–23, 28–29
parathyroid hormone-related protein, neoplasm, **166,** 260–261
Parathyroid hormone
astrocyte reaction to, **148,** 146–147
biochemical features, **166,** 231–232
chicken, gene structure, **166,** 235
effect on chondrocyte differentiation, **159,** 271, 318–319
human, gene structure, **166,** 235
-parathyroid hormone related protein receptor, **166,** 266–267
Parathyroid hormone-related protein
amphibia, **166,** 263–264
assay method, **166,** 236–242
bird, **166,** 262–263
brain, **166,** 251–252
breast, neoplasm, **166,** 259
cardiovascular system, **166,** 242–244
central nervous system, **166,** 251–252
chemical structure, **166,** 231, 232
chemistry, **166,** 232–234
chicken, gene structure, **166,** 234–236
connective tissue, **166,** 246–247
detection method, **166,** 236–242
distribution, **166,** 242
endocrine tumor, **166,** 261
fetus, **166,** 253–255
fish, **166,** 264–266
functions, **166,** 242
gene expression, **166,** 241–242
gene structure, **166,** 234–236
human, gene structure, **166,** 234–236
immune system, **166,** 252–253
immunohistochemistry, **166,** 236–238
integument, **166,** 249–250
kidney, neoplasm, **166,** 256
lactation, **166,** 251
lung, neoplasm, **166,** 257
lymphoid system, neoplasm, **166,** 261–262
mammary gland, **166,** 244–246
mouse, gene structure, **166,** 234–236
neoplastic cell, **166,** 255–262
nonmammalian vertebrate, **166,** 262–266
ovary, **166,** 250–251
parathyroid gland, neoplasm, **166,** 260–261
pregnancy, **166,** 250, 253–255
radioimmunoassay, **166,** 238–241
rat, gene structure, **166,** 234–236
reproductive system, **166,** 250–251
gynecological tumors, **166,** 259–260
skin, **166,** 249–25O
neoplasm, **166,** 256–257

species comparison, **166,** 234–236
testis, **166,** 250
urogenital system, **166,** 247–249
vascular wall, **166,** 243, 244
vasoactive factor, **166,** 243, 244
vertebrate cell, **166,** 231–268
Paraventricular nucleus, avian salt glands, **144,** 182, 188–189, 195
Paraxonemal body
Euglena, **148,** 284–286
isolation, **148,** 271
photoreceptors in, **148,** 257
signal generation from, **148,** 269
ultrastructural organization, **148,** 246–247
Parenchymal cells
avian salt glands and, **144,** 143
blood supply, **144,** 148–149
hormonal control, **144,** 193, 196
innervation, **144,** 158–159, 162–163, 168
receptive systems, **144,** 180, 189
effect of microgravity and clinorotation, **170,** 48–50
symplast and, **142,** 268, 270, 279, 291, 296, 300
Parietal epithelial cells
average volumes, **161,** 151–154
quantity estimation in average glomeruli, **161,** 148–151
surface areas, **161,** 151–154
Parkinson's disease, with dementia, PHF-Tau protein, **171,** 195–196
Parotid gland, ecto-ATPase levels, **158,** 172
Pars intermedia cells
differentiation, **159,** 52
hormone synthesis, **159,** 52
periodic acid–Schiff-positive cells
acid–base regulation, **169,** 14
calcium regulation, **169,** 10, 12
somatolactin synthesis, **169,** 1, 3, 5, 8–10
regulation of activity
bioamines, **159,** 53
peptides, **159,** 53
steroid hormones, **159,** 54–55
Particles
intracellular movement, role of nonmuscle tropomyosin, **170,** 18–19
intramembrane
in eyespot apparatus of flagellate algae, **148,** 239, 246
lens
cortical and nuclear fiber cells, **163,** 338–340
epithelial cell junctions, **163,** 337
microparticles, polyacrylamide, as antigen-delivering carriers, **167,** 141
prion, associated disease, role of chaperones, **174,** 171–173
Particulate tracers, Peyer's patch dome epithelium uptake of, **167,** 177–122
Parvalbumin
effect of electrostimulation, **170,** 168–169
muscle fiber type-specific distribution, **170,** 157–158
proteins in striated muscle and, **142,** 127–128
Patau syndrome
associated craniofacial abnormalities, **158,** 216, 223
incidence in live births of, **167,** 265
Pathogenesis
extracellular matrix, **146,** 261, 267, 280, 287
modification, **146,** 287, 299–300, 305–313
stress, **146,** 303, 305
urease in plant cells, **145,** 83–84, 98
Pathogenesis-related protein
Arabidopsis thaliana, **144,** 59, 74, 78
as elicitor-releasing factors, **148,** 10, 12
gene expression, **148,** 26
promoters, elicitor-responsive regions in, **148,** 28
Pathogens
Arabidopsis thaliana, see Arabidopsis thaliana
intracellular calcium messenger system and, **142,** 307, 338
Peyer's patch M cells
as primary entry site for invasion by, **167,** 138–139
uptake, **167,** 117–122
plant cell recognition, **148,** 1–2
Pathology
effect of brain transplants
avian reflex epilepsy, **175,** 291–293
Fepi/JA chimera construction, **175,** 294
NMR analysis, *see* Nuclear magnetic resonance
retinal pigmented epithelium, **146,** 75

Paxillin
complexes formed by FAK with, **167,** 166–167
phosphorylation by focal adhesion kinase/Src complex, **168,** 91–93
as potential FAK substrate, **167,** 172, 173
tyrosine phosphorylation in FAK-deficient cells, **167,** 177–178
vinculin *in vitro* interaction with, **167,** 162, 163
PCM, *see* Pericentriolar material
PCNA
imported to nucleus, **162B,** 164–165
perturbed localization, **162B,** 171
at replication foci, **162B,** 316–317
PCR, *see* Polymerase chain reaction
PDGF, *see* Platelet-derived growth factor
Pea
cotyledons, clathrin-coated vesicles
assembly and disassembly of, **167,** 14–18
composition of, **167,** 12–13
protein transport and, **167,** 7–8
uncoating of, **167,** 22
V-ATPase activity, **167,** 24–25
effect of microgravity
on biochemical content and enzyme activity, **171,** 38–39
on microviscosity and lipid peroxidation, **171,** 41
on root statocytes, **171,** 20–21
environmental stimuli in, **145,** 293–294
flowering plants, **143,** 305–306, 308
intracellular calcium messenger system, **142,** 309–310, 315, 329–330, 333
nodule development, **156,** 3, 21, 26–28, 31, 33–35, 40, 46
non-nodulating mutants, **156,** 40–41
plastids
amino acid synthesis, **145,** 191, 193–194, 199–200
carbohydrate oxidation, **145,** 201
fatty acid synthesis, **145,** 184, 187–188, 190
starch synthesis, **145,** 177, 179–181
translocation, **145,** 168–171
seedlings, microgravity experiments, **156,** 323, 325, 326, 328
supernodulating mutants, **156,** 42
symplast and, **142,** 281, 284, 286, 294
urease in plant cells and, **145,** 67–69, 74, 82, 90, 99
Peaks, intensity and shape in NMR, **145,** 11–12
Peanut, urease in plant cells, **145,** 67
Peanut agglutinin, **156,** 168, 172, 175
Pectin
biosynthesis and transport, **176,** 170–171
detection in various systems, **171,** 91
epitopes
distribution in plant cell wall, **171,** 87, 89
modulation, **171,** 88
esterified and acidic, distribution, **176,** 164–165
extracellular matrix
characterization, **146,** 263–264, 266, 268, 270
immunology, **146,** 285
modification, **146,** 294–295, 299, 303, 307
synthesis, **146,** 274, 283
role in pollen tube growth, **171,** 90–91
role in primary cell walls, **171,** 87–88
self-incompatibility, **143,** 260
structure, **171,** 85–86
urease in plant cells, **145,** 94
Pectinase, *see* Polygalacturonase
Pectinesterase
moss, effect of clinostating, **171,** 31
role in cell wall changes under microgravity, **170,** 63
Pectocellulosic layer, outer
cellulose microfibrils, **176,** 165–167
esterified and acidic pectins, **176,** 164–165
Pelagia noctiluca, **156,** 285
Pelamis, avian salt glands and, **144,** 139
P elements, *Drosophila* muscles, **143,** 64–65, 103, 117, 139
Pelomyxa
chloroplasts, **144,** 270
locomotion, **150,** 54–55
Pelvis, mouse
neonatal treatment with sex hormones and, **139,** 33–36
sexual dimorphism and, **139,** 31–33
Penicillin-binding protein, in *Cyanophora paradoxa* cyanelles, **151,** 34
Penicillum griseofulvam, mitosis and, **144,** 235
Penis
glans, **139,** 25–26

os, **139,** 23–25
perinatal exposure to antihormones and, **139,** 23–26
Pennaria, instability of homogeneous state, reduction fields, **154,** 349
Pentasaccharides, **156,** 21–22
Pentose phosphate pathways, oxidative, **145,** 150, 189, 193, 201, 203
Pentylenetetrazol, biological clocks in retina, **146,** 105
PEPC, *see* Phosphoenolpyruvate carboxylase
Pepper
chromoplast differentiation
carbohydrate metabolism, **163,** 209–211
carotenogenic gene expression, **163,** 214–216
genetic factors, **163,** 193
light regulation, **163,** 199
transcription activity, **163,** 195–197
plastid envelope membrane budding, **163,** 206
Pepsin, eggshells and biomineralization, **145,** 237
Peptide chains, clathrin-coated vesicles
description, **167,** 10–13
uncoating, **167,** 22–23
Peptide histidine isoleucine, control of amphibian metamorphosis, **145,** 137
Peptide histidine methionine, **138,** 170
Peptide receptor
cloning of cnidarian, **167,** 55–56
insect, cloning
diuretic hormone receptors, **164,** 241–242
orphan receptors, **164,** 239–241
Peptides
astrocyte reaction to
atrial natriuretic, **148,** 128–128, 132–134
brain natriuretic, **148,** 128–128, 132
calcitonin-gene related, **148,** 146–147
C-type natriuretic, **148,** 128, 132–133
calcitonin-gene-related, phenotypic expression affected by, **153,** 275
cnidarian
potential endocrinological role of, **167,** 82
protection against degradation of biologically active, **167,** 49
intestinal, in mucosal immune system, **167,** 97
intracellular calcium messenger system and, **142,** 309
intramedullary regulatory, effect of ET, **171,** 283
natriuretic, effect on endothelin release, **171,** 293
opiate, effects, **138,** 148–149
POMC-derived
in central nervous system, **170,** 83
in immune system, **170,** 85–87
proteins in striated muscle and, **142,** 83–84, 89, 121, 125
short, plant analog, **174,** 271
sperm-activating, **163,** 251
thymic, neurotransmission, **149,** 269
Peptide-tyrosine-tyrosine, **138,** 127
Peptidoglycans
in bacterial cell walls, **161,** 267–269
ring, bacterial flagellar motor, **147,** 100–102, 105
Peptidyl-glycine, in neuropeptide biosynthesis, **167,** 58
Percolation theory, gel transition of protein fiber networks, **150,** 140–141, 145–146
Percoll gradient, **166,** 149
Perforin, expression in granulated metrial gland cells, **153,** 123–124
Perfusion
reactors based on, **142,** 167–169, 175–176
symplast and, **142,** 275–276, 280
vertebrate cell culture technology and, **142,** 188, 190
Periarticular calcification, crystal structure in, **172,** 147
Pericentrin, with MTOC properties, **175,** 114
Pericentriolar material
lepidopteran
male eupyrene meiosis, **152,** 19, 22
structural features, **152,** 1
microtubule nucleation, **169,** 26–28, 31–34
structure, **169,** 32–33
Perichromosome region, mitotic cells, **162B,** 31–36
Peridinin, chloroplasts and, **144,** 286, 288–290
Peridinium balticum
chloroplasts and, **144,** 288–290, 292
photoreceptive apparatuses, **148,** 249–250

Peridinium gatunense, photosynthetic accessory pigments, **148,** 275
Perikarya, avian retina and, **146,** 2–3
Perilymph, blood barrier and, glucose transport, **172,** 23–24
Perimeningeal fluid, ependymin, **165,** 146
Perinatal period, hormonal imprinting during, **155,** 24–29
Perineurium
GLUT1 in, **172,** 24–25
as permeability layer, **172,** 24
Periodontal ligament, regenerative cementogenesis and
established root surfaces *in vitro,* **142,** 28–29, 31, 36–37, 39–42
established root surfaces *in vivo* 44-46, **142,** 49–50
growing root surfaces, **142,** 11, 15, 17, 22
Periodontal regeneration, **142,** 1–2, 42–43
Periodontitis, regenerative cementogenesis and, **142,** 37, 45, 47
Peripheral cells, microgravity experiments, **156,** 330
Peripheral nervous system
blood barrier
glucose transport, **172,** 24–25
structure, **172,** 24
Drosophilia, **139,** 134–140
ganglia, environmental effects in patterning, **138,** 304–310
vertebrate neuronal cell models, **153,** 236–237
Peripherin
retinal precursor cells and, **146,** 147
in vitro binding to nuclear components, **164,** 99–100
Periventricular zone, ependymin, granular cells, **165,** 140–142
Perlecan, as lamina densa component, **173,** 108–109
Peronospora parasitica, plant-pathogen interactions, **144,** 65, 76–78
Peroxidase
cell wall bound
effects of suppressors on, **172,** 80–81
in lignin polymerization, **172,** 271–273
in lignin biosynthesis, **176,** 114–117
role in cell wall changes under microgravity, **170,** 61–62
in wheat seedling, effect of microgravity, **171,** 39
Peroxidation, lipid, effect of microgravity, **171,** 40–41
Peroxide, generation
cell wall-bound ATPase and, **172,** 81
effects of suppressors on, **172,** 80–81
in fungal signal transduction cascade model, **172,** 82–84
Peroxisome-proliferator-activated receptor
ligand recognition, **160,** 331
structure, **160,** 331
tissue distribution, **160,** 332
Peroxisomes
endosymbiont hypothesis and, **141,** 258, 329–330
hepatocyte
carnitine acetyltransferase, **160,** 305, 319–321
catalase, **160,** 314, 317
conferences, **160,** 304–305
enzymes, **160,** 312–313
fatty acid oxidation, **160,** 318, 325
functional definition, **160,** 304
magnesium-ATPase, **160,** 305, 307, 313, 321–324
pH gradient, **160,** 324
plasmalogen synthesis, **160,** 318
proliferators, **160,** 303, 315–316, 325, 327
age-dependent sensitivity, **160,** 333–334
hepatocarcinogenIcity, **160,** 328
rat strain sensitivity, **160,** 334–335
recovery, **160,** 340
sex differences on response, **160,** 329–333
protein import, **160,** 332
round-to-tubular-type transformation, **160,** 329, 340
serine:pyruvate aminotransferase, **160,** 318–319
structure
core, **160,** 314–315
diameter in rat, **160,** 306
fibril bundles, **160,** 316
limiting membrane, **160,** 312–314
marginal plates, **160,** 315
matrix plates, **160,** 315
microscopy, **160,** 317
peroxisome matrix, **160,** 307–308, 311–312
tubular projections, **160,** 315–316

supressors
evidence for endogenous suppressor, **160,** 325, 341
nicardipine, **160,** 328
phenobarbital, **160,** 326–328
trifluoperazine, **160,** 328
targeting signals, **160,** 335–337
thyroid hormone response, **160,** 327
higher plant plastids and, **145,** 196, 198
Persistence, cell, computation, **163,** 63
Perturbation analysis
biological clocks in retina and, **146,** 96–98, 115
retinal precursor cells and, **146,** 150
Pertussis toxin, fish chromatophores and, **143,** 227, 240
Petromyzon marinus, **138,** 112
Petunia
chalcone synthase and
control of activity, **147,** 254–259
control of gene, **147,** 266–267, 276
functions of products, **147,** 251–252
endosymbiont hypothesis and, **141,** 304
extracellular matrix and, **146,** 270
flowering plants and, **143,** 312
higher plant mitochondrial genomes and, **141,** 157, 159–160, 162–163
abnormal phenotypes, **141,** 130–133
recombination, **141,** 151–154, 156
structure, **141,** 136, 139, 141–143, 146–148
self-incompatibility, **143,** 280, 282, 284–286, 288–289
Petunia hybrida, **140,** 27, 43, 47; **141,** 144–145, 158
Peyer's patch, *see also* Dome epithelium
description, **167,** 94–96
intestinal, M cells in
characteristics
general, **167,** 111–112
variations among species and locations, **167,** 112–117
clinical aspects
intestinal immune response induction and, **167,** 142–143
as potential entry sites for oral vaccines, **167,** 139–142
as targets for enteropathogenic microorganisms, **167,** 138–139
functions
antigen interaction with surface of, **167,** 122–126
antigen and tracer transport, **167,** 117–122
cytoskeleton, **167,** 128–130
transcytosis of antigens, **167,** 126–128
identification, **167,** 100–107
alkaline phosphatase, **167,** 102
cross-correlation of labeling patterns, **167,** 104–107
cytoskeleton, **167,** 104
lectins, **167,** 103–104
monoclonal antibodies, **167,** 102
ultrastructure, **167,** 100–102
locations outside of gut
BALT, **167,** 136–137
NALT/DALT/CALT, **167,** 137–138
tonsils, **167,** 135–136
lymphoid cells in gut wall
immunological functions, **167,** 92
localization, **167,** 92–96
overview, **167,** 91–92
in patchwork arrangement with enterocytes, **167,** 130–131
macrophage function *versus* different compartments of, **167,** 132
transport capacity *versus* ordinary gut epithelium, **167,** 142
PF4, therapeutic use in vascular disease, **161,** 32
PGDF, *see* Platelet-derived growth factors
PGLA microspheres, *see* Poly(DL-lactic coglycolic acid) microspheres
PGlu-His-Gly, inhibition of colonic epithelial proliferation by, **167,** 227
P-Glycoprotein
–ABC protein, forms, **171,** 157–158
aspects of cellular expression
alkalinization of cytoplasm, **171,** 149–153
ATP transport, **171,** 155–156
drug accumulation rate, **171,** 144–149
membrane potential, **171,** 139–141, 153–155
cell overexpressing, depolarization, **171,** 137–139
and CFTR, and SUR, homology, **171,** 125–127
in classic MDR phenomenon, **156,** 137
drug pump model
ATPase activity and stoichiometry, **171,** 132–134

P-Glycoprotein (*continued*)
kinetics, **171,** 134–135
photolabeling, **171,** 136
thermodynamics, **171,** 135
encoding gene, **156,** 78–79; **171,** 124–125
function
associated hypotheses, **171,** 124–125
direct and indirect models, **171,** 127–128
historical context, **171,** 130–132
multidrug resistance
function, **156,** 82–86
as marker, **156,** 87–88
regulation, **156,** 86–87
reversal, **156,** 88–92
multidrug resistance mediated by, **156,** 79–92
collateral sensitivity, **156,** 81
cross resistance, **156,** 81
drugs, **156,** 79–81
overproduction, **171,** 123–124
structure, **156,** 82
pH
cytoplasmic streaming requirements, **155,** 115
cytoskeletal dynamics and, **144,** 86, 105, 108
dependency of azuki bean EXGT, **173,** 176–177
effect on pollen tube tip growth, **176,** 179
eggshells and biomineralization and, **145,** 241
environmental stimuli in plant cells and, **145,** 290, 298, 300
intracellular calcium messenger system and, **142,** 313–317
nuclear magnetic resonance and, **145,** 19–20, 41, 45
ΔpH, effect on drug transport, **171,** 153–155
pH_i, effect on weak base accumulation, **171,** 149–153
role of V-ATPase, **174,** 89–90
symplast and, **142,** 268, 273, 275–276
transdifferentiation in medusae and, **142,** 237
urease in plant cells and, **145,** 66, 92–93
PHA, *see* Phytohemagglutinin
Phaeodactylum tricornatum, morphogenesis
colloidal silica formation, **150,** 321, 324
nonsilicified protoplasmic masses, **150,** 260
Phaeophyceae, eyespot reflection in, **148,** 259, 269
Phaeophytes, chloroplasts and, **144,** 277–278, 282
Phagocytes
chloroplasts and, **144,** 277, 290, 292
cytoskeletal dynamics and, **144,** 111
depletion, dendritic cell, **153,** 52
Phagocytosis
as autoimmune response, **149,** 125
avian retina and, **146,** 6
in bacteria–macrophage interactions, **161,** 302–304
bacteria resistant to, **161,** 304
chemotactic component, **161,** 304–306
coiling, **161,** 303
effect of cytokines, **170,** 106–107
growth factor-induced retinal regeneration and, **146,** 50
Leydig cells, **149,** 122
by macrophages
M cell particle uptake parallel to, **167,** 127
particulates and cell debris, **167,** 132
by M cells of antigens and tracers, **167,** 118–122
melanocytes and, **146,** 194
regenerative cementogenesis and, **142,** 37
retinal pigmented epithelium and, **146,** 75
role of POMC products, **170,** 88, 94–95
Sertoli cell
mechanism, **154,** 109, 116
processing of residual bodies, **154,** 111, 114, 116
Sertoli-germ cell communication network and, **147,** 71, 73–74
steroid effects, **149,** 120
testicular macrophages, **149,** 123
zipper interaction, **161,** 303
Phagokinetic track assay, as migration analysis method, **155,** 58
Phalloidin, **142,** 83, 111; **156,** 203, 209–210, 254; **166,** 16
Pharbitis, symplast and, **142,** 273
Pharmacodynamics, exogenously administered TGF-β, **167,** 212

Pharmacology
avian salt glands and, **144,** 161–164, 167–169, 205
calcium channels and, **147,** 193–194
fish chromatophores and, **143,** 226–228, 231, 236, 240, 247
nicotinic, control of rhythmic behaviors and, **147,** 17–18
Phase-based angiography, nuclear magnetic resonance and, **145,** 26
Phase-contrast microscopy, polytene chromosome, **165,** 66–68
Phase delay, biological clocks in retina
Bulla, **146,** 129, 134, 136–137
concepts, **146,** 100
entrainment pathway, **146,** 106, 111–112
pacemaker mechanisms, **146,** 118, 121–122
Phase microscopy, retinal precursor cells and, **146,** 153, 168
Phaseolus, symplast and, **142,** 266, 268
Phaseolus lunatus, cytokinin-induced habituation, **152,** 271
Phaseolus vulgaris
cytokinin-induced habituation, **152,** 271–272
higher plant mitochondrial genomes and, **141,** 133–134
plastids and, **145,** 175, 200–201
urease in plant cells and, **145,** 76
Phase–response curve, biological clocks in retina
Bulla, **146,** 129, 131, 134–137
concepts, **146,** 96–97
entrainment pathway, **146,** 102–103, 105–106, 110–111
pacemaker mechanisms, **146,** 116, 122
Phase shifting, biological clocks in retina
Bulla, **146,** 118, 128–134, 136–137, 139
entrainment pathway, **146,** 102–103, 105, 107–115
Phase transition, as model of actin filament bundle formation, **175,** 65–68
Phelum pretense, **140,** 12
Phe-Met-Arg-Phe-NH_2, *see* FMRFamide
Phenobarbital, effect on clofibrate-induced peroxisomes, **160,** 326–328
Phenocopying, in *Acetabularia,* environmental effects on, **150,** 237–238
Phenolamine receptor
cloning, **164,** 236
subtypes, pharmacological profiles, **164,** 223
Phenolamines, functions in insect CNS, **164,** 221–223
Phenolic acids, in plant cell wall, dynamic aspects
bonds with polysaccharides, **151,** 249–252
cytology, **151,** 247–249
hydroxycinnamoyl–polysaccharides
biological roles, **151,** 257–259
biosynthesis, **151,** 252–253
oxidative coupling, **151,** 253–255
photodimerization, **151,** 256–257
occurrence, **151,** 247–249
quinone methides, **151,** 255–256
Phenolic compounds, in plant cell wall, dynamic aspects
conjugates, **140,** 45–48
cutin, **151,** 259–260
glycolipids, bacterial, **161,** 288, 290
occurrence, **151,** 261–262
phenol, **151,** 229
soluble extraprotoplasmic phenolics, **151,** 260–261
sporopollenin, **151,** 261
structure, **151,** 229
suberin, **151,** 259–260
Phenolics, extracellular matrix and
characterization, **146,** 266, 271–272
modification, **146,** 305, 309–312
synthesis, **146,** 278–281, 283
Phenol oxidases, in lignin polymerization, **172,** 271–273
Phenotype
animal mtDNA and, **141,** 222, 227
Arabidopsis thaliana and, **144,** 54
bacteria, **144,** 66–69
biochemistry, **144,** 74
fungi, **144,** 64–66
genetic model, **144,** 61–63
models, **144,** 77–78
nematodes, **144,** 71–72
pathogens, **144,** 56–57, 59–60
viruses, **144,** 69–71
bacterial flagellar motor and, **147,** 109
cell, desmosomal cadherin, **165,** 179–180

Phenotype (*continued*)
chalcone synthase and, **147,** 254, 257, 259
cytoskeletal dynamics and, **144,** 120
dendritic cells, human, comparisons of subtypes, **153,** 60–61
diversity of peripheral glial cells, **138,** 294–295
Drosophila muscles and, **143,** 63–64
actins, **143,** 114
contractile protein genes, **143,** 99, 101, 103–104, 134, 136
development, **143,** 65, 78, 85, 87
interactions, **143,** 127, 131–132
tropomyosin, **143,** 117
Z-disc proteins, **143,** 123, 125
endosymbiont hypothesis and, **141,** 235, 243, 251
expression, and neuronal plasticity
adult, **153,** 275–279
developing, **153,** 271–275
mechanism, **153,** 270–271
neonatal, **153,** 271–275
extracellular matrix protein-regulated expression, **150,** 129
flowering plants and, **143,** 300, 304–305, 314–316, 321
glial, microenvironmental control, **138,** 301–302
granulated metrial gland cells, rodent, **153,** 113–119
growth factor-induced retinal regeneration and, **146,** 53
habituated, reversal in plant cells, **152,** 273–275
higher plant mitochondrial genomes and, **141,** 129–135, 164
melanocytes and, **146,** 195
genetics of melanoma, **146,** 245, 250
pigment-cell genetics, **146,** 206–209, 211, 215, 218, 222
mitochondrial genomes in fungi and, **141,** 109, 115
mitosis and, **144,** 228
Nar, **156,** 43–44
proteins in striated muscle and, **142,** 106
regenerative cementogenesis and, **142,** 35, 37–38, 40–41
retinal pigmented epithelium and, **146,** 76
retinal pigment epithelial
cytomechanics of expression regulation, **150,** 119–120
destabilized, **150,** 125
expression on basement membrane proteins, **150,** 126–127
retinal precursor cells and, **146,** 149–150, 153
experimental analysis, **146,** 155, 166
microenvironmental factors, **146,** 176, 179
photoreceptor default hypothesis, **146,** 169, 172
photoreceptor subtypes, **146,** 174–175
self-incompatibility and, **143,** 257–259, 279–280, 289
spasmodic and *oscillator* mouse, **176,** 231–232
transcriptional regulation and, **143,** 5
transdifferentiation in medusae and, **142,** 215–216
urease in plant cells and, **145,** 81, 85, 88, 94
vertebrate cell culture technology and, **142,** 192
Phenoxybenzamine, avian salt glands and, **144,** 168
Phenoxyradicals, in lignin polymerization, **172,** 271, 273
Phenylalanine, flowering plants and, **143,** 309, 311
Phenylalanine ammonia lyase
chalcone synthase and, **147,** 246, 261, 265
and CHS, activities after pollen-tapetum fractionation, **140,** 50
encoding gene in plants, **148,** 26–27
expression
correlation with lignification, **172,** 250–251
environmental stress, **172,** 250
pathogen stress, **172,** 250–251
extracellular matrix and
modification, **146,** 295, 300, 304, 309–310
synthesis, **146,** 278–281
flowering plants and, **143,** 312
genes, **172,** 250
inhibitors, effects *in planta,* **172,** 251–252
localization, **172,** 249
in phenylpropanoid pathway, **172,** 247; **176,** 102–105
polymorphism, **172,** 249–250
role in cell wall changes, **170,** 61–62
structure, **172,** 249

L-Phenylalanine mustard, **156,** 130
Phenylalkylamines, intracellular calcium messenger system and, **142,** 320
Phenylarsine oxide, **167,** 177
Phenylepinephrine, effect on thymus, **156,** 189
L-3-Phenyllactyl-Leu-Arg-Asn-NH_2, *see* Antho-RNamide neuropeptide
Phenylphosphorodiamidate, urease in plant cells, **145,** 75–77, 82, 91, 95–96
Phenylpropanoid pathway
  chalcone synthase and, **147,** 256, 261, 268
  in lignin biosynthesis, **172,** 246
    caffeoyl-CoA 3-*O*-methyltransferase, **172,** 258
    cinnamate 4-hydroxylase, **172,** 253–254
    cinnamyl alcohol dehydrogenase, **172,** 264–269
    CoA-esterification, **172,** 258
    4-coumarate-CoA ligase, **172,** 259–262
    coumarate 3-hydroxylase, **172,** 254–255
    ferulate 5-hydroxylase, **172,** 255
    O-methyltransferase, **172,** 255–258
    phenylalanine ammonia lyase, **172,** 247–252
    tyrosine ammonia lyase, **172,** 252
  manipulation
    chemical inhibitors and, **172,** 276–277
    compensation effects, **172,** 275
    enzyme specificity and, **172,** 275–276
    future directions, **172,** 278–279
    mutations and, **172,** 277–278
  metabolism, extracellular matrix and, **146,** 278–281, 301, 310
  products as lignin precursors, **176,** 102–105
  regulation, **172,** 274
Pheomelanins
  biology, **146,** 193
  pigment-cell genetics, **146,** 199, 208–211
Pheromones, sexual induction in *Volvox carteri* and, **139,** 219–220
PHF, *see* Paired helical filaments
Phlobophene, chalcone synthase and, **147,** 272, 274
Phoenix dactylifera generative cell in, **140,** 191, 192, 193, 194, 195, 196–197, 202, 203
Phorbol esters
  fish chromatophores and, **143,** 240
  immunologic studies in microgravity, **156,** 311–314, 333
  intracellular calcium messenger system and, **142,** 332
  transcriptional regulation and, **143,** 14, 16, 23–24, 28, 41
    function, **143,** 2
    inducible activation, **143,** 7, 9–10
  transdifferentiation in medusae and, **142,** 247
Phorbol myristate acetate, tissue factor induction by, **152,** 67–68
Phosphatase, role in Tau proteins
  phosphorylation regulations, **171,** 180–181
  regulations, **171,** 179–180
  Tau dephosphorylation, **171,** 179
Phosphatase inhibitors
  angiogenic effects, **159,** 137–138
  in studying FAK function in epithelial cells, **167,** 177
Phosphate
  eggshells and biomineralization and, **145,** 231, 240, 244
  incorporation in apatite, **172,** 158
  inorganic, *see* Inorganic phosphate
  nuclear magnetic resonance and, **145,** 36
Phosphate translocator, higher plant plastids, **145,** 168–170, 179–180, 189, 193
Phosphatidylinositides
  integrin signal transduction, **168,** 97–99
  metabolism in signal transduction, **168,** 126–127
  plants, **168,** 127, 168
Phosphatidylinositol
  cycle modulation, **138,** 252–254
  pathway, electromagnetic field effect, **158,** 282–283
  signaling in yeast cells, **139,** 323–325
  signal transduction in plants, **158,** 16, 43–44, 53
  turnover, intracellular calcium messenger system and, **142,** 321–328, 338
Phosphatidylinositol bisphosphate, in plant defense signaling, **172,** 71, 73
Phosphatidylinositol 4,5-bisphosphate
  cleavage, **142,** 321
  effect on profilactin, **156,** 244
  in signal transduction, **144,** 102
  turnover, **142,** 325–327

Phosphatidylinositol 3-kinase
activation pathway, **157,** 229
binding of other kinases, **157,** 230
C19 binding, **157,** 229
inhibitors, **157,** 227–228
phosphorylation, **157,** 230–231
Ras binding, **157,** 230
role
cytoskeleton, **157,** 229
lymphocyte signaling, **157,** 216, 227–231
receptor internalization, **157,** 228–229
structure, **157,** 227
substrates, **157,** 227
Phosphatidylinositol 4-kinase, in lamina–pore complex, **162B,** 394
Phosphatidylinositol 4-phosphate, intracellular Ca^{2+} messenger, **142,** 321, 323, 327
Phosphatidylinositol 4-phosphate kinase, intracellular Ca^{2+} messenger system and, **142,** 321–323, 328
L-Phosphinothricin, flowering plants and, **143,** 301
3′-Phosphoadenosine 5′-phosphosulfate, eggshells and biomineralization and, **145,** 231, 237
Phosphocreatine, NMR and, **145,** 42–45
Phosphodiesterase
cAMP level and, **139,** 315
retinal precursor cells and, **146,** 147–148
Phosphoenolpyruvate carboxylase
phosphorylation, **149,** 50, 92–93
CAM plants, **149,** 59–60, 67–68
consensus sequences, **149,** 61–63
C_3 plants, **149,** 60–61, 67
C_4 plants, **149,** 60, 68
nitrate role, **149,** 86
physiology, **149,** 67–68
salt tolerance, **165,** 15
Phosphoenolpyruvate carboxylase-kinase, phosphorylation, **149,** 50, 62, 64–68
Phosphoenolpyruvate carboxylase-phosphatase, phosphorylation, **149,** 64–65
Phosphofructokinase
avian salt glands and, **144,** 145
Turing biochemical mechanism affected by, **154,** 365
Phosphoglycerate, higher plant plastids and, **145,** 169, 189–190, 196
3-Phosphoglyceric acid, higher plant plastids, **145,** 168, 177, 190, 196
Phosphoinositide cascade, **139,** 277
Phosphoinositols
avian salt glands and, **144,** 203–205
control of rhythmic behaviors and, **147,** 11
Phospholamban
effect of electrostimulation, **170,** 167–168
muscle fiber type-specific distribution, **170,** 158
proteins in striated muscle and, **142,** 123–124
Phospholipase, in plant defense transmembrane signaling, **172,** 71, 73–74
Phospholipase A, intracellular calcium messenger system and, **142,** 326, 338
Phospholipase A_2, stimulation by endothelin, **171,** 281–282
Phospholipase C
cytoskeletal dynamics and, **144,** 102–103
fish chromatophores and, **143,** 228, 244
intracellular calcium messenger system and, **142,** 334, 338
phosphatidylinositol turnover, **142,** 321–323, 326, 328
receptors, **142,** 311
isoforms, **157,** 218
phosphorylation, **157,** 211
role in lymphocyte signaling, **157,** 216–221
stimulation by endothelin, **171,** 279–280
Phospholipase Cγ, microgravity, **156,** 314–316
Phospholipase D, intracellular calcium messenger system and, **142,** 326
Phospholipids
avian salt glands and, **144,** 144, 169, 203
calcium channels and, **147,** 216
coagulation affected by, **152,** 54–55
cytoskeletal dynamics and, **144,** 98, 102
intracellular calcium messenger system and, **142,** 317, 338
phosphatidylinositol turnover, **142,** 322, 326
protein kineses, **142,** 332–334
karyosphere and, **144,** 41
NMR analysis, **145,** 2, 19, 37, 39, 42, 45
in phototactic signal transduction, **148,** 297
symplast and, **142,** 288
transdifferentiation in medusae and, **142,** 247

Phosphomonoesters, nuclear magnetic resonance and, **145,** 42
Phosphonium compounds, steady-state accumulation, **171,** 141–144
Phosphoprotein phosphatase, **139,** 317–318
Phosphoproteins
insulin-sensitive
pp120 as, **173,** 267–269
Shc as, **173,** 269–272
Tau proteins as, **171,** 172–174
Phosphoribosyl pyrophosphatetransferase, **156,** 122
Phosphorus
eggshells and biomineralization and, **145,** 231, 235, 240
NMR and, **145,** 2
imaging, **145,** 34–35, 37, 41–42, 44–45
techniques, **145,** 19, 23–24, 33
Phosphorylation
animal mtDNA and, **141,** 174
avian salt glands and, **144,** 169, 205
biological clocks in retina and, **146,** 110
calcium channels and, **147,** 201, 205, 207
centrosomal components, effects on microtubule behavior, **164,** 272–273
connexins, **157,** 333–335, 339, 344
control of rhythmic behaviors and, **147,** 2–3, 10
cytoskeletal dynamics and, **144,** 96, 103, 108, 117
dephosphorylation
intracellular calcium messenger system and, **142,** 324, 337
mitosis and, **144,** 248–249, 252
Drosophila muscles and, **143,** 99, 105, 107–108
effect on prolactin activity, **157,** 66–67
endosymbiont hypothesis and, **141,** 308–309
environmental stimuli in plant cells and, **145,** 264, 276, 282, 286
extracellular matrix and, **146,** 281, 311–312
fish chromatophores and, **143,** 241–242, 245
FRNK on serine residues, **167,** 170–171
growth factor-induced retinal regeneration and, **146,** 59
higher plant plastids and, **145,** 190
histone, **162A,** 213–214
induced disassembly of speckled domain, **162B,** 313
integrins in regulating pp120, **167,** 164–165
intracellular calcium messenger system and, **142,** 322, 329, 332–337
kinesin regulation by, **164,** 300–301
kinetoplastid mtDNA and, **141,** 70, 80
mitosis and, **144,** 218, 247, 249–251
myofibrillogenesis and, **143,** 165
NPC components, **162B,** 293–294
paxillin and tensin in FAK-deficient cells, **167,** 177–178
regulation of cytoplasmic dynein, **164,** 293–295
regulation of FAK kinase activity
by another kinase possibly, **167,** 168–169
by autophosphorylation, **167,** 167
Src-mediated, **167,** 174
regulation of transcription factors, **157,** 244–245
reversible, in plants, *see* Plant enzymes
role in actin cytomatrix, **139,** 291–292
role in microtubule stability, **157,** 6, 11–14, 18, 20–21, 23
role in nuclear lamin dynamics, **162B,** 154–158
Sertoli-germ cell communication network and, **147,** 75
signal transduction in lymphocytes, **157,** 211, 221–222, 228, 230–231, 233–234, 238, 240–242, 245, 257
sites on lamin C-terminus, **162B,** 144–145
sites in Tau proteins, **171,** 172–174
state of Tau protein, relationship to neuronal injuries, **171,** 203
Tau proteins, regulations, **171,** 180–181
TGF-β regulating pRB state of, **167,** 194
transcriptional regulation and, **143,** 24–25, 36–37
Phosphotyrosine, integrins in increasing level of, in FAK, **167,** 168
Phosvette, **166,** 107
Phosvitin, **166,** 107
Photoaffinity labeling, glucan-binding proteins, **148,** 17–18
Photodinesis, environmental stimuli in plant cells
in light-induced intracellular movement, **145,** 252–259
motile systems, **145,** 259–263
signal transduction, **145,** 263–268
Photography, sequential, retinal precursor cells and, **146,** 169–170

Photolabeling, in drug pump model, **171,** 136
Photolyases, **156,** 123
Photomultiplier tube, bacterial flagellar motor end, **147,** 115
Photoperiod, effects on testicular macrophages, **149,** 121–122
Photoreceptor cells
avian retina and, **146,** 40
cellular processes, **146,** 26, 28
normal development, **146,** 2–3, 5–6
retinogenesis *in vitro,* **146,** 10, 16–17
biological clocks in retina and
entrainment pathway, **146,** 101–102, 107, 109
opisthobranch retina, **146,** 86, 88–93, 95
environmental stimuli in plant cells and, **145,** 256, 264, 270–274, 299
in flagellate algae, **148,** 273–274
cryptochromes as, **148,** 286–289
photosynthetic accessory pigments as, **148,** 274–276
retinal pigments as, **148,** 276–286
growth factor-induced retinal regeneration and, **146,** 50, 61
ocular tissues, **146,** 57–59
transdifferentiation, **146,** 52–55
insect
phototransduction, **176,** 314–316
plasma membrane
organization, **176,** 311–313
protein mobility, **176,** 313–314
intracellular calcium messenger system and, **142,** 308
Na^+/K^+-ATPase
in cockroach, **176,** 325–327
developmental redistribution, **176,** 332–334
in fly, **176,** 329–332
in honeybee, **176,** 325
in insects, **176,** 323–324
in locust, **176,** 325–327
maintenance and distribution
in fly, **176,** 337–338
in honeybee, **176,** 334–337
in moth, **176,** 327–329
in neurosensory epithelium, **173,** 209–210
rescue
control for RPE cell transplantation studies, **138,** 25–26
first RPE cell transplantation results, **138,** 13–17
mechanism of action of photoreceptor cell rescue, **138,** 26–33
optimal conditions for RPE cell transplantation results, **138,** 24–25
structural and functional characteristics of rescued photoreceptor cells, **138,** 17–24
retinal, RPE role in function, **171,** 226–227
retinal pigmented epithelium and, **146,** 75, 79–80
retinal precursor cells and, **146,** 145–150
chick embryo cells, **146,** 151–155
default hypothesis, **146,** 167–174
experimental analysis, **146,** 162–167
microenvironmental factors, **146,** 176–183
molecular polarity, **146,** 158–163
structural polarity, **146,** 155–158
subtypes, **146,** 174–176
survival, effects of RPE cell age on ability to support, **138,** 34–39
survival-promoting activity, retinal precursor cells and, **146,** 182
Photorespiration
flowering plants, mutants, **143,** 303–306
higher plant plastids, **145,** 194–199
Photosynthesis
angiosperm leaf cells, effect of microgravity, **171,** 36–38
apparatus, cyanelle genes for, **151,** 45–49
chloroplasts and, **144,** 259–260, 270, 277, 293
cryptomonads, **144,** 273, 275–276
dinoflagellates, **144,** 285–286, 288, 290–291, 293
euglenoids, **144,** 282, 284
red algae, **144,** 261, 263–265
endosymbiont hypothesis, **141,** 233–234
future prospects, **141,** 331–332
mitochondrial 310, **141,** 313
molecular biology, **141,** 262–263, 268
plastids, **141,** 253–254, 261, 271, 273, 278–279
environmental stimuli in plant cells and, **145,** 251
chloroplasts, **145,** 270–271, 276
photodinesis, **145,** 253–254, 256, 259, 264, 266, 268

flowering plants, **143,** 299, 305, 314
higher plant plastids, **145,** 150
amino acid synthesis, **145,** 191, 197–198
CO_2 fixation, **145,** 172–173, 175–176
fatty acid synthesis, **145,** 182, 189
genome, **145,** 160–162
metabolism, **145,** 170
structure, **145,** 150, 153, 155, 157–158, 166
intracellular calcium messenger system and, **142,** 332
symplast and, **142,** 264
Photosystem I, cyanelle genes for, **151,** 45
Photosystem II
ciliates and, **141,** 37
cyanelle genes for, **151,** 45–46
Phototaxis
chloroplasts and, **144,** 284, 291
in flagellate algae, **148,** 229–231, 300–302
chromophytes, cryptochromes in, **148,** 288–289
cryptophyceae, photosynthetic accessory pigments in, **148,** 274–275
dinophyceae, photosynthetic accessory pigments in, **148,** 275–276
Euglena
cryptochromes in, **148,** 287–288
retinal pigments in, **148,** 284–286
eyespot apparatuses, **148,** 232–235
dinoflagellate, **148,** 247–251
type A, **148,** 235–240
type B, **148,** 240–243
type C, **148,** 243–247
eyespot reflection function, **148,** 257–269
green algae, retinal pigments in, **148,** 278–284
late potential difference, **148,** 291–293
photoreceptive apparatuses, components, **148,** 269–273
photoreceptors, **148,** 273–274
cryptochromes as, **148,** 286–289
photosynthetic accessory pigments as, **148,** 274–276
retinal pigments as, **148,** 276–286
primary potential difference, **148,** 291–292
regenerative response, **148,** 291–292
signal generation, **148,** 251–257
signal transduction, **148,** 290
Euglena, **148,** 298, 300
green algae, **148,** 290–299
structural aspects, **148,** 231–232
Phototransduction
in insect photoreceptors, **176,** 314–316
retinal precursor cells and, **146,** 147–148
retinal regeneration and, **146,** 58
Phragmatpbia fuliginosa, male eupyrene meiosis in, **152,** 21, 34
Phragmoplasts
cell model systems, **175,** 9–11
future studies, **175,** 18–19
isolated phragmoplasts, **175,** 12–13
permeabilized cells, **175,** 11–12
conventional cell plates and spindles, **140,** 170–172
mitosis and, **144,** 220
Phragmosomes, environmental stimuli in plant cells, **145,** 292–293
Phrynocephalus, karyosphere, **144,** 37
O-Phthalaldehyde-mercaptoethanol, **138,** 53, 58–59
Phycobilins, chloroplasts
in chloroplasts, **144,** 259–260
cryptomonads, **144,** 273, 275
in dinophysoid chloroplasts, **144,** 295
green algae, **144,** 266, 270
heterokont algae, **144,** 285, 290, 292
red algae, **144,** 261, 263–265
Phycobiliproteins, endosymbiont hypothesis and, **141,** 254–255, 263
Phycobilisomes
chloroplasts and, **144,** 263, 265–266, 273
cyanelle genes for, **151,** 47–48
Phycocyanin, chloroplasts and, **144,** 261, 263
Phycoerythrin, chloroplasts and, **144,** 261, 263, 273, 290
Phycomitrium turbinatum, environmental stimuli in, **145,** 287
Phycomyces, symplast and, **142,** 280
Phycomyces blakesleanus, mitochondrial genomes in fungi and, **141,** 90
Phyllomedusa sauvagei, **138,** 106
Phylogenetic trees, endosymbiont hypothesis and
future prospects, **141,** 332
inferred tree, **141,** 239
lineages of life, **141,** 237–239, 241, 245–246

Phylogenetic trees (*continued*)
mitochondria, **141,** 308–309, 317, 319–320, 323
nuclear genome, **141,** 249
plastids, **141,** 268
true tree, **141,** 239
Phylogeny
endosymbiont hypothesis and
algae, **141,** 273
gene structure, **141,** 308, 310–315, 317–318
gene transfer, **141,** 278
genome organization, **141,** 290, 294, 297
lineages of life 238, **141,** 240, 243
mitochondria, **141,** 327
nuclear genome, **141,** 248, 251–253
organelles, **141,** 330
Plasmodium, **141,** 326
plastids, **141,** 253–255, 261, 278
rRNA sequences, **141,** 319–324
sequences, **141,** 268–273
insects, **174,** 8–10
mouthpart type distribution, **174,** 10–11
Phylum, endosymbiont hypothesis and, **141,** 289–291, 293–298, 326
Physalia, **156,** 284
Physarum
cytoplasmic streaming affected by action potential, **155,** 121
fragmins, **149,** 161–162
microfilaments, *see* Microfilaments, *Physarum* as model
mitosis and, **144,** 221, 224, 234
Physarum polycephalum
actin, **156,** 213–218
cytoskeleton, **156,** 202–213, 253–254
actomyosin, superprecipitation, **156,** 247–253, 254
connectin, **156,** 245–247
cytoplasmic streaming, **156,** 200–202, 247–253
cytoskeletal dynamics and, **144,** 107, 115
flagellate, **156,** 232–233, 254
fragmin, **156,** 230–232
life cycle, **161,** 70–72
microgravity experiments, **156,** 343
mitochondrial fusion
defective strains, **161,** 76–78
differentiation during sclerotization and sporulation, **161,** 72–75
genetic element controlling, **161,** 81–84
isolation, **161,** 76–78
mF plasmid-mediated constitutive recombination, **161,** 91–95
nuclear alleles controlling, **161,** 84–86
during sporulation and spore germination, **161,** 75–76
mitosis and, **144,** 226, 235
mt-nuclear fusion, **161,** 61–62, 96–97
myosin, **156,** 219–221
myxameba, **156,** 231–232
profilin, **156,** 227–228
protein, 52-kDA, **156,** 240
synchronous cell cycle, **162B,** 381–382
titin, **156,** 246
Physcomitrella patens, environmental stimuli in, **145,** 287–288
Physical disector method
characterization, **161,** 122–125
disector/fractionator combinations, **161,** 128–129
Physiology
and cellular aspects of GSI, **140,** 457–459
genetics, **140,** 452–457
effects of MCH, **138,** 151–152
electrophysiology
Sertoli-germ cell communication network and, **147,** 39
symplast structure, **142,** 265–274, 297
relationship with glycine receptor isoforms, **176,** 210–212
Phythium, mitochondrial genomes in fungi and, **141,** 91–92
Phytoalexins
accessibility induction in, **172,** 58
accumulation
elicitor-mediated, **148,** 20–21, 23–24
in necrotic area, **148,** 1
Arabidopsis thaliana and, **144,** 59, 73, 75
biosynthesis by chalcone synthase, **148,** 27
chalcone synthase and, **147,** 233, 237
control of gene, **147,** 271
functions of products, **147,** 245–248, 250
defense response, in soybean, **148,** 17
effects of suppressors on, **172,** 67
elicitors in, **172,** 57

induction by orthovanadate, **172,** 68–69
intracellular calcium messenger system and, **142,** 307, 328
oligogalacturonide induction, **148,** 12
in plant-parasite interactions, **172,** 56–57
synthesis, elicitor-mediated, **148,** 26
Phytochrome
chalcone synthase and, **147,** 246–247
environmental stimuli in plant cells and, **145,** 288, 299
photodinesis, **145,** 254, 256, 258–259, 266, 268
extracellular matrix and, **146,** 300, 313
intracellular calcium messenger system and, **142,** 308–309, 327
Phytoestrogens, perinatal exposure to, **139,** 12–13, 39
Phytohemagglutinin, **143,** 10, 16, 30; **152,** 68; **156,** 166, 170–172, 175
Phytohormones
abscisic acid, **166,** 39
auxin, **166,** 39
cytoskeleton, cell elongation, **166,** 37, 38–39
environmental stimuli in plant cells and, **145,** 299–300
ethylene, **166,** 39
gibberellin, **166,** 39
in habituated plant cells
autonomy, **152,** 275–277
concentration changes, **152,** 267–269
kinetin, **166,** 39
nodulation, **156,** 44–46
Phytopathogens
elicitors in, **172,** 57, 58–59
host plant accessibility and, **172,** 58–59
qualities, **172,** 81–82
suppressor production, **172,** 59–61
suppressors in spore germination fluids, **172,** 62–65
Phytophthora
activity of glucan suppressors in, **172,** 65
mitochondrial genomes in fungi and, **141,** 91–92
suppressor production, **172,** 59–60
Phytophthora megasperma
effect on 4CL gene expression, **148,** 27
elicitor binding to, **148,** 19
intracellular calcium messenger system and, **142,** 307
Phytophthora megasperma f. sp. *glycinea*
binding properties, **148,** 16
elicitor recognition in, **148,** 9–10
Phytotoxins, intracellular calcium messenger system and, **142,** 307
Pick bodies, in neurodegenerative disorders, **171,** 187–188
Pick's disease, with Tau 55 and 64, **171,** 201
Piebaldism, melanocytes and, **146,** 220
Pigment-aggregating agents, fish chromatophores and, **143,** 234, 236–237, 239–240, 245
Pigmentation
chalcone synthase and, **147,** 245–247, 254, 257, 271–272
depigmentation, melanocytes, **146,** 219–224
melanocytes and, **146,** 191
biology, **146,** 192–196
development, **146,** 214–224
function, **146,** 202–214
genetics, **146,** 196
mouse coat-color loci, **146,** 197–202
melanoma and
genetics, **146,** 229, 235–236, 250–251
growth factors, **146,** 237–239
oncogenes, **146,** 241–243
role of tyrosinase gene, **171,** 232–235
Pigment cells
avian retina and, **146,** 11–12
sensory, induction in ascidian tissue formation, **176,** 279–281
Pigment epithelial-derived factor, cytokine production and, **146,** 79
Pigments
copigments, chalcone synthase and, **147,** 245
pink, facultative methylotroph, urease in plant cells, **145,** 93, 96–99
Pigs, Peyer's patches in
dome epithelium, **167,** 107*f*
ontogeny *versus* pattern of, **167,** 94
Pimozide, hormonal control of amphibian metamorphosis and, **145,** 134
Pimpinella anisurum, microgravity experiments, **156,** 327, 328
Pineal gland
biological clocks in retina and, **146,** 84–85, 139
retinal precursor cells and, **146,** 181–182

Pine seedlings, microgravity experiments, **156,** 321–322, 324, 332
Pinocembrin, chalcone synthase and, **147,** 236
Pinus, pollen
scheme of isolation and purification of wings from, **140,** 56
surface waxes isolated from, **140,** 39
wing material isolated, **140,** 62
Pinus elliotti, effect of microgravity, **171,** 32
Pinus mugo, pollen from, **140,** 38, 39, 47
Pinus sylvestris
chalcone synthase and, **147,** 236–237, 243, 247
effect of microgravity, **171,** 31–32
extracellular matrix and, **146,** 295
PIP_2, *see* Phosphatidylinositol 4,5-bisphosphate
PIPAS cells, *see* Pars intermedia periodic acid, –Schiff-positive cells
Pisatin
effects of elicitors and suppressors on, **172,** 65–66
effects of mucin-type suppressors on, **172,** 69–70
Pistil receptivity
and embryo sac isolation, **140,** 571–573
embryo sac and gamete analyses, **140,** 573
pistil receptivity and *in vitro* pollination, **140,** 571–572
and *in vitro* pollination, **140,** 571–572
Pisum, environmental stimuli in plant cells and, **145,** 261
Pisum sativum
cell wall changes under microgravity, role of Ca^{2+}, **170,** 65
intracellular calcium messenger system and, **142,** 309
microgravity experiments, **156,** 323, 325, 326, 328
plastids, **145,** 155, 181
protoplast wall regeneration, **170,** 45
Pituitary adenylatecyclase-activating polypeptide, **148,** 129, 142–143
Pituitary-adrenal axis, avian salt glands and, **144,** 189, 192–192
Pituitary cells
graft, **159,** 75, 78
steroid receptors, **159,** 82–83
Pituitary gland
cell proliferation
age-related changes, **157,** 45
circadian variations in cell division, **157,** 44
factors affecting proliferation, **157,** 46–52
identification of mitotic cells, **157,** 44
streptozotocin-induced diabetic mice, **157,** 53–54
control of amphibian metamorphosis and, **145,** 105, 140
adrenocortical hormones, **145,** 122, 124
prolactin, **145,** 128–130, 132–134, 136–137
thyroid hormone, **145,** 112–115, 117, 119
differentiation of cells, **157,** 41
effect on testicular macrophages, **149,** 121
fish chromatophores, **143,** 236
GnRH-encoding gene in teleosts and, **147,** 165, 167, 169–171, 175, 185
POMC gene, **170,** 82–83
treatment of tumors, **157,** 47–48
tumorigenesis, **157,** 41, 43
Pituitary transcription factor-1, tissue-specific, **162A,** 102–103
PKC, *see* Protein kinase C
Placenta
confined mosaicism, **167,** 287
ecto-ATPase levels, **158,** 169
endomitosis, **165,** 68–69
human, blood barrier in
bidirectional glucose transport, **172,** 29
glucose transport, **172,** 26–29
structure, **172,** 25–26
ploidy level
embryonal (trophoblast), **165,** 54, 55
maternal (decidua), **165,** 54, 55
polyteny, **165,** 68–69
rat, blood barrier in
gap junctions, **172,** 31
glucose transport, **172,** 29–31
structure, **172,** 29
survival of aneuploidy fetus with normal cells in, **167,** 289
Placenta growth factor
stimulation of vascular cells, **161,** 13–14
in tumor angiogenesis, **161,** 13–14
Placental ribonuclease inhibitor, inhibition of endothelial cell growth, **161,** 21

Placeopectern megellanicus, animal mtDNA and, **141,** 173, 203–204
Plakoglobin, **165,** 187
Planarian, transdifferentiation in medusae and, **142,** 217–218
Plant cells
aneuploidy mechanisms, **158,** 117–118
cell wall extracellular matrix, **166,** 79–81
centrosome-like structures, **175,** 97–101
chromosomal aberrations, **158,** 67–68
cell culture age effects, **158,** 69–70
chromosome distribution, **158,** 110–112
explant source effects, **158,** 68–69
heterochromatin role in breakage, **158,** 112–117
types, **158,** 71
cytokinesis of secondary MTOCs, **175,** 107–111
cytoskeleton, interaction with nuclear matrix, **162B,** 90–95
embryogenesis, **158,** 70–71, 91–92
environmental stimuli, *see* Environment, stimuli in plant cells
establishment, **158,** 67–68
gene expression, DNA methylation role, **158,** 116–117
haploidy mechanisms, **158,** 117–119
MAR sequences, **162B,** 127–128
meiosis
chromosome disjunction, **158,** 96–99
chromosome pairing, **158,** 95–96
commitment in carrot, **158,** 91–92, 94
cytological features, **158,** 123
embryo sac, **158,** 103–104
homeotic transformation, **158,** 92
homologous chromosome segregation, **158,** 87–93
hormonal triggering, **158,** 123
impairment of entry, **158,** 93–94
multipolar spindle, **158,** 100–102
nuclear DNA content, **158,** 105–108
pollen, **158,** 103
proembryogenic masses, **158,** 103–104
RAPD analysis, **158,** 109
RFLP analysis, **158,** 109
role of heat shock proteins, **158,** 123–125
somatic reduction, **158,** 86–91
synaptonemal complex, **158,** 95–96
variant phenotypes of carrot, **158,** 100
variations, **158,** 93–102
microtubule–cytoplasmic interactions, model, **175,** 104–107
microtubule-organizing center, **175,** 94–97
mitosis
amitosis, **158,** 76–77
C-metaphase, **158,** 79–80, 83
cytokinesis failure, **158,** 84–86
endopolyploidy, **158,** 73–76
evolution, **158,** 125–126
indefinite growth, **158,** 78–79
nuclear envelope breakdown, failure, **158,** 72–79
phases, **158,** 72
prophase chromosome reduction, **158,** 77–78
reductional grouping, **158,** 80, 82–83
spindle assembly failure, **158,** 79–86
split spindle, **158,** 83–84
model systems
cortical microtubules, **175,** 6–9
cytoplasmic streaming, **175,** 4–6
future studies, **175,** 18–20
microtubule-organizing center, **175,** 13–14
miniprotoplast, **175,** 15–18
phragmoplast, **175,** 9–13
spindle, **175,** 14–15
nuclear matrix
isolation protocols, **162B,** 79–88
structural domains, **162B,** 88–90
organogenisis, **158,** 70–71
polyploidy
chromosomal reduction, **158,** 119–120
colchicine effects, **158,** 120
evolutionary importance, **158,** 117, 119
mechanisms, **158,** 117–120
premitotic, primary and secondary MTOCs, **175,** 101–104
regenerated plants
genome segregation, **158,** 127
single recessive mutations, **158,** 127–128
variation in genetic recombination frequency, **158,** 128–129
somaclonal variation, **158,** 66
totipotency, **158,** 65–66, 121–123
tumorous, habituation in
auxin–cytokinin interactions, **152,** 270

Plant cells (*continued*)
cell properties
hormonal, **152,** 275–277
molecular genetic, **152,** 277–279
genetic tumor as example
auxin effects, **152,** 285–286
cellular transfer DNA genes, **152,** 287–290, 293
characterization, **152,** 279–280
cytokinin effects, **152,** 286–287
hormonal aspects, **152,** 265–266
induction
by auxin, **152,** 265–266, 292
by cytokinins, **152,** 265–266, 292
by hormones, **152,** 265–266, 292
in vitro, **152,** 282–285
in vivo, **152,** 282
interconversion
molecular basis, **152,** 287–291, 293
system, **152,** 283–285
in *Nicotiana,* **152,** 280–282
protoplast fusion, **152,** 282–283
tumor-specific genes, **152,** 290–291
induction
by antiauxins, **152,** 269
genetic background, **152,** 271–272
hormone concentration changes, **152,** 267–269
inoculum size, **152,** 270–271
temperature, **152,** 271–272
mechanism, **152,** 265–267
reversibility, **152,** 273–275, 293–294
urease in, *see* Urease, in plant cells
Plant defenses, *see also* Plants, –parasite interactions
accessibility and, **172,** 58–59
Arabidopsis thaliana, **144,** 53–55, 73, 76–77
genes, *Arabidopsis thaliana,* **144,** 60, 75
in host-parasite interactions, **172,** 55–56
nuclear protein phosphorylation and, **172,** 75–76
phytoalexin theory and, **172,** 56–57
role of cell wall in, **172,** 76–77
transmembrane signaling and, **172,** 70–76
Plant enzymes, regulation by protein phosphorylation
CAM plants, **149,** 50, 59–60, 67–68
chloroplast stromal enzyme, **149,** 49–53
C_3 plants, **149,** 60–61, 67
C_4 plants, **149,** 60, 68
cytosolic enzymes, **149,** 85–88
future studies, **149,** 93
H^+-ATPase, **149,** 91
HMG-CoA reductase, **149,** 88–90, 93
malate synthase, **149,** 90–91
mechanisms, **149,** 47–49, 91–93
NADH:nitrate reductase, **149,** 79–82, 92
PEPC, **149,** 59–68, 92
pyruvate, P_i dikinase, **149,** 49–53
pyruvate dehydrogenase complex, **149,** 54–58, 92–93
quinate:NAD^+ oxoreductase, **149,** 82–84
sucrose phosphate synthase, **149,** 68, 76, 79, 92
modulation af activity *in vivo,* **149,** 68–72
sucrose phosphate synthase-kinase, **149,** 72–74
sucrose phosphate synthase-phosphatase, **149,** 74–76
Plant giant chromosome, **165,** 78–79
Plant hormones, *see* Phytohormones
Plants, *see also specific plants*
actin identification in, **155,** 108
and animals
recognition and gamete fusion in, **140,** 408–414
sperm capacitation in, **140,** 409–411
arabinogalactan proteins
analogs of animal extracellular matrix components
adhesion proteins, **174,** 272
fibronectin, **174,** 271–272
integrins, **174,** 270–275
vitronectin, **174,** 271–272
comparison to animal proteoglycans and mucins, **174,** 268–270
breeding, potential applications, **140,** 477–479
cell development, role of arabinogalactan-proteins, **171,** 100–101
cell differentiation, effect of microgravity and clinostating, **171,** 53–56
cell surface
cell wall, **174,** 198–199
cell wall synthesis sites, **174,** 200
cuticle, **174,** 200
middle lamella, **174,** 199–200
molecular probing with antibodies, **171,** 83–85

plasma membrane–cell wall interface, **174,** 197–198
plasma membrane–cytoskeleton interaction, **174,** 196–197
cellular role of V-ATPase
metabolic processes, **174,** 89–92
space filling, **174,** 87–89
cell wall
architectural models, **173,** 160–162
classification, **173,** 159–160
construction
cellulose microfibril synthesis, **173,** 168–169
cleavage of load-bearing crosslinks, **173,** 192–194
dynamic aspects, **173,** 167–168
EXGT-mediated molecular grafting, **173,** 194–195
xyloglucan–cellulose microfibril interaction, **173,** 195–196
loosening, cleavage model, **173,** 169–171
molecular grafting model, **173,** 171–172
cell wall polymers, probing with antibodies
enzymes, **171,** 100
extensin, **171,** 95–99
glycine-rich proteins, **171,** 99–100
lignin, **171,** 94–95
matrix polysaccharides, **171,** 92–94
pectin, **171,** 85–92
proline-rich proteins, **171,** 99
cell walls
–microbe interactions, **171,** 107–109
structure and components, **171,** 79–80
clathrin-coated vesicles in
acidification, **167,** 23–26
composition
assembly proteins and adaptors, **167,** 13–18
heavy and light chains, **167,** 10–13
receptors, **167,** 18–21
function
endocytosis, **167,** 2–5
membrane recycling, **167,** 5–6
protein sorting and transport, **167,** 6–8
historical background, **167,** 1–2
isolation, **167,** 8–9
prospects and unresolved problems, **167,** 26–28
uncoating, **167,** 22–23
cyanelle genome, **151,** 35–37
cytokinesis in, **173,** 217–218
defense response elicitors, **148,** 1–2, 29–30
elicitor-receptor model, **148,** 29
endogenous, **148,** 5, 12–14
galacturonides, **148,** 13
jasmonic acid, **148,** 14
methyl jasmonate, **148,** 14
oligouronides, **148,** 13
salicylic acid, **148,** 14
exogenous, **148,** 4–6
chitosan, **148,** 11–12
general, **148,** 8–12
race-specific, **148,** 6–8
perception, **148,** 15–19
primary signals, **148,** 3
regulation, **148,** 26–29
secondary signals, **148,** 2–3
signal transduction, **148,** 19–25, 29–30
development, role of AGPs
early studies, **174,** 245–246
pattern formation in roots and coleoptiles, **174,** 255–258
reproductive phase of development, **174,** 247–255
somatic embryogenesis, **174,** 258–260
effect of microgravity
cell adaptation strategy, **171,** 56–58
space vehicle cabin environment, **171,** 5
techniques and instruments, **171,** 5–7
weightlessness, definition, **171,** 4
extracellular matrices, **140,** 300–304
cell wall, **140,** 300–301
stylar wall, **140,** 301
role of in pollination, **140,** 303–304
structure and chemical composition, **140,** 301–303
growth in symplast, *see* Symplast, plant growth in
leguminous plants, **156,** 1–51
microgravity effects, **156,** 303, 305, 319–333, 344–348
mitochondrial genomes, *see* Genomes, mitochondrial
myosin identification in, **155,** 103–108, 130
nod factors, response to, **156,** 46–49
nodulation
genetic control, **156,** 38–40
mutants, **156,** 40–43

Plants (*continued*)
nodulin genes, **156,** 32–38
nuclear organization, **153,** 203
organelles, effect of microgravity and clinostating, **171,** 46–53
-parasite interactions, *see also* Elicitors; Suppressors
accessibility in, **172,** 57–59
bacterial, **172,** 61
phytoalexin theory in, **172,** 56–57
plant resistance in, **172,** 55–56
specificity in, **172,** 63, 64–65, 84–85
suppressors in, **172,** 59–61
pathology in *Arabidopsis thaliana, see Arabidopsis thaliana*
potato, organ culture, effect of microgravity, **171,** 42
recognition and gamete fusion in, **140,** 408–414
reproductive biology, trends
advances in compatibility and male-female interactions, **140,** 567
experimental fertilization, **140,** 578–581
attempts to obtain artificial zygotes, **140,** 579
plant transformation using sexual route, **140,** 579–581
pollen selection, **140,** 578–579
pistil receptivity and embryo sac isolation, **140,** 571–573
embryo sac and gamete analyses, **140,** 573
pistil receptivity and *in vitro* pollination, **140,** 571–572
pollen, **140,** 567–571
cellular organization and quality, **140,** 569–570
pollen gene expression and genetic overlap, **140,** 567–568
pollen wall as living structure, **140,** 568–569
sperm cells, **140,** 570–571
progamic phase, **140,** 573–576
pollen tube chemotropism, **140,** 575–576
pollen tube growth, **140,** 574–575
pollen tube wall formation, **140,** 573–574
syngamic phase, **140,** 576–578
cytoplasmic DNA inheritance, **140,** 577–578
MGU, preferentiality and synchronization of double fertilization, **140,** 576–577
preprogrammed fertilization, **140,** 578–579
conclusions
data from double fertilization and cytoplasmic DNA inheritance, **140,** 587
data from gametes and syngamic system, **140,** 586–587
data from pollen and progamic phase, **140,** 586
data from self-incompatibility and mechanisms, **140,** 587–588
examples of progress, **140,** 565–566
self-incompatibility and male-female interactions, **140,** 581–586
generalities, **140,** 581–582
molecular analysis of self-incompatibility, **140,** 582–585
S genes and S products of pollen, **140,** 585–586
–*Rhizobium* interactions
host specificity, **156,** 30–32
plant hormones, **156,** 45–46
signal exchange, **156,** 49–51
signal exchange, **156,** 49–51
structural–functional organization in altered gravity
algae, **171,** 25–30
angiosperms, *see* Angiosperms
ferns, **171,** 31
gymnosperms, **171,** 31–32
moss, **171,** 30–31
specialized graviperceptive cells, **171,** 10–11, 20–22
tip-growing cells, **171,** 22–25
in study of space flight biology, **171,** 2–4
transdifferentiation in medusae and, **142,** 215–216
transformation using sexual route, *see* Sexual route, plant transformation using
wheat, effect of microgravity, **171,** 41
whole-chromosome hybridization studies, **153,** 200
Planula larva
as cnidarian developmental stage, **167,** 38

neuropeptides and metamorphosis of *Hydractinia echinata*, **167,** 54–55
Plasma, control of amphibian metamorphosis and
adrenocortical hormones, **145,** 119, 122–123, 125
growth hormone, **145,** 138–139
prolactin, **145,** 131, 133–136
thyroid hormone, **145,** 111, 115–117, 119
Plasma cells, Peyer's patches and
B lymphocytes and lymphoblasts as precursors, **167,** 93
in lamina propria, **167,** 98
as percentage of lymphoid cells in dome epithelium, **167,** 133
Plasma-free media, in assays of polymorphonuclear leukocytes, **139,** 160–161
Plasmalemma
in diatom valve morphogenesis, **150,** 326–327
extracellular matrix and, **146,** 273, 275, 285, 307
in protoplast wall, effect of microgravity, **170,** 43
symplast and, **142,** 278, 299–300
integration of activity, **142,** 289, 291, 293–294, 296
proton pumps, **142,** 274–275
Plasma membrane receptor, binding, **166,** 10–11
Plasma membranes
anchorage of cytoplasmic intermediate filaments to, **164,** 102–109
antigens, **171,** 105
avian salt glands and, **144,** 129, 157, 196, 203–204
secretory mechanism, **144,** 169, 171
secretory tissue, **144,** 135, 139–140, 143–144
axonal assembly, **165,** 272
bacterial flagellar motor and, **147,** 139
biological clocks in retina and, **146,** 133
–cell wall interface, **174,** 197–198
chemiosmotic mechanisms, **165,** 2
chloroplasts and, **144,** 280, 283, 292
cryptomonads, **144,** 275
green algae, **144,** 266, 268
red algae, **144,** 261
clathrin-coated vesicles
acidification of ligand–receptor complexes, **167,** 23–26
in endocytosis, **167,** 2–4
membrane recycling, **167,** 5–6
receptors, **167,** 18–21
cytoskeletal dynamics and, **144,** 96, 98–100, 118
–cytoskeleton interaction, **174,** 196–197
egg
brown alga, **163,** 278
CD-4-like proteins, **163,** 269–272
echiurian worm, **163,** 278–279
FcR proteins, **163,** 269–272
mammalian, **163,** 269–273
PH-30/fertilin cognate, **163,** 273
recognition events, **163,** 283–284
sea urchin, **163,** 273–278
sperm-binding factor, **163,** 274–275
sperm receptor, **163,** 275–278
table listing, **163,** 270–271
endosymbiont hypothesis and, **141,** 258
environmental stimuli in plant cells and, **145,** 265, 276, 280, 296, 299–300
external, **140,** 190–191
extracellular matrix and, **146,** 307, 310
fusion with secretory vesicle, regulation
aggregation, **176,** 173–174
pathway, **176,** 173
H^+-ATPase, **165,** 23–24, **176,** 179
hormonal control of amphibian metamorphosis and, **145,** 110
intracellular calcium messenger system and, **142,** 338
concentration, **142,** 312–315, 318, 320
G proteins, **142,** 309–311
phosphatidylinositol turnover, **142,** 321–322, 325, 327–328
protein kineses, **142,** 332, 337
receptors, **142,** 307–308
mitosis and, **144,** 220
myofibrillogenesis and, **143,** 169, 174
Na^+/K^+-ATPase distribution
fly, **176,** 337–338
honeybee, **176,** 334–337

Plasma membranes (*continued*)
neuron, **165,** 273
photoreceptor cell, insect
organization, **176,** 311–313
protein mobility, **176,** 313–314
plant, antigens, **171,** 105
in pollen tube, **176,** 137–138
preparation, **162A,** 490–491
retinal pigment epithelium
alterations, effect of, **152,** 252–253
location, **152,** 230–231
properties, **152,** 227, 229
retinal precursor cells and, **146,** 147
salt tolerance
chloride transport, **165,** 23
ion transport, **165,** 17–24
plasma membrane H^+-ATPase, **165,** 23–24
potassium uptake, **165,** 17–21
sodium efflux, **165,** 21–24
sodium uptake, **165,** 17–21
Sertoli-germ cell communication network and, **147,** 81
function, **147,** 40–41, 72, 76
structure, **147,** 36–37
sperm cells, **140,** 201–202
Plasmids
ciliates and, **141,** 9
higher plant, **145,** 164–165
higher plant mitochondrial genomes and, **141,** 134, 136–137, 147, 152, 159
mitochondrial
circular, **161,** 68
genomes in fungi and, **141,** 89, 110
linear, **161,** 68–69
mF (mitochondrial fusion-promoting)
constitutive homologous recombination with mtDNA, **161,** 91–95
DNA polymerase genes, **161,** 90–91
genetic organization, **161,** 86–87
and mitochondrial sex, **161,** 100–101
terminal structure, **161,** 89–90
table, **161,** 54–55
terminal repeat sequences, **161,** 67–68
urease in plant cells and, **145,** 92
Plasminogen activator, transcriptional regulation and, **143,** 19
Plasmodesmata, symplast and, **142,** 270, 292
Plasmodium
actin, **156,** 213–218
cytoskeleton, **156,** 202–213, 253–254
caffeine drops, **156,** 208–210
cytoplasmic streaming, **156,** 199–202, 247–253
endosymbiont hypothesis and, **141,** 325–326, 328
fragmin, **156,** 230–232
mitosis and, **144,** 224, 234
Physarum, **156,** 199–255
profilin, **156,** 227–229
protein, 52-kDA, **156,** 240
small sandwich, *Physarum,* **149,** 195–198
Plasmodium falciparum, binding to sialic acids, **175,** 182
Plasmolysis
in diatom valve morphogenesis, **150,** 337–338
environmental stimuli in plant cells and, **145,** 261
extracellular matrix and, **146,** 303, 307, 310
Plasticity
NC cells, **138,** 271–273
neuronal, *see* Neuronal plasticity
retinal pigment epithelium cells, **171,** 245–247
Plastids
ATP/ADP sources, **163,** 207–208
chlorochromoplasts, **163,** 185
chloroplasts and, **144,** 268, 276, 284, 293–294
dinoflagellates, **144,** 286, 290–292
heterokont algae, **144,** 277, 280–281
ciliates and, **141,** 37
endosymbiont hypothesis and, **141,** 233–234, 236–239, 253–254, 278–279
algae, **141,** 273–275
eukaryotes, **141,** 275–277
future prospects, **141,** 331–333
gene expression, **141,** 306, 308
gene structure, **141,** 308, 311, 315
gene transfer, **141,** 277–278
genome organization, **141,** 292–293, 297–298
hypotheses of origin, **141,** 255, 257–258
mitochondria, **141,** 280–281
molecular biology, **141,** 262–268

nuclear genome, **141,** 251–253
organelles, **141,** 329–330
phylogenetic diversity, **141,** 254–255
Plasmodium, **141,** 326
rRNA sequences, **141,** 320–321, 323
secondary symbioses, **141,** 256, 258–262
sequences, **141,** 268–273
envelope membrane vesiculation, **163,** 206–207
environmental stimuli in plant cells and, **145,** 263
evolution, cyanelle genome study and, **151,** 30, 53–55
fusion and translocation factor, **163,** 207
gerontoplasts, **163,** 185
higher plant, **145,** 149–150, 204
fatty acid synthesis, **145,** 182–190
metabolism
amino acid synthesis, **145,** 190–201
carbohydrate oxidation, **145,** 201–203
CO_2 fixation, **145,** 171–176
mitochondrial genomes and, **141,** 135, 149
starch synthesis, **145,** 176–182
structure
chloroplast genome, **145,** 153–160
evolutionary origns, **145,** 164–167
genome interactions, **145,** 160–164
types of plastid, **145,** 150–154
translocation, **145,** 167–171
NADPH sources, **163,** 208
nonphotosynthetic, starch synthesis, **163,** 209–210
in pathogenic tissues, **163,** 185–187
plant, effect of microgravity, **171,** 48–49
in symbiotic tissue, **163,** 185–187
Platelet-activating factor
neurotransmission, **149,** 269
as procoagulant response regulator, **152,** 64
Platelet activation-dependent granule external membrane protein, **175,** 184–186
Platelet-derived growth factor
activity in embryogenesis
angiogensis, **172,** 116–117
central nervous system, **172,** 113–114
early stages, **172,** 111–113
heart, **172,** 114–115
lung, **172,** 115–116
patch mouse mutation and, **172,** 110–111
targeted gene disruption and, **172,** 117–118
antibodies, for vascular disease control, **161,** 30
autocrine signaling in, **172,** 104, 105
cellular activities and, **172,** 95–96, 102–103
distribution in embryogenesis
central nervous system, **172,** 108
early developmental stages, **172,** 104–105
heart, **172,** 109
kidney, **172,** 110
limb, **172,** 108–109
lung, **172,** 109
neural crest, **172,** 105
somites, **172,** 105, 108
dominant negative mutants in, **172,** 103
effect on arterial smooth muscle cell culture proliferation, **169,** 208–210
effects on nuclear transport, **151,** 214
endothelial cell growth factor, stimulation of vascular cells, **161,** 14
IFN-β effects combined with, **167,** 231
intracellular signaling molecules and, **172,** 100–103
ligands
dimerization, **172,** 96–97
extracellular localization, **172,** 98
genetic conservation, **172,** 97
isoforms, **172,** 96–97
post-transcriptional modification, **172,** 97–98
splice variants, **172,** 98
melanocytes and, **146,** 218, 240–241
migration induction by, **155,** 75–77
mitogenic activity, **172,** 95, 96
paracine signaling in, **172,** 104, 105
retinal pigmented epithelium and, **146,** 76–77
role in chondrogenesis, **159,** 318
in stimulating FAK tyrosine phosphorylation, **167,** 171–172
synthesis in arterial smooth muscle cell culture, **169,** 219–221
Platelet-derived growth factor receptor
chimeric constructs, **172,** 102–103
chromosomal organization, **172,** 99–100
dimeric forms, **172,** 100, 101

Platelet-derived growth factor receptor (*continued*)
SH2 protein binding and, **172,** 100
tyrosine kinase activity of, **172,** 99
Platelet factor 4
inhibition of angiogenesis, **159,** 141–142
inhibition of endothelial cell growth, **161,** 17–18
Platelets
adenosine nucleotide release, **158,** 158
ADP-induced aggregation, **158,** 158, 160–161, 197
aggregation, diabetes effect, **158,** 197
ecto-ATPase levels, **158,** 158
fish chromatophores, **143,** 248
cytoskeleton, **143,** 224
iridophores, **143,** 206–213, 216–219
motility control, **143,** 226, 235
melanocytes, **146,** 198
role in wound healing, **169,** 152–153, 155
RVD response, swelling-initiated signal transduction pathway, **161,** 225
TGF-β production, **167,** 210
Platypoecilus, melanocytes and, **146,** 233
PLC, *see* Phospholipase C
Pleodorina, **139,** 198, 210, *see also* Algae, volvocine
Pleurodeles waltii, **138,** 108, 148, 274
Plexus, *see* Nerve plexus, density of
plO5, transcriptional regulation and, **143,** 31, 36–38
Ploidy level, placenta, **165,** 54, 55
Plumbago, sperm dimorphism in, **140,** 133
PMF, *see* Pulsed magnetic field
PML, *see* Promyelocytes
PMN, *see* Polymorphonuclear leukocyte
PNA, *see* Peanut agglutinin
pNcollagen III
on fine collagen fibrils, **173,** 111–112
role in collagen fibril growth *in vitro,* **173,** 114–116
Poa pratensis, **140,** 9
Podocoryne, transdifferentiation in medusae and, **142,** 227, 230, 236, 249
Podocoryne carnea, transdifferentiation in medusae, **142,** 216, 222–224, 252
Podophyllotoxins, **156,** 81
Podospora anserina, mitochondrial genomes, **141,** 12, 36–37, 47–48, 50, 93–97, 108–109, 113, 115
Point resolved spectroscopy, and NMR, **145,** 21
Polar body biopsy, preimplantation embryo analysis, **153,** 4
Polarity
in actin filament bundle formation *in vivo,* **175,** 71–72
avian retina and, **146,** 12, 34
cell, role of Rab proteins, **176,** 42–44
cytoskeleton, **166,** 7–9
muscle cells, **166,** 8
single cells, **166,** 7–8
in embryonic induction, definition, **150,** 413
epithelial cell, animal, **166,** 8–9
growth factor-induced retinal regeneration and, **146,** 53, 66
maternal, gene systems, **174,** 18
neuronal
axon, **165,** 247
dendrite, **165,** 247
microtubule cytoskeleton, **165,** 248
recombination phenomenon, **161,** 65
segment, gene expression, **174,** 18–19
Polarization
cytokine production and, **146,** 79
depolarization, *see* Depolarization
hyperpolarization, *see* Hyperpolarization
polymorphonuclear leukocytes, **139,** 170–174
repolarization, biological clocks in retina and, **146,** 126
retinal pigment epithelium cells, **171,** 237–243
retinal precursor cells and, **146,** 147, 153, 155–163, 178
Poliovirus
M cell uptake of type 1, **167,** 140
preferential adherence to M cell apical membranes, **167,** 123
vaccines, M cell uptake of oral, **167,** 140
Pollen
Arabidopsis thaliana and, **144,** 62
chalcone synthase and, **147,** 245–246, 251–252, 258
flowering plants and, **143,** 321
gene expression, molecular evidence
gene characterization and identification, **140,** 7–12
gene transcription during development, **140,** 5–7

haploid transcription, evidence for, **140,** 3–5
promoter structure, **140,** 12–16
germination, **176,** 135–137
hydrated and autoclaved, gymnosperms, isolation, **140,** 54–55
hydration, activation and germination, **140,** 77–78
interactions
cellular organization and quality, **140,** 569–570
pollen gene expression and genetic overlap, **140,** 567–568
pollen wall as living structure, **140,** 568–569
sperm cells, **140,** 570–571
outer wall
carotenoids, flavonoids and phenolic conjugates
anther tapetum in biosynthesis of wall phenolics, **140,** 49–50
biological functions of pollen wall substances, **140,** 50–51
carotenoids, **140,** 41–43
flavonoids, **140,** 43–45
phenolic conjugates, **140,** 45–48
pollen volatiles and surface lipids, **140,** 36–41
Pinus
scheme of isolation and purification of wings from, **140,** 56
surface waxes isolated from, **140,** 39
wing material isolated, **140,** 62
and pollen tube, vegetative cell walls in
cell walls, **140,** 192–194, 196, 198–200
external plasma membranes, **140,** 190–191
organelles, **140,** 193
vegetative nucleus, **140,** 191–193
and progamic phase, data from, **140,** 586
rejection
interspecific, **158,** 1
intraspecific, *see* Self-incompatibility
selection, **140,** 399–400
and experimental fertilization, **140,** 578–579
interactions, **140,** 578–579
self-incompatibility and, **143,** 257–258, 289–290
gametophytes, **143,** 280–283, 285, 287–288
sporophytes, **143,** 259–264, 270, 277–279
specific genes
allergenic proteins encoded by, **140,** 23–25
characterization and identification, **140,** 7–12
expression of two groups during pollen development, **140,** 20–22
function of pollen-and anther-specific genes, **140,** 29–32
future directions, **140,** 32–33
localization of pollen- and anther-specific proteins, **140,** 28–29
localization of pollen- and anther-specific transcripts, **140,** 28
localization of pollen-specific transcript, **140,** 25–27
regulating, **140,** 19
temporal expression patterns, **140,** 22–23
temporal expression patterns of, **140,** 22–23
and stigma quality and protocol of pollination
nuclease activity, **140,** 438–439
pollination protocol, **140,** 439–440
and tubes, localization of cytoskeletal elements in
actin filaments in pollen grains, **140,** 96–102
overview, **140,** 96
pollen grains, **140,** 96–100
pollen tubes, **140,** 100–102
microtubules, **140,** 102–111
overview, **140,** 102
pollen grains, **140,** 102–103
pollen tubes, **140,** 103–107
myosin and kinesin
kinesin-like protein, **140,** 111–112
myosin, **140,** 111
pollen tubes, **140,** 111–112
as vectors of transforming DNA
direct DNA transfer, **140,** 432–435
pollen and stigma quality and protocol of pollination
nuclease activity, **140,** 438–439
pollination protocol, **140,** 439–440

Pollen (*continued*)
results and controls of transformation experiments, **140,** 435–438
strategies for transformation, **140,** 432–435
volatiles and surface lipids, **140,** 36–41
Pollen coat protein, extracted and separated, **140,** 553
Pollen tubes
cell wall, structure, **176,** 163–164
inner callosic lining, **176,** 167–168
outer pectocellulosic layer, **176,** 164–167
changes, **140,** 410
chemotropism, **140,** 575–576
evolution, **140,** 129–130
tube growth, **140,** 574–575
tube wall formation, **140,** 573–574
cytoskeleton, **176,** 140
actin-binding proteins, **176,** 143–145
actin biochemical characteristics, **176,** 141–143
actin filament distribution, **176,** 146–149
microtubule-associated proteins, **176,** 143–145
microtubule distribution, **176,** 146–149
tubulin biochemical characteristics, **176,** 141–143
endomembrane system organization
endoplasmic reticulum, **176,** 138–139
Golgi apparatus, **176,** 139–140
plasma membrane, **176,** 137–138
secretory vesicles, **176,** 140
entry and fertilization in selected species, **140,** 362–364
extension, role of stylar matrix
conclusions, **140,** 314
evidence for active biological role of, **140,** 304–313
latex bead translocation, **140,** 304–305
model for pollen tube extension via stylar matrix, **140,** 312–313
occurrence of substrate adhesion molecule vitronectin in plants, **140,** 307–312
stylar matrix of Vicia *faba,* **140,** 305–307
sexual reproduction in flowering plants, **140,** 297–304
growth, **176,** 135–137
growth, role of pectin, **171,** 90–91
intermicrotubular connection, **140,** 109–111
kinesin-like protein, **140,** 111–112
localization, **140,** 103–107
microtubule orientation, **140,** 107–109
myosin, **140,** 111
pathway, **140,** 434–437
polarized growth, affecting factors
fluctuated tip growth, **176,** 174–175
boron, **176,** 179–180
Ca^{2+} as secondary messenger, **176,** 175–177
cell wall properties, **176,** 177–178
flavonols, **176,** 180
H^+-ATPase, **176,** 179
K^+, **176,** 179
oligosaccharides as signals, **176,** 180–181
pH, **176,** 179
turgor pressure, **176,** 178–179
oscillatory and pulsatory growth, **176,** 174
Pollen wall, and sporopollenin
biochemical studies on sporopollenin biosynthesis, **140,** 55–61
application of inhibitors, **140,** 57–58
tracer experiments, **140,** 58–61
biosynthesis and structure of sporopollenin, **140,** 52–66
isolation and purification procedures of exines, **140,** 53–55
isolation of exines by acetolysis method, **140,** 53
isolation of exines from hydrated and autoclaved pollen, **140,** 54–55
isolation of exines using 4-methylmorpholine N-oxide, **140,** 54
isolation of sporopollenin by extraction hydrolysis method, **140,** 54
isolation of sporopollenin from separated wings of *Pinus* pollen, **140,** 55
chemical studies of degradation and spectroscopic analysis
degradation experiments, **140,** 61–63
spectroscopic analyses, **140,** 63–66
conclusions, **140,** 66–67
domains, **140,** 35–36

occurrence of in higher plants
tapetal and peritapetal cell walls, **140,** 52–53
Ubisch bodies, **140,** 52
viscin threads, **140,** 53
outer wall as accumulation site of chemicals of high diversity
carotenoids, flavonoids and phenolic conjugates
biological functions of pollen wall substances, **140,** 50–51
carotenoids, **140,** 41–43
flavonoids, **140,** 43–45
phenolic conjugates, **140,** 45–48
significance of anther tapetum for biosynthesis of pollen wall phenolic, **140,** 49–50
pollen volatiles and surface lipids, **140,** 36–41
Pollination
post-pollination, **140,** 136–141, 397–398
protocol, **140,** 439–440
role of stylar wall in, **140,** 303–304
in vitro
in maize, **140,** 394–396
and pistil receptivity, **140,** 571–572
in vitro, in analysis of environmental stress
heat shock response in reproductive tissues, **140,** 400–402
influence of temperature stress at different phases of *in vitro* reproductive process, **140,** 397–400
kernal development, **140,** 399
pollen selection, **140,** 399–400
postpollination, **140,** 397–398
stress on reproductive tissues, **140,** 398–399
stress responses to high temperature, **140,** 391–392
temperature stress on *in vivo* reproductive process, **140,** 392–394
control of male and female quality, **140,** 396–397
in vitro pollination in maize, **140,** 394–396
Polyadenylation
animal mtDNA and, **141,** 174, 189, 191
endosymbiont hypothesis and, **141,** 298
kinetoplastid mtDNA and, **141,** 77, 81, 84
signal motif, AATAAA, **140,** 8, 10
Polyamines
cytochemical methods
cellular content and fluorescence yield, **138,** 60–64
characterization, **138,** 68–71
sensitivity and precision, **138,** 69–71
specificity, **138,** 68–69
conclusions, **138,** 71
formaldehyde fluorescamine methods
development and early consideration, **138,** 52–53
identification of FF-reactive compound, **138,** 53–58
OPT method, **138,** 58–59
polyamine immunocytochemistry
accessibility of polyamines, **138,** 66–68
polyamine antibodies, **138,** 64–65
tissue fixation, **138,** 65–66
endogenous, compartmentalization and possible functions
association with chromatin
chromosomes, **138,** 76–77
condensed nuclei, **138,** 73–76
evidence for close association with condensed DNA, **138,** 73
interactions with DNA, chemical aspects, **138,** 78
conclusions, **138,** 83
cytoplasmic localization, **138,** 78–83
general aspects, **138,** 72–73
localization
compartmentalization and functions
association with chromatin, **138,** 73–78
conclusions, **138,** 83
cytoplasmic localization, **138,** 78–83
general aspects, **138,** 72–73
cytochemical methods
cellular content and fluorescence yield, correlation, **138,** 60–64
characterization of polyamine cytochemical methods, **138,** 68–71
conclusions, **138,** 71
formaldehyde fluorescamine methods
development and early consideration, **138,** 52–53
identification of FF-reactive compound, **138,** 53–58

Polyamines (*continued*)
OPT method, **138,** 58–59
polyamine immunocytochemistry, **138,** 66–68
polyamines, **138,** 51–52
summary and perspectives, **138,** 83–84
Polygalacturonase
moss, effect of clinostating, **171,** 31
role in cell wall changes under microgravity, **170,** 62–63
Polygenome nucleus, polytene chromosome, transformation, **165,** 92–98
Polyglutamylation
5-fluorouracil, **156,** 118–119
methotrexate, **156,** 111–112, 114–115
Poly-Ig receptor
dome *versus* villus enterocytes, **167,** 131
lack of expression by dome epithelial cells, **167,** 126
Polykaryocyte, ultrastructural study, **165,** 99
Poly(DL-lactic coglycolic acid) microspheres, **167,** 141
β-Polymerase, encoding gene, **156,** 79
Polymerase chain reaction
amplification specificity, **140,** 518–519
cloning of G-protein-coupled receptors by, **164,** 234
degenerate oligonucleotide-primed, in whole-chromosome hybridization
clinical studies, **153,** 217–218
method, **153,** 196–197
detection of gene mutations, **158,** 235–236
Drosophila muscles and, **143,** 64, 120, 124
flowering plants and, **143,** 298, 313
in fluorescent *in situ* hybridization, **153,** 20
gene expression analysis, **158,** 238
GnRH-encoding gene in teleosts and, **147,** 174, 181
intracellular calcium messenger system and, **142,** 310
melanocytes and, **146,** 244
in preimplantation diagnosis of genetic defects, **151,** 2–5, 17–19
self-incompatibility and, **143,** 277
SLG-homologous transcripts in *Brassica* anthers
anthers and leaves, **140,** 519–521
specificity, **140,** 518–519
whole-chromosome hybridization
clinical studies, **153,** 217
degenerate oligonucleotide-primed, **153,** 196–197, 217–218
interspersed repetitive sequence, **153,** 195–196, 217
linker-adaptor, **153,** 197
method, **153,** 195–198
random primed, **153,** 196–197
Polymerases, tracking, localization, **162A,** 131–137
Polymerization
actomyosin thread models, **149,** 167
chalcone synthase and, **147,** 274
cytoskeletal dynamics and, **144,** 85, 110, 116
current concepts, **144,** 97, 102–103
cytoskeletal components, **144,** 105, 107
theories, **144,** 86, 89
depolymerization
cytoskeletal dynamics and, **144,** 109–110
environmental stimuli in plant cells and, **145,** 279, 285–286, 296
mitosis and, **144,** 243, 245
myofibrillogenesis and, **143,** 164–170, 172
Drosophila muscles and, **143,** 97, 112, 139
environmental stimuli in plant cells and, **145,** 279, 285, 287
intracellular calcium messenger system and, **142,** 307
karyosphere and, **144,** 9
in lignin biosynthesis
final step, **176,** 109–111
role of laccase, **176,** 111–114
role of peroxidase, **176,** 114–117
mechanism reevaluation, **162A,** 126–127
microfilament system, **149,** 148, 151
MIP-1α, **167,** 199–200
mitosis and, **144,** 235, 237
myofibrollogenesis and
sarcomere formation, **143,** 181
thick filament proteins, **143,** 176
thin filament proteins, **143,** 158–172
protoplasmic streaming, **149,** 190–191
regenerative cementogenesis and, **142,** 22

Sertoli-germ cell communication network and, **147,** 75
Tetrahymena 14nm filament protein, effects on citrate synthase, **164,** 18–20
vertebrate cell culture technology and, **142,** 180–181, 198
Polymers
biopolymer, cholesteric ordering, *in vitro,* **166,** 84–92
extracellular matrix and
characterization, **146,** 264–266, 271
immunology, **146,** 284
modification, **146,** 294–295, 300, 310
synthesis, **146,** 272, 282–284
growth factor-induced retinal regeneration and, **146,** 60
melanocytes and, **146,** 193, 247–250
Mg-polymer, *Physarum* actin, **156,** 217–218
plant cell wall, probing with antibodies
enzymes, **171,** 100
extensin, **171,** 95–99
glycine-rich proteins, **171,** 99–100
lignin, **171,** 94–95
matrix polysaccharides, **171,** 92–94
pectin, **171,** 85–92
proline-rich proteins, **171,** 99
Polymorphisms
ciliates and, **141,** 10
DNA, **167,** 280–282
length, collagen α(IV) chains, **173,** 90–92
long terminal repeats, transcriptional regulation and, **143,** 33, 39–44
mitochondrial genomes in fungi and, **141,** 109, 111
restriction fragment length
analysis of somatic segregation in carrots, **158,** 109
identification of disease genes, **158,** 229–230
melanocytes and, **146,** 240, 245–246
and self-incompatibility in plants, **158,** 11
S gene alleles, **158,** 51
single-strand conformation, detection of gene mutations, **158,** 235–236
Polymorphonuclear leukocytes
ecto-ATPase levels, **158,** 157–158
role in wound healing, **169,** 155–156
in vitro, random motility
assays, **139,** 177–182
asymmetric migration, **139,** 180
conditions for, **139,** 158
direct microscopy, **139,** 177–179
discrepancies in, **139,** 157–159
filter membrane, **139,** 180–182
results, **139,** 158
cell-biological components, **139,** 162–176
active forward protrusion and, **139,** 165, 167
cell viability and, **139,** 168–169
membrane fluidity and, **139,** 176
membrane ratchet and, **139,** 166, 167–168
membrane recycling models and, **139,** 164, 166–167
membranous structures and, **139,** 169
motor of ameboid cell and, **139,** 162–163
polarization and, **139,** 170–174
pseudopod formation and, **139,** 171–173
rollingballoonmodelsand, **139,** 163–164, 166
spreading and, **139,** 174–175
streaming and, **139,** 176
surface wave theories and, **139,** 165, 167
traction and, **139,** 168–170
zeiosis and, **139,** 175–176
general considerations for, **139,** 157–159, 182
methodology and concepts, **139,** 159–162
attachment to artificial substrata and, **139,** 162, 170
chemokinesis and, **139,** 161–162
early, **139,** 159–160
separated polymorphonuclear leukocytes and plasma-free media and, **139,** 160–161
rates, **139,** 158, 160
Polynucleosomes, salt-soluble, **162A,** 221–225
Polynucleotide ligase, **156,** 124
Polyomavirus
binding to sialic acids, **175,** 172
tumorigenesis in transgenic mice bearing gene for, **164,** 62

Polyorchis, transdifferentiation in medusae and, **142,** 222–223
Polyorchis penicillatus, see also Hydromedusae
biosynthesis of neuropeptides in, **167,** 74–77
RFamide neuropeptides isolated from, **167,** 48
Polyovular follicles, sex hormones and, **139,** 8–10
Polypeptides
in AGP biosynthesis and degradation, **174,** 235–236
animal mtDNA and, **141,** 224
assembly
adaptor complex, **167,** 14–18
GTP in association with Golgi membranes, **167,** 21
–carbohydrate linkage, in AGPs, **174,** 230–232
chloroplasts and, **144,** 272
clathrin, **167,** 12–13
component of AGPs
cDNA-deduced sequence, **174,** 218–230
conformation, **174,** 230
general features, **174,** 217–218
cytoskeletal dynamics and, **144,** 109
Drosophila muscles and, **143,** 104–105, 107, 115, 125
endosymbiont hypothesis and, **141,** 282, 296, 301
environmental stimuli in plant cells and, **145,** 261, 263, 299
fish chromatophores and, **143,** 241
higher plant mitochondrial genomes and, **141,** 134
higher plant plastids and, **145,** 149
metabolism, **145,** 168, 173, 182, 197, 200
structure, **145,** 155–157
intracellular calcium messenger system and, **142,** 308, 310, 336–337
lamina-associated, **164,** 118
mitosis and, **144,** 245, 250
myofibrillogenesis and, **143,** 173
myosin heavy chain, in motor proteins, **176,** 154–157
NIC96p, characterization, **162B,** 286–287
pituitary adenylate cyclase-activating, astrocyte reaction to, **148,** 129, 142–143
proteins in striated muscle and, **142,** 87, 127
self-incompatibility and, **143,** 269, 284, 289
transcriptional regulation and, **143,** 34, 39
Polypeptide toxin, ω-Aga-l, calcium channels and, **147,** 212
Polyphaga, karyosphere and, **144,** 25
Polyphosphoinositides
messenger system, calcium oscillation in mammalian eggs by, **152,** 191
metabolism
functional association with membrane-bound ATPase, **172,** 76
in fungal signal transduction cascade model, **172,** 82–84
in plant defense transmembrane signaling, **172,** 71–73
Polyphyletic origins, endosymbiont hypothesis and
eukaryotes, **141,** 273
future prospects, **141,** 331
mitochondria, **141,** 281, 327
plastids, **141,** 253–254, 257, 278–279
sequences, **141,** 269, 271
Polyploidy
definition, **167,** 264
effect in humans, **158,** 224
mitosis and, **144,** 225, 230–231, 235–237, 239
plants
chromosomal reduction, **158,** 119–120
colchicine effects, **158,** 120
evolutionary importance, **158,** 117, 119
mechanisms, **158,** 117–120
Poly-L-proline, myofibrillogenesis and, **143,** 161, 163
Polyps, *see Hydra*
Polysaccharides
eggshells and biomineralization and, **145,** 231
extracellular matrix and
characterization, **146,** 262–267, 269
immunology, **146,** 284–285
modification, **146,** 292, 295, 303, 310
synthesis, **146,** 272–276, 282–283
lectins in detection of M cell, **167,** 103–104
pectic network, **173,** 167
plant cell plasma membrane recycling and matrix deposition of, **167,** 5

plant cell wall
effect of clinorotation, **170,** 57–60
effect of space flight, **170,** 54–56
Polysaturated fatty acid, effects on volume regulation, **161,** 224–225
Polytene chromosomes
Diptera, **165,** 53
dissociation, **165,** 93–95
Drosophila muscles and, **143,** 63, 105, 107–108, 119
endochromsome, **165,** 94–95
human trophoblast, **165,** 68–69
mammalian cell, **165,** 53–107
interphase nucleus markers, **165,** 70–75
nonclassic form characteristics, **165,** 75–81
sex chromatin, **165,** 70–72
nucleolus-like body, **165,** 81–87
oligotene fibril, **165,** 93–94, 95–96
phase-contrast microscopy, **165,** 66–68
polygenome nucleus, transformation, **165,** 92–98
trophoblast
cell morphology, **165,** 56–69
cellular reproduction, **165,** 54–75
classic *vs.* nonclassic, **165,** 79–81
Diptera-rat similarity, **165,** 76–77
mammal-Diptera differences, **165,** 77–79
morphology, **165,** 54–75
nonclassic features, **165,** 77–79
trophoblast giant cell, **165,** 54–56
trophoblast giant cell
differentiation, **165,** 103–104
genome multiplication, **165,** 103–104
placenta, **165,** 102–103, 105–106
significance, **165,** 102–106
Polytene nucleus
cycle
G phase or endoprophase, **165,** 69–70
S phase or endointerphase, **165,** 69–70
stages, **165,** 69–70
trophoblast giant cell, ultrastructural organization, **165,** 81–89
POM121
characterization, **162B,** 271–272
similarity with yeast POM152, **162B,** 290
POM152, NPC protein, **162B,** 290
Pomacentrus lepidogenys, chromatophores and, **143,** 209
POMC, *see* Proopiomelanocortin
Ponticulin
cytoskeletal dynamics and, **144,** 107–108, 111
mutants, 3D motion analysis, **163,** 92–93
Pores
in ADH-sensitive epithelia, **163,** 36–37
antibiotic, osmotic conductance (P_{os}), **163,** 3–5, 30–34
aquapores, osmotic conductances (P_{os}), **163,** 5–6
–lamina complex, 89–90, 194, 209–210; **162B,** 394, *see also* Nuclear pore complexes
leaky
bimodal theory, **163,** 21–26
hydrodynamic theory, **163,** 8–10
single file theory, **163,** 16–18
surface forces, **163,** 9–10
ultrafiltration theory, **163,** 8
and membrane-spanning proteins, **163,** 37–38
semipermeable
bimodal theory, **163,** 19–21
hydrodynamic theory, **163,** 6–8
osmotic conductance (P_{os}), **163,** 3–5
single file theory, **163,** 11–16
Porosity, cell wall, **176,** 177–178
Porphyra purpurea, cyanelles
plastid genome, **151,** 37
translation apparatus protein genes, **151,** 44
Porphyra yezoensis, chloroplasts and, **144,** 263
Porphyridium aeruginosum, chloroplasts and, **144,** 264
Porphyrin, Harderian gland secretion
biosynthesis, **168,** 36
estrogen effects, **168,** 54
levels in rodents, **168,** 35–37
melatonin effects, **168,** 51
pituitary hormone effects, **168,** 50–51
structure, **168,** 36
thyroid hormone effects, **168,** 50–52
Portuguese man-of-war, nematocyst discharge, **156,** 284
Positional cloning
disease gene identification, **158,** 227–228
gene sequence identification, **158,** 233–235
Positional information, in embryonic induction, definition, **150,** 412

Position independence, domains, **162A,** 425–427
Positron emission tomography, nuclear magnetic resonance and, **145,** 41
Postinhibitory rebound, control of rhythmic behaviors and, **147,** 13
Post-pollination, **140,** 136–141, 397–398
Postreplication repair, **156,** 125
Post-transcriptional regulation
  animal mtDNA and, **141,** 197
  endosymbiont hypothesis and, **141,** 290, 302, 304–305
  higher plant mitochondrial genomes and, **141,** 161
Post-translational processing, endosymbiont hypothesis and, **141,** 248, 299, 315–316, 323, 329
Potassium
  avian retina and, **146,** 6
  avian salt glands and, **144,** 132, 139, 192, 201
  biological clocks in retina and, **146,** 102–103, 110–111, 127, 131
  calcium channels and, **147,** 206, 211
  control of rhythmic behaviors and, **147,** 13
  cytoskeletal dynamics and, **144,** 109
  effect on pollen tube tip growth, **176,** 179
  extracellular matrix and, **146,** 312
  fish chromatophores and, **143,** 205, 231, 239
  glial cells, **149,** 275
  glutamate metabolism and, **147,** 294, 305
  in insect retina, **176,** 316–319
  intracellular calcium messenger system and, **142,** 307, 311, 320–321, 327
  K^+/Cl^- cotransporter
    protein phosphorylation/dephosphorylation effects, **161,** 233–234
    in RVD, **161,** 186–189
  K^+/H^+ exchanger, in RVD, **161,** 197
  in neuromediation
    acetylcholine, **149,** 239
    ion channels, **149,** 227–231
  nuclear magnetic resonance and, **145,** 20
  retinal precursor cells and, **146,** 147, 160
  and sodium, concentration in reproductive tract fluids, **149,** 21–22
  sperm function regulation, **149,** 17–20
  symplast and, **142,** 276–277
  volume-sensitive channels
    cell types affected by, **161,** 181–182
    properties, **161,** 182–184
Potassium channels, neuroblastoma cells, tamaxifen and, **139,** 42
Potato
  ecto-ATPase levels, **158,** 176
  environmental stimuli in, **145,** 297
  flowering plants and, **143,** 311
  intracellular calcium messenger system and, **142,** 328
  organ culture, effect of microgravity, **171,** 42
  plant-pathogen interactions in, **144,** 72
  plastids and, **145,** 177, 179
  spindle tuber viroid, *Arabidopsis thaliana,* **144,** 71
  symplast and, **142,** 282
  urease in plant cells and, **145,** 90
Potentials
  action, *see* Action potentials
  electric, symplast, **142,** 265–269, 271–273
  end-plate, calcium channels, **147,** 194, 203–206, 209
  excitatory postsynaptic potentials, **167,** 42
  membrane, *see* Membrane potential effects
  radial, symplast and, **142,** 272–273
  surface, symplast and, **142,** 267, 269
Potentiation, long-term
  interphase chromatin reorganization by, **150,** 163–164
  mechanism, **153,** 267
Pottia intermedia, protonema, apical cells, effect of microgravity, **171,** 24–25
Pouch assay
  angiogenesis quantitation, **159,** 117–118
  effector media, **159,** 117–118
  polyvinyl acetate disk implantation, **159,** 117
pp50, transcriptional regulation and, **143,** 37–38
$Pp60^{STC}$ tyrosine kinase, in discovery of $pp125^{FAK}$, **167,** 164
pp120
  ATPase activity, **158,** 148, 151
  phosphorylation, **158,** 148, 196
  role in signal transduction, **173,** 267–269
$Pp125^{FAK}$, *see* Focal adhesion kinase
PPB, absence, **140,** 177–178
PPDK, *see* Pyruvate, P_i dikinase

PrA, *see* Proteinase A
Prader-Willi syndrome
association with UPD, **167,** 289
chromosomal deletions, **160,** 84
gene imprinting role, **160,** 85
Prasinophyceae, eyespot reflection in, **148,** 259, 269
Prasinophytes, chloroplasts and, **144,** 266, 288
Pratylenchus penetrans, plant-pathogen interactions in, **144,** 72
PRB, *see* Retinoblastoma, susceptibility gene product
PrB, *see* Proteinase B
Precementoblasts, regenerative cementogenesis and, **142,** 13, 27, 41
Precipitation, hormonal control of amphibian metamorphosis, **145,** 124
Precursor cells, retinal, *see* Retinal precursor cells
Precursor proteins, *see* Preprohormones
Predators, fish chromatophores and, **143,** 193
Predentine, regenerative cementogenesis and, **142,** 9, 11–13, 19, 27
Preformation, and epigenesis, antithesis between, **154,** 313–322
Pregastrulation movements, in embryonic induction, **150,** 391–397
Pregnancy
incidence of mosaicism in, **167,** 287
nondisjunction as leading cause of loss, **167,** 289
parathyroid hormone-related protein, **166,** 250, 253–255
tamoxifen and, **139,** 43
Pregnenolone, control of amphibian metamorphosis and, **145,** 123
Preimplantation development, human conceptus
cell cycle regulation, **164,** 161–164
cell number/integrity, **164,** 148–155
chromosomal abnormalities, **164,** 142–143
oocyte, **164,** 143–144
spermatozoa, **164,** 144–145
culture conditions effects, **164,** 165–166
culture media, **164,** 168–170
cytoplasmic maturation effects, **164,** 145–146
cytoskeleton role in multinuclear blastomeres, **164,** 160–161
DNA content, **164,** 155–160
endometrial synchronization, **164,** 166–168
and genomic activation, **164,** 172–173
incidence of aneuploidy, **167,** 267–268
markers for conceptus developmental potential, **164,** 173–174
nucleocytoplasmic ratio effects, **164,** 172
oocyte maturation effects, **164,** 140–142
oxidative stress effects, **164,** 170–171
Preinfection threads, **156,** 47
Prekinetochore, and centromere/kinetochore region, **162B,** 36–38
Prematurely condensed chromosomes, mitosis and, **144,** 238, 241–243
Premelanocytes
biology, **146,** 196
genetics, **146,** 235
Premelanosomes
biology, **146,** 195
fish chromatophores and, **143,** 195, 197
Prenucleolar bodies, assembly, **162B,** 310
Prenyllipids, in chromoplasts, **163,** 188–189
Preoptic recess organ, hormonal control of amphibian metamorphosis, **145,** 122–123
Preosteoblastic cells, regenerative cementogenesis and, **142,** 25, 40–41
Prephenate, flowering plants and, **143,** 311
Preproenkephalin, Sertoli-germ cell communication network and, **147,** 69, 77
Preprogrammed fertilization, pollen selection, **140,** 578–579
Preprohormones
Antho-RPamides
in biosynthesis of, **167,** 68–69, 69–71
diversity, **167,** 77–78
biosynthesis in higher animals, **167,** 57–58
cnidarian
in Antho-RFamide biosynthesis, **167,** 58–64, 64–67
in Antho-RPamide I biosynthesis, **167,** 68–69
in Antho-RPamides II-IV biosynthesis, **167,** 69–71
for Glu-Gln-Pro-Gly-Leu-Trp-NH_2 peptide, **167,** 55
versus higher animals, **167,** 78

Preprohormones (*continued*)
in metamorphosis-inducing peptides, **167,** 71
Pol-RFamide, **167,** 74–75, 76*t*
Preprophase bands
electron-dense vesicle and granule, **175,** 108
environmental stimuli in plant cells and, **145,** 293
Presequence binding factor, as cytosolic chaperone, **174,** 140–141
Pressure
hydrostatic, symplast and, **142,** 263, 293, 297, 300
osmotic, cytoskeletal dynamics, **144,** 89
turgor, effect on pollen tube tip growth, **176,** 178–179
Presynaptic element, **149,** 220
Presynaptic membrane, **149,** 220
Presynaptic receptor, serotoninergic, **149,** 250
Primates, whole-chromosome hybridization studies, **153,** 199–200
Priming, animal mtDNA and
regulation, **141,** 226–229
replication, **141,** 223
Primordial germ cells, programmed cell death in, **163,** 139
Prion particles, associated disease, role of chaperones, **174,** 171–173
PRL, *see* Prolactin
Probenazole, systemic acquired resistance induction, **168,** 326
Probenecid, **156,** 110
Probing
in lignin detection, **176,** 95
lignin structure *in situ,* **176,** 97–98
Procaine, effect on sea urchin egg stiffness, **157,** 122–123
Procephalon, unique features, **174,** 13–16
Processing enzymes, in cnidarian neuropeptide biosynthesis
of Antho-RFamides, **167,** 59–62
of Antho-RPamide I, **167,** 68
in Antho-RPamides II-IV biosynthesis, **167,** 69
description, **167,** 78–82
in prohormone conversion to active peptides, **167,** 57–58
Processing sites, cnidarian neuropeptide sequences, **167,** 78–82
Prochloron
chloroplasts and, **144,** 260, 270, 272
higher plant plastids and, **145,** 166
Prochlorophytes
chloroplasts and, **144,** 260, 272
endosymbiont hypothesis and, **141,** 255, 269
higher plant plastids and, **145,** 166–167
Procoagulants
adhesion receptor regulation, **152,** 75–77
expression modulation, **152,** 86–89
and malignancy, **152,** 82–86
monocyte–macrophage
cytokine modulation, **152,** 61, 63
mechanism, **152,** 57–59
protein antigen induction, **152,** 60
properties, **152,** 77–82
Procollagens, regenerative cementogenesis and, **142,** 11, 13
Proctolin
characterization, **164,** 225–226
control of rhythmic behaviors and, **147,** 10, 13, 18
Proenkephalin, transcriptional regulation and, **143,** 16
Proenkephalin-related peptides
immunohistochemical localization of enkephalins, **138,** 147–148
opiate peptides, effects, **138,** 148–149
Profilactin, **156,** 244
Profilin
binding to monomeric actin, **139,** 279
cytoskeletal dynamics and, **144,** 101–103, 108
dissociation, **156,** 244
effect on actin filament assembly, **139,** 292–299
isolation, **156,** 227–229
myofibrillogenesis and, **143,** 161, 163, 169
in pollen, **140,** 96
regulation by PIP_2, **156,** 315
Progamic phase, interactions
and pollen, data from, **140,** 586
pollen tube chemotropism, **140,** 575–576
pollen tube growth, **140,** 574–575
pollen tube wall formation, **140,** 573–574
Progenitor cells
defective adhesion and excess proliferation of CML ph^{1+}, **167,** 195
hemopoietic
factors regulating, **167,** 189–192

lactoferrin as regulator of, **167,** 228
LAI suppression of normal, **167,** 234–235
MIP-1α
in cell cycle regulation, **167,** 200–206
cell mobilization by, **167,** 242–243
in myelosuppression resulting from chemotherapy, **167,** 239–240
proximity to stromal components, **167,** 194–195
TGF-β effects on, **167,** 210–214
TNF-α and, **167,** 229
Progenote, endosymbiont hypothesis and, **141,** 235, 238, 246
Progesterone
control of amphibian metamorphosis and, **145,** 121, 123
effect on decidual cell reaction, **160,** 6–7, 10, 12
effect on female human genital system, **139,** 21
gene transcription, **160,** 18–19
neonatal exposure to sex hormones and antihormones and, **139,** 10–11
regulation of pars intermedia cells, **159,** 54–55
sensitivity of postnatal rats, **159,** 84–86
uptake in androgenized rats, **160,** 8, 13–14
Progesterone receptor
accessory factors, **162B,** 362–363
chromatin acceptor sites, **162B,** 346–348
Harderian gland, **168,** 42
RBF
as acceptor protein, **162B,** 338
in avian oviduct, **162B,** 348–353
binding to c-*myc* promoter region, **162B,** 355–356
RBF-1, in rat tissues, **162B,** 354
Proglucagon, fish
gene
evolution, **168,** 192–193
homology between species, **168,** 192–193
structure, **168,** 192–193
processing, **168,** 192
transcription, **168,** 198, 200
Proglumide, control of rhythmic behaviors and, **147,** 19
Prohormone convertase, cloning, **167,** 57
Prohormones, *see* Preprohormones
Projectin, *Drosophila* muscles and, **143,** 124–125, 138
Projection, cytoskeletal dynamics and, **144,** 97–99
Prokaryotes
animal mtDNA and, **141,** 193, 195–196, 198
replication, **141,** 225
transcription, **141,** 221
auxin-regulated genes, **152,** 133–134
chloroplasts and, **144,** 259–261, 263, 283, 294
cryptomonads, **144,** 273, 276
dinoflagellates, **144,** 286, 292–293
green algae, **144,** 266, 268
ciliates and, **141,** 43, 45, 59
endosymbiont hypothesis and, **141,** 233, 235–237
future prospects, **141,** 332
lineages of life, **141,** 238, 240, 243–244, 246
mitochondria, **141,** 298, 306, 309, 312–313
nuclear genome, **141,** 247
organelles, **141,** 330
plastids, **141,** 255, 258, 267–268, 278
growth, effect of altered gravity, **171,** 56
H^+-ATPase peripheral studies, **174,** 99–100
higher plant plastids and, **145,** 161, 164, 166
microgravity experiments, **156,** 339, 342
mitochondrial genomes in fungi and, **141,** 120
mitosis and, **144,** 226
Prolactin
avian salt glands and, **144,** 198
biological activity, **157,** 33, 66
cellular synthesis, *see* Mammosomatotroph; Mammotroph
control of amphibian metamorphosis and, **145,** 113, 138–139
larval growth, **145,** 128–132
levels, **145,** 133–137
control of calcium metabolism, **159,** 62
differentiation of cells, **159,** 50
effect on Harderian gland secretion, **168,** 50–51
effect on release
acetylcholine, **159,** 65
γ-aminobutyric acid, **159,** 64–65

Prolactin (*continued*)
catecholamines, **159,** 64–65
histamine, **159,** 65
opioids, **159,** 64
prolactin-releasing hormone, **159,** 62–63
serotonin, **159,** 65
thyrotropin-releasing hormone, **159,** 63, 74
vasoactive intestinal peptide, **159,** 63–64
estrogen effect on cells, **159,** 65–67, 73, 85–86, 91
evolution, **159,** 61, 91
gene structure, **159,** 61
GnRH-encoding gene in teleosts and, **147,** 185
inverse relationship to growth hormone, **159,** 73–74, 91–92
isoforms, **157,** 65–66
precursor, **159,** 61
proteases, **157,** 65
release, **159,** 40, 58
secretion in androgenized rats, **160,** 4–5
secretion in pituitary cells, **159,** 4, 58–59
feedback mechanisms, **159,** 65–66
storage in granules, **159,** 66
secretion of variants
charge, **157,** 64
glycosylation, **157,** 64–65
intracellular age, **157,** 63–64
phosphorylation, **157,** 66–67
physiological significance, **157,** 65–67, 77
size, **157,** 62, 64–65
seminal vesicle epithelium response, **160,** 287–288
size, **157,** 33, 64–65
solubility, **157,** 63
somatolactin homology, **169,** 1–2, 4–5
stress response, **159,** 58
Prolactin receptor, **159,** 58
Prolactin-releasing hormone
effect on prolactin release, **159,** 62–63
synthesis, **159,** 62–63
Proliferation
avian retina, **146,** 27–29, 36, 42
cellular, *see* Cell proliferation
and differentiation, osteoblast, **162A,** 255–256
marker Ki67, expression in dome epithelium, **167,** 134
metabolism of differentiation and morphogenesis, **154,** 360–362
rates, microgravity experiments, **156,** 319, 342–343
Proliferative vitreoretinopathy, cytokine production and, **146,** 75, 78, 80
Prolifin, *Physarum,* **149,** 156, 162
Proline
in Antho-RPamide I, **167,** 68
flowering plants, **143,** 321–323
in higher plants, **165,** 11–12
metabolic cycle, **165,** 9
nematocyst discharge, **156,** 284, 286
proteins rich in, extracellular matrix
characterization, **146,** 267–269
immunology, **146,** 285
modification, **146,** 292, 295, 303, 311
urease in plant cells, **145,** 68–69
Proline isomerase, activity of cyclophilins, **162A,** 34
Prolylhydroxylase, extracellular matrix and, **146,** 276–277, 285, 311
Promoters
animal mtDNA and, **141,** 202, 229–230
regulation, **141,** 227
replication, **141,** 223
transcription, **141,** 218–221
ASTP, **162B,** 362
chalcone synthase and
control of activity, **147,** 254, 256, 259
control of gene, **147,** 262–271
trans -acting factors, **147,** 274–277
Drosophila muscles and, **143,** 94, 103, 108–109, 115, 132
endosymbiont hypothesis and, **141,** 267
gene expression, **141,** 298–299, 301, 333, 336
genome organization, **141,** 289, 297
extracellular matrix and, **146,** 295, 297, 301, 304
GnRH-encoding gene in teleosts and, **147,** 174–175
higher plant mitochondrial genomes and, **141,** 161
histone gene, as model, **162A,** 255–256
H-strand
animal mtDNA and, **141,** 218, 223, 226–227
endosymbiont hypothesis and, **141,** 299

kinetoplastid mtDNA and, **141,** 78–79
L-strand
animal mtDNA and, **141,** 218, 223, 226–227
endosymbiont hypothesis and, **141,** 299–300
melanocytes and, **146,** 214, 216, 243
mitochondrial genomes in fungi and, **141,** 108, 115
ocs, motif in chalcone synthase control, **147,** 263
plastid, **163,** 196–197
for regulation of transcription specificity, **140,** 12–16
retinal precursor cells and, **146,** 182–183
S/MAR cooperation, **162A,** 421–423
transcriptional regulation and, **143,** 7, 13–14, 28, 31
proteins, **143,** 46–47
viruses, **143,** 43–44
Promyelocytes
fused to retinoic acid α receptor, **162A,** 80,169–170
restoration in leukemia, **162A,** 80
Pronase, transdifferentiation in medusae and, **142,** 237–238, 241, 244
Proopiomelanocortin
control of amphibian metamorphosis and, **145,** 125
corticosteroid response, **159,** 86–88
encoding gene
structure, **170,** 80–82
in vertebrate cells and organs, **170,** 82–85
evolution of precursors, **159,** 10
fragmentation, **170,** 82
inhibition by estrogen, **159,** 47
mRNA levels, **159,** 47
peptides from
in central nervous system, **170,** 83
in immune system, **170,** 85–87
role in invertebrate phagocytosis, **170,** 94–95
processing, **159,** 9–10, 46
products
in inflammatory response, **170,** 118–121
interaction with nitric oxide, **170,** 117–118
role in cytotoxicity, **170,** 95–97
role in immune response, **170,** 87–93, 121–122
role in inflammatory response, **170,** 121–122
role in stress response, **170,** 87–93, 121–122
role in immune response, **159,** 78
secretion
brain neurons, **159,** 45–46
pituitary cells, **159,** 4, 9
Sertoli-germ cell communication network and, **147,** 58
Proopiomelanocortin peptides
central effects of alpha-MSH, **138,** 144–145
immunahistochemical localization of POMC-derived peptides, **138,** 139–144
Propagation, vegetative, effect of microgravity, **171,** 41
Propagators, vertebrate cell culture technology and, **142,** 170–171
Prophase, **140,** 179
Proplastids, higher plant, **145,** 160, 162–163
Propranolol, avian salt glands and, **144,** 168
Prorocentrum, photosynthetic accessory pigments in, **148,** 275
Prorocentrum micans, chloroplasts and, **144,** 291
Prosaposin, *see* Glycoprotein, gp1, sulfated
Prostacyclin, inhibition of smooth muscle cell growth, **161,** 26
Prostaglandin
decidual cell reaction, induction, **160,** 16–21
effects on arterial smooth muscle cell culture
phenotype, **169,** 203, 206
proliferation, **169,** 216–217
insect, characterization, **164,** 230–231
procoagulant expression modulation by, **152,** 88
regenerative cementogenesis and, **142,** 38
transcriptional regulation and, **143,** 10
uterine synthesis, **160,** 17–18, 20, 29
Prostasome
ecto-ATPase levels, **158,** 168–169
transport, **158,** 168–169
Prosystemin mRNA, accumulation in plant defense responses, **148,** 13–14
Protamine
Drosophila muscles and, **143,** 77
inhibition of angiogenesis, **159,** 140

Proteasomes, in degradation of immature CFTR protein, **172,** 204–205
Protectants, ultraviolet, chalcone synthase and, **147,** 233, 245–247
Proteinase
chaperone-assisted, *Escherichia coli*
Clp protease, **174,** 155–157
FtsH protease, **174,** 157
La protease, **174,** 155
chaperone-assisted, in mitochondria
Clp homologs, **174,** 158–159
ClpP proteolytic subunit homologs, **174,** 158–159
FtsH homologs, **174,** 159–160
La protease homologs, **174,** 157–158
environmental stimuli in plant cells and, **145,** 261, 301
Sertoli-germ cell communication network and, **147,** 45, 53, 80
stress proteins as, **174,** 152–154
transcriptional regulation and, **143,** 18
urease in plant cells and, **145,** 84, 98
Proteinase A, biosynthetic routes, **139,** 67–68, 70
Proteinase B, biosynthetic routes, **139,** 67, 68–69
Proteinase inhibitors, in coated vesicle isolation from plant cells, **167,** 9
Protein bodies, clathrin-coated vesicles and
origin in legume seeds, **167,** 6–7
synthesis and transport of proteins destined for, **167,** 19–20
unresolved problems, **167,** 28
Protein fiber networks
boundaries of immiscibility function, **150,** 146–147
gel formation, **150,** 140
matrix-driven translocation
definition, **150,** 140
percolation, **150,** 145–146
physical mechanisms, **150,** 143–146
process, **150,** 141–143
wetting, **150,** 143–145
model system, **150,** 140–141
properties, **150,** 139–140
Protein folding
matrix chaperone sequential action in, **174,** 152
process, **174,** 128–129
role of chaperones, **174,** 151–152
Protein kinase
avian salt glands and, **144,** 205
biological clocks in retina and, **146,** 108, 111
calcium-dependent
control of plant enzyme, **149,** 49–50, 64, 74, 82–83
intracellular calcium messenger system and, **142,** 332–335
and neurotransmission, **149,** 236
calmodulin-dependent, activation of mammalian eggs, **152,** 210–211
cAMP-dependent, **139,** 316–317
control of rhythmic behaviors and, **147,** 2–3
cyclic AMP-dependent protein kinase, **139,** 316–317; **172,** 195, 197, 199
cytoskeletal dynamics and, **144,** 97
effect of Ca^{2+}, **139,** 290–291
environmental stimuli in plant cells and, **145,** 281–282, 286
fish chromatophores and, **143,** 241
intracellular calcium messenger system and, **142,** 332–327, 338
mitogen-activated, *see* Mitogen-activated protein kinase
mitosis and, **144,** 227, 249–252
in nuclear matrix, **162A,** 27
p34cdc2, in lamina disassembly, **162B,** 154–156
in phototactic signal transduction, **148,** 297–298
proteins in striated muscle and, **142,** 123
self-incompatibility and, **143,** 263, 270, 275
transdifferentiation in medusae and, **142,** 247–249
Protein kinase A
intracellular calcium messenger system and, **142,** 333, 335–336
lamin phosphorylation, **162B,** 157–158
TPK2, **165,** 32–33
TPK3, **165,** 32–33
TPKI, **165,** 32–33
transcriptional regulation and, **143,** 24, 29
transdifferentiation in medusae and, **142,** 249
Protein kinase C
activation by lipopolysaccharide, **161,** 297–298

in calcium activation of mammalian eggs, **152,** 211–212
calcium channels and, **147,** 201, 221
cytoskeletal dynamics and, **144,** 102
dependent and independent pathways, **138,** 217–219
effect of electromagnetic field, **158,** 315–316
fish chromatophores and, **143,** 230, 241, 243–244
intracellular calcium messenger system and, **142,** 321, 327, 332–333, 335
lipid activators, **157,** 228
melanocytes and, **146,** 236–237, 245
microgravity, **156,** 302, 311–318, 348, 358
phosphorylation, **157,** 228
post-transitional modification, intermediate filament protein role, **154,** 44–46
role in angiogenesis, **159,** 121, 125
role in apoptotic processes, **163,** 115
role in oncogenesis, **157,** 338
tamoxifen and, **139,** 42
transcriptional regulation and, **143,** 10, 16, 24–25, 36, 47
transdifferentiation in medusae and, **142,** 247, 249, 257

Protein phosphorylation
in activation of plant defense genes, **172,** 75–76
–dephosphorylation
and K^+/Cl^- cotransporter, **161,** 233–234
and Na^+/H^+ exchanger, **161,** 234–236
and $Na^+/K^+/2Cl^-$ cotransporter, **161,** 230–233
elicitor-mediated, in plant cells, **148,** 23–24
P110, **162B,** 202–204
pathological Tau proteins, in Alzheimer's disease
abnormal phosphorylation, **171,** 188–189
sites in PHF-Tau and normal Tau, **171,** 189
Tau protein concepts, **171,** 189–191
in phototactic signal transduction, **148,** 294–295, 297–298
in plant defense transmembrane signaling, **172,** 74–76
Rab proteins, **176,** 48–49
Rag proteins, **157,** 139–140
reversible, in plants, *see* Plants, control of enzyme activity by phosphorylation
role in glycine receptor modulation, **176,** 214–215
in striated muscle and, **142,** 85, 98, 121, 123–124

Proteins
ABF2, yeast mitochondrial gene expression affected by, **152,** 149
acidic, instability of homogeneous state, electrophoretic analysis, **154,** 341
actin-associated, **139,** 278–287
actin-bundling, in *Dictyostelium discoideum,* **175,** 52–56
actin network interaction with, **148,** 44
adhesion, plant analog, **174,** 272
advanced glycosylation end products, **152,** 70
allergenic, encoding by pollen specific genes, **140,** 23–25
amyloid precursor, in Tau protein aggregation, **171,** 193–194
androgen-binding, Sertoli-germ cell communication network and
function, **147,** 45, 51, 61
germ cell depletion, **147,** 64–66
germ cell *in vitro,* **147,** 67–71
mechanisms, **147,** 72, 74–75, 78
animal mtDNA and, **141,** 174, 180, 208–209, 229–231
genes, **141,** 181–189
nucleotide bias, **141,** 204–206
regulation, **141,** 226, 229
replication, **141,** 225
rRNA, **141,** 189–191
sequences, **141,** 204
transcription, **141,** 218–222
tRNA, **141,** 193, 199
anther- and pollen-specific, localization
intracellular targeting, **140,** 28–29
localization of transcripts in transgenic plants, **140,** 28
Arabidopsis thaliana and, **144,** 59–60, 68, 70, 74, 78
assembly
FAK and, **167,** 165
focal adhesion during mitosis, integrins in, **167,** 162–163
focal adhesions and stress fibers, FAK and, **167,** 176
associated with cell division, **167,** 291

Proteins (*continued*)
auxin-binding, analysis, **152,** 110–111
avian salt glands and, **144,** 161, 192, 205
secretory mechanism, **144,** 169, 172
secretory tissue, **144,** 141, 143
bacterial flagellar motor and model comparisons, **147,** 147–148
models, **147,** 13O, 132–134, 136, 143–145
structure, **147,** 104, 106–110
basement membrane, and retinal pigment epithelial cells
phenotype expression, **150,** 126–127
transdifferentiation, **150,** 128–129, 131
basic, instability of homogeneous state, electrophoretic analysis, **154,** 341–342
bifunctional regulatory, control of plant enzyme activity, **149,** 51–53
biological clocks in retina and *Bulla,* **146,** 129, 132, 134, 137–130
entrainment pathway, **146,** 106–108, 110
pacemaker mechanisms, **146,** 115, 117–123
BPF-1, in plant defense gene expression modulation, **148,** 28–29
Brassica pollen grains, **140,** 551–554
Brassica stigma
history, **140,** 527–528
homology modeling of female glycoproteins, **140,** 544–551
other families of glycoproteins homologous with S locus-specific glycoproteins, **140,** 533–543
primary structure of S locus-specific glycoproteins, **140,** 544–551
in Ca^{2+} activation of mammalian eggs, **152,** 212
Ca^{2+} channels and, **147,** 205
caldesmon-like, *Physarum,* **149,** 156
and cardiac myocytes
function, **150,** 83
growth, **150,** 80
physiology affected by synthesis rate, **150,** 76–78
Ca^{2+}-regulatory
effect of electrostimulation, **170,** 167–169
muscle fiber type-specific distribution, **170,** 156–158
cationic, bactericidal activity, **161,** 308–309
cell cycle-dependent, **162B,** 385–387
cellular, diethylstilbestrol effect on, **139,** 11–12
chalcone synthase and, **147,** 277
control of activity, **147,** 253–254
control of gene, **147,** 262–264, 266, 269–270
enzymatic activity, **147,** 235–237, 241, 244–245
trans-acting factors, **147,** 272–273, 275–276
chloroplasts and, **144,** 263, 282, 289
ciliates and, **141,** 4, 11, 58–59
genetic code, **141,** 53, 55
Paramecium aurelia, **141,** 12, 35–38
classical matrix, **162A,** 303–309
coding genes
endosymbiont hypothesis and gene expression, **141,** 298, 304
gene structure, **141,** 308–309
gene transfer, **141,** 325
genome organization, **141,** 289–290, 293, 295–297
lineages of life, **141,** 239
mitochondrial, **141,** 288
nuclear genome, **141,** 247, 250
plastids, **141,** 267
higher plant mitochondrial genomes and, **141,** 157–159, 162
kinetoplastid mtDNA and, **141,** 70–73
composition of plant lamina, **162B,** 98–100
contractile, in interphase nucleus, **150,** 164–167
control of rhythmic behaviors and, **147,** 4
C-reactive
host defense function, **170,** 236–238
regulation in transgenic mouse, **170,** 249
cyanelle genome
biosynthetic pathway, **151,** 49–50
import, **151,** 51–52
routing, **151,** 52–53
translation apparatus, **151,** 42–45
cystine-rich, extracellular matrix and, **146,** 269
cytoplasmic, interaction with insulin, **173,** 261–265

cytoskeletal dynamics and, **144,** 85–86, 89, 120
current concepts, **144,** 96–97
cytoskeletal components, **144,** 104–109
Dictyostelium, **144,** 110–111
mechanism, **144,** 116–117
in cytoskeleton-based motor proteins
dynein-like proteins, **176,** 159–160
kinesin-like proteins, **176,** 157–159
in cytoskeleton–membrane connections, **176,** 152–154
decreased synthesis rate during cardiac atrophy, **150,** 81
degradation, *see* Proteolysis
doubling time, effects on H1 histone, **148,** 101–102
Drosophila muscles and, **143,** 65
actins, **143,** 109–114
components, **143,** 137–138
contractile protein genes, **143,** 132–137
development, **143,** 77–85, 87
interactions, **143,** 126–132
isoforms, **143,** 138–139
muscle protein genes, **143,** 125–126
myofibrils, **143,** 94–99
myosin heavy chains, **143,** 99–104
myosin light chains, **143,** 104–108
paramyosin, **143,** 108–109
tropomyosin, **143,** 114–117
troponin, **143,** 117–122
Z-disc proteins, **143,** 122–125
eggshells and biomineralization and, **145,** 220–221, 226, 231, 234–235, 239
endosymbiont hypothesis and, **141,** 234
gene expression, **141,** 298–299, 301–302, 304, 306–307
gene structure, **141,** 308–313
gene transfer, **141,** 278–279, 324–325
genome organization, **141,** 292–295, 298
lineages of life 242, **141,** 246
mitochondrial 281-282, **141,** 329
nuclear genome, **141,** 247–249
organelles, **141,** 329
Plasmodium, **141,** 326
plastids, **141,** 253, 263, 265, 267–268
sequences, **141,** 268–273
environmental stimuli in plant cells and, **145,** 299–301
chloroplasts, **145,** 281, 285–287
photodinesis, **145,** 259, 261, 263
wounding, **145,** 291, 295
extracellular matrix
characterization, **146,** 262–263, 267–270
cytomechanics of adhesion, **150,** 121–123
immunology, **146,** 284–285
modification, **146,** 291–292, 295, 299–300
pathogenesis, **146,** 309–313
phenotypic expression, role in, **150,** 129
stress, **146,** 303–304
synthesis, **146,** 273–278
fish chromatophores and, **143,** 199, 222, 224, 241–243, 249
flowering plants and, **143,** 297
amino acids, **143,** 311, 314, 316, 321, 325
ammonium, **143,** 307–308
nitrate, **143,** 299, 301
in fluid mosaic model of cell membrane, **148,** 43
focal contact, cytomechanics of adhesion, **150,** 120–121
GAL4, transcription initiation role, **154,** 19
genes, **139,** 227, 239–243
in genetic tumors of habituated plant cells, **152,** 290
germ cell-induced, Sertoli-germ cell communication network and, **147,** 69–70, 72
glucan-binding, photoaffinity labeling, **148,** 17–18
glutamate metabolism and, **147,** 286, 298, 301, 306
glycine-rich
extracellular matrix and, **146,** 267, 269–270, 295, 297, 303
overexpression as cause of tumorous growth, **152,** 279
GnRH-encoding gene in teleosts and, **147,** 165, 173, 176, 183, 185–186
growth-associated, neurite initiation and elongation role, **153,** 262–263
growth factor-induced retinal regeneration and, **146,** 58
GTP-binding
in phototaceic signal transduction, **148,** 296–297

Proteins (*continued*)
role in calcium oscillation in mammalian eggs, **152,** 191–194
heat shock
DnaJ homologs, **174,** 139–140, 147–148
DnaK, DnaJ, and GrpE, **174,** 131–132
Drosophila muscles and, **143,** 113–114
Hsp70, **148,** 176–177, 208; **167,** 22; **172,** 205, 207
Brownian ratchet model, **174,** 146–147
as cytosolic chaperone, **174,** 139
force-generated motor, **174,** 147
mitochondrial, reaction cycle, **174,** 149
induction, **148,** 176, 178
prompt, **148,** 201
synthesis during cardiac hypertrophy, **150,** 79–80
in tolerant zone, **148,** 213–214
higher plant mitochondrial genomes, **141,** 162–164
abnormal phenotypes, **141,** 131–134
gene location, **141,** 157–159
structure, **141,** 137, 146
higher plant plastids and, **145,** 204
amino acid synthesis, **145,** 190–194, 199
fatty acid synthesis, **145,** 182–187
genomes, **145,** 153, 155–158, 160–162, 164
metabolism, **145,** 167–169, 172, 175, 177
structure, **145,** 151, 166
homeodomain, role in differentiation, **148,** 103–104
hormonal control of amphibian metamorphosis and, **145,** 117, 122–123, 129–130, 138
inhibitory, brain, **139,** 278, 279
interactions with apatite in enamel, **172,** 142
internal matrix, immunolocalization, **162B,** 105–111
interphotoreceptor retinoid-binding, retinal precursor cells, **146,** 173, 180
intracellular Ca^{2+} messenger system, **142,** 305
calmodulin, **142,** 328–330
concentration, **142,** 317–319
G proteins, **142,** 308–311
protein kineses, **142,** 334–337
receptors, **142,** 306–308
ionic strength effect on enzyme activity, **160,** 102–103
karyosphere and, **144,** 1, 40, 42–45
oocytes of insects, **144,** 19
vertebrates, **144,** 25, 28, 32, 37–39
kinetoplastid mtDNA and, **141,** 66, 69, 76, 82
Kit, melanocytes and, **146,** 215–216, 218–219
liver regulating, Sertoli-germ cell communication network, **147,** 55, 72, 76
load-induced synthesis during cardiac hypertrophy, **150,** 79
mannan-binding, host defense function, **170,** 238
mannose-binding, specificity, **175,** 187–189
melanocytes and, **146,** 193
development, **146,** 215–216, 219
genetics of melanoma, **146,** 235, 245, 247–248, 250
pigment-cell genetics, **146,** 196–197, 204, 206–209, 212–213
membrane
capping, **148,** 47
cytoskeletal association, **148,** 50–53
cytoskeletal hauling, **148,** 47–50
platelet activation-dependent granule external protein, **175,** 184–186
retinal precursor cells and, **146,** 160
methods of visualization
actin, actin filaments and microfilaments, **140,** 86–90
myosin and kinesin, **140,** 93–96
tubulin and microtubules, **140,** 90–93
microsomal, RPE-specific, expression, **171,** 236
mitochondrial genomes in fungi and, **141,** 96, 98, 107
mitosis and
chromosomes, **144,** 241, 243–247, 250–251
dissociation of events, **144,** 218, 231
mobility in photoreceptor plasma membrane, **176,** 313–314
in monocyte/macrophage procoagulant activity, **152,** 60

mot, bacterial flagellar motor
models, **147,** 130, 133–134, 136, 145
structure, **147,** 108–100
mouse epididymal
characterization, **154,** 148
immunolocalization, **154,** 148
with MTOC properties
chromosomal protein, **175,** 112
inner centromere protein, **175,** 115
intermediate filament-like protein, **175,** 112–114
keratin-like protein, **175,** 112–114
nuclear protein, **175,** 111–115
multifunctional
elongation factor-1α, **164,** 25–27
Tetrahymena 14nm filament protein, **164,** 4–9
neurite elongation and initiation role, **153,** 237–238, 243
neurofilament
ADP-ribosylation, **154,** 43
expression subject to feedback control mechanism, **154,** 63
glycosylation, **154,** 42
NMR analysis, **145,** 7, 31
NSP1 pore protein, macromolecular exchanges, **151,** 200
nuclear, import
across nuclear envelope, **151,** 191–193
antibody effects, **151,** 199–200
via NPC, **162B,** 244–248
wheat germ agglutinin effects, **151,** 199–200
nuclear, localization signal binding, **151,** 197–199
nuclear matrix-associated, **154,** 28, 31; **162B,** 387–396
parA, in tobacco mesophyll protoplasts, **152,** 122–124
parB, in tobacco mesophyll protoplasts, **152,** 123–124
penicillin-binding, in *Cyanophora paradoxa* cyanelles, **151,** 34
in phototactic signal transduction, **148,** 294–297
Physarum microfilament system, **149,** 156
physiological significance of, **139,** 287–288
plant cell coated vesicles and
assembly, **167,** 13–18
receptors and sorting, **167,** 20
in plant defense
elicitor identification, **148,** 29
gene expression regulation, **148,** 28
pore, role in nuclear transport, **151,** 199–200
precursors to neuropeptides, *see* Preprohormones
present at replication foci, **162B,** 315–318
production by testicular macrophage, **149,** 111–112
proline-rich, extracellular matrix and
characterization, **146,** 267–269
immunology, **146,** 285
modification, **146,** 292, 295, 303, 311
proteolipid, in preimplantation diagnosis of genetic defects, **151,** 2, 18
RAP1, physical association with nucleus, **154,** 33
Rb, nuclear function, **154,** 12, 14
regenerative cementogenesis and, **142,** 23, 37–38, 40–41
regulation of FAK activity by cytoplasmic, **167,** 170–171
retinal pigmented epithelium and, **146,** 77–79
retinal precursor cells and, **146,** 145, 147, 160, 173, 175, 182
ribosmal gene complement of cyanelle DNA, **151,** 44
role in cell repair after heat shock, **148,** 205–209
secretory, in sperm maturation, **154,** 105–106, 126–128
self-incompatibility, **143,** 257–258, 289
gametophytes, **143,** 284, 286, 288
sporophytes, **143,** 261–265, 269–270, 272, 274–275, 277
Sertoli-germ cell communication network, **147,** 80
function, **147,** 40–41, 45, 50–52
germ cell depletion, **147,** 66
germ cell in vitro, **147,** 67–69
mechanisms, **147,** 75–78
stage-dependent changes, **147,** 60
structure, **147,** 33, 36–37
signal-mediated import through *Xenopus,* **151,** 191
snRNP-associated, **162A,** 71–75
sperm membrane, in epididymal transit, **154,** 179

Proteins (*continued*)
S protein, role in coronavirus, **175,** 171–172
stress, **148,** 171
in cell repair, **148,** 206–209, 215–216
gene expression, **148,** 211
in increased cell stability, **148,** 190–202
in increased primary stability of cell functions, **148,** 203–205, 215
induction by extreme heat, **148,** 175–179
as proteases, **174,** 152–154
in tolerant zone, **148,** 214
in striated muscle, **142,** 61–62, 127–129
cardiac muscle, **142,** 102
antibody use, **142,** 115–119
structural proteins, **142,** 110–115
thick-filament proteins, **142,** 103–107
thin-filament proteins, **142,** 107–110
sarcolemma, **142,** 119–120, 124–127
sarcoplasmic reticulum, **142,** 119–124
skeletal muscle, **142,** 62
86-kDa protein, **142,** 81–82, 94
antibody use, **142,** 100–102
C protein, **142,** 79–80
F protein, **142,** 82–83
H protein, **142,** 82–83
MM-creatine kinase, **142,** 81
M protein, **142,** 80–81
myomesin, **142,** 80–81
myosin, **142,** 62–79
ribosomes, **142,** 83
sarcomere, **142,** 93–100
skelemins, **142,** 83
thin-filament proteins, **142,** 83–93
X protein, **142,** 82–83
structural role in plant cell wall
glycine-rich proteins, **171,** 99–100
proline-rich proteins, **171,** 99
substrate, **139,** 291
synthesis in entrainment model, **146,** 134
targeting to mitochondria
chaperones at ribosomes, **174,** 136–138
posttranslational preprotein import, **174,** 135–136
preprotein interaction with cytosolic chaperones, **174,** 138–139
telomere-binding, physical association with nucleus, **154,** 34
thermostability, heat shock-induced increase, **148,** 179–182
threonine-rich, extracellular matrix and, **146,** 268
thyroid hormone RBF-like, **162B,** 355
transcriptional regulation, **143,** 1–3, 47–48
biochemistry, **143,** 23, 25
genes, **143,** 11–13, 18–19
inducible activation, **143,** 7, 10–11
molecular cloning, **143,** 25–31, 33–39
proteins, **143,** 45–47
structure, **143,** 21–22
viruses, **143,** 39, 41–43
transdifferentiation in medusae and, **142,** 214, 227, 250
translocation
inner membrane components, **174,** 144–145
mitochondrial preprotein proteolytic maturation, **174,** 145–146
outer membrane components, **174,** 142–144
transport from cytoplasm to nucleus, **162B,** 207–210
ubiquitous, genes encoding, **139,** 225–246
urease in plant cells and, **145,** 65–66, 98–99
biochemical genetics, **145,** 84–86, 89, 91–92
elimination, **145,** 80–82
metabolic origins, **145,** 67–68, 71, 77
nickel metabolism, **145,** 95
vacuolar
biosynthesis of, **139,** 63–75
degradation of, **139,** 75–76
endocytosis and, **139,** 79–81
enzyme turnover and, **139,** 76–77
fructose 1,6-bis-phosphatase and, **139,** 77–79
H^+-ATPase, **139,** 81–89
vertebrate cell culture technology, **142,** 146–147, 179, 184, 197–198
vertebrate NPC, **162B,** 266–267
viral, **162A,** 492–493; **162B,** 210–213
volume changes and osmotic pressure, **160,** 103–105, 134
Volvox-specific, encoding genes, **139,** 243–246
glycoprotein SSG 185 and, **139,** 245–246
sex-inducer gene and, **139,** 243–245

yeast mitochondrial gene expression effects
categorization, **152,** 154–155
function, **152,** 156
post-translational activity, **152,** 163–164
regulation, **152,** 166–167
reverse transcriptase activity, **152,** 153
yeast NPC, **162B,** 276–278,290
Protein sorting
plant cell coated vesicles and, **167,** 6–8
in yeast, regulation by Rab GTPase, **176,** 27–28
Protein synthesis
apoptosis-coincident, **163,** 116–117
and unbalanced growth and H1 accumulation, **148,** 97–100
Protein targeting
and association with nuclear functional domains, **162B,** 303–327
to replication foci, **162B,** 318–322
Protein transport, plant cell coated vesicles and, **167,** 6–8
Protein-tyrosine kinase
activation by lymphocyte receptors, **157,** 210, 217
apoptosis role, **157,** 213
B cell expression, **157,** 212
binding to antigen recognition activation motif, **157,** 234–235, 238–239
Csk
phosphorylation of protein kinases, **157,** 242
regulation of CD45, **157,** 242–243
families, **157,** 211–212
Lck
role in lymphocyte signaling, **157,** 235–236
role in T cell development, **157,** 213–214
phosphorylation, **157,** 241
posttranslational modification, **157,** 211
role in lymphocyte signal transduction, **157,** 182, 193, 199
substrates, **157,** 210–211
T cell signaling, **157,** 213–214
translocation, **157,** 217–218
ZAP-70, role in lymphocyte signaling, **157,** 234–236
Proteoglycans
animal, comparison with plant AGPs, **174,** 268–270
definition, **174,** 202–203
eggshells and biomineralization and, **145,** 221, 226, 231, 237
retinal precursor cells and, **146,** 154, 178–179
synthesis in arterial smooth muscle cell culture, **169,** 225–226
Proteolysis
arabinogalactan proteins, **174,** 235–240
AGP turnover, **174,** 238–239
carbohydrate component, **174,** 236–238
hydrolytic enzymes, **174,** 239–240
polypeptide component, **174,** 235–236
ciliates, **141,** 6
Drosophila muscles and, **143,** 124–125, 136
intracellular calcium messenger system and, **142,** 307
in mitochondria, role of chaperones, **174,** 152–154
TGF-β precursor, **167,** 209–210
transcriptional regulation and, **143,** 30, 39
transdifferentiation in medusae and, **142,** 236, 239, 245
vertebrate cell culture technology and, **142,** 149
Proteus vulgaris microgravity experiments, **156,** 342
Prothrombinase complex
assembly, **152,** 79–80
viral infection effects, **152,** 81
Prothyrotropin-releasing hormone, tissue distribution, **159,** 39
Protistan
mitochondrial endosymbiont hypothesis and, **141,** 305–306
mtDNA, endosymbiont hypothesis and, **141,** 294–298, 316
Protists, endosymbiont hypothesis and, **141,** 251, 278, 283, 286, 324, 332
Protocadherin
functional diversity, **165,** 161–162
structural similarities, **165,** 161–162
Protodyn experiment, for cell wall formation, **170,** 42
Protofilament, proteins in striated muscle and, **142,** 96, 98
Protonema cells, moss, apical cells, effect of microgravity, **171,** 23–25
Proton motive force
bacterial flagellar motor
models, **147,** 141–142

Proton motive force (*continued*)
motor energetics, **147,** 123, 127, 129
structure, **147,** 111, 113–114
symplast, **142,** 291, 296
Proton pump
functional reconstitution
lipid manipulation, **174,** 106
protein component manipulation, **174,** 106
solubilization of tonoplast V-ATPase, **174,** 103–104
tonoplast preparation, **174,** 102–103
V-ATPase into liposomes, **174,** 105–106
V-ATPase purification, **174,** 104–105
symplast and, **142,** 271, 273–277, 285–287, 299–300
Protons
environmental stimuli in plant cells and, **145,** 290
urease in plant cells and, **145,** 83
Proton well, bacterial flagellar motor and, **147,** 148
Proto-oncogenes
expression in early stages of cardiac hypertrophy, **150,** 79–80
intracellular calcium messenger system and, **142,** 310
latent potential, **167,** 232–233
melanocytes and, **146,** 228, 230, 234, 240, 244–247
in neoplastic change, **167,** 232–233
retinal regeneration and, **146,** 56
in transgenic mice
bcr/abl, **164,** 47–49
c-erbB-2, **164,** 49–51
c-fos, **164,** 46–47
c-myc, **164,** 45–46
E2A-PBX homeobox gene, **164,** 52–54
growth factor/growth factor receptor genes, **164,** 49–52
neu, **164,** 49–51
pathologic changes induced by, table, **164,** 42–44
ras, **164,** 47
transforming growth factor-α, **164,** 51–52
vertebrate cell culture technology and, **142,** 160
Protoplasmic drops, in analysis of *Physarum* microfilament system, **149,** 177–179
Protoplasmic streaming, *Physarum,* **149,** 190–191
Protoplastids, endosymbiont hypothesis and, **141,** 277–278
Protoplasts
cell wall regeneration, **170,** 42–47
chromosomal aberrations in culture, **158,** 114–115
coated vesicles in plant
in endocytosis, **167,** 4–5
in plasma membrane recycling, **167,** 5–6
cytokinesis failure, **158,** 85
effect of microgravity, **171,** 44–46
environmental stimuli in plant cells and
light, **145,** 260, 266–267, 278, 282
wounding, **145,** 292, 295
fusion, in genetic tumors, **152,** 282–283
higher plant mitochondrial genomes and, **141,** 151, 153
intracellular calcium messenger system and, **142,** 308, 332
calmodulin, **142,** 329, 332
concentration, **142,** 315, 320
phosphatidylinositol turnover, **142,** 325, 327
protein kineses, **142,** 332
miniprotoplasts
cytokinesis, **175,** 15–16
for cytoskeleton studies, **175,** 16–18
future studies, **175,** 19–20
reductional grouping, **158,** 83
Prototheca wickerhamii, endosymbiont hypothesis and, **141,** 296–297, 324
Protozoa
binding to sialic acids, **175,** 182–183
ciliated
basal body cage, **173,** 211–213
basal body proximal structures, **173,** 211
centrosomes, **173,** 214
ciliary motility, **173,** 213
ciliates and, **141,** 4, 12, 36
cytoskeletal dynamics and, **144,** 98
endosymbiont hypothesis and, **141,** 252, 273, 275, 294, 330
hormone receptors
characteristics, **155,** 3–7
imprinting, **155,** 2, 21
membrane poll, **155,** 12
offspring generations, **155,** 17

microgravity experiments, **156,** 342, 343
plasma membrane structure, **155,** 22
replication band, **153,** 137–167
Provirus, transcriptional regulation and, **143,** 17, 40
PRPP, *see* Phosphoribosyl pyrophosphatetransferase
Prunus persica, self-incompatibility, **143,** 282–283
Prymnesiophytes, chloroplasts and, **144,** 277–278, 280, 282, 289–290
psb G, ciliates and, **141,** 36–38
Pseudo-genes, GnRH-encoding gene in teleosts, **147,** 182–183
Pseudomonas, plant-pathogen interactions in, **144,** 57, 73, 76
Pseudomonas aeruginosa
binding to sialic acids, **175,** 182
in cystic fibrosis, **172,** 194
spaceflight experiments, **156,** 341
Pseudomonas syringae, plant-pathogen interactions, **144,** 66–69, 74–75, 78
Pseudopod, amoeboid
formation
ameboid cells and, **139,** 171–173
and chemoattractant gradients, **163,** 94–97
during translocation, **163,** 83
retraction
and chemoattractant gradients, **163,** 94–97
during translocation, **163,** 83–85
Psoriasis, dendritic cell role in, **153,** 90
PstI sites, mitochondrial genomes in fungi and, **141,** 110, 116
Pterins, as photoreceptors in algae, **148,** 287
Pteris, environmental stimuli in, **145,** 274–275
PTK, *see* Protein-tyrosine kinase
Puccinia graminisi, intracellular calcium messenger system and, **142,** 307
Pulse duration, biological clocks in retina and, **146,** 121
Pulse sequences, in NMR, **145,** 2O, 24, 45
Pumiliotoxin B, calcium channels and, **147,** 205
Pumpkin, urease, **145,** 67–69, 72
Purine insertase, **156,** 123–124
Purinergic neurotransmission, **149,** 254–255
Purinergic receptor, **149,** 254–255
Purines
fish chromatophores and, **143,** 220
urease in plant cells and, **145,** 73–79
Puromycin, environmental stimuli in plant cells and, **145,** 291
Putrescine
γ-aminobutyric acid metabolism and, **147,** 312
Sertoli-germ cell communication network and, **147,** 49
Pylaiella littoralis
chloroplasts and, **144,** 281
cyanelles, plastid genome, **151,** 37
Pyramimonas parkae
photoreceptive apparatuses, **148,** 250
phototactic signal generation, **148,** 252
Pyrenoids, chloroplasts and, **144,** 263, 288, 290
Pyrenomonas salina, endosymbiont hypothesis and, **141,** 276–277
Pyridine derivatives, calcium channels and, **147,** 206
Pyridoxal 5′-phosphate, γ-aminobutyric acid metabolism, **147,** 289, 313–315, 317
Pyridylaminooligosaccharides, in characterization of EXGT, **173,** 174–176
Pyrogen, endogenous, **161,** 269
Pyrogenicity, muramyl dipeptide, **161,** 269
H^+-Pyrophosphatase, **167,** 24
Pyrophosphate, incorporation in apatite, **172,** 159
Pyrus serotina, self-incompatibility, **143,** 282
Pyruvate
higher plant plastids and, **145,** 171, 187–190, 203
P_i dikinase, phosphorylation, **149,** 49–53
Sertoli-germ cell communication network and, **147,** 55
Pyruvate dehydrogenase complex
higher plant plastids and, **145,** 187, 189
phosphorylation, **149,** 92–93
components, **149,** 54–55
intact systems, **149,** 57–58

Q

Qualitative analysis
base and amino acid changes, **140,** 504
nucleotide changes among *SLR* cDNAs, **140,** 505

Quantitative analysis
connections between DNA-containing components, **140,** 131–132
exceptions to rule
post-pollination connections/ associations, **140,** 136–141
scale of sperm dimorphism, **140,** 141–142
prevalence and implications of male germ unit, **140,** 133–136
sperm dimorphism in *Plumbago,* **140,** 133
Quercetine, **156,** 46
Quinate : AD^+ oxoreductase, phosphorylation, **149,** 82–84
Quinine, MDR reversal, **156,** 89, 91
Quinone methides, in plant cell wall, dynamic aspects, **151,** 255–256
Quinuclidinyl benzilate, avian salt glands, **144,** 161–162
Quox 7, expression, **138,** 280–281

R

R7 cells, fate in *Drosophilia* development, **139,** 146–148
Rabaptin-5, as Rab effector protein, **176,** 55–56
Rabbit
bronchus-associated lymphoid tissue in, **167,** 136–137
conjunctiva-associated lymphoid tissue in, **167,** 138
M cell studies in
antigen and tracer uptake, **167,** 118
apical cytoplasm of, **167,** 113*f*
identified in appendix, **167,** 100
lectin histochemistry, **167,** 103–104
Peyers patches in, lymph sinuses, **167,** 95–96
trophoblast, cell morphology, **165,** 61, 62
Rab effector proteins
Rabaptin-5 as, **176,** 55–56
Rabin3 as, **176,** 57
Rabphilin-3a, **176,** 53–55
Rab escort protein, role in Rab isoprenylation, **176,** 45–47
Rab GTPase
amino terminus variability, **176,** 15–16
carboxy terminus variability, **176,** 15–16
effector region, **176,** 14–15
functional specificity, **176,** 17–18
function in yeast secretory pathway, **176,** 25–27
intracellular localization, **176,** 20–25
linkage to membrane transport machinery, **176,** 57–58
membrane association, **176,** 44–47
molecular switch, **176,** 16–17
posttranslational modification, **176,** 44–47
regulation of yeast functions, **176,** 27–28
role in 19.5S complex, **176,** 62–63
role in hydrolysis, **176,** 5–14
role in nucleotide binding, **176,** 5–14
Rab3 GTPase
in regulated exocytic vesicles, **176,** 22–23
role in regulated secretion, **176,** 33–35
Rabin3, as Rab effector protein, **176,** 57
Rabphilin-3a, as Rab effector protein, **176,** 53–55
Rab1 protein
in endoplasmic reticulum, **176,** 20–21
in Golgi apparatus, **176,** 20–21
role in ER–Golgi interface trafficking, **176,** 30–31
Rab2 protein, role in ER–Golgi interface trafficking, **176,** 30–31
Rab3a protein, interaction with Rabin3, **176,** 57
Rab3b protein, post-Golgi transport intermediates, **176,** 22
Rab3d protein, interaction with Rabin3, **176,** 57
Rab4 protein, role in early endocytic events, **176,** 36–39
Rab5 protein
role in early endocytic events, **176,** 36–39
role in homotypic membrane fusion, **176,** 42
Rab6 protein, role in intra-Golgi transport, **176,** 31–32
Rab7 protein, role in late endocytic pathway, **176,** 39–40
Rab8 protein, post-Golgi transport intermediates, **176,** 21–22
Rab8a protein
role in cell polarity, **176,** 43–44
role in constitutive exocytosis, **176,** 35–36
Rab9 protein, role in endosome–Golgi communication, **176,** 32–33
Rab10 protein, post-Golgi transport intermediates, **176,** 21–22

Rab11 protein, role in recycling endosomes, **176,** 39
Rab13 protein
  post-Golgi transport intermediates, **176,** 22
  role in cell polarity, **176,** 44
Rab17 protein, role in cell polarity, **176,** 43–44
Rab18 protein, role in cell polarity, **176,** 44
Rab20 protein, role in cell polarity, **176,** 44
Rab proteins
  C terminus, isoprenylation, **176,** 44–47
  early endosomal proteins, **176,** 23–24
  on late endocytic compartments, **176,** 24–25
  membrane association and GTPase cycle
    role of GDP dissociation inhibitors, **176,** 49–51
    role of GTPase-activating proteins, **176,** 52–53
    role of guanine nucleotide exchange factors, **176,** 51–52
  –yeast SNARE interaction, genetic interactions, **176,** 59–60
  nonubiquitous expression, **176,** 18–20
  phosphorylation, **176,** 48–49
  role in cell polarity, **176,** 42–44
  role in homotypic membrane fusion, **176,** 41–42
  role in transport vesicle formation, **176,** 40–41
  Schizosaccharomyces pombe, **176,** 29–30
  ubiquitous expression, **176,** 18
Race specificity, *Arabidopsis thaliana* and, **144,** 54, 56–58, 65, 73
Radial cells
  avian retina and, **146,** 5, 42
    cellular processes, **146,** 30, 34–37
    retinogenesis *in vitro,* **146,** 16, 19
  fiber organization in columns, **163,** 314
Radialization, avian retina and, **146,** 42
Radial potential, symplast and, **142,** 272–273
Radial spokes, flagellar dynein regulation, **155,** 170–171
Radiation, spaceflight experiments, **156,** 355–357
Radioactivity
  biological clocks in retina and, **146,** 117
  Sertoli-germ cell communication network and, **147,** 34, 39
Radioautography
  calcium channels and, **147,** 203–204
  DNA fiber
    basic principles, **154,** 265–269
    critical data, **154,** 269–274
    double-pulse labeling prior to, **154,** 274–280
    method, **154,** 263
    replicon model, **154,** 296
  retinal pigmented epithelium and, **146,** 76
  Sertoli-germ cell communication network and, **147,** 59
Radiofrequency, in NMR
  applications, **145,** 38, 44
  excitation of nuclei, **145,** 4–5
  at resonance, **145,** 9–11
  techniques, **145,** 21, 24, 31
Radioimmunoassay
  control of amphibian metamorphosis and
    adrenocortical hormones, **145,** 119, 125
    growth hormone, **145,** 138
    prolactin, **145,** 13O, 133, 137
    thyroid hormone, **145,** 107, 115, 117, 119
  control of rhythmic behaviors and, **147,** 4, 9
  environmental stimuli in plant cells and, **145,** 265
Radioprotectant drugs, spaceflight experiments, **156,** 305
Rag proteins
  genomic organization, **157,** 139
  markers of B cell differentiation, **157,** 139–140
  posttranslational modification, **157,** 139–140
  role in B cell lineage commitment, **157,** 162
  role in immunoglobulin recombination, **157,** 139–140, 197
Raman spectroscopy, in analysis of calcifications, **172,** 137
Ramipril, inhibition of smooth muscle cell growth, **161,** 24
Rana aurora, premetamorphic forebrain–spinal cord system, **138,** 97
Rana calmitans, hormonal control of metamorphosis and, **145,** 107
Rana cascadae, premetamorphic forebrain–spinal cord system, **138,** 97

Rana catesbeiana
arterial strips, **138,** 137
CGRP concentration, **138,** 166
control of metamorphosis and
adrenocortical hormones, **145,** 119, 123–125, 127
growth hormone, **145,** 140
prolactin, **145,** 135
thyroid hormone, **145,** 107, 113, 116
effect of GnRH, **138,** 99
growth factor-induced retinal regeneration and, **146,** 55
innervation, **138,** 130
magnocellular preoptic area, **138,** 120
MT injection, **138,** 125
NPY-containing neurons, distribution, **138,** 127
olfactory bulb, **138,** 157
optic chiasma, **138,** 107
premetamorphic forebrain–spinal cord system, **138,** 97
PRL from pituitary, **138,** 103
telencephalon, **138,** 113
vasoconstrictor effects, **138,** 169
Rana erythraea, VSCP-1 and VSCP-5 peptides, **138,** 174
Rana esculenta
AVT antidiuretic effect, **138,** 125
brain, **138,** 154, 168
carnosine immunoreactivity, **138,** 171
GnRH-containing perikarya, **138,** 93
hypothalamohypophyseal system, **138,** 164
mesencephalon, **138,** 147
neuropeptide concentrations in hypothalamus, **138,** 116
pituitary, **138,** 120
sympathetic ganglia, **138,** 95
thalamus and hypothalamus, **138,** 163
thyroxin levels in plasma, **138,** 99
TRH levels, **138,** 100
ventral infundibulum, **138,** 165
Rana ornativentris, hormonal control of metamorphosis and, **145,** 118
Rana perezi
ACTH cells, **138,** 110
hormonal control of metamorphosis and, **145,** 113, 115, 140
TSH cells, **138,** 103
Rana pipiens
brain, **138,** 97, 174
brain and retina, **138,** 112
control of metamorphosis and, **145,** 106, 108–109
growth hormone, **145,** 140
prolactin, **145,** 128
thyroid hormone, **145,** 112, 116–117
dorsal thalamus and optic tectum, **138,** 101
hypothalamus, **138,** 163
melanophore expansion, **138,** 152
mesencephalon, **138,** 147
TRH levels, **138,** 100
Rana ridibunda
ACTH secretion, **138,** 124
AVT-ir nerve fibers, **138,** 123
brain, **138,** 143, 145, 163
corticosteroid secretion, **138,** 137
hormonal control of metamorphosis and, **145,** 115, 124–127
internal zone of median eminence, **138,** 130
karyosphere and, **144,** 32–36
lateral amygdala, **138,** 141
MCH-ir neurons, **138,** 149
neural lobe, **138,** 95
neurointermediate lobe, **138,** 125, 132
nucleus preopticus, **138,** 108
optic chiasma, **138,** 107
PACAP isolation, **138,** 170
pars intermedia, **138,** 105
pituitary cells, **138,** 111
POMC cDNA, **138,** 139
preoptic and infundibular extracts, **138,** 109–110
PYY isolation, **138,** 127
SRIF-ir cell distribution, **138,** 115
SRIF-ir perikarya, **138,** 113
thyroxin levels, **138,** 103
thyroxin levels in plasma, **138,** 99
Rana rigosa, granuliberin-R, **138,** 174
Rana temporaria
CCK levels in optic tectum, **138,** 154
infundibular nuclei, **138,** 147
karyosphere and, **144,** 47
nuclear matrix, **144,** 41
vertebrates, **144,** 27–28, 30, 33, 35
mesencephalic anteroventral tegmentum, **138,** 165
neural lobe, **138,** 123
neurointermediate lobe, **138,** 119
NPY immunoreactivity, **138,** 131

pancreatic polypeptide, **138,** 127
SRIF-ir cells, **138,** 115
thyroxin levels in plasma, **138,** 99
Rana tigrina
arterial strips, **138,** 137
cardiac stimulation, **138,** 126
TRH in skin, **138,** 100
Ran/TC4, GTPase, role in nuclear import of proteins, **162B,** 246–248
Ranunculus scleratus, nuclei level, **140,** 243
RAP-1, 92-kDa protein, **162A,**202
Rapeseed, microgravity experiments, **156,** 320
Raphanus
higher plant mitochondrial genomes and, **141,** 142
self-incompatibility, **143,** 260, 262–263
Raphanus sativus
coated vesicles in, **167,** 2
higher plant mitochondrial genomes and, **141,** 133, 142, 145, 149
Raphidophytes, chloroplasts and, **144,** 277–278, 280–282
Ras protein
activation pathway, **157,** 155, 223–226
biological activity, **157,** 221–222
encoding gene, in adenylate cyclase regulation, **139,** 311–313
GTPase-activating protein, **157,** 222–224
GTP binding, **157,** 221, 223
kinase binding, **157,** 230
protein kinase cascade, **157,** 222
role in lymphocyte signaling, **157,** 216, 221–226
Rat
FAK expression in central nervous system of, **167,** 175–176
genital abnormalities in, **139,** 3
perinatal exposure to antihormones and, **139,** 14–17, 21–26
Long-Evans, RPE cells, **138,** 34–35
macrophage frequency in dome epithelium cells, **167,** 132
nongenital abnormalities in
prenatal exposure to phytoestrogens and, **139,** 39
perinatal exposure to sex hormones and antihormones and, **139,** 26–29
parathyroid hormone-related protein, gene structure, **166,** 234–236
trophoblast, cell morphology, **165,** 56–68
RBF, *see* Receptor binding factor
RB tumor suppression proteins, regulation of cell cycle, **157,** 249
Reacidification, symplast and, **142,** 275
Reactors, packed-bed, vertebrate cell culture technology, **142,** 187–192
Reaggregation
avian retina and, **146,** 23, 40–41
normal development, **146,** 2
retinogenesis *in vitro,* **146,** 9–13, 19
retinal precursor cells and, **146,** 150, 173, 183
Rearrangements
animal mtDNA and, **141,** 180
endosymbiont hypothesis and, **141,** 292
higher plant mitochondrial genomes and, **141,** 150–151, 153–157
mitochondrial genomes in fungi and, **141,** 113–117
Receptor binding factor
as acceptor protein for progesterone receptor, **162B,** 338
estrogen, in rabbit endometrium, **162B,** 354–355
progesterone
in avian oviduct, **162B,** 348–353
binding to c-*myc* promoter region, **162B,** 355–356
Receptor binding factor-1
localization in oviduct, **162B,** 350–354
potential binding to MAR-like elements, **162B,** 359
progesterone, in rat tissues, **162B,** 354
Receptor binding factor-2, extranuclear matrix binding, **162B,** 352–353
Receptors
in clathrin-coated vesicles
acidification of ligand, **167,** 23–26
description, **167,** 18–21
unresolved problems, **167,** 27
cnidarian neuropeptide, **167,** 55–56
FAK activation by nonintegrin, **167,** 171–172
in gut epithelium
in antigen binding to M cell membranes, **167,** 124
binding sIgA to M cells, **167,** 126
blocking to prevent pathogen invasion via M cells, **167,** 122
in endocytosis of antigens, **167,** 127
imaging by NMR, **145,** 30–31

Receptors (*continued*)
intracellular transport, **166,** 11–12
mediated endocytosis, **166,** 109–110
MIP-1α, **167,** 206–207
systems, sex hormones effect on, **139,** 6–7
T cell, types of, **167,** 96–97
TGF-β, **167,** 214–215
TNF-α, **167,** 229
Receptor tyrosine kinase
insect data, **164,** 197–199
properties, **164,** 195–197
signal transduction, **164,** 199–200
Reciprocal inhibition, control of rhythmic behaviors and, **147,** 13
Recognition, and gamete fusion in animals and plants
intergametic recognition during fertilization process, **140,** 413–414
sperm capacitation in animals and plants
male gamete changes, **140,** 410–411
pollen tube changes, **140,** 410
in vitro fertilization
in animals, **140,** 408–409
for higher plants by mixing gametes, **140,** 411–413
Recognition markers, sialic acids as, **175,** 155–156
Recognition sequences, prohormone, **167,** 58
Recombination
animal mtDNA and, **141,** 175, 225
chromosomal, in transgenic mice, **164,** 73–74
ciliates and, **141,** 3, 8, 10, 42, 59
DNA repair, **156,** 125–126
Drosophila muscles and, **143,** 63, 65, 71
endosymbiont hypothesis and, **141,** 291–293
flowering plants and, **143,** 315
higher plant mitochondrial genomes and, **141,** 129, 147–157, 164
abnormal phenotypes, **141,** 130–131, 133–135, 138–147
kinetoplastid mtDNA and, **141,** 67
mitochondrial
genetics, **161,** 64–65
and mitochondrial sex, **161,** 100–101
mobile introns, **161,** 65–67
mtDNA and mF plasmid, **161,** 91–95
and mt-nuclear fusion in zygotes, **161,** 96–97
polarity, **161,** 65
mitochondrial genomes in fungi and, **141,** 121
DNA diversity, **141,** 91, 93, 103, 105
generation of mtDNA diversity, **141,** 107, 111–116
reciprocal, higher plant mitochondrial genomes, **141,** 147, 150
vertebrate cell culture technology and, **142,** 169, 176, 194, 196
Reconstitution, functional, proton pump
lipid manipulation, **174,** 106
protein component manipulation, **174,** 106
solubilization of tonoplast V-ATPase, **174,** 103–104
tonoplast preparation, **174,** 102–103
V-ATPase into liposomes, **174,** 105–113
V-ATPase purification, **174,** 104–105
Reconstruction
cytoskeletal functions *in vitro,* **175,** 3–4
in demembered cell models of Characeae, **155,** 116–117
Recording, cell motility
three-dimensional analysis, **163,** 70–73
two-dimensional analysis, **163,** 50–53
Recoverin, retinal precursor cells and, **146,** 147
Rectal glands, avian salt glands and
blood supply, **144,** 149, 154
hormonal control, **144,** 196–197
innervation, **144,** 157, 163, 165
receptive systems, **144,** 177, 185, 187
secretory mechanism, **144,** 169, 171, 176
secretory tissue, **144,** 131, 140, 147–148
stimulus-secretion coupling, **144,** 201, 203
Red algae, chloroplasts and
biochemistry, **144,** 263–264
cryptomonads, **144,** 273, 275–276
euglenoids, **144,** 283, 285
heterokont algae, **144,** 281
pigments, **144,** 261
structure, **144,** 261–263
Red pigment concentrating hormone, control of rhythmic behaviors, **147,** 11, 15–16
Reduced folate system, **156,** 109–110
Reduced gravity, *see* Microgravity
5α-Reductase, inhibitors of effect on mouse pelvis
perinatal treatment with, **139,** 29–30
sex hormones and, **139,** 8

Reductional grouping
cultured plant cells, **158,** 80, 82–83
protoplasts, **158,** 83
Reduction fields, and instability of homogeneous state, **154,** 347–355
Reflection, eyespot, function in algae, **148,** 257, 269
Reflection confocal laser scanning microscopy, in eyespot reflection analysis, **148,** 261, 266–267, 269
Reflex arc, simplicity of cnidarian *versus* mammalian monosynaptic, **167,** 42
Reflexion coefficient
defined, **163,** 6, 8
leaky pores, **163,** 8–9
Regeneration
guided, and cementogenesis, **142,** 46–49
Hydra and hydrozoan polyps, capacity for, **167,** 38–39
spontaneous, cementogenesis, **142,** 44–46, 49
transdifferentiation in medusae and, **142,** 217
Regenerative cementogenesis, *see* Cementogenesis, regenerative
Regional induction, embryonic, definition, **150,** 407–408
Region of interest, NMR analysis, **145,** 29, 42–43, 45
Reg mutants, germ/soma differentiation in *Volvox carteri* embryos and, **139,** 204–206
Regulation
associated signals, cell cycle, integration, **162A,** 255–256
cell cycle, **167,** 192–194
FAK activity
by cytoplasmic proteins, **167,** 170–171
by integrins, **167,** 168–170
N-terminal domain in, **167,** 167
by FAK of focal adhesion assembly, **167,** 177
hemopoeitic stem cells, controlled within stem cell compartment, **167,** 191
by integrins of pp120 phosphorylation, **167,** 14–15
negative feedback, principles of, **167,** 187*f,* 189
rhizobial gene expression, **156,** 13–14, 16
Regulatory volume decrease
Ca_i^{2+} in, **161,** 215–216
Ca^{2+} influx, **161,** 215
characterization, **161,** 178–181
Cl^- channels, volume-sensitive
cell types, **161,** 181–182
properties, **161,** 184–185
cytoskeleton role, **161,** 203–205
in Ehrlich ascites tumor cells, swelling-initiated signal transduction pathway, **161,** 225–229
and free Ca^{2+} changes, **161,** 217–218
in human platelets, swelling-initiated signal transduction pathway, **161,** 225
K^+ channels, volume-sensitive
cell types, **161,** 181–182
properties, **161,** 182–184
K^+/Cl^- cotransport in, **161,** 186–189
K^+/H^+ exchanger, **161,** 197
long-term, **161,** 197
major changes during, **161,** 179
short-term, **161,** 197
swelling-activated taurine channel, **161,** 199–201
Regulatory volume increase
characterization, **161,** 178–181
Cl^-,HCO_3^- exchanger, **161,** 194–196
cytoskeleton role, **161,** 205–206
long-term, **161,** 197
major changes during, **161,** 179
Na^+/H^+ exchangers, **161,** 194–196
$Na^+/K^+/2Cl^-$ cotransporter
bumetanide binding, **161,** 193–194
characterization, **161,** 189–191
glide symmetry model, **161,** 191–192
identification and purification, **161,** 192
mechanisms, **161,** 191–194
occlusion of ions by, **161,** 194
primary structure, **161,** 191–194
RVI-after-RVD protocol, **161,** 178
short-term, **161,** 197
Reinfection, cementogenesis and, **142,** 47
Reissner's fiber, ependymin, **165,** 143–144
Rel, transcriptional regulation
domain in NF-κB, **143,** 45
molecular cloning, **143,** 25–31, 35–37, 39
structure, **143,** 21–22
viruses, **143,** 39, 41
Relaxation
polar, cytoskeletal dynamics and, **144,** 91, 93
times in NMR, **145,** 9–11, 31–32, 50

Relaxed selection hypothesis, for maternal age effect in aneuploidy, **167,** 278
Released cells, purification, **140,** 215–218
Renal glomerulus
absolute or total quantities, **161,** 125–127
Cavalieri principle, **161,** 127–128
disector/fractionator combination methods, **161,** 128–129
fractionator principle, **161,** 128–129
cell biology, **166,** 181–222
conventional methodological aspects, **166,** 183–187
anionic sites, **166,** 185–186
basement membrane, **166,** 184–185
foot process, **166,** 187
mesangial matrix, **166,** 184–185
slit diaphragm, **166,** 187
deep-etching method, **166,** 193–203
anionic sites, **166,** 196–198
basement membrane, **166,** 193–196
mesangial matrix, **166,** 198–200
slit diaphragm, **166,** 200–203
endothelial cell, **166,** 182
epithelial cell, **166,** 182
experimental alteration, **166,** 204–216
anionic site, **166,** 214–216
basement membrane, **166,** 204–214
mesangial matrix, **166,** 204–214
slit diaphragm, **166,** 214–216
freeze-substitution method, **166,** 188–193
anionic sites, **166,** 189–191
basement membrane, **166,** 188–189
foot process, **166,** 191–193
slit diaphragm, **166,** 191–193
function, **166,** 183–203
mesangial cell, **166,** 182
quantitative morphological assessment, **161,** 114–115
quick-freezing method, **166,** 187–188
sampling for stereology
sample distribution, **161,** 132–133
sample orientation, **161,** 133
sample size, **161,** 133–134
stereological methods
average volumes of cell types, **161,** 151–154
for glomerular number, **161,** 134–141
for glomerular volume, **161,** 141–143
individual glomeruli, **161,** 162–165
for kidney volume, **161,** 141–143
length density, **161,** 119–221
length of capillaries, **161,** 154–155
matrix volume density, **161,** 160–161
mean length of capillaries, **161,** 158
number of capillaries, **161,** 156–157
number of cell types, **161,** 148–151
numerical density, **161,** 121–122
surface area of capillaries, **161,** 155–156, 158
surface areas of cell types, **161,** 151–154
surface density, **161,** 117–119
thickness of glomerular basement membrane, **161,** 161–162
total number of cells in, **161,** 143–148
total number of each cell type, **161,** 151
volume density, **161,** 116–117
volume density of capillary lumina, **161,** 158–159
volume density of urinary space, **161,** 158–159
volume of an embedded kidney, **161,** 141
structure, **166,** 183–203
terminology, **161,** 115–116
in vivo cryotechnique
glomerular changes by abnormal hemodynamics, **166,** 219–222
glomerular morphology, **166,** 216–222
normal blood circulation ultrastructure, **166,** 217–219
volume estimation
indirect methods, **161,** 129–130
point-sampled intercepts, **161,** 130–131
selector method, **161,** 131–132
Renilla köllikeri, see also Sea pansies; 64–67; **167,** 48
Renin angiotensin system
avian salt glands, **144,** 192
intraadrenal, effect of endothelin, **171,** 284
Reovirus
adherence to M cells, **167,** 124
binding to sialic acids, **175,** 172–175
REP, *see* Rab escort protein
Repair
MAR/ORI DNA refractory to, **162A,** 353–354
occurrence on nuclear matrix, **162A,** 316–317
in vitro synthesis, **162A,** 139–140

Repetition time, in NMR, **145,** 11–12, 31
Replication
animal mtDNA and, **141,** 174–175, 201–203, 209, 229–231
elongation, **141,** 224–225
initiation, **141,** 222–224
regulation, **141,** 225–229
transcription, **141,** 218
Arabidopsis thaliana and, **144,** 70–71
ciliates and, **141,** 3, 59–60
structure, **141,** 4–9, 11
DNA, *see* DNA, replication
endosymbiont hypothesis and, **141,** 248, 300, 329
enhancers, model, **162A,** 315–316
higher plant mitochondrial genomes and, **141,** 137, 145–147, 164
kinetoplastid mtDNA and, **141,** 66, 69, 73, 82–83
late, AT-rich sequences, **148,** 91–94, 112
mitochondrial genomes in fungi and, **141,** 111
mitosis and
chromosomes, **144,** 240, 243, 248–249
dissociation of events, **144,** 227–228, 231, 238
mitotic events, **144,** 219, 225
timing, genes
allele inactivation, **160,** 70
H19 gene, **160,** 70, 73
Igf2 gene, **160,** 70,73
viral, **162A,** 486–488
Replication band, ciliated protozoa
characterization, **153,** 137–138, 166
chromatin replication models, **153,** 161–166
cytochemical studies, **153,** 154–155
early studies, **153,** 140–143
functional characteristics, **153,** 143–148
hypotrichs
characterization, **153,** 138–139
chromatin replication models, **153,** 162–165
early studies, **153,** 143
function, **153,** 148
immunochemical studies, **153,** 155–161
macronuclei
characterization, **153,** 137–139
chromatin replication models, **153,** 162–165
cytochemical studies, **153,** 155
function, **153,** 144, 146, 148
immunochemical studies, **153,** 156–157, 159–160
ultrastructure, **153,** 149–150, 152
micronuclei
characterization, **153,** 138
function, **153,** 144
immunochemical studies, **153,** 156–157, 159
microscopy, **153,** 140, 151
properties, **153,** 167
structure, **153,** 137–138, 166
ultrastructure, **153,** 148–154
Replication factories, *see also* Nucleolar factories
associated with nuclear matrix, **162A,** 461–462
DNA polymerase concentration, **162A,** 126–127
visualization, **162A,** 132–136
Replication fork barrier, **154,** 263, 292–293
Replication fork rate
concept, **154,** 263
replicon model, **154,** 295, 299
replicon termination, **154,** 293
Replication granules, correspondence to *in vivo* clustersomes, **162A,** 44–53
Replication initiator protein, sites in MAR/ORIs, **162A,** 333–335
Replication origins, *see* Origins of replication
Replicators, chromosomal, identification, **162A,** 464–466
Repolarization, biological clocks in retina and, **146,** 126
Reproduction
barriers in flowering plants, **140,** 450–452
definition, **140,** 450–452
interspecific, **140,** 450, 451
intraspecific, **140,** 450, 451
types of systems, **140,** 452
cnidarian
neuropeptides and, **167,** 54–55
overview, **167,** 38
phase of plant development, role of AGPs, **174,** 247–255
sexual, *see* Sexual reproduction
in vitro, pollination in maize, **140,** 394–396

Reproduction (*continued*)
in vitro, temperature stress at different phases, **140,** 397–400
kernal development, **140,** 399
pollen selection, **140,** 399–400
postpollination, **140,** 397–398
stress on reproductive tissues, **140,** 398–399
in vivo, influence of temperature stress on, **140,** 392–394
control of male and female quality, **140,** 396–397
Reproductive system
male tract, and sperm maturation, *see* Sperm, maturation
parathyroid hormone-related protein, **166,** 250–251
gynecological tumors, **166,** 259–260
tissues
heat shock response in, **140,** 400–402
stress on, **140,** 398–399
vertebrate tract, cell death in, **163,** 141
Reptiles
avian salt glands and, **144,** 129, 160, 187
blood supply, **144,** 149
secretory tissue, **144,** 131–132, 139, 147–148
karyosphere and, **144,** 25, 36–37
Rescued photoreceptor cells, **138,** 17–24, 26–33
Resetting hypothesis, biological clocks in retina and, **146,** 121
Resialylation, cell surface glycans, **175,** 150–151
Residual bodies, Sertoli-germ cell communication network, RB/CES, **147,** 34–35, 37, 68, 71, 74
Residues
Antho-RFamide
sea anemones, and biosynthesis of, **167,** 59
sea pansies, and biosynthesis of, **167,** 64–67
Antho-RPamide I, sea anemones, and biosynthesis of, **167,** 68–69
Antho-RPamides II-IV biosynthesis, **167,** 69–71
as cnidarian preprohormone processing sites, **167,** 78–82
MIP-1α cysteine, **167,** 199
Resistance
alkylation, glutathione S-transferase, **156,** 134–137
associated genes, *see* Major resistance genes
cellular, cancer chemotherapy, **156,** 77–137
disease, *see* Disease resistance
drug, *see* Drug resistance
nonhost, *Arabidopsis thaliana,* **144,** 58, 69
Resorption, regenerative cementogenesis and, **142,** 44, 46–47, 49–51
Respiration
avian salt glands, **144,** 144, 176
–heat discharge, in angiosperms, effect of microgravity, **171,** 40
intracellular calcium messenger system and, **142,** 318
symplast and
auxin, **142,** 285
electrophysiological structure, **142,** 270–272
integration of activity, **142,** 285–296, 299
proton pumps, **142,** 274, 276
Respiratory burst
bacteria resistant to, **161,** 307
bactericidal, **161,** 306
Respiratory chain proteins, endosymbiont hypothesis and, **141,** 308–311
Respiratory inflammation, in cystic fibrosis, **172,** 211
Responsive element, vitamin D, enhancer function, **162A,** 261–263, 266–269
Restenosis, therapeutic strategies, **161,** 29–30
Resting zygote, volvacacean, **139,** 199
Restorative force, biological clocks in retina and, **146,** 134, 137
Restriction fragment length polymorphism
analysis of somatic segregation in carrots, **158,** 109
Arabidopsis thaliana, **144,** 62–65, 67, 76
higher plant plastids and, **145,** 159
identification of disease genes, **158,** 229–230
melanocytes and, **146,** 240, 245–246
urease in plant cells and, **145,** 86
Restriction maps, higher plant mitochondrial genomes, **141,** 135–144, 164

Resveratrol, chalcone synthase and, **147,** 237
Rete testis, structure and function, **154,** 131–137
Retina
avian
histogenesis, **146,** 40–42
normal development
lamination, **146,** 1–2
model, **146,** 2–4
regenerative retinogenesis, **146,** 6–8
retinogenesis, **146,** 4–5
retinogenesis *in vitro*
cell adhesion, **146,** 21–26
cell sorting, **146,** 8–9
histotypical differentiation, **146,** 13, 15–23
primary lamination, **146,** 36–40
radial cell columns, **146,** 34–37
reaggregation, **146,** 10–14
rosetted spheroids, **146,** 27–28
stratospheroids, **146,** 28–34
in avian neurogenesis
acetylcholinesterase in study, **151,** 150–152
butylcholinesterase
downregulation during *in vitro* regeneration, **151,** 151, 153–156
in mature retina, **151,** 146
biological clocks in, **146,** 83–86, 139
Bulla, **146,** 128–129
rhythm generation, **146,** 134–139
transmembrane loop, **146,** 129–133
entrainment pathway, **146,** 100–101, 115
efferent regulation, **146,** 108–114
membrane potential, **146,** 102–103
photoreceptor, **146,** 101–102
protein synthesis, **146,** 106–108
second messengers, **146,** 103–106
free-running rhythms, **146,** 96–97
opisthobranch retina
functional organization, **146,** 88–93
ocular rhythm, **146,** 86–88
pacemaker localization, **146,** 93–95
oscillator, **146,** 99–100
output pathway, **146,** 123, 126–128
pacemaker mechanisms, **146,** 115
intracellular processes, **146,** 116–125
membranes, **146,** 115–116
pathway analysis, **146,** 97–99
blood barrier
glucose transport, **172,** 15–17
structure, **172,** 14–15
cell transplants, evolution
photoreceptor cell transplants, **138,** 8
separated layers and suspensions, **138,** 7
techniques developed to deliver retinal cells to eye, **138,** 5–6
whole retinal transplants, **138,** 6–7
cognin, avian retina and, **146,** 24
detachment, cytokine production and, **146,** 75
gap junctions
connexin distribution, **157,** 349–350, 354
coupling, **157,** 354
freeze-fracturing, **157,** 350–354
morphology, **157,** 349
horizontal cells, **157,** 349–350, 354
inner layer, growth factor-induced regeneration, **146,** 58
insect
ion concentration, **176,** 316–319
light-induced ion movements, **176,** 316–319
organization
ommatidium in honeybee compound eye, **176,** 309–311
photoreceptor cell plasma membrane, **176,** 311–313
protein mobility in photoreceptor plasma membrane, **176,** 313–314
light adaptation, **157,** 354
melanocytes and, **146,** 192, 194, 198
neural, *see* Neural retina
RCS dystrophic rats, RPE cell transplantation into, **138,** 34–36
regeneration, growth factor-induced, **146,** 49, 69
development
differentiation, **146,** 50–52
optic cup, **146,** 50
fibroblast growth factors, **146,** 67–69
induction by growth factors, **146,** 59–60
chicken embryo, **146,** 61–67
intraocular delivery, **146,** 60–61
ocular tissues, **146,** 55–59
transdifferentiation
adult vertebrates, **146,** 52–54
vertebrate embryos, **146,** 54–55

Retina (*continued*)
vasculature in RPE cell transplanted retinas, **138,** 33–34
vessel preparation, **138,** 33
Retinaldehyde, retinal precursor cells and, **146,** 181
Retinal dystrophy, in RCS rat, **138,** 40
Retinal ganglion cells
growth factor-induced retinal regeneration and, **146,** 62–64
retinal precursor cells and, **146,** 148, 151, 171
Retinal pigment epithelium
avian, **146,** 41
cellular processes, **146,** 28–34
normal development, **146,** 2, 4, 6–8
retinogenesis *in vitro,* **146,** 10
cell fate choice regulation
by soluble factors, **150,** 125–128
by substratum, **150,** 125–128
cell–substratum adhesiveness, **150,** 123–124
cells
plasticity in culture, **171,** 245–247
polarization, **171,** 237–243
cytokine production by, **146,** 75–80
deficiency of in RCS dystrophic rats, **138,** 2–3
development during embryogenesis, **171,** 227–228
differentiated state, effects of environmental changes, **171,** 243–245
gene expression patterns, **171,** 228–232
in glucose transport, **172,** 15–16
growth factor-induced retinal regeneration, **146,** 49, 59–62, 66–67
development, **146,** 50–52
differentiation, **146,** 52–55
fibroblasts, **146,** 67–69
ocular tissues, **146,** 58–59
human, transplants
diseases possibly treated by, **138,** 41
studies, first, **138,** 41–42
mammalian, regeneration
basement membrane effects, **152,** 253
cytologic changes, **152,** 226–227
cytoskeletal changes, effects, **152,** 255
extracellular matrix effects, **152,** 253–254
growth factor effects, **152,** 253–254
interactions, **152,** 223–224, 254
laser lesions, **152,** 224–225
mechanisms, **152,** 255
models, **152,** 225–226
plasma membrane alterations, **152,** 252–253
polarity, **152,** 227–232
in rabbits, **152,** 251
sodium iodate retinopathy
cytology, **152,** 234, 236–251
as model, **152,** 232–235
medium conditioned by, cytokine production and, **146,** 79
melanocytes and, **146,** 194
molecular marking during differentiation, **171,** 235–237
Na^+,K^+-ATPase polarity, **165,** 174–177
nuclear magnetic resonance and, **145,** 38
as permeability barrier, **172,** 15
phenotype change, role of oncogenes, **171,** 255
retinal precursor cells and, **146,** 148, 153, 177–178, 180–182
role in retinal photoreceptor function, **171,** 226–227
structure, **171,** 225–226
tissue type specification, **171,** 228–232
TPA effects, **150,** 124–125
transdifferentiation
amphibian RPE, **171,** 248–249
cellular mechanics, **150,** 128–131
chick RPE *in vitro,* **171,** 249
in medusae and, **142,** 217–218
process, **150,** 119–120, 122
rat RPE *in vitro,* **171,** 249–254
transplantation, **171,** 247
applications to human eye disease, **138,** 39–42
first human RPE transplantation studies, **138,** 41–42
techniques developed for possible human eye transplantation, **138,** 39–41
various diseases possibly treated by human RPE transplants, **138,** 41
conclusions
achievements, **138,** 42–43
challenges, **138,** 43–44
technology
closed-eye method, **138,** 9–10

open sky method, **138,** 9
transcleral approach, **138,** 10
transplantation, animal models
effects of RPE cell age, **138,** 34–39
photoreceptor cell rescue
characteristics of rescued photoreceptor cells, **138,** 17–24
control for RPE cell transplantation studies, **138,** 25–26
first RPE cell transplantation results, **138,** 13–17
mechanism of action of photoreceptor cell rescue, **138,** 26–33
optimal conditions for RPE cell transplantation results, **138,** 24–25
prevention of neovascularization in RCS dystrophic rats, **138,** 33–34
RPE cell isolation and culture, **138,** 10–13
tyrosinase gene transcription regulation, **171,** 233–235
tyrosinase and melanin synthesis, **171,** 232–233
Retinal precursor cells
chick embryo cells, **146,** 151–155
development, **146,** 149–150
differentiation, **146,** 145, 150
experimental analysis
commitment, **146,** 165–167
molecular polarity, **146,** 158–163
responsiveness to light, **146,** 162–165
structural polarity, **146,** 155–158
microenvironmental factors, **146,** 176
extracellular matrix, **146,** 178–180
growth factors, **146,** 176–178
neurotransmitters, **146,** 181–182
retinoids, **146,** 180–181
tissue extracts, **146,** 182–183
photoreceptor default hypothesis, **146,** 167–174
photoreceptors, **146,** 145–149
photoreceptor subtypes, **146,** 174–176
Retinectomy, growth factor-induced retinal regeneration, **146,** 61, 67, 69
Retinoblastoma
gene, as inhibitory regulator of cell proliferation, **167,** 233
melanocytes and, **146,** 228, 233, 248–249
protein
associated with nuclear matrix, **162B,** 30–31
lamin A binding domain, **162B,** 395–396
lamin-binding, **162B,** 172
retinal precursor cells and, **146,** 179
susceptibility gene product, as cell cycle regulator, **167,** 193–194
Retinoic acid
amphibian gene expression regulation, **169,** 128–129
chondrocyte
dose response, **159,** 305–306
effect on differentiation, **159,** 271, 274–275, 277, 300, 303, 305–306
effect
acute promyelocytic leukemia cells, **159,** 182–183
alkaline phosphatase, **159,** 305
induced heparin-binding factor, role in chondrogenesis, **159,** 317
neurite initiation and elongation role, **153,** 255
nuclear receptors, **159,** 182–183
receptor expression in mesenchyme, **159,** 274–275
retinal precursor cells and, **146,** 180–181
Retinoic acid α receptor, fused to PMLs, **162A,** 80,169–170
Retinoic acid receptor-β-galactosidase fusion protein, **164,** 75–76
Retinoic acid syndrome
animal models, **158,** 260
craniofacial abnormalities, **158,** 260, 262
Retinoids, retinal precursor cells and, **146,** 176, 180–181
Retinol, retinal precursor cells and, **146,** 180–181
Retinol-binding protein, Sertoli-germ cell communication network, **147,** 50
Retinospheroids, avian retina
cellular processes, **146,** 24, 27, 32, 34, 36
histiotypic differentiation, **146,** 40–42
normal development, **146,** 2
retinogenesis *in vitro,* **146,** 9, 11, 13, 15–17, 19–21
rosettes, **146,** 2, 10
Retrograde actin escalator, growth cone, **165,** 258–259

Retrograde surface transport
actin treadmilling, **150,** 51–54
definition, **150,** 41
future research, **150,** 62–65
hypothesis, alternative, **150,** 46–48
mechanism, **150,** 43–46
membrane raking hypothesis, **150,** 48–51
Retroviruses
application in gene therapy, **158,** 264
avian retina and, **146,** 35
growth factor-induced retinal regeneration and, **146,** 52
higher plant mitochondrial genomes and, **141,** 149, 158
integration, at multiple sites, **162A,** 436–437
melanocytes and, **146,** 205, 211
mitochondrial genomes in fungi and, **141,** 107
retinal precursor cells and, **146,** 150
transcriptional regulation and, **143,** 27, 39, 43
Reverse transcription
higher plant mitochondrial genomes and, **141,** 149, 158–159
mitochondrial genomes in fungi and, **141,** 107–109
Reversible protein phosphorylation, *see* Plant enzymes
Reversion, chromoplast, **163,** 205
Revertants, higher plant mitochondrial genomes and, **141,** 134, 149–151
Rev-responsive element, HIV-1, **162B,** 210–213
RF-amide neuropeptides
antisera, in visualizing cnidarian neurotransmitter substances, **167,** 47–48
apparent ubiquity in cnidarians, **167,** 48–49, 77
as candidates for earliest neurotransmitters, **167,** 77
in hydrozoans, **167,** 74–77
neuromuscular transmission and, **167,** 51
sequence similarity of cnidarian neuronal substances to, **167,** 44
transdifferentiation in medusae
DNA replication, **142,** 255
initiation, **142,** 238–239, 243, 247–248
potential, **142,** 224, 229–230, 235
RFLP, *see* Restriction fragment length polymorphism
Rhabdomyosarcoma, proteins in striated muscle and, **142,** 101–102
Rhamnogalacturonans, extracellular matrix and, **146,** 264–265, 285
Rheumatic disease, dendritic cell role in, **153,** 88–89
RHIB, *see* Retinoic acid-induced heparin-binding factor
Rhizobium
chalcone synthase and, **147,** 248–250
extracellular matrix and, **146,** 269
NGR234, **156,** 12–13
host-specific modifications, **156,** 25
Nod factors, **156,** 22
–plant interactions
host specificity, **156,** 30–32
plant hormones, **156,** 45–46
signal exchange, **156,** 49–51
Rhizobium etli, **156,** 7
Rhizobium fredii
host-specific modifications, **156,** 24–27
nodD copies, **156,** 12
Nod factors, **156,** 22
nolJ gene, **156,** 16
Rhizobium galageae, **156,** 12
Rhizobium leguminosarum
effect on flavonoid production in host plants, **156,** 9
host plants for, **156,** 1
host-specific modifications, **156,** 24–25
nodD gene, **156,** 12–13
NodD proteins, **156,** 16
nodE gene, **156,** 30
Nod metabolites, **156,** 19
nodN gene, **156,** 28
repressor presence, **156,** 17
Sym plasmid, **156,** 7
Rhizobium meliloti
binding of NolR repressor, **156,** 11–14
extracellular signal strength, **156,** 18
host-specific modifications, **156,** 24–26
inoculation with, **156,** 9
modulation of host plants, **156,** 1
nodD1 gene, **156,** 16
Nod factors, **156,** 19
Nod metabolites, **156,** 19
nodM mutant, **156,** 31
nolR repressor gene, **156,** 8

sulfated Nod factors, **156,** 49
transport functions, **156,** 28
Rhizobium tropicii, **156,** 13, 15, 22
Rhizoids, *Chara globularis,* effect of microgravity, **171,** 23
Rhizopus niveus, self-incompatibility and, **143,** 287
Rhodamine, environmental stimuli in plant cells, **145,** 277, 283, 289, 291
Rhodella violacea, cyanelles, plastid genome, **151,** 37
Rhodophytes, endosymbiont hypothesis and, **141,** 263, 271–272, 275–276, 279
Rhodopsin
cytoskeletal dynamics and, **144,** 107
insect, cloning, **164,** 242–243
as photoreceptor in algae, **148,** 277–278, 281–286
retinal precursor cells and, **146,** 175, 179, 181–182
Rhopalia, ocelli and statocysts in, **167,** 39–40
Rhynia, symplast and, **142,** 264
Ribonuclease
activity of S proteins, biological relevance of, **140,** 472–475
self-incompatibility, **143,** 287–288, 290
Ribonuclease protection assay, gene expression analysis, **158,** 237–238
Ribonucleoproteins
export/import, **162B,** 248–249
fibrogranular network, **162A,** 81–82
in interchromatinic regions, **162A,** 6–8
karyosphere, **144,** 15, 42–43, 45
matrix channels overloaded with, **162B,** 439–440
nuclear matrix domain, **162B,** 103–112
small nuclear
associated proteins, **162A,** 71–75
distribution throughout cell cycle, **162B,** 33–36
NuMA association, **162B,** 18–19
and speckled domain, **162B,** 311
Ribonucleotide reductase, **156,** 129
Ribosomal DNA
ciliates and, **141,** 48
endosymbiont hypothesis and, **141,** 272, 276
higher plant plastids and, **145,** 159
karyosphere and, **144,** 11, 17, 37
repeat size, **162B,** 427–430
Ribosomal RNA
animal mtDNA and, **141,** 174, 189–191, 193, 203, 208–209
genome content, **141,** 179–180
regulation, **141,** 226–229
transcription, **141,** 221
chloroplasts, **144,** 265, 276, 284–285, 293–294
ciliates and, **141,** 1, 53, 58–59
genes, **141,** 45–46
large rRNA, **141,** 48, 50–52
Paramecium aurelia, **141,** 12–13, 35–36, 41, 43
small rRNA, **141,** 47–49
structure, **141,** 7–8
cyanelle genome organization, **151,** 39–41
endosymbiont hypothesis
algae, **141,** 273–275
eukaryotes, **141,** 276
gene expression, **141,** 298, 303, 307–308
gene structure, **141,** 308, 311, 316–319
gene transfer, **141,** 325
genome organization, **141,** 289–291, 293–297
lineages of life, **141,** 237–242, 245–246
mitochondrial 280- 282, **141,** 288, 327
nuclear genome, **141,** 247–251
Plasmodium, **141,** 326
plastids, **141,** 253, 264, 268–273
rRNA sequences, **141,** 319–324
formation in nucleolar matrix, **162B,** 112–114
higher plant mitochondrial genomes and, **141,** 148–149, 157, 164
higher plant plastids, **145,** 155–156, 159, 162, 166–167
kinetoplastid mtDNA and, **141,** 69–70, 73–74, 77–78
large subunit
ciliates and, **141,** 46, 48, 50–53, 59
endosymbiont hypothesis and
gene expression, **141,** 298
gene structure, **141,** 308
gene transfer, **141,** 325
genome organization, **141,** 293–297
lineages of Iffe, **141,** 238, 242
mitochondria, **141,** 288, 316
nuclear genome, **141,** 247–248
Plasmodium, **141,** 326
plastids, **141,** 263, 269, 276
rRNA sequences, **141,** 319, 323

Ribosomal RNA (*continued*)
mitochondrial genomes in fungi and, **141,** 120
DNA diversity, **141,** 91, 96–97, 100, 103, 105–106
generation of mtDNA diversity, **141,** 107–109, 118
small subunit
ciliates and, **141,** 46–50, 52, 59
endosymbiont hypothesis and gene structure, **141,** 308, 316
gene transfer, **141,** 325
genome organization, **141,** 294–297
lineages of life 238-239, **141,** 241
mitochondria, **141,** 288
nuclear genome, **141,** 251
Plasmodium, **141,** 326
plastids, **141,** 263, 269, 272, 276
rRNA sequences, **141,** 319-321, 323–324
kinetoplastid mtDNA and, **141,** 69
mitochondrial genomes in fungi and, **141,** 120
DNA diversity, **141,** 91, 95, 97, 102–103, 105
generation of mtDNA diversity, **141,** 109, 111–112
Ribosomes
animal mtDNA and, **141,** 174, 183, 191
avian salt glands and, **144,** 135, 141, 172
chloroplasts and, **144,** 263, 275, 277
ciliates and, **141,** 35–36, 38, 41, 55, 58
as contaminants in isolation of plat cell coated vesicles, **167,** 8
endosymbiont hypothesis and gene expression, **141,** 306–307
gene structure, **141,** 311, 313
genome organization, **141,** 293, 295, 298
lineages of life, **141,** 242
mitochondria, **141,** 280, 282, 288
nuclear genome, **141,** 247–251
Plasmodium, **141,** 326
plastids, **141,** 259, 267–268, 276, 278
genes, in *Volvox carteri,* **139,** 224–225
GnRH-encoding gene in teleosts and, **147,** 185
higher plant mitochondrial genomes and, **141,** 134, 157–158, 163–164
higher plant plastids and, **145,** 151, 155, 157–158, 171
karyosphere and, **144,** 18
kinetoplastid mtDNA and, **141,** 71
mitochondrial genomes in fungi and, **141,** 96, 98
Sertoli-germ cell communication network and, **147,** 73
transdifferentiation in medusae and, **142,** 229
urease in plant cells and, **145,** 84
Ribulose-bisphosphate, higher plant plastids and, **145,** 171, 173
Ribulose-bisphosphate carboxylase, salt tolerance, **165,** 15
Rice
plastids and, **145,** 153, 155, 157, 200
urease in plant cells and, **145,** 67, 80
Ring canals
cytoplasmic streaming, **173,** 220–221
formation in *Drosophila* development, **175,** 44–46
Ring–spoke complex, NPC, **162B,** 261–263
RNA
Arabidopsis thaliana, **144,** 70, 74
avian salt glands, **144,** 141, 143, 158, 172
chalcone synthase and, **147,** 254, 257
chloroplasts and, **144,** 285
ciliates and, **141,** 1, 7–8, 13, 60
complementary, Sertoli-germ cell communication network, **147,** 59
denaturation, **158,** 237
Drosophila muscles and, **143,** 112, 117–118
contractile protein genes, **143,** 94, 100
myosin light chains, **143,** 105, 107
editing
animal mtDNA and, **141,** 198
endosymbiont hypothesis and
gene expression, **141,** 304–305
gene structure, **141,** 309
gene transfer, **141,** 325
genome organization, **141,** 294
mitochondria, **141,** 281
higher plant mitochondrial genomes and, **141,** 129, 162–163
kinetoplastid mtDNA and, **141,** 65, 68, 84
developmental control, **141,** 81–82
maxicircle genes, **141,** 70, 74
efflux, import signals, **151,** 193–195
endosymbiont hypothesis and, **141,** 235, 240, 250, 267
gene expression, **141,** 300, 302, 305, 307
mitochondria, **141,** 290, 294, 328

export/import, **162B,** 248–249
extracellular matrix and, **146,** 292, 299
extraction, **158,** 237
flowering plants and, **143,** 307, 314
GnRH-encoding gene in teleosts and, **147,** 172, 185–186
guide, *see* Guide RNA
higher plant mitochondrial genomes and, **141,** 129, 137, 149, 160–161
higher plant plastids and, **145,** 151, 158–159, 162, 164–165, 175
hormonal control of amphibian metamorphosis and, **145,** 121
and instability of homogeneous state
metabolic field, **154,** 359–360
intermediate filament proteins, gene regulatory function, **154,** 5, 69
intracellular calcium messenger system and, **142,** 333
karyosphere and, **144,** 1, 11, 19, 30, 42
kinetoplastid mtDNA and, **141,** 65–66, 81–82, 84
maxicircle genes, **141,** 69–70, 73–74
maxicircle transcripts, **141,** 74–79
melanocytes and, **146,** 240
messenger, *see* Messenger RNA
metabolism
on nuclear matrix, **162A,** 99–105
and transport, **162B,** 184–186
mitochondrial, *see* Mitochondrial RNA
myofibrillogenesis, **143,** 154, 176
nascent, chains, **162B,** 419
Northern blot analysis, **158,** 237
pre-rRNA, nucleolar, **162B,** 306–307
primed DNA synthesis, **162A,** 550–552
primer, synthesis inhibition, **162A,** 554–556
processing
sites and organization, **162B,** 310–314
and splicing, **162A,** 159–165
purification of messenger RNA, **158,** 237
in quantitative regulation of tissuespecific genes, **148,** 88
retention in nuclear matrix, **162A,** 85–86
ribosomal, *see* Ribosomal RNA
role in nuclear structure, **162A,** 97–99
in *Saccharomyces cerevisiae* gene expression
enzyme-dependent processing, **152,** 168
processing, **152,** 150–158, 168–169
regulation, **152,** 165–166
translation, **152,** 159–162
turnover, **152,** 158–159
Sertoli-germ cell communication network and, **147,** 65, 67, 74
in situ hybridization, **158,** 238
splicing
coordinated with transcription, **162A,** 39
on nuclear matrix, **162A,** 103–104
synthesis during myocardial growth
messenger, **150,** 78
ribosomal, **150,** 78
in temporal expression experiments, **148,** 83–84
transcription, electromagnetic field effect, **158,** 294, 308–312
α polymerase, **158,** 311
Drosophila melanogaster, **158,** 308–309
Escherichia coli, **158,** 312–313
HL-60 cells, **158,** 310–311
NusA, **158,** 312
oncogenes, **158,** 310–312
Saccharomyces cerevisiae, **158,** 310
transcriptional regulation and, **143,** 11, 17, 48
transcript tracks and domains, **162A,** 78–79
transdifferentiation in medusae and, **142,** 237
transfer, *see* Transfer RNA
U snRNA, import signals, **151,** 195–196
in whole-chromosome hybridization
detection methods, **153,** 181
development of technique, **153,** 174
probe labeling methods, **153,** 179–180
in situ hybridization technique, **153,** 176
RNA polymerase I, nucleolar factories, **162A,** 137–139
RNA polymerase II
large subunit, and matrin p50, **162A,** 34–37
nuclear distribution, **162A,** 155
transcription foci, **162A,** 136–137
RNA polymerases
animal mtDNA and, **141,** 230
regulation, **141,** 226, 228
transcription, **141,** 220–222
cell cycle dependence, **162B,** 391–393
chalcone synthase and, **147,** 256, 262
endosymbiont hypothesis and
lineages of life, **141,** 240, 242–243
mitochondria, **141,** 299, 301, 306, 326
nuclear genome, **141,** 248–250
plastids, **141,** 263, 266

RNA polymerases (*continued*)
higher plant mitochondrial genomes and, **141,** 161
higher plant plastids and, **145,** 155–156, 164, 166
immobilized on nuclear matrix, **162A,** 158–159
proteins in striated muscle and, **142,** 107
in *Saccharomyces cerevisiae* gene expression, **152,** 149
transcriptional regulation and 1, **143,** 18
RNase, animal mtDNA and, **141,** 224, 226, 228–230
RNP, *see* Ribonucleoproteins
Robertsonian translocation, effect on gametes, **158,** 226
Rodents, Peyer's patches in, **167,** 94
Rods
growth factor-induced retinal regeneration and, **146,** 61
outer segments
retinal pigmented epithelium and, **146,** 75
retinal regeneration and, **146,** 59
retinal precursor cells and, **146,** 147–148, 161
microenvironmental factors, **146,** 178–179, 181, 183
photoreceptor subtypes, **146,** 174–175
signals inducing, retinal precursor cells, **146,** 173
Rolling balloon model, ameboid movement, **139,** 163–164, 166
Root cap, microgravity experiments, **156,** 331–332
Root cementum, *see* Cementogenesis, regenerative
Root hairs
curling, **156,** 2–3
deformation, **156,** 2–3, 39
emerging, rhizobia, **156,** 2
nod factors, response to, **156,** 46
nodulins, **156,** 32–37
rhizobia, action at, **156,** 2–3, 39
Roots
angiosperm, effect of microgravity
on elongation and differentiation, **171,** 34–35
on meristem cells, **171,** 32–34
growth, microgravity experiments, **156,** 322–327
pattern formation, role of AGPs, **174,** 255–258
Root sheath, regenerative cementogenesis, **142,** 9, 13
Root surfaces, regenerative cementogenesis
biological problems, **142,** 49–51
clinical conditions, **142,** 43–44
collagen, **142,** 31–37
culture material, **142,** 28–31
factors, **142,** 37–43
guided regeneration, **142,** 46–49
spontaneous regeneration, **142,** 44–46
Rosa canine, pollen and pollenkitt, chemicals, **140,** 37–38
Rosa rugosa, pollen and pollenkitt, chemicals, **140,** 37–38
Rosettes
avian retina and, **146,** 25–28, 41
thymocytes, **156,** 173–174
Rotary-motive force, bacterial flagellar motor, **147,** 122
Rotation cultures, avian retina and, **146,** 8–10, 21–22, 40
Rotation threshold, bacterial flagellar motor end, **147,** 141–142
Rotaviruses, binding to sialic acids, **175,** 173–174
Royal College of Surgeons rat strain, and retinal transplants, **138,** 1–3, 24–25
RPE, *see* Retinal pigment epithelium
RS domain, *see* Arginine/serine-rich domain
R-type cells, biological clocks in retina and, **146,** 92, 114
Rubisco
chloroplasts and, **144,** 263, 276, 281, 291
cyanelle genes of photosynthetic apparatus, **151,** 48–49
higher plant plastids and
amino acid synthesis, **145,** 196–198
CO_2 fixation, **145,** 172–176
large subunit (LSU), **145,** 16O, 172
small subunit (SSU), **145,** 16O, 168, 172, 175
structure, **145,** 160
translocation, **145,** 168
Ruffling
in actin assembly, **150,** 64
definition, **150,** 38, 41

rearward traction forces correlated to, **150,** 41
in retrograde surface transport, **150,** 48
Rusts, mitochondrial genomes in fungi and, **141,** 104
Ruthenium red, agravitropism induction in plants, **168,** 161–162
RVD, *see* Regulatory volume decrease
RVI, *see* Regulatory volume increase
Ryanodine, calcium channels, **147,** 209
Ryanodine receptor
in calcium activation of mammalian eggs, **152,** 197–198
proteins in striated muscle, **142,** 122–123
in striated muscle, **142,** 122–123

S

Saccharides
chains, self-incompatibility, **143,** 284
dome *versus* villus enterocyte composition of terminal, **167,** 131
M cell
in bacterial adherence to, **167,** 124–126
site-specific variations in, **167,** 115
Saccharomyces
cytoskeletal dynamics and, **144,** 97
mitosis and, **144,** 222
Saccharomyces cerevisiae
19.5S complex, role of Rab GTPase, **176,** 62–63
animal mtDNA and, **141,** 230
cytokinesis, role of nonmuscle tropomyosin, **170,** 20–22
cytoskeletal dynamics and, **144,** 103
endosymbiont hypothesis and gene expression, **141,** 298, 300–302
gene structure, **141,** 311
genome organization, **141,** 290
mitochondria, **141,** 282
flowering plants and, **143,** 307
function regulation by Rab GTPase, **176,** 27–28
mitochondrial gene expression
mechanism, **152,** 145, 167
post-translational factors, **152,** 162–164, 169
regulation, **152,** 164–167
RNA
messenger, **152,** 151–162, 165–166, 168–169
modification, **152,** 151, 168
processing, **152,** 151–158, 168
regulation, **152,** 165
ribosomal, **152,** 150–151, 168
splicing, **152,** 153
transfer, **152,** 150–151, 165, 168
translation, **152,** 159–162
turnover, **152,** 158–159
RNA polymerase role, **152,** 149
transcription, **152,** 146–150, 167–169
transcription factors, **152,** 149–150
transcription promoters, **152,** 146–149
translation, **152,** 159–162
mitochondrial genomes, **141,** 12, 40, 42, 50, 53
mitochondrial genomes, in fungi
DNA diversity, **141,** 93, 98–103, 105
generation of mtDNA diversity, **141,** 107–113, 115–119
mitosis and, **144,** 226, 228, 230, 250–251
mt-nuclear fusion after mitochondrial fusion, **161,** 61–62
precursor protein processing, **167,** 81
Rab–SNARE interaction, genetic interactions, **176,** 59–60
salt stress, **165,** 27–34
salt tolerance, **165,** 3
secretory pathway, Rab GTPase function, **176,** 25–27
signal transduction, **165,** 27–34
signal transduction systems in, **139,** 309–332, *see also* Signal transduction systems and yeast
tropomyosin gene, **170,** 5
vacuole fusion, **176,** 63–64
vacuole in, **139,** 59–114
Ypt1p, role as SNARE complex regulator, **176,** 60–62
Saccharomyces pombe, tropomyosin gene, **170,** 5–6
Saccus dorsalis, ependymin, **165,** 143–144
Saccus vasculosus, ependymin, **165,** 143–144
SAF-A
120-kDa protein, **162A,** 202
in DNA looping, **162A,** 305
mediated DNA loop formation, **162A,** 543–544
Sail-sheet cultures, vertebrate cell culture technology and, **142,** 160–161
Saintopin, **156,** 108

Salicylic acid
as endogenous elicitor of plant defense, **148,** 14
systemic acquired resistance induction
binding protein, **168,** 317
biosynthesis, **168,** 316–317
catalase inhibition, **168,** 317–319
conjugation, **168,** 317
constitutive levels, **168,** 316
discovery, **168,** 314–315
translocation in tobacco, **168,** 316
Salinity
historical aspects, **165,** 2
nonspecific osmotic effect, **165,** 3
specific toxicity of ions, **165,** 3
Salivary gland
avian salt glands and, **144,** 165, 167
duct-associated lymphoid tissue in monkeys, **167,** 137–138
S allele-associated proteins, identification of, **140,** 459–461
S alleles
genealogical tree of, **140,** 476
generation of new, **140,** 509–511
self-incompatibility, **143,** 257–258, 289–290
gametophytes, **143,** 280, 282–283, 287
sporophytes, **143,** 259–260, 264–265, 277–278
temporal and tissue-specific expression of, **140,** 461–463
Salmonella
microgravity experiments, **156,** 340
species in oral vaccine development, **167,** 140
Salmonella typhimurium
bacterial flagellar motor and, **147,** 116, 123
function, **147,** 112–114
structure, **147,** 100, 104, 109
M cells and
nonuniformity of adhesion to, **167,** 117
phagocytosis of, **167,** 127–128
preferential adherence to, **167,** 123
uptake of entire, **167,** 120*t,* 122
Salt concentration, **165,** 2, 13–17
Salt glands, *see* Avian salt glands
Salt stress
Escherichia coli, **165,** 25–27
Saccharomyces cerevisiae, **165,** 27–34
signal transduction, **165,** 25–35
whole-plant level, **165,** 35–37
Salt tolerance
animals compared, **165,** 2
Arabidopsis thaliana, **165,** 3
cation uptake system, **165,** 17–21
defense responses, **165,** 1–39
enzyme, **165,** 13–14
Escherichia coli, **165,** 3
genetic analysis, **165,** 3
glutathione reductase, **165,** 15
halophytic archaebacteria, **165,** 14
Knal gene, **165,** 37
osmolyte synthesis, tobacco mutants, **165,** 13
osmotic toxicity, **165,** 4–6
phosphoenolpyruvate carboxylase, **165,** 15
plant levels, **165,** 2
plasma membrane
chloride transport, **165,** 23
ion transport, **165,** 17–24
plasma membrane H^+-ATPase, **165,** 23–24
potassium uptake, **165,** 17–21
sodium efflux, **165,** 21–24
sodium uptake, **165,** 17–21
responses at cellular level, **165,** 4–35
ribulose-bisphosphate carboxylase, **165,** 15
Saccharomyces cerevisiae, **165,** 3
specific ion toxicity, **165,** 4–6
stomatal opening regulation, **165,** 35
toxicity, **165,** 4–35
toxicity targets, **165,** 1–39
vacuole, **165,** 24–25
Salt toxicity
mineral nutrition, **165,** 6
osmolyte synthesis, **165,** 6–13
Escherichia coli, **165,** 7–8
higher plants, **165,** 11–13
Saccharomyces cerevisiae, **165,** 8–11
osmolyte transport, **165,** 6–13
Escherichia coli, **165,** 7–8
higher plants, **165,** 11–13
Saccharomyces cerevisiae, **165,** 8–11
primary mechanism, **165,** 5–6
Salvage pathway, DNA precursors, **162B,** 389

Salyut spaceflights, microgravity experiments
microorganisms, **156,** 340–342, 343
plants, **156,** 310, 321–328, 333
Samanea saman, intracellular calcium messenger system and, **142,** 327
Sampling methods
for renal glomeruli
sample distribution, **161,** 132–133
sample orientation, **161,** 133
sample size, **161,** 133–134
for stereology
sample distribution, **161,** 132–133
sample orientation, **161,** 133
sample size, **161,** 133–134
SAMs, *see* Substrate adhesion molecules
Sand dollar, sperm-binding proteins, **163,** 279
SAP, self-incompatibility, **143,** 277
SAR, *see* Scaffold-attached regions; Systemic acquired resistance
Sarcolemma, proteins in striated muscle
cardiac muscle, **142,** 107–110, 112, 114–115
contraction stimulus, **142,** 119–120
dystrophin, A-CAM, and T tubules, **142,** 124–127
skeletal muscle, **142,** 66, 71, 87, 92, 99–101
Sarcoma, *mdr1* gene expression, **156,** 88
Sarcomeres
cytoskeletal dynamics, **144,** 89, 96, 115, 119
Drosophila muscles and
contractile protein genes, **143,** 95, 97–98, 126, 136
development, **143,** 66, 71, 75, 87, 94
myosin heavy chains, **143,** 107
tropomyosin, **143,** 17
Z-disc proteins, **143,** 122
formation, **143,** 181–184
thick filament proteins, **143,** 177–178
thin filament proteins, **143,** 160, 163–164, 172
myofibrillogenesis and, **143,** 153
proteins in striated muscle and, **142,** 93–100
cardiac muscle, **142,** 106, 110–115, 118
thick-filament proteins, **142,** 65, 79, 82–83
thin-filament proteins, **142,** 85, 88
Sarcophyton glaucum, animal mtDNA and, **141,** 180, 193, 199
Sarcoplasmic reticulum
Drosophila muscles and, **143,** 98, 115, 125
proteins in striated muscle and, **142,** 81, 92, 102, 107, 119–124
Sarsia, transdifferentiation in medusae and, **142,** 222
SATB1
binding to A;nd;T-rich sequences, **162A,** 544
tissue-specific, **162A,** 303–305
Satellite cells, proteins in striated muscle and, **142,** 71, 73, 79, 97
Saturation
endothelin receptor subtypes, **171,** 271–273
NMR and, **145,** 26, 29–30
Saturation transfer, in NMR, **145,** 33
Sauromalus obesus, avian salt glands and, **144,** 149
SC-10, inhibition of *Chlamydomonas* gametogenesis, **168,** 292
Scaffold association, S/MAR structural features leading to, **162A,** 410–416
Scaffold-attached regions, *see also* Matrix-attached regions
association with topoisomerase II, **162A,** 128
in cell type-specific gene expression, **162A,** 280
genomic distribution, **162B,** 422–424
low-salt extraction, **162A,** 199
underwound status, **162A,** 443–444
Scaffold–matrix attachment regions, *see* S/MAR elements
Scanning electron microscopy
in analysis of calcifications, **172,** 137–138
avian salt glands and, **144,** 180, 182
eggshells and biomineralization, **145,** 219, 224, 239–240
Scanning transmission electron microscopy, NPCs, **162B,** 228–230
Scatter assay, as migration analysis method, **155,** 63
Scatter factor–hepatocyte growth factor
biological activity, **160,** 238–239
embryogenesis role, **160,** 240–241
hepatic development role, **160,** 241
mutagenesis, **160,** 236
pathology role, **160,** 241–242

Scatter factor–hepatocyte growth factor (*continued*)
posttranslational processing, **160,** 235
structure, **160,** 234–236
tissue distribution, **160,** 239–240
variants, **160,** 236
Scatter factor–hepatocyte growth factor receptor, *see* c-Met receptor
Scattering, fish chromatophores and, **143,** 203, 207–208, 217
Schizophyllum commune, mitochondrial genomes in fungi, **141,** 117
Schizosaccharomyces pombe
chromosome structure, **152,** 34
endosymbiont hypothesis and, **141,** 282, 290
mitochondrial genomes in fungi and
DNA diversity, **141,** 93, 97–100, 102
generation of mtDNA diversity, **141,** 108–110, 113, 117
mitosis and, **144,** 226, 228, 231, 250–251
Rab proteins, **176,** 29–30
Schwann cell myelin protein, **138,** 297–299
Sclerotization, mitochondrial differentiation during, **161,** 72–75
SCO, *see* Subcommisural organ
Scute, extracellular matrix, **166,** 77–78
Scyliorhinus, **138,** 151
Scyphozoa
centralization and complexity of nervous system, **167,** 47
life cycle of, **167,** 38
SDS–PAGE
higher plant plastids and, **145,** 168
intracellular calcium messenger system and, **142,** 310–311
in study of Sertoli-germ cell communication network, **147,** 66, 69, 72, 76
urease in plant cells and, **145,** 89
SDZ PSC833, **156,** 91
Sea anemones
centralization and complexity of nervous system, **167,** 47, 77–78
as cnidarians, **167,** 38
G-protein-coupled receptors, **167,** 55–56
metamorphosis, neuropeptides inducing, **167,** 71–74
nematocyst discharge, **156,** 284, 286
neurons, transmitters and multifunctionality of, **167,** 50–51
neuropeptides in
Antho-RFamide biosynthesis in, **167,** 58–64
C-terminal sequences, **167,** 49–50
neuromuscular transmission, **167,** 51–54
Sea pansies, neuropeptides in
Antho-RFamide, **167,** 48, 64–67
neuromuscular transmission, **167,** 51–54
Seasonal effects, control of rhythmic behaviors and, **147,** 12
Sea urchin, sperm, flagellar motility, role of Ca^{2+}, **173,** 27
Secale cereale, **140,** 12
Second messengers
avian salt glands and, **144,** 200–201
biological clocks in retina and, **146,** 84, 103, 106, 139
control of rhythmic behaviors and, **147,** 2–3, 10
fish chromatophores and, **143,** 227, 240, 250
and hormone receptors, **155,** 8–11
intracellular calcium messenger system, **142,** 305–306, 311, 321, 328, 338
in nematocyst discharge, **156,** 289–291
Sertoli-germ cell communication network and, **147,** 79, 81
transcriptional regulation and, **143,** 48
Secretin, astrocyte reaction to, **148,** 142–143
Secretin receptor, properties, **164,** 212–213
Secretion
avian salt glands and
blood supply, **144,** 151–154
duct system, **144,** 145–148
enzymes, **144,** 144–145
fine structure, **144,** 133–140
hormonal control, **144,** 196, 199
hypertrophy, **144,** 140–144
innervation, **144,** 155, 157–163, 165, 167
mechanism, **144,** 168–176
receptive systems, **144,** 177–190
stimulus-secretion coupling, **144,** 200–205
zoology, **144,** 130–133
effect of endothelin
glucocorticoid from zona fasciculata reticularis, **171,** 277–278
mineralocorticoid from zona glomerulosa, **171,** 275–277
steroid from interrenal cells, **171,** 278–279
eggshells and biomineralization and, **145,** 241

glandular, cholesteric supramolecular order, **166,** 83–84
hormonal control of amphibian metamorphosis and, **145,** 116, 123–128, 136–137
hypersecretion, hormonal control of amphibian metamorphosis, **145,** 136
sperm maturation role, **154,** 105
germ cells, **154,** 126, 128
principal cells of epididymis, **154,** 149–170
immobilin, **154,** 157–158, 164–168
SPG-2, **154,** 155, 157, 164, 166
Sertoli cells
proteins, **154,** 122–123
sulfated glycoprotein-1, **154,** 123–124
sulfated glycoprotein-2, **154,** 124–127
vas deferens, **154,** 177–178
Secretory granules, possible roles of polyamine in, **138,** 81–83
Secretory immunoglobulin-A, Peyer's patches and, **167,** 126
Secretory system
in pollen tube
endoplasmic reticulum, **176,** 138–139
Golgi apparatus, **176,** 139–140
secretory vesicles, **176,** 140
regulated, role of Rab3 GTPase, **176,** 33–35
yeast, Rab GTPase function, **176,** 25–27
Secretory vesicles, ADP ribosylation factor in formation of, **172,** 228
Sectoring, cell, in 2D analysis systems, **163,** 68
Sedimentation
microgravity environment, **156,** 344–345
NMR analysis, **145,** 14
Seedlings, effect of microgravity
angiosperms, **171,** 32, 40–41
wheat peroxidase level, **171,** 39
Seed plants, *see* Nonflowering seed plants
Seeds
higher plant plastids and, **145,** 151, 184, 186–187, 204
urease in plant cells and, **145,** 65–66, 98
biochemical genetics, **145,** 84, 89, 91
elimination, **145,** 80–83
metabolic origins, **145,** 67–73, 75, 77–78
nickel metabolism, **145,** 93, 97
Segmentation
in *Drosophila*
clypeolabral region, **174,** 26–27
dorsal ridge, **174,** 27
head gap genes, **174,** 23–26
labial segments, **174,** 20–23
maternal components, **174,** 19–20
maternal polarity, **174,** 18
maxillary segments, **174,** 20–23
segment-polarity genes, **174,** 18–19
Drosophilia, **139,** 123–125
evolution, **174,** 27–29
specific stages of genes, **174,** 48–52
Segment diversity, genetic control, **174,** 30–33
Segment polarity
gene expression, **174,** 18–19
genes, in pattern formation in *Drosophilia* epidermis, **139,** 126–128
Segregation
ooplasmic, in ascidian eggs, **176,** 253–255
postmitotic, secondary MTOCs, **175,** 107–111
Selaginella, environmental stimuli in, **145,** 269–271
E-Selectin, sialoglycoprotein ligand as counterreceptor, **175,** 192
L-Selectin, counterreceptors, **175,** 190
P-Selectin, sialoglycoprotein ligand as counterreceptor, **175,** 190–192
Selectins
carbohydrate specificity, **175,** 186–189
counterreceptors, **175,** 189–192
role in leukocyte–endothelial interaction, **175,** 183–186
Self-compatibility, in plants, **143,** 263–264
Self-incompatibility
gametophytic systems, *see* Gametophytic self-incompatibility systems
homomorphic, *see* Homonorphic self-incompatibility
and male-female interactions
generalities, **140,** 581–582
molecular analysis of self-incompatibility, **140,** 582–585
S gene and S products identified from pistil, **140,** 582
S gene and S products in sporophytic self-incompatibility (SSI), **140,** 584–585
S genes and S products of pollen, **140,** 585–586
and mechanisms, data from, **140,** 587–588

Self-incompatibility (*continued*)
molecular analysis, **140,** 582–585
S gene and S products identified from pistil, **140,** 582
S gene and S products in sporophytic self-incompatibility, **140,** 584–585
plants
cell–cell interactions
induction of RNA synthesis, **158,** 41–42
receptor identification, **158,** 15–16
signal transduction, **158,** 16–17, 43–44, 46–47, 52–53
classification of systems
floral morphology, **158,** 3–4
heteromorphic system, **158,** 4–5
homomorphic systems, with gametophytic control of pollen, **158,** 5–9
incompatibility phenotype of pollen, **158,** 3–4
developmental onset, **158,** 17
discovery, **158,** 2
diversity of systems, **158,** 49–50
evolution of systems, **158,** 9–11
genetics, *see also* G genes; S genes; Z genes
heredity, **158,** 2–3, 11
loci number in systems, **158,** 6–8
polymorphisms, **158,** 11
population genetics, **158,** 11–13
grasses, **158,** 6, 10, 46–49
incidence among species, **158,** 2
models
Papaveraceae, **158,** 41–45
Poaceae, **158,** 47–48
Solanaceae, **158,** 36–37
outcome, environmental manipulation, **158,** 17
phosphorylation pattern of proteins, **158,** 46–47
physiological response, **158,** 17–18
bioassay, **158,** 18–20
gametophytic systems, **158,** 19–21
sporophytic systems, **158,** 18–19
sporophytic, *see* Sporophytic self-incompatibility
Self-renewal, cellular
hemopoietic growth factors and intensified chemotherapy in limiting toxicity, **167,** 241
hemopoietic stem cell proliferation regulation, **167,** 194–196
MIP-1α and, **167,** 205–206
Seminal vesicle
development, **160,** 269
fertility role, **160,** 268
lymphocyte content, **160,** 282–283
Seminal vesicle epithelium
castration effects, **160,** 267, 284–286, 294
cell culture, **160,** 290–293
endocytosis, **160,** 267–268
horseradish peroxidase, **160,** 279, 281, 292
physiological roles, **160,** 281–283
substrates, **160,** 279
testosterone effects, **160,** 289–290
hormone response
estrogen, **160,** 288
prolactin, **160,** 287–288
testosterone, **160,** 284–286, 294
vasoactive intestinal peptide, **160,** 289, 294
membrane recovery, **160,** 283–284
protein secretion
measurement, **160,** 275, 277
testosterone regulation, **160,** 285–286
ultrastructure and aging, **160,** 267, 269–271, 273
Seminiferous tubules
epithelium, Sertoli-germ cell communication network, **147,** 25, 27, 37, 80–81
function, **147,** 40, 57
germ cell depletion, **147,** 63, 65
mechanisms, **147,** 73–74
stage-dependent changes, **147,** 58–61
intermediate region, **154,** 130–131
Senecio vulgaris, plant-pathogen interactions in, **144,** 58–59, 64
Sensory neurons, cnidarian
centralization in *Hydra* species, **167,** 44–45
description, **167,** 40
light- and gravity-sensitive, **167,** 39–40
Separation, immunophenotypic, dendritic cell, **153,** 52
Sequence domains, Rab GTPase effector region, **176,** 14–15
Sequences
amino acid, *see* Amino acid sequence

animal mtDNA and, **141,** 173–175, 207–210
control region, **141,** 201–202
direct repetition, **141,** 203–204
genome content, **141,** 176, 180
nucleotide bias, **141,** 204–206
protein genes, **141,** 181, 183–184, 187–189
rRNA, **141,** 189–191
tRNA, **141,** 193, 195, 198–199
Arabidopsis thaliana and, **144,** 63, 71, 74
chloroplasts and, **144,** 260, 291, 293–295
cryptomonads, **144,** 276
euglenoids, **144,** 284
green algae, **144,** 272
heterokont algae, **144,** 281
ciliates and, **141,** 1, 4, 58, 60
genetic code, **141,** 52, 55
Paramecium aurelia, **141,** 12–45
rRNA genes, **141,** 47, 49–52
structure, **141,** 6–7, 10–11
cnidarian neuropeptide
high copy number of immature, **167,** 78
prohormone recognition, **167,** 58
cytoskeletal dynamics and, **144,** 97–98, 105–109
Drosophila muscles and, **143,** 65, 81, 126, 138
actins, **143,** 109, 111
contractile protein genes, **143,** 101, 103, 126
myosin light chains, **143,** 107–108
troponin, **143,** 118–119, 121
Z-disc proteins, **143,** 124–125
eggshells and biomineralization and, **145,** 239, 241
endosymbiont hypothesis and, **141,** 234
algae, **141,** 273–275
eukaryotes, **141,** 276
gene expression, **141,** 298–306, 308
gene structure, **141,** 308–317
gene transfer, **141,** 277–278, 325–326
genome organization, **141,** 288–295, 297
lineages of life 238-244, **141,** 246
mitochondria, **141,** 280, 282, 327
nuclear genome, **141,** 248–251
plastids, **141,** 253, 262–263, 268–273, 279
rRNA, **141,** 319–324
fish chromatophores and, **143,** 249
flowering plants and, **143,** 311, 313–314, 323
higher plant mitochondrial genomes and, **141,** 135, 160–164
gene location, **141,** 157–159
recombination, **141,** 148–153
structure, **141,** 135, 137, 140–143, 145
higher plant plastids and
metabolism, **145,** 168, 171, 185, 193–194, 201
structure, **145,** 155, 157–158, 160, 165–167
hormonal control of amphibian metamorphosis and, **145,** 115–116, 126–127, 130, 138
intracellular calcium messenger system and, **142,** 306
calmodulin, **142,** 328–329
G proteins, **142,** 309–311
protein kineses, **142,** 333, 335
kinetoplastid mtDNA and, **141,** 65, 67–68, 81–84
maxicircle genes, **141,** 69–74
maxicircle transcripts, **141,** 75–80
mitochondrial genomes in fungi and, **141,** 89, 119–120
DNA diversity, **141,** 93, 96–98, 100–101, 103, 105–106
generation of mtDNA diversity, **141,** 107–112, 116, 118
myofibrillogenesis and
isoform change, **143,** 154–155
thick filament proteins, **143,** 178
thin filament proteins, **143,** 163, 165–168, 173–174
nuclear magnetic resonance and
applications, **145,** 35, 44
techniques, **145,** 20, 24–25, 27, 29
proteins in striated muscle and, **142,** 75, 91, 96, 123
self-incompatibility and, **143,** 258
gametophytes, **143,** 284–289
sporophytes, **143,** 268–270, 272, 274–276
small single copy, plastids, **145,** 155, 158
TGF-β isoform shared homology, **167,** 210
transcriptional regulation and, **143,** 1–2, 6, 47
biochemistry, **143,** 23
genes, **143,** 13, 19

Sequences (*continued*)
molecular cloning, **143,** 29–30, 33–34, 36–38
structure, **143,** 19, 21–22
urease in plant cells and, **145,** 86, 89, 91, 98
Sequoia sempervirens, **140,** 322
SER, *see* Smooth endoplasmic reticulum
Serial endsymbiosis theory, **141,** 332
Serine proteinase
inhibition by acute phase proteins, **170,** 240–242
in neovascularization, **161,** 4
Serine–threonine kinase receptor, characterization, **164,** 200–201
Ser-Lys-Leu motif, peroxisomal targeting, **160,** 335–336
Serotonin, *see also* 5-Hydroxytryptamine
biological clocks in retina
Bulla, **146,** 131–132
entrainment pathway, **146,** 101, 108, 110–112, 114
control of amphibian metamorphosis and, **145,** 127–128, 137
in control of rhythmic behaviors, **147,** 12–15, 18
fish chromatophores and, **143,** 236
fluorescence visualization, **157,** 282–283, 292
functions, **157,** 296
hypoxia effect on release, **157,** 289
inhibition of smooth muscle cell growth, **161,** 25–26
modulation of fish gill function, **157,** 297
neurotransmission, **149,** 247–252
pharmacology, **149,** 247, 250–252
retinal precursor cells and, **146,** 155
secretion by neuroendocrine cells, **157,** 278, 280, 282–283, 292
unsuccessful attempts to isolate in *Hydra* neurons, **167,** 43–44
Serotoninergic receptor, **149,** 249–250
Sertoli cells
cytoplasmic organelles, **160,** 176–178
cytoskeletal elements, **160,** 179–181
ectoplasmic specializations, **160,** 187–189, 204
functions, **160,** 193, 206
nutrient provision to developing spermatogenic cells, **160,** 194–195
phagocytosis, **160,** 187, 196–197
protein secretion, **160,** 200–204
spermiation, **160,** 204–206
steroid hormone synthesis, **160,** 197–200
tubular fluid, ionic composition maintenance, **160,** 195–196
germ cell
Sertoli cell junctions, **160,** 184, 186–188, 208
translocation, **160,** 165, 191–193
–germ cell adhesion molecules, structure, **147,** 36
–germ cell communication network, **147,** 25–28, 79–81
function, **147,** 37–38
Sertoli cell barrier, **147,** 40–44
Sertoli cell products, **147,** 44–58
synchrony, **147,** 38–40
germ cell depletion, **147,** 62–67
germ cell *in vitro,* **147,** 67–72
mechanisms, **147,** 72–79
stage-dependent changes, **147,** 58–62
structure
devices, **147,** 28–32
morphoregulatory mechanisms, **147,** 32–37
GLUT1, **172,** 32
intercell junctions
occluding junctions, **160,** 183–184
permeability, **160,** 185
receptor-mediated endocytosis, **160,** 204
tissue remodeling, **160,** 191–192
types, **160,** 183
metabolism, **160,** 194–195
microtubules and spermatogenesis, **160,** 180–181, 205
morphology
amphibians, **160,** 168
birds, **160,** 166
cephalopods, **160,** 173
Chondrichthyes, **160,** 167
crustaceans, **160,** 171
Cyclostomata, **160,** 166–167
earthworms, **160,** 171
echinoderms, **160,** 174
insects, **160,** 171–172
jellyfish, **160,** 169
lizards, **160,** 165–166
Mollusca, **160,** 172–173
nematodes, **160,** 170

prosobranchs, **160,** 173
sea anemones, **160,** 169
sponges, **160,** 169
Teleostei, **160,** 167–168
nucleus structure, **160,** 175–176
S-100 localization, **160,** 181
seasonal changes, **160,** 165–166
sperm maturation role
endocytosis
adsorptive, **154,** 117–118
fluid-phase, **154,** 116–118
receptor-mediated at base of cells, **154,** 120–122
in secondary lysosomes, **154,** 118–120
function, **154,** 106
phagocytosis of residual bodies, **154,** 109, 111, 114, 116
secretory functions in
proteins, **154,** 122–123
sulfated glycoprotein-1, **154,** 123–124
sulfated glycoprotein-2, **154,** 124–127
ultrastructure, **154,** 107–115
testosterone response, **160,** 181
tubulobulbar complex, **160,** 183, 189–191
Serum, acute phase protein alteration, **170,** 244
Serum albumin, advanced glycosylation end products, **152,** 70
Serum responsive factor, transcriptional regulation and, **143,** 14
Serum sickness nephritis, **166,** 204–207
Sesbania rostrata, **156,** 5, 47
Severin, **144,** 107, 111; **156,** 234
Sex
behavior, perinatal exposure to sex hormones and antihormones, **139,** 26–28
inducer genes, codon bias and, **139,** 254
mitochondrial, *see* Mitochondria, sexuality
parasitic origins, **161,** 99–100
Sex chromatin, interphase nucleus marker, **165,** 70–72
Sex chromosomes
gonosomes, *see* Gonosomes
mammalian
comparative studies, **154,** 246, 248
dosage compensation, **154,** 206, 214–216
evolution, **154,** 202–206
function, **154,** 191
gene content, **154,** 196–198
gonadal differentiation role
higher vertebrates, **154,** 220–221
hormone function, **154,** 216–218
marsupials, **154,** 219–220
regional localization, **154,** 226–229
testis determination, **154,** 221–222
organization, **154,** 191–195
testis-determining factor
function, **154,** 222
regional localization, **154,** 226–229
Y chromosome as determinant, **154,** 223–226
testis-determining factor *SRY* gene
cloning, **154,** 233–237
demonstration of process, **154,** 236, 238
evolution of organization, **154,** 241–244
variations, **154,** 195–202
X chromosome inactivation
comparative studies, **154,** 211–212
cytological changes, **154,** 209–210
dosage compensation role, **154,** 214–216
evolutionary aspects, **154,** 211–212
gene expression, **154,** 207–209
mechanism, **154,** 206–207, 209–210
sex chromosome evolution, **154,** 212–214
Y chromosome as sex determinants, **154,** 223–226
Sex-determining genes
characteristics, **154,** 191–195
comparison, **154,** 244, 248
dosage compensation, **154,** 206–209, 214–216
function, **154,** 246
gene dosage effects, **154,** 214
gonadal differentiation
developmental steps, **154,** 247
higher vertebrates, **154,** 220–221
hormone function, **154,** 216–218
marsupials, **154,** 219–220
testis determination, mammalian, **154,** 221–222
inactivation, **154,** 209–213
mapping studies, **154,** 196–202, 204, 206, 246

Sex-determining genes (*continued*)
sexual dimorphisms
examples, **154,** 216
in marsupials, **154,** 219–220
and testis-determining factor
function, **154,** 222
genetic research, **154,** 229–233
regional localization, **154,** 226–229
SRY gene, **154,** 233–245
Y chromosome role, **154,** 223–226
Sex differences, acute phase proteins, **170,** 235
Sex hormones, *see also specific sex hormones*
early exposure to, **139,** 1–57
effect on female human genital system, **139,** 21
effect on *in vitro* cell growth, **139,** 7–8
effect on receptor systems, **139,** 6–7
enzyme inhibitors and, **139,** 8
general considerations for, **139,** 43–45
genital abnormalities induced by, **139,** 3–13
molecular changes and, **139,** 11–12
nongenital abnormalities induced by, **139,** 26–39
enzyme levels and, **139,** 28–29
mouse pelvis and, **139,** 33—36
5α-reductase inhibitor and, **139,** 29–30
sexual behavior and, **139,** 26–28
sexual dimorphism of mouse pelvis and, **139,** 31–33
sexually dimorphic brain nuclei and, **139,** 26–28
phytoestrogens and, **139,** 12–13
polyovular follicles and, **139,** 8–10
thymic epithelial cells, **156,** 186
Sexing, human cells
postnatal samples, **153,** 21
preimplantation embryos, **153,** 22–25
clinical application, **153,** 24–25
early studies, **153,** 22–23
patient strategies, **153,** 25
research needs, **153,** 22
single-color FISH, **153,** 24
studies, **153,** 23–24
prenatal samples, **153,** 22
Sexual dimorphisms
examples, **154,** 216
in marsupials, **154,** 219–220
mouse pelvis, **139,** 31–33
Sexual induction, in *Volvox carteri*
cAMP and, **139,** 220–221
extracellular matrix in, **139,** 221–223
inducer action and, **139,** 219, 223–224
pheromone and, **139,** 219–221
Sexual reproduction
extracellular matrix in animal systems, **140,** 299–300
in flowering plants, **140,** 297–304
volvacacean, **139,** 199
Sexual route, in plant transformation
conclusions, **140,** 442–443
microspores
approaches to gene transfer in microspores, **140,** 427–431
discussion, **140,** 431
pollen grains as vectors of transforming DNA
direct DNA transfer, **140,** 432–435
pollen and stigma quality and protocol of pollination
nuclease activity, **140,** 438–439
pollination protocol, **140,** 439–440
strategies for transformation, **140,** 432–435
techniques, **140,** 425–426, 579–581
transformation *in situ* of female sexual organs, **140,** 441–442
S-fimbriae
expression by *Escherichia coli,* **175,** 176–177
specificity to glycolipids, **175,** 177
SFO, *see* Subfornical organ
S genes
alleles, **158,** 5–8, 10–11, 23–24, 33
distribution between populations, **158,** 14
encoding of specificity, **158,** 51
frequencies, **158,** 12
number in population, **158,** 13
number in species, **158,** 14–15
polymorphism, **158,** 51
receptor-binding site association, **158,** 15
cloning, **158,** 3, 17
Brassicacea, **158,** 23–24
Papaveraceae, **158,** 40–41

Poaceae, **158,** 49
Solanaceae, **158,** 31–33
evolution, **158,** 34, 50–51
genomic organization of family, **140,** 496–498
identification of related genes, **158,** 24–25
molecular basis of recognition and response of products, **140,** 554–558
molecular characteristics
cloning of DNA encoding S proteins, **140,** 465–466
molecular organization of S locus, **140,** 469–471
primary structural features of S proteins, **140,** 466–469
conserved residues, **140,** 467
extreme sequence diversity, **140,** 466–467
variable and hyperactive residues, **140,** 467
molecular evolution of family
Brassica S gene family, **140,** 503–509
SLG expression in *Brassica* stigmas, **140,** 512–514
SLR1 subfamily, **140,** 503–506
SLR2 subfamily, **140,** 506–507
evolution of globin genes, **140,** 499–502
evolution of immunoglobulin genes, **140,** 502–503
generation of new S alleles, **140,** 509–511
overview, **140,** 498–499
molecular organization, **140,** 469–471
mutants, **140,** 454–456
tripartite mode of, **140,** 456
mutations, **158,** 10
pollen components, **158,** 29, 49, 51–52, 54
products
identified from pistil, **140,** 582
pollen, **140,** 585–586
in sporophytic self-incompatibility, **140,** 584–585
products, sporophytic self-incompatibility systems
historical background, **140,** 525–527
molecular basis of recognition and response, **140,** 554–558
proteins of *Brassica* pollen grains, **140,** 551–554
proteins of *Brassica* stigma, **140,** 527–551
history, **140,** 527–528
homology modeling of female glycoproteins, **140,** 544–551
other families of glycoproteins homologous of S locus-specific glycoproteins, **140,** 533–543
primary structure of S locus-specific glycoproteins, **140,** 544–551
screening in Poaceae, **158,** 46
self-incompatibility, **143,** 258, 290
effect of outside mutation, **140,** 457
gametophytes, **143,** 280–281, 283, 289
sporophytes, **143,** 265, 278
Solanaceae
RNase activity, **158,** 34–36
sequence homology, **158,** 33–34
structure, **158,** 33
structure, **140,** 511–512; **158,** 10
transformation, **158,** 35
SGLT1, in dietary absorption of glucose, **172,** 5–6
Shark rectal glands, *see* Rectal glands
Sharpey's fibers, regenerative cementogenesis and, **142,** 3–4, 17
Shc protein, role in signal transduction, **173,** 269–272
SH2 domain, *see* Src homology 2 domain
Shear sensitivity, vertebrate cell culture technology, **142,** 166–167, 178, 196
Sheep, Peyer's patches in, ileal, **167,** 94–95
Shell membranes, eggshells and biomineralization and, **145,** 219–224
Shiga toxin, structure, **161,** 287
Shigella flexneri, M cells and
pilus plasmids in adherence to, **167,** 124
preferential adherence to, **167,** 123
uptake of entire, **167,** 120*t,* 122
Shikimate pathway
flowering plants, **143,** 310–313, 316
in lignin biosynthesis, **172,** 246
S homology, sequence expression in *Brassica* anthers
overview, **140,** 517–518
PCR detection of SLG-homologous transcripts, **140,** 518–519, 519–521
Shoots, angiosperm, effect of microgravity
on apical meristem cells, **171,** 35–36
on primary structure, **171,** 36
Short-circuit current, avian salt glands
hormonal control, **144,** 197
innervation, **144,** 161, 163

Short-circuit current (*continued*)
secretory mechanism, **144,** 175–176
stimulus-secretion coupling, **144,** 201, 204
SH2 proteins, in platelet-derived growth factor signal mediation, **172,** 100
SHPTP2 tyrosine phosphatase, mediation of FAK dephosphorylation, **167,** 171–172
Shrimp, oocytes, sperm binding, **163,** 263
Shrinkage necrosis, **163,** 109
Shuttle streaming, **156,** 200
Sialic acids
contactinhibin receptor, **175,** 154–155
function, **175,** 138
galactose residue masking, **175,** 144–145
hepatic galactose receptor, **175,** 145–146
Kupffer cell galactose receptors, **175,** 147
metabolism, **175,** 142–143
peritoneal macrophage galactose receptor, **175,** 147–153
properties, **175,** 138–142
as recognition determinants, **175,** 155–156
for bacteria, **175,** 175
binding site, **175,** 204–207
CD22, **175,** 197–200
CD33, **175,** 202–203
characterization, **175,** 193–196
for coronaviruses, **175,** 171–172
counterreceptors, **175,** 207–208
for *Escherichia coli,* **175,** 175–178
for *Helicobacter pylori,* **175,** 179–181
for influenza A virus, **175,** 162–169
for influenza B virus, **175,** 162–169
for influenza C virus, **175,** 169–170
myelin-associated glycoprotein, **175,** 200–202
for other bacteria, **175,** 181–182
for paramyxoviruses, **175,** 170–171
for polyomavirus, **175,** 172
for protozoa, **175,** 182–183
for reoviruses, **175,** 172–175
role of modifications, **175,** 208–210
selectfor family, **175,** 183–192
sialoadhesin, **175,** 196–197
for *Streptococcus* strains, **175,** 178–179
structural features, **175,** 203–204
for viruses, **175,** 157–162
structure, **175,** 137–138
Sialoadhesin family
CD22, **175,** 197–200
CD33, **175,** 202–203
characterization, **175,** 193–196
myelin-associated glycoprotein, **175,** 200–202
sialic acid binding
binding site, **175,** 204–207
counterreceptors, **175,** 207–208
role of modifications, **175,** 208–210
structural features, **175,** 203–204
species differences, **175,** 196–197
Sialylation
cell surface glycans, **175,** 150–151
CFTR and, **172,** 227
cyclic AMP regulation of, **172,** 228
trans Golgi network and, **172,** 227–228
α-2,6-Sialyltransferase, regulation of CD22 binding, **175,** 199–200
Sias
interaction with influenza A and B viruses, **175,** 162–169
masking effects, **175,** 144–145, 148–150, 153, 155–156
as receptor determinant for paramyxoviruses, **175,** 170–171
Sickle cell anemia, preimplantation diagnosis, **151,** 3
SIgA, *see* Secretory immunoglobulin-A
Signaling
autocrine, platelet-derived growth factors and, **172,** 104–105, 112
gustatory, insect, **164,** 231
nuclear localization
function, **151,** 192–193
identification, **151,** 191–192
masking
by NRD transcription factors, **151,** 203–206
by steroid receptors, **151,** 203–206
number, effect of, **151,** 201–202
transcriptional regulation and, **143,** 38
olfactory, insect, **164,** 231
paracine, in platelet-derived growth factors, **172,** 104, 105
phosphatidylinositol in yeast cells, **139,** 323–325
visual, insect, **164,** 232
Signal-to-noise ratio, in NMR
applications, **145,** 32, 44–45
data averaging, **145,** 6
improvement, **145,** 9
techniques, **145,** 15, 23

Signal transduction
abiotic stress interactions, **165,** 25–35
and activation of transcription
regulation, **157,** 246–247
structure, **157,** 247
adaptor protein role, **164,** 199–200
agonist-induced Ca^{2+}-dependent volume
changes, **161,** 218–219
in apoptotic cell death, **163,** 115–116
Arabidopsis thaliana and, **144,** 56, 62,
68, 73
avian salt glands and, **144,** 131, 188
in bacteria-induced macrophage
activation, **161,** 296–299
B cell, role of CD22, **175,** 198–199
Ca^{2+}–calmodulin in, **161,** 214–217
and chemical neuromediators
aminoacidergic, **149,** 255–266
catecholaminergic, **149,** 240–249
cholinergic, **149,** 236–241
process, **149,** 232–236
purinergic, **149,** 254–255
serotoninergic, **149,** 247–252
crosstalk in, FAK and, **167,** 171–172
cytokine pathways
early events, **170,** 260–262
immediate early events, **170,** 262–264
cytoskeleton
actin filament, **166,** 14–15
cytoplasmic anchoring, **166,** 12
desensitization, **166,** 10–11
dynamics, **144,** 101–103
endocytosis, **166,** 10–11
exocytosis, **166,** 12–14
intracellular signaling, **166,** 9–10
plasma membrane receptor binding,
166, 10–11
receptor intracellular transport, **166,**
11–12
steroid hormone, **166,** 11
dependence on mode of cell swelling,
161, 229–230
eicosanoid role, **161,** 220–222
environmental stimuli in plant cells and,
145, 252, 298–299, 301
chloroplasts, **145,** 279–283
photodinesis, **145,** 257, 263–268
Escherichia coli, **165,** 25–27
fish chromatophores and, **143,** 238–245
in flagellate algae phototaxis, **148,** 290
Euglena, **148,** 298, 300
green algae, **148,** 290–299
focal adhesion in, as site of, **167,** 162
G-protein-coupled receptors, **164,**
210–211
by growth factors, **173,** 247–248
by hormones, **173,** 247–248
and inositol phosphate cycle, **161,**
219–220
insect
gustatory signaling, **164,** 231
olfactory signaling, **164,** 231
visual, **164,** 232
intracellular
effect on cytoskeleton, **166,** 9–10
modulation of calcium and related
metabolism, **138,** 254–261
intracellular calcium messenger system
and, **142,** 305–306, 338
G proteins, **142,** 308, 311
phosphatidylinositol turnover, **142,** 326
protein kineses, **142,** 335–336
kinetochore role, **162B,** 46–48
lamin role, **162B,** 167
macromolecular crowding effects, **161,**
211–214
from membrane receptors to Ras/MAP
kinase pathway, GRB2 mediation of,
167, 175
microtubule/nucleus system, **175,** 119
myofibrillogenesis and, **143,** 169
network for insulin, **173,** 244–247
organelle-specific stress, **174,** 161–163
and plant defense gene activation, **148,**
19–25, 29–30
receptor protein classes, schematic, **164,**
203
receptor tyrosine kinases, **164,** 199–200
for regulation of flagellar motility, **173,**
21–23
role of MAG, **175,** 201
role of pp120, **173,** 267–269
role of Shc, **173,** 269–272
Saccharomyces cerevisiae, **165,** 27–34
salt stress, **165,** 25–35
self-incompatibility and, **143,** 263, 270,
289–290
stretch-activated channels, **161,** 207–208
selective channels, **161,** 210–211
swelling-initiated
in Ehrlich Ascites tumor cells, **161,**
225–229
in human platelets, **161,** 225

Signal transduction (*continued*)
termination of signal, **161,** 229
transcriptional regulation and, **143,** 10, 16, 24
transdifferentiation in medusae and, **142,** 249, 257
and yeast
calcium-related cascade and, **139,** 325–326
cAMP cascade and
adenylate cyclase regulation in, **139,** 311–314
cAMP-dependent protein kinase in, **139,** 316–317
cAMP-responsive gene expression in, **139,** 318–319
downstream of cAMP pathway in, **139,** 318
isolation of mutants related to, **139,** 310–311
phosphodiesterases in, **139,** 315
phosphoprotein phosphatases in, **139,** 317–318
regulation of intracellular cAMP level in, **139,** 315–316
regulation of metabolism by, **139,** 319–320
cyclic AMP cascade and, **139,** 310–319
general considerations for, **139,** 310–311
meiosis regulation and, **139,** 322–323
phosphatidylinositol signaling in, **139,** 323–325
regulation of cell division and, **139,** 320–322
tyrosine kinase in, **139,** 326–327
Silica
deposition vesicles, in diatom morphogenesis
definition, **150,** 253
formation, **150,** 269
shell, modeling, **150,** 275–277, 279, 283
structure, **150,** 254
in diatom morphogenesis
colloidal properties, **150,** 320–326
deposition vesicle, **150,** 253–254, 269
diffusion, **150,** 322–323
diffusion-limited precipitation, **150,** 271–275
formation, **150,** 320–322
precipitation, **150,** 271–288, 345
receptor role in DNA synthesis, **150,** 327
requirements, **150,** 324–326
reserves, **150,** 338–339
shell
deposition vesicle, **150,** 275–277, 279, 283
diffusion-limited precipitation, **150,** 275–288
sources and sink distribution in models, **150,** 281–284
silica antenna, diatom as, **150,** 338–340
surface chemistry in sintering, **150,** 295–296
transport model, **150,** 341–342
valve
ionophores, **150,** 328–329
receptors, **150,** 327–328
transport vesicles, **150,** 328, 335–336
fiber cultures, **142,** 156, 158–159
Silicalemma
definition, **150,** 253
formation, **150,** 254–255, 343
Silurus, fish chromatophores and, **143,** 195, 231
Silurus asotus, fish chromatophores and, **143,** 193, 237
Simian virus, transformed rat fibroblasts, **142,** 159–160
Simian virus 40
DNA
intermediates, **162A,** 508–516
minichromosomal, **162A,** 520–521
and T-antigen, **162A,** 517–518
life cycle, **162A,** 488–490
model system for DNA replication, **162A,** 495–498
myofibrillogenesis and, **143,** 168
T-antigen
NLS, **162B,** 259–260
regulation of DNA replication, **162A,** 498–501
transcriptional regulation and, **143,** 2, 39, 43–44
tumorigenesis in transgenic mice bearing gene for, **164,** 54–60
virions
accumulation during infection, **162A,** 522–523
assembly stages, **162A,** 520–528

association with chromatin, **162A,** 530–531
formation during course of infection, **162A,** 525–528
Sinapyl alcohol, *see* Monolignols
Single-headed myosin, **156,** 219, 221–222, 224
Single-photon emission-computed tomography, proteins in striated muscle, **142,** 115–116
Single-strand conformation polymorphism, detection of gene mutations, **158,** 235–236
Sintering
bone, **172,** 162–164
in diatom morphogenesis, *see* Diatom morphogenesis, chemical basis
synthetic bone graft materials, **172,** 164
Sinuses, lymph, **167,** 95–96
Sinus rhomboidalis, lateral and caudal regions, **175,** 254–255
Siphonocladus, environmental stimuli, **145,** 295
Siphonophores, colonies of, **167,** 38
Siphonous green algae, environmental stimuli in, **145,** 294–296
Sister chromatid exchange
formation, replication model, **154,** 293
visualization, **156,** 126
Size
cell nucleus, change in, **162A,** 10
chromatin domain, and transcriptional activity, **162A,** 430–431
DNA loops, **162B,** 407–411
Skelemin, in striated muscle, **142,** 83
Skeletal muscle
cytoskeletal dynamics, **144,** 96, 106
differentiation
mechanical stimulation, **150,** 77–78
response to mechanical loading, **150,** 72
environmental stimuli in plant cells, **145,** 260–261, 263, 277
FAK localization in *Xenopus,* **167,** 175
fiber type transitions
mechanism, **170,** 188–192
related to energy metabolism, **170,** 152–153
related to myosin isoform, **170,** 153–155
fiber type transitions, factors
aging, **170,** 186–188
denervation, **170,** 176–177
development, **170,** 158–162
electrostimulation-induced fast-to-slow transitions, **170,** 163–170
electrostimulation-induced slow-to-fast transitions, **170,** 177–178
exercise training, **170,** 171–172
hindlimb suspension, **170,** 181–182
hormone action, **170,** 184–186
immobilization in shortened position, **170,** 180–181
innervation, **170,** 162–163
mechanical overloading, **170,** 172–175
microgravity, **170,** 182–183
myotonia, **170,** 170–171
spinal cord transection and isolation, **170,** 177
thyroid hormone action, **170,** 183–184
gene expression, fiber type-specific programs, **170,** 155–158
metabolic enzyme-based fiber types, **170,** 148–150
myofibrillogenesis
differential assembly, **143,** 179–180
isoform change, **143,** 154–155
sarcomers formation, **143,** 182–183
thick filament proteins, **143,** 178
thin filament proteins, **143,** 157–174, 176
myosin ATPase-based fiber types, **170,** 144–148
myosin heavy chain isoforms, **170,** 150–152
proteins
antibody use, **142,** 100–102
cardiac muscle, **142,** 102–103, 108–111
sarcolemma, **142,** 124–125
sarcomere, **142,** 93–100
sarcoplasmic reticulum, **142,** 120, 122, 124
thick-filament proteins, **142,** 62–83
thin-filament proteins, **142,** 83–93
soleus
myosin, **142,** 68, 70, 74–79
thick-filament proteins, **142,** 80–82, 87–88
Z-disk, intermediate filament associations, **164,** 109–112

Skeleton, lamina densa, and collagen IV, **173,** 120–123
SK&F 104,864, against MDR phenotypes, **156,** 105
Skin, parathyroid hormone-related protein, **166,** 249–250, **166,** 256–257
Skylab, cell biology experiments, **156,** 354
Sliding velocity, flagellar dynein regulation, **155,** 170–171
Slime mold
  microgravity experiments, **156,** 343
  myosin, **156,** 223–225
Slit diaphragm, polyethylenimine administration, **166,** 214–215
S locus-specific glycoprotein
  primary structure, **140,** 544–551
  self-incompatibility
    gametophytes, **143,** 283–290
    sporophytes, **143,** 263–270, 272–280
Slow-twitch oxidative fibers, proteins in striated muscle, **142,** 120
S/MAR elements
  coinciding
    with DNase I sites, **162A,** 397
    with hypersensitive sites, **162A,** 398–404
  as domain borders, **162A,** 425–435
  functions, **162A,** 440–444
  as integration targets, **162A,** 435–437
  mediation of long-term stability, **162A,** 430–431
  prototype, **162A,** 407–410
  role
    in cultured cells and in differentiation, **162A,** 431–435
    in nucleosomal positioning, **162A,** 398–400
  structural features leading to scaffold association, **162A,** 410–416
  and transcriptional regulation, **162A,** 416–425
SMI1, in matrix attachment, **162A,** 305
Smoking, nondisjunction and, **167,** 283
Smooth muscle
  cardiac, *see* Cardiac muscle
  dense bodies, intermediate filaments associations, **164,** 109–112
  ecto-ATPase levels, **158,** 164–165
  proteins in, **142,** 94, 98, 124
  transdifferentiation in medusae and, **142,** 257
    DNA replication, **142,** 252–255
    potential, **142,** 224–230, 235
Smooth muscle cells
  arterial
    atherosclerosis
      mouse model, smooth muscle cell changes, **169,** 229–230
      pathogenesis role, **169,** 184, 187–190, 226, 231
    cell culture
      attachment substrates, **169,** 192–193
      cell lines, **169,** 191
      clones, **169,** 191–192
      coculture with other cell types, **169,** 192
      enzymatic isolation of cells, **169,** 190–191
      explantation technique, **169,** 190
      extracellular matrix components, effects on phenotype, **169,** 197–207
      organ culture, **169,** 193
      phenotypic changes, **169,** 193–197
      proliferation induction, **169,** 207–218
      secretion of extracellular matrix components, **169,** 223–226
      synthesis of growth-promoting agents, **169,** 219–223
    development
      differentiation, **169,** 185–187
      embryogenesis, **169,** 184–185
      fibronectin expression, **169,** 186–187
      growth factors, **169,** 185
      markers, **169,** 186–187
    manipulation of behavior
      angiotensin-converting enzyme inhibitors, **169,** 228
      calcium channel blockers, **169,** 227–228
      cyclooxygenase inhibitors, **169,** 227
      gene therapy, **169,** 229
      heparin, **169,** 227
      recombinant toxins, **169,** 228
      vastatins, **169,** 228
  cardiac, ecto-ATPase levels, **158,** 159–160
  development, **161,** 1–5
  disease states, therapeutic strategies, **161,** 27–32
    antibodies to growth factors, **161,** 30–31

antisense oligonucleotides, **161,** 31
drugs, **161,** 32
toxins, **161,** 31–32
inhibitors
ACE inhibitors, **161,** 24–25
eicosanoids, **161,** 26
heparan sulfate, **161,** 23–24
heparin, **161,** 23–24
and hyperplasia diseases, **161,** 29–30
nitrous oxide, **161,** 26–27
serotonin, **161,** 25–26
somatostatin, **161,** 26
myofibrillogenesis, **143,** 153, 155, 160–161, 174
stimulators
angiotensin II, **161,** 14
apolipoprotein (a), **161,** 14–15
betacellulin, **161,** 12–13
fibroblast growth factors, **161,** 6–7
heparin-binding epidermal growth factor-like growth factor, **161,** 10–12
insulin-like growth factors, **161,** 14
interleukin-1, **161,** 14
lipoprotein (a), **161,** 14–15
platelet-derived growth factor, **161,** 9–10
vascular, ecto-ATPase levels, **158,** 159–160
Snakes, avian salt glands and, **144,** 131, 139
SNAP, in CFTR recycling, **172,** 219–220
SNARE protein
in CFTR recycling, **172,** 219–220
–Rab interaction, genetic interactions, **176,** 59–60
role in vesicle transport, **176,** 57–58
Ypt1p role in regulation, **176,** 60–62
SNSC hair bundles, **156,** 288, 292, 293, 294
SOD, *see* Superoxide dismutase
Sodium
acrosomal exocytosis requirements for, **149,** 20–21, 39
avian retina and, **146,** 6
avian salt glands and, **144,** 129, 154, 192, 205
innervation, **144,** 155, 157, 161
receptive systems, **144,** 178, 183–185, 187
secretory mechanism, **144,** 173, 175–176
secretory tissue, **144,** 131–133, 135, 139, 141
stimulus-secretion coupling, **144,** 200–201
bacterial flagellar motor and, **147,** 99, 154
biological clocks in retina and, **146,** 102–103, 131
calcium channels and, **147,** 205, 207
control of rhythmic behaviors and, **147,** 12
efflux, **165,** 21–24
fertilization requirements for, **149,** 20–22
fish chromatophores and, **143,** 239
hormonal control of amphibian metamorphosis and, **145,** 135
in insect retina, **176,** 316–319
ion channels, and neuromediation, **149,** 227–231
ionic changes leading to acrosomal exocytosis, **149,** 37–38
ionic regulation of sperm function, **149,** 20–22
calcium channels, **149,** 31–37
functional evidence, **149,** 27–28
Na^+/Ca^{2+} exchanger role, **149,** 29
Na^+/H^+ exchanger, **149,** 30–31
Na^+/K^+-ATPase role, **149,** 28–29
nuclear magnetic resonance and, **145,** 20
retinal precursor cells and, **146,** 147, 153, 155, 160
sperm capacitation requirements for, **149,** 18–21, 39
functional evidence, **149,** 22–25
Na^+/Ca^{2+} exchanger role, **149,** 26–27
Na^+/K^+-ATPase role, **149,** 25–26
Sodium–calcium exchanger
acrosomal exocytosis role, **149,** 29
sperm capacitation regulation, **149,** 14, 26–27
Sodium channels
epithelial, possible CFTR regulation of, **172,** 231
modulation by CFTR, **172,** 213–214
Sodium chloride
avian salt glands and, **144,** 178, 182, 184–185, 187
reabsorption in cystic fibrosis, **172,** 196, 210–211
Sodium iodate retinopathy
cytology, **152,** 234, 236, 238–251
as model of retinal pigment epithelium regeneration, **152,** 232–235

Sodium-motive force, bacterial flagellar motor end, **147,** 113
Sodium–potassium ATPase, structure, **159,** 11
Sodium–potassium–chloride cotransporter
bumetanide binding, **161,** 193–194
cytoskeleton role in RVI, **161,** 205–206
identification and purification, **161,** 192
mechanism, **161,** 191–194
occlusion of ions by, **161,** 194
primary structure, **161,** 191–194
protein phosphorylation/dephosphorylation effects, **161,** 230–233
in RVI, **161,** 189–191
Sodium–proton exchanger
acrosomal exocytosis role, **149,** 30–31, 38
protein phosphorylation/dephosphorylation effects, **161,** 234–236
in RVI, **161,** 194–196
sperma capacitation role, **149,** 30–31
Software, visualization, in computer imaging, **162A,** 50–53
Soil amoeba, myosin, **156,** 221–223
Soil bacteria, *Rhizobium*–plant interactions, **156,** 1–51
Solanaceae, extracellular matrix and, **146,** 305
Solanum, self-incompatibility and, **143,** 280, 283–284, 286, 289
Solanum tuberosum
plastids and, **145,** 174
protoplast culture, effect of microgravity, **171,** 44–45
protoplast wall regeneration, **170,** 44
Solation, cytoskeletal dynamics and, **144,** 86, 88
Solid-state perspective, *see also* Diffusionist perspective
application to transportants, **162B,** 207–210
cytoplasmic mRNA distribution, **162B,** 199–200
models
assessment, **162B,** 189–192
evaluation, **162B,** 205–207
generalizability, **162B,** 192–193
on mRNA transport, **162B,** 184–186
Solubilization, tonoplast V-ATPase, **174,** 103–104
Soluble factors, Sertoli-germ cell communication network and, **147,** 75–79
Somateria mollissima, avian salt glands, **144,** 130, 141
Somatic chromosomes, reduction
cultured plant cells, **158,** 90–91
induction by colchicine, **158,** 89
plants, **158,** 87–90
Somatic–germ cells, differentiation in asexual *Volvox carteri* embryos
asymmetric division and cell size in, **139,** 208–211
cell type-specific genes in absence of asymmetric division in, **139,** 214–215
cytological control of division symmetry in, **139,** 216–218
differential gene expression in, **139,** 211–214
model for genetic control of, **139,** 206–208
mutations as disruption in, **139,** 204–206
Somatic hybrids, higher plant mitochondrial genomes and, **141,** 151–152
Somatogastric system, control of rhythmic behaviors and, **147,** 11
Somatolactin
discovery, **169,** 2–3
functions
acid–base regulation, **169,** 14–16, 21
background adaptation, **169,** 19
calcium regulation, **169,** 11–13
fat metabolism, **169,** 16–18
maturation, **169,** 10–11
reproduction, **169,** 10–11
stress response, **169,** 13
gene expression, **169,** 8
glycosylation, **169,** 1, 4, 21
hormonal characteristics, **169,** 21
immunocytochemical localization, **169,** 5, 8
secretion
hypothalamic control, **169,** 19–20
Pit-1 regulation of expression, **169,** 20
sequence homology
growth hormone, **169,** 1–2, 4–5
interspecies homology, **169,** 2–4
prolactin, **169,** 1–2, 4–5
synthesis site, **169,** 1, 3, 5, 8–10
Somatostatin
astrocyte reaction to, **148,** 129, 147–154
avian salt glands and, **144,** 154
conservation between species, **159,** 37
control of amphibian metamorphosis and, **145,** 140

expression in Harderian gland, **168,** 45
inhibition of smooth muscle cell growth, **161,** 26
messenger RNA levels, **159,** 9
neuronal phenotypic expression affected by, **153,** 272, 276–278
neuron distribution, **159,** 37–38
precursor, **159,** 36–38
secretion in fish, **159,** 7
synthesis, **159,** 36–37
Somatostatin receptor, Harderian gland, **168,** 42–43
Somatotroph
growth hormone synthesis, **157,** 34, 38
ontogenesis, **157,** 35–37, 43–44
Somatotropin-release inhibiting factor
immunohistochemical localization, **138,** 113–117
neuroendocrine effects, **138,** 117
role in apoptotic processes, **163,** 118
Somites
platelet-derived growth factor distribution in, **172,** 105, 108
programmed cell death in, **163,** 138–139
Sorbitol, **165,** 11–12
Sorting
avian retina and, **146,** 27, 41
MACS, **166,** 149
Sounding rockets, **156,** 352–353
Soybean
flowering plants and, **143,** 322
intracellular calcium messenger system and
G proteins, **142,** 310
phosphatidylinositol turnover, **142,** 322, 325–326, 328
protein kineses, **142,** 334
receptors, **142,** 307
lectin, thymocytes and, **156,** 172
nodulation, **156,** 3, 22, 27, 31, 34–37, 39, 42, 46
non-nodulating mutants, **156,** 41
plastids and, **145,** 184, 192
supernodulating mutants, **156,** 42
urease in plant cells and, **145,** 65–66, 98–99
biochemical genetics, **145,** 84–92
elimination, **145,** 81–84
metabolic origins, **145,** 67–69, 72–73, 75–79
nickel metabolism, **145,** 92–97
Soyuz spaceflights, microgravity experiments, **156,** 321, 323, 325–326
SP120, interaction with matrix proteins, **162A,** 304
Spa1, spindle pole body functions, **169,** 66
Spaceflights
biology
experimental equipment complexity, **171,** 5–7
fundamental tasks in, **171,** 2–4
effect on plant cell wall polysaccharide synthesis, **170,** 54–56
experiments, immune cells, **156,** 305–316
gravity, **156,** 301–303, 305
microgravity effects
on animals, **156,** 335
on plants, **156,** 313–314, 316, 319, 321–326, 328
space vehicle cabin environment, **171,** 5
techniques and instruments, **171,** 5–7
weightlessness, definition, **171,** 4
MIR, microgravity experiments, **156,** 335
orbital spaceflight, **156,** 301–302, 353–357
spacecraft, **156,** 353–357
weightlessness as active factor in, **171,** 2
Spacelab
animals, **156,** 334, 335, 336
in-flight centrifuge, **156,** 356
microgravity experiments, **156,** 354
microorganisms, **156,** 340, 342, 343
plants, **156,** 306, 312, 323, 325–328, 330
Space radiation, spaceflight experiments, **156,** 355–357
Space shuttle, **156,** 349, 354
SPARC, Sertoli-germ cell communication network and, **147,** 50
Spastic syndromes
human hyperplexia, **176,** 230–231
oscillator mouse, **176,** 231–232
spasmodic mouse, **176,** 231–232
spastic mouse, **176,** 233–234
SPB, *see* Spindle pole body
Spc110, spindle pole body functions, **169,** 65–66
Special chromatin structures
enhancer blocking assay, **162A,** 428–430
lacking enhancing activity, **162A,** 426–427
Species
BALT variations among, **167,** 136–137
M cells
identification in different, **167,** 100
variations among, **167,** 112–114

Species (*continued*)
NALT variations among, **167,** 137
Peyer's patches in, development among, **167,** 94–95
Species differences, acute phase proteins
bovine, **170,** 234
canine, **170,** 235
equine, **170,** 234
feline, **170,** 235
goat, **170,** 234
guinea pig, **170,** 232
hamster, **170,** 234
mink, **170,** 235
mouse, **170,** 231–232
ovine, **170,** 234
porcine, **170,** 234
rabbit, **170,** 232
rat, **170,** 231
Species specificity
avian retina and, **146,** 36
growth factor-induced retinal regeneration and, **146,** 60
sea urchin sperm receptor, **163,** 276
sperm–egg interactions, **163,** 252
Specific adhesion hypothesis, avian retina and, **146,** 23
Specific ion toxicity, salt tolerance, **165,** 4–6
Speckled domains
bright spots in, **162B,** 311–314
containing splicing machinery components, **162A,** 160–164
location and function, **162A,** 73–78
Spectrin
brain, **139,** 284–287
cytoskeletal dynamics and, **144,** 108
in cytoskeleton–membrane connections, **176,** 150–151
in membrane-cytoskeletal interactions, **148,** 50–52
nonerythroid spectrins, **156,** 241
retinal precursor cells and, **146,** 160
in striated muscle, **142,** 92–93, 101–102, 110
Spectrometry, in analysis of calcifications, **172,** 139
Spectroscopy
in analysis of calcifications, **172,** 137, 139
depth resolved surface, in pathology, **145,** 21
electron, in analysis of calcifications, **172,** 139
nuclear magnetic resonance, *see* Nuclear magnetic resonance
point resolved, and NMR, **145,** 21
proton, pathology and, **145,** 2, 19, 43
Raman, in analysis of calcifications, **172,** 137
sporopollenin, **140,** 63–66
Sperm
abnormalities, conceptus development-related, **164,** 144–145
aneuploidy
incidence, **167,** 269–270
paternal age and, **167,** 281–282
ATP induction of acrosome reaction, **158,** 168
capacitation
acrosomal exocytosis, *see* Exocytosis, acrosomal
in animals and plants, **140,** 409–411
male gamete changes, **140,** 410–411
pollen tube changes, **140,** 410
associated changes, **149,** 2–5
Ca^{2+}-ATPase regulation, **149,** 14–16, 39
calcium role, **149,** 9–11, 14–17, 31–36
ionic changes, **149,** 38
Na^+/H^+ exchanger role, **149,** 30
potassium role, **149,** 17–20
process, **149,** 1–2, 39
sodium requirements
functional evidence, **149,** 22–25, 39
ionic regulation, **149,** 18–21
Na^+/Ca^{2+} exchanger, **149,** 26–27
Na^+/K^+-ATPase, **149,** 25–26
chemotaxis, **163,** 251
cytoskeleton and generative cell, *see* Generative cell and sperm cytoskeleton
decapacitation factors, **149,** 5–7
dimorphism
functional significance of, **140,** 143–144
in *Plumbago,* **140,** 133
scale, **140,** 141–142
ecto-ATPase levels, **158,** 168
–egg interactions
fusion events, **163,** 284–285
gamete surface components, **163,** 253
lectin-like, **163,** 268–269, 281
recognition events, **163,** 280–284
signaling events, **163,** 285
species specificity, **163,** 252

sperm receptors
egg outer coat, **163,** 280–282
egg plasma membrane, **163,** 283–284
flagellar motility, **173,** 23–25, 27
function, **140,** 188–189
calcium regulation, **149,** 9–17, 33
chlortetracycline as monitor, **149,** 3–7
ionic changes, **149,** 37–38
Na^+/H^+ exchanger, **149,** 30
Na^+/K^+-ATPase, **149,** 28–29
potassium regulation, **149,** 17–20
sodium regulation, **149,** 20–22
invertebrate, acrosome reaction, **175,** 37–40
mammalian, special adaptations, **173,** 9–13
maturation, **154,** 105–106
conditions required, **154,** 184
efferent ducts, epithelial nonciliated cells of, **154,** 131, 137–146
endocytosis in, *see* Endocytosis
epididymis, *see* Epididymis
germ cells, **154,** 126, 128–130
membrane, modification during epididymal transit, **154,** 178–184
rete testis epithelial cells, **154,** 131–137
secretion role, *see* Secretion
seminiferous tubule intermediate region, **154,** 130–131
Sertoli cells, *see* Sertoli cells
vas deferens, **154,** 177–178
nucleus, **140,** 202
peptides activating, **163,** 251
plasma membrane, **140,** 201–202
pollen interaction, **140,** 570–571
produced by Cnidarian medusa, **167,** 38
sperm nucleus, **140,** 202
sperm-to-sperm connections, functional significance, **140,** 142–143
true cells or naked nuclei, **140,** 130–131
Spermatocytes, karyosphere and, **144,** 1, 8
Spermatogenesis
double
definition, **152,** 2
development in Lepidoptera, **152,** 36
MIP-1α as regulator of DNA synthesis in, **167,** 208, 242
Spermatozopis
eyespot apparatus, **148,** 233–234
photoreceptive apparatuses, components, **148,** 272
photosynthetic accessory pigments in, **148,** 281–283
phototaxis, calcium requirements, **148,** 294
Sperm-binding proteins
egg plasma membrane
brown alga, **163,** 278
CD4-like proteins, **163,** 269–272
echiurian worm, **163,** 278–279
FcR proteins, **163,** 269–272
mammalian, **163,** 269–273
PH-30/fertilin cognate, **163,** 273
recognition events, **163,** 283–284
sea urchin, **163,** 273–278
sperm-binding factor, **163,** 274–275
sperm receptor, **163,** 275–278
table listing, **163,** 270–271
egg surface
abalone vitelline layer, **163,** 262–263
in amphibians, **163,** 263–264
ascidian vitelline coat, **163,** 258–259
echinoderm egg jelly, **163,** 259–262
in fishes, **163,** 263–264
penaeoidean shrimp, **163,** 263
recognition events, **163,** 280–282
table listing, **163,** 260–261
zona pellucida
ZP2, **163,** 267–268
ZP3, **163,** 264–267
identification criteria, **163,** 255
Spermidine
Sertoli-germ cell communication network and, **147,** 49
spermine concentrations in rat prostatic lobes, **138,** 62
Spermination, Sertoli cell role, **160,** 204–206
Spermine, Sertoli-germ cell communication network and, **147,** 49
Spheroid assay, as migration analysis method, **155,** 62
Spheroids
avian retina and, **146,** 41
cellular processes, **146,** 27–28, 32, 36
retinogenesis *in vitro,* **146,** 9, 11–12, 15–17, 19

Spheroids (*continued*)
rosetted, avian retina and, **146,** 40–41
cellular processes, **146,** 25, 27–32, 34, 36
retinogenesis *in vitro,* **146,** 11, 15, 17, 20
Spider web cultures, vertebrate cell culture technology and, **142,** 153–155
Spinacea oleracea, **140,** 201–202; **143,** 299
Spinach
intracellular calcium messenger system and, **142,** 328, 330, 332, 334–335
plastids and
amino acid synthesis, **145,** 192, 194
CO_2 fixation, **145,** 172, 175
fatty acid synthesis, **145,** 184–187
structure, **145,** 162–163, 167
translocation, **145,** 168, 170
Spinal cord, as model of reduced neuromuscular activity, **170,** 177
Spinal nerve, amphibian development, **169,** 99–100
Spin coupling, in NMR, **145,** 8–9
Spin decoupling, in NMR, **145,** 8–9
Spin density, nuclear magnetic resonance and, **145,** 26
Spindle pole body
centrosome homolog in yeast, **169,** 63
microtubule nucleation, **169,** 69
mitosis and, **144,** 226
morphology, **169,** 63–64
organization during cell cycle, **169,** 64–65
protein components, **169,** 65–68
replication, **169,** 69–70
separation, **169,** 70–71
Spindles
abnormally functioning, mitosis and, **144,** 230, 232
cell model systems, **175,** 14–15
elongation, kinesin-like proteins in, **164,** 305
midbody, composition, **162B,** 54–55
midzone, during metaphase-to-anaphase, **162B,** 48–54
mitotic, evolution, **158,** 126
orientation in cell division and development, **173,** 216
and phragmoplast-cell plates, conventional, **140,** 170–172
poles, centrosomes as, **162B,** 56–59
structure, lepidopteran, *see* Lepidoptera, spindle structure
Spine apparatus, **139,** 271–272, 277–278
Spin echo, in NMR, **145,** 11–12, 25, 27
Spirodela polyrrhiza, effect of microgravity, **171,** 41
Spleen
dendritic cells, **153,** 67
hormonal control of amphibian metamorphosis and, **145,** 131
myofibrillogenesis, **143,** 161
nuclear magnetic resonance and, **145,** 31
Spliceosomes
and coiled bodies, **162B,** 326
NuMA association, **162B,** 18–19
Splicing
alternative for NuMA, **162B,** 6–8
component distribution and accumulation, **162A,** 162–164
RNA
coordinated with transcription, **162A,** 39
on nuclear matrix, **162A,** 103–104
Tau protein genes, **171,** 170–171
trans-splicing, endosymbiont hypothesis, **141,** 267, 281, 304
Splicing factors
coiled bodies containing, **162A,** 71–73
localization, **162B,** 311–314
non-snRNP, **162A,** 104–105
snRNP and non-snRNP, **162A,** 159–161
in speckled domains, **162A,** 75–78
U2AF
diffuse pattern in nucleoplasm, **162A,** 160
non-snRNP splicing factor, **162A,** 71–75
Sponge
transdifferentiation in medusae and, **142,** 216
vertebrate cell culture technology and, **142,** 154, 156–157
Spontaneous abortions
aneuploidy and
incidence of, **167,** 265–267
increased maternal age reducing likelihood of, **167,** 278
mosaicism in, **167,** 287–288
Spores
germination fluids, suppressors, **172,** 62–65
microgravity experiments, **156,** 339–340, 352

S receptor kinase (*continued*)
conserved cysteine residues, **158,** 26
expression in development, **158,** 27
expression in pollen, **158,** 29
gene structure, **158,** 26
linkage to S locus, **158,** 25–27, 31, 52
role in self-incompatibility, **158,** 27–28, 52
self-incompatibility, **143,** 270, 274–277, 279–280, 290
sequence, **158,** 25–26
signal transduction in self-incompatibility, **158,** 17, 52
species distribution, **158,** 50–51
SRIF, *see* Somatotropin-releasing inhibiting factor
S RNase
Rosaceae, **158,** 37–38
Solanaceae
catalytic domain, **158,** 34
role in pollen inhibition, **158,** 35–36
substrate specificity, **158,** 34
SRP1p, yeast homolog of importin, **162B,** 287–289
SSCP, *see* Single-strand conformation polymorphism
S sequences, expression
SLG expression in *Brassica* stigmas, **140,** 512–514
SLG gene and promoter constructs in transgenic plants, **140,** 514–517
S subfamilies, matrix of percentage of amino acid identification, **140,** 504
Stability
internal matrix, **162B,** 104–105
long-term, mediated by S/MARs, **162A,** 430–431
Stable tubule only peptide
induction of microtubule stability, **157,** 20
role in microtubule cold adaptation, **157,** 20–21
structure, **157,** 20
Stablin, melanocytes and, **146,** 208
Stachydrine, **156,** 9
Staining
apoptotic cells, **163,** 120
cnidarian neurons
immunocytochemical method, **167,** 44
methylene blue, **167,** 40
DAPI, mitochondrial nuclei, **161,** 61
EDTA regressive staining, **162A,** 4–8, 15–21
Stall torque, bacterial flagellar motor and, **147,** 110, 116, 126–127, 155
Staphylococcus aureus, spaceflight experiments, **156,** 340–341
Starch
higher plant plastids and, **145,** 167, 172, 176–182, 203–204
synthesis
in chromoplasts, **163,** 209–211
in nonphotosynthetic plastids, **163,** 209–210
Starch synthase, higher plant plastids and, **145,** 177, 181
Starfire, **156,** 352–353
Start sites, nascent strand, identification, **162A,** 466–473
STAT, *see* Signal transducer and activator of transcription
Static head equilibrium, bacterial flagellar motor and, **147,** 126
Statocysts, structure, **167,** 40
Statocytes
microgravity experiments, **156,** 328–330
plant, endoplasmic reticulum volume, effect of microgravity, **171,** 50–51
root cap
effects of microgravity, **171,** 11, 20–22
structural characteristics, **171,** 10–11
Staurosporine, **156,** 333; **168,** 292–293
Steady-state free precession sequences, NMR and, **145,** 29
Steele–Richardson–Olszewski syndrome, with Tau 64 and 69, **171,** 199–200
steel factor, melanocytes and, **146,** 215–216, 222, 247
Stele factor, **156,** 45, 47
Stem cells
assays for inhibitory regulators of, **167,** 197
avian retina and, **146,** 28–31, 35, 39
defective mice, **157,** 143
differentiation, **157,** 129
dome epithelial cells as possibly derived from, **167,** 108
hematopoietic microenvironment, **157,** 143
hematopoietic ontogeny, **157,** 130–132
hemopoietic
Pre-CFU-S
recent recognition of more primitive, **167,** 190
reversible inhibition of by pEEDCK, **167,** 188
proliferation regulation of, **167,** 194–196

search for MIP-1α receptors on, **167,** 206–207
in structure of developing cell populations, **167,** 190–192
identification, **157,** 130
introducing into *Hydra* epithelium, **167,** 39
migration, **157,** 132
ploidy level, **158,** 122
prevalence, **157,** 130
regenerative cementogenesis and, **142,** 38, 40
self-renewal, **157,** 129
transdifferentiation in medusae and, **142,** 229
Stenosis, proteins in striated muscle and, 141:118
Stephanodiscus hantzschia, morphogenesis
microfilaments, **150,** 306
silicalemma, **150,** 343
Stereocilia, actin filament bundles, **175,** 35–37
Stereological methods, renal glomeruli
cell type average volume, **161,** 151–154
for glomerular number, **161,** 134–141
for glomerular volume, **161,** 141–143
individual glomeruli, **161,** 162–165
for kidney volume, **161,** 141–143
length of capillaries, **161,** 154–155
matrix volume density, **161,** 160–161
mean glomerular volume, **161,** 141–143
mean length of capillaries, **161,** 158
number of capillaries, **161,** 156–157
number of cell types, **161,** 148–151
surface area of capillaries, **161,** 155–156, 158
surface area of cell types, **161,** 151–154
systematic uniform random schemes for fields selection, **161,** 137
thickness of glomerular basement membrane, **161,** 161–162
total number of cells in, **161,** 143–148
total number of each cell type, **161,** 151
volume density of capillary lumina, **161,** 158–159
volume density of urinary space, **161,** 158–159
volume of embedded kidney, **161,** 141
Stereotropism, vertebrate cell culture technology, **142,** 154
Sterility, cytoplasmic, male
endosymbiont hypothesis and, **141,** 292
higher plant mitochondrial genomes and, **141,** 129, 164
abnormal phenotypes, **141,** 130–134
structure, **141,** 141, 149–150, 152–153, 156
Steroid hormones
direct *versus* cascade models, **162B,** 340–344
nuclear matrix role, **162B,** 344–345
responsive osteocalcin gene expression, **162A,** 265–269
simplistic model, **162B,** 366–367
thymic epithelial cells, **156,** 186
Steroid receptor
accessory factors, **162B,** 359–364
characterization, **164,** 202–205
hormone-responsive element-binding domain, **159,** 161, 167
imprinting
by aromatic hydrocarbons, **155,** 32–33
pathological, **155,** 31
inducibility, **155,** 16
ligand-bound, regulation of gene expression, **162B,** 340–341
memory effect, **159,** 162, 171–173
nuclear acceptor sites, **162B,** 345–348
nuclear localization signal masking, **151,** 206–207
signal transduction, **159,** 161–162
subcellular localization
cytoplasm, **159,** 167–169
nuclear matrix, **159,** 170–171
nucleolus, **159,** 170
nucleus, **159,** 167–170
target gene activation, **159,** 162
transcriptional regulation and, **143,** 7, 38
as transcription factors, **162B,** 338–340
Steroids
avian salt glands and, **144,** 191–192
control of amphibian metamorphosis and, **145,** 119–121, 124, 126–128
GnRH-encoding gene in teleosts and, **147,** 170, 175, 187
gonadal, sex hormones and antihormones effect on, **139,** 10–11
secretion, effect of endothelin, **171,** 278–279
Sertoli-germ cell communication network and, **147,** 44, 58, 73
sexual, tumor secreting, role of endothelin, **171,** 295
and testicular macrophages, **149,** 115–116, 120
transcriptional regulation and, **143,** 18

Stick insect, microgravity experiments, **156,** 335, 336, 356
Stilbenes
  chalcone synthase and, **147,** 247
  derivatives, **139,** 1
Stilbene synthase, chalcone synthase and, **147,** 236–237, 242–245
Stillbirths, incidence of trisomy in, **167,** 265
Stimulated echo acquisition mode, in NMR, **145,** 21, 45
Stimulators
  CFU-S, from marrow macrophages, **167,** 192
  SK&F107647, **167,** 219
  TGF-β, on certain hemopoietic cell populations, **167,** 210
Stimulus-secretion coupling, avian salt glands and, **144,** 200–205
Stoichiometry, ATPase, in drug pump model, **171,** 132–134
Stomata, structure and transpiration, effect of microgravity, **170,** 53–54
Stomatal opening regulation
  drought tolerance, **165,** 35
  salt tolerance, **165,** 35
Stomotoca, transdifferentiation in medusae and, **142,** 222–223, 236
STOP, *see* Stable tubule only peptide
Stop mutations, in CFTR, **172,** 200–201
Stratospheroids, avian retina
  cellular processes, **146,** 28–34, 36
  normal development, **146,** 2
  retinogenesis *in vitro,* **146,** 10–13, 15–17, 19–21
Streaming
  cytoplasmic, *see* Cytoplasmic streaming
  environmental stimuli in plant cells and
    chemicals, **145,** 297
    chloroplasts, **145,** 283–284
    low temperature, **145,** 288–290
    photodinesis, **145,** 252–254, 256–257, 260–261, 264–266
    primary streaming, **145,** 252
    secondary streaming, **145,** 252
    wounding, **145,** 295
  protoplasmic, *see* Protoplasmic streaming
Streptococcus
  bacterial flagellar motor, **147,** 106, 114
    dynamic measurements, **147,** 115–117
    model comparisons, **147,** 152, 155
    motor energetics, **147,** 123, 126–127, 129
  binding to sialic acids, **175,** 178–179
Streptomyces griseus, transdifferentiation in medusae and, **142,** 237
Streptomycin, **145,** 159; **156,** 290
Stress
  abiotic, drugs, **165,** 31
  environmental, *see* Environmental stress
  extracellular matrix and, **146,** 261, 301–305, 312–313
  mechanical, in embryonic tissue, *see* Embryonic tissue, mechanical stress in
  organelle-specific, signaling pathways, **174,** 161–163
  osmotic, activation of sucrose phosphate synthase, **149,** 71–72
  oxidative
    in *Drosophila melanogaster,* **165,** 37
    effects on conceptus development, **164,** 170–171
  regulation of chromoplast development, **163,** 203
  salt
    Escherichia coli, **165,** 25–27
    Saccharomyces cerevisiae, **165,** 27–34
    signal transduction, **165,** 25–35
    whole-plant level, **165,** 35–37
  water-related, extracellular matrix and, **146,** 301, 303
Stress fibers
  actin filament polarity, **156,** 251
  adhesive plaque connection, **150,** 63
  FAK and assembly of
    MASMC evidence against, **167,** 176
    in mouse FAK-deficient cells, **167,** 177–178
  rearward traction forces on, **150,** 41
  related structures
    myofibrillogenesis, **143,** 159–160, 166, 177, 179
    proteins in, **142,** 95, 105, 108
  retrograde surface transport, **150,** 47
  in tissue-cultured cells, **156,** 211–213
Stress-force feedback, in embryonic tissue
  cell columnarization models, **150,** 23–24
  definition, **150,** 2
Stress proteins
  in cell repair, **148,** 206–209, 215–216

gene expression, **148,** 211
in increased cell stability, **148,** 190–202
in increased primary stability of cell functions, **148,** 203–205, 215
induction by extreme heat, **148,** 175–179
as proteases, **174,** 152–154
in tolerant zone, **148,** 214
Stress response
by chaperone genes, **174,** 161
evolution, **170,** 97–98
link to immune response, **170,** 98
mediating molecules, **170,** 121–122
role of cytokines, **170,** 107–108
role of nitric oxide, **170,** 116–117
role of POMC products, **170,** 87–93
Stress tolerance
fast growth, **165,** 31–32
higher plants, **165,** 34–35
SNFI pathway, **165,** 33
Stretch-activated channels
cytoskeleton role, **161,** 211
nonselective cation channels, **161,** 208–210
selective channels, **161,** 210–211
Stretch-inactivated channels, **161,** 211
Stretching
–contraction cell response, in embryonic tissue
cell columnarization, **150,** 23
morphomechanical situations, **150,** 28–29
effect on muscle fiber type transitions, **170,** 172–175
Striated muscle
myofibrillogenesis, **143,** 153, 155, 175
proteins, *see* Proteins, in striated muscle
transdifferentiation in medusae and, **142,** 252, 256
initiation, **142,** 238–241, 245, 247, 249–251
isolation of tissues, **142,** 222–223
potential, **142,** 223–231, 233–235, 238
tissue organization, **142,** 218, 220
Stromal cells
activation by TNF-α, **167,** 229
adhesion molecule synthesis, **157,** 145
cloning, **157,** 144–145
growth factors, **157,** 145
proliferation regulation of hemopoietic progenitor cells and, **167,** 195
role in B cell development, **157,** 143–145, 148–149
Strongylocentrotus purpuratus
animal mtDNA and, **141,** 180, 187–188
endosymbiont hypothesis and, **141,** 289, 299
Strontium, incorporation in apatite, **172,** 159
Strophanthidin, biological clocks in retina and, **146,** 102
Structural organization, GSP, **140,** 151–158
Structure–function relationship, plant, in altered gravity
algae, **171,** 25–30
angiosperms, *see* Angiosperms
ferns, **171,** 31
gymnosperms, **171,** 31–32
moss, **171,** 30–31
organelle rearrangements, **171,** 46–53
specialized graviperceptive cells, **171,** 10–11, 20–22
tip-growing cells, **171,** 22–25
Strychnine, binding by glycine receptor, **176,** 212–213
Stylar matrix, in pollen tube extension
evidence for active biological role
latex bead translocation, **140,** 304–305
model for pollen tube extension via stylar matrix, **140,** 312–313
stylar matrix of *Vicia faba,* **140,** 305–307
substrate adhesion molecule vitronectin in plants, **140,** 307–312
sexual reproduction in flowering plants, **140,** 297–304
extracellular matrix in animal systems, **140,** 299–300
extracellular matrix in plants, **140,** 300–304
cell wall, **140,** 300–301
stylar wall, **140,** 301
Stylocephalus, mitosis and, **144,** 221–222
Stylonychia
mitochondrial genomes, **141,** 1
replication band
cytochemical studies, **153,** 154
ultrastructure, **153,** 150, 152
Subcommisural organ
effects of serotonin, **159,** 25–26
ontogeny, **159,** 26
structure, **159,** 26
vascularization, **159,** 26

Suberin
extracellular matrix and, **146,** 283, 305
in plant cell wall, dynamic aspects, **151,** 259–260
Suberization, as cell wall lignification model, **176,** 99–100
Subfertility syndrome, in males, nondisjunction, **167,** 286
Subfornical organ
avian salt glands and, **144,** 195, 197
efferent projections, **159,** 27
function, **159,** 27
structure, **159,** 26
Subfractionation
plant nuclear matrix, **162B,** 95–123
in situ nuclei, **162A,** 512–516
Subjective day, biological clocks in retina and, **146,** 96–97, 131, 136–137
Sublimons, higher plant mitochondrial genomes and, **141,** 152–153
Subnucleus
fiber attachments, **162A,** 142–143
SV40 minichromosomes and virion-contained DNA distribution, **162A,** 523–525
SV40 replication in conjunction with, **162A,** 531–533
T-antigen subclass distribution, **162A,** 502–508
Suborbital flight, **156,** 352–353
Subsilicea fragilarioides, sintering, **150,** 298
Substance P
astrocyte reaction to, **148,** 129, 135–137
avian salt glands and, **144,** 199
effects, **138,** 160–161
neuromodulation, **149,** 270–271
precursor, **159,** 46
Substrate, FAK, **167,** 172–175
Substrate adhesion molecules, **140,** 299–300, 307–312
Substratum mechanics, and cell differentiation
cell type conversion, **150,** 119
cellular mechanics of transdifferentiation process, **150,** 128–132
cytomechanics, **150,** 120–125
retinal pigment epithelial cells
cell fate choice regulation, **150,** 125–128
transdifferentiation process, **150,** 119–120
Subumbrella, transdifferentiation in medusae, **142,** 218, 220, 222–223
Succinate
γ-aminobutyric acid metabolism and, **147,** 311, 317
urease in plant cells and, **145,** 71
Succinate dehydrogenase, avian salt glands and, **144,** 145, 147
Succinic semialdehyde
γ-aminobutyric acid metabolism and, **147,** 311, 317
glutamate metabolism and, **147,** 300
Succinic semialdehyde dehydrogenase, γ-aminobutyric acid metabolism, **147,** 311, 316–318
Sucrose
avian salt glands and, **144,** 183–184
destabilization of clathrin-coated vesicles by, **167,** 22
higher plant plastids and, **145,** 167, 171–172
as major osmolyte in higher plant, **165,** 11–12
as regulatory metabolites of plant enzymes, **149,** 86–87
Sucrose phosphate synthase, phosphorylation
amino acids, **149,** 87–88
carbohydrates, **149,** 86
light/dark signals, **149,** 69–70
mannose activation, **149,** 70–71
in metabolism, **149,** 92–93
modulation of activity *in* vivo, **149,** 68–72
multisite nature, **149,** 72
nitrate, **149,** 86
osmotic stress activation, **149,** 71–72
physiology, **149,** 78–79
in plants, **149,** 68
regulation, **149,** 76–77
species variation, **149,** 77–78
Sucrose phosphate synthase-kinase, phosphorylation
properties, **149,** 72–73
site sequences, **149,** 74
Sucrose phosphate synthase-phosphatase
activating, **149,** 74–75
inactivating, **149,** 75–76
Sucrose phosphate synthase protein phosphatase, inhibition, **149,** 71, 76
Sugars
eggshells and biomineralization and, **145,** 235

ethanol-soluble, in pea plant, effect of microgravity, **171,** 38
extracellular matrix and
characterization, **146,** 265–266
modification, **146,** 300, 303
synthesis, **146,** 272, 275–276, 283
Suicide assays
cytotoxics using S-phase, **167,** 204–205
description, **167,** 197–198
as major tool for assessing cell proliferation, **167,** 243
myelo-protection model and, **167,** 204
Suillus, mitochondrial genomes in fungi, **141,** 104–106
Sulfate, activated, pathway, **165,** 15–16
Sulfolipids, bacterial, **161,** 288–289
Sulfolobus solfataricus, endosymbiont hypothesis and, **141,** 246, 248
Sulfonylurea, flowering plants and, **143,** 323
Sulfonylurea receptor, and P-glycoprotein, and CFTR, homology, **171,** 125–127
Sulfur
endosymbiont hypothesis and, **141,** 240–241, 249
flowering plants and, **143,** 299
kinetoplastid mtDNA and, **141,** 71
Sunflower, microgravity experiments, **156,** 322
Superantigens
intraepithelial lymphocytes and, **167,** 97
microbial toxins, table, **161,** 277
Superhelical tension, complete relaxation, **162B,** 407–408
Superhelicity, S/MAR, **162A,** 410–416
Superior cervical ganglion neurons, phenotypic plasticity, **153,** 271
Supernodulating mutants, **156,** 42–43
Superoxide dismutase, **167,** 225–226
Superprecipitation, *Physarum* actomyosin, **156,** 247–254
Suppression
cosuppression, in control of chalcone synthase expression, **147,** 258–259
immunosuppression
melanocytes, **146,** 240
Sertoli-germ cell communication network, **147,** 44
transcriptional regulation, **143,** 16, 18
urease in plant cells and, **145,** 80
Suppressor cells, decidual, pregnant uterus, **153,** 109
Suppressors
accessiblity-inducing activities, **172,** 62–65
chemical-nature of, **172,** 61
effects on ATPase
cell wall-bound, **172,** 77–80
membrane-bound, **172,** 67–70
effects on GTP-binding protein, **172,** 73–74
effects on peroxide generation, **172,** 80–81
effects on phytoalexins, **172,** 67
effects on plant defense transmembrane signaling, **172,** 70–76
–enhancer genetics, **139,** 149–151
in host-parasite specificity, **172,** 63, 64–65, 84–85
interactions with elicitors, **172,** 65–67
mucin-type, **172,** 69–70
in *Mycosphaerella,* **172,** 62–63
origin of term, **172,** 58–59
phytopathogen production of, **172,** 59–61
pisatin accumulation and, **172,** 65–66
possible modes of action, **172,** 84
signal transduction cascade model, **172,** 82–84
Supramembranal ring, bacterial flagellar motor end, **147,** 100–101, 105, 131, 135, 140
Supramolecular structure, collagen VII, **173,** 101–102
Supraoptic nucleus, avian salt glands, **144,** 182, 195
SUR, *see* Sulfonyl urea receptor
Suramin, inhibition of ecto-ATPases, **158,** 157–158
Surface angle hypothesis, embryonic tissue, mechanical stress
amphibians, **150,** 3–7
stretched cell group formation, **150,** 25
Surface area
glomerular capillaries, **161,** 155–156, 158
glomerular cell types, **161,** 151–154
Surface-attached growth, vertebrate cell culture technology, **142,** 151–152
Surface density
parameter specification, **161,** 117–118
renal glomeruli, **161,** 117–119
Surface forces, in leaky pores, **163,** 9–10
Surface lipids, and pollen volatiles, **140,** 36–41

Surface potential, symplast and, **142,** 267, 269
Surface proteins, Gram-positive bacteria, **161,** 272–274
Surface pump, symplast and, **142,** 273–275, 299
Surface receptor, macrophage, **161,** 291–293
Surface tension, in embryonic tissue
  cellular level, **150,** 19, 22–23
  mechanisms, **150,** 11–12
  subcellular level
    relaxation of tension, **150,** 12–19
    stretching, **150,** 19–21
  tissue level, **150,** 19, 22–23
Surface transport, *see* Retrograde surface transport
Surface wave theory, ameboid movement, **139,** 165, 167
Surface wax, isolation from *Pinus* pollen and wing material, **140,** 39
Surfing, in retrograde surface transport, **150,** 48
Surirella, morphogenesis, silica deposition vesicle formation, **150,** 269
Survival
  growth factor-induced retinal regeneration and, **146,** 49, 55–56
  melanocytes and, **146,** 247
  selective, retinal precursor cells and, **146,** 170
Susceptibility
  Arabidopsis thaliana, **144,** 68–69, 72–73
  in plant–parasite interactions, **172,** 57–59, 62–65
Sutures
  comparative anatomy
    fiber length, **163,** 321–322
    fiber width, **163,** 322–323
  formation, **163,** 314
  line-type, **163,** 317–319
  star-type, **163,** 319–321
  umbilical, **163,** 317
  Y-type, **163,** 319
SV40, *see* Simian virus 40
Sviatro, nodulation, **156,** 17
Sweat duct, reabsorptive, CFTR function in, **172,** 196, 210–211
Swelling
  associated pressure, cytoskeletal dynamics, **144,** 89
  paraflagellar, role in eyespot organization, **148,** 246
Switch complex, bacterial flagellar motor and, **147,** 109–110, 145
Sycamore, intracellular calcium messenger system and, **142,** 323
Sychomedusae, bidirectionalality of nerve net synapses, **167,** 43
Symbionts, chloroplasts
  cryptomonads, **144,** 275–276
  dinoflagellates, **144,** 286, 288–289, 292–293
  euglenoids, **144,** 283
  green algae, **144,** 268, 270
  heterokont algae, **144,** 278, 280
  prokaryotes and eukaryotes, **144,** 260
Symbiosis
  chloroplasts, **144,** 261, 283, 294
  Rhizobium–plant interactions, **156,** 1, 9
Sympathetic ganglion
  axotomy effects
    frogs, **159,** 215–220
    mammals, **159,** 208–215
  composition, **159,** 196
  frog chracteristics, **159,** 196–197
  reinnervation
    effect of deafferentation, **159,** 195, 222, 236–246
    heterogenic
      frog, **159,** 241–243, 246–247, 249–250
      mammals, **159,** 222, 236–241, 247–250
  sympathetic neuron classification
    amphibians, **159,** 201
    mammals, **159,** 200
  synapse, *see* Ganglionic synapse
Sympathetic nervous system, fish chromatophores, **143,** 228–230, 238
Symplast, in plant growth
  auxin
    Lockhart equations, **142,** 277–279, 283–285
    mechanical properties, **142,** 279–282
    water relations, **142,** 282–283
  electrophysiological structure, **142,** 265
    cowpea hypocotyl, **142,** 267–270
    elongation growth, **142,** 265–267
    two-pump hypothesis, **142,** 270–274
  integration of activity
    apoplast canal, **142,** 291–294
    auxin-enhanced uptake, **142,** 294–297
    dynamic regulation, **142,** 297–299

models, **142,** 287–291
respiration, **142,** 285–287
plasmalemma, **142,** 299–300
proton pumps
surface pump, **142,** 273–275
xylem pump, **142,** 275–277
water levels, **142,** 263–265
Synapses
γ-aminobutyric acid metabolism and, **147,** 315–316
avian retina and, **146,** 1–2, 19–20
avian salt glands and, **144,** 158–159
biological clocks in retina and, **146,** 92, 102, 104
and blood-brain barrier, **149,** 278
calcium channels and, **147,** 193, 199–200, 221, 227
electron microscopy, **147,** 214, 216
location, **147,** 203, 206
motor end plate, **147,** 194–195
chemical, *see* Chemical synapses
cnidarian chemical
bidirectionality of, **167,** 42–43
contacts with multiple neurons, **167,** 40–41
cnidarian electrical, confirmation of, **167,** 43
control of rhythmic behaviors and, **147,** 2–3, 19–20
ion channels, **147,** 3, 5, 10–11
neuromodulation, **147,** 13–15
electrical synapses, **149,** 219; **167,** 43
ganglionic, *see* Ganglionic synapses
glutamate metabolism and, **147,** 285–286, 304–305, 309
mixed, **149,** 220
and neuronal plasticity
formation mechanism, **153,** 263–264
invertebrate neurons, **153,** 268–270
postsynaptic potentials, **153,** 265–267
vertebrate neurons, **153,** 264–268
and neuropeptides, **149,** 271–272
neurotransmission, **149,** 219–222
noradrenaline, **149,** 243
reciprocal, **149,** 219
retinal precursor cells and, **146,** 147, 149, 154
Synapsin
regulation of MT–microfilament interactions, **173,** 228–229
in regulation of neuromediator release, **149,** 220–221
role in neurogenesis, **159,** 17
role in neuron survival, **163,** 152–153
Synaptic cleft, and neurotransmission
amino acids, **149,** 260–262
cholinergic, **149,** 238–239
in neuronal interconnections, **149,** 219–220
Synaptic complex, **149,** 220
Synaptic plasticity, dendritic spines, **139,** 269–271
Synaptic vesicles
acetylcholine content, **149,** 238
amino acid release, **149,** 260, 262
ATP release, **149,** 255
catecholamines
release, **149,** 243
storage, **149,** 244
taurine release, **149,** 258
Synaptonemal complex
karyosphere and, **144,** 7, 40, 43–45
oocytes of insects, **144,** 9, 11, 17, 21, 23
vertebrates, **144,** 36
lepidopteran, in meiosis
female, **152,** 14
male eupyrene, **152,** 20
Synaptophysian
calcium binding, **159,** 17
role in neurogenesis, **159,** 17
Synaptosomes
γ-aminobutyric acid metabolism and, **147,** 312, 314, 318
glutamate metabolism and, **147,** 295, 298, 302, 305, 307–309
Synchronization, biological clocks in retina and, **146,** 107, 133
Synchrony, Sertoli-germ cell communication network and, **147,** 38–41, 59–61
Syncytiotrophoblasts
in human placenta
as blood barrier, **172,** 26
GLUT1 in, **172,** 27
in rat placenta
as blood barrier, **172,** 29
GLUT1 in, **172,** 30, 31
Syncytium
mitosis and, **144,** 231, 234
neuronal, in cnidarians, **167,** 43

Synemin
myofibrillogenesis and, **143,** 182
in striated muscle, **142,** 98–99
Synergid
angiosperm lacking, pollen tube arrival, **140,** 370
female germ unit
cell wall and filiform apparatus, **140,** 250–256
cytoplasmic features, **140,** 256–258
histochemistry and cytochemistry of, **140,** 258–260
micromanipulation, **140,** 414–419
role of in double fertilization, **140,** 360–368
Synergistic effects, interferon classes on colony formation, **167,** 231
Syngamic phase, interactions
cytoplasmic DNA inheritance, **140,** 577–578
MGU, in double fertilization, **140,** 576–577
preprogrammed fertilization, **140,** 578–579
Synlactin, control of amphibian metamorphosis and, **145,** 131
Synovial fluid dendritic cells, human, **153,** 63–64
Systemic acquired resistance
agricultural applications, **168,** 311–312, 329
Arabidopsis thaliana, **144,** 59, 77–78
characterized plants, table, **168,** 306–307
chemical induction
3-aminobutyric acid, **168,** 326
benzo-1,2,3-thiadiazole carbothioic acid-S-methyl ester, **168,** 327–328
chitosan, **168,** 324
criteria for inducers, **168,** 323
2,6-dichloroisonicotinic acid, **168,** 326–327
elicitin, **168,** 324–325
fatty acids, **168,** 324
nicotinamide, **168,** 325
phosphate salts, **168,** 323–324
probenazole, **168,** 326
structural requirements, **168,** 328
WL 28325, **168,** 326
comparison to immunization, **168,** 304
discovery, **168,** 303–304
effects
cytology, **168,** 312–313
enzyme induction, **168,** 312–313
electrical signaling, **168,** 322–323
ethylene signaling, **168,** 320–321
jasmonic acid signaling, **168,** 321–322
model, **168,** 308
pathogenesis-related proteins
genes, **168,** 308–309, 329
inducers, **168,** 309–310
plant transformation studies, **168,** 311–312
types, **168,** 309–311
rationale, **168,** 303
salicylic acid signaling
binding protein, **168,** 317
biosynthesis, **168,** 316–317
catalase inhibition, **168,** 317–319
conjugation, **168,** 317
constitutive levels, **168,** 316
discovery, **168,** 314–315
translocation in tobacco, **168,** 316
specificity against pathogens, **168,** 304–305
systemin signaling
constitutive levels, **168,** 320
induction, **168,** 319
translocation, **168,** 319–320
time requirements, **168,** 305
Systemin, systemic acquired resistance induction
constitutive levels, **168,** 320
induction of protein, **168,** 319
translocation, **168,** 319–320

T

T3, *see* Triiodothyronine
T4, *see* Thyroxine
Tachykinins
immunohistochemical localization of, **138,** 156–160
substance P. effects of, **138,** 160–161
Tail bud, programmed cell death in, **163,** 137–138
Talin
FAK C-terminal in binding to, **167,** 167
myofibrillogenesis and, **143,** 159

in striated muscle, **142,** 92–93, 99–100, 114–115
as structural element of focal adhesion, **167,** 162
Tamoxifen
action, **139,** 40–43
calmodulin and, **139,** 42
critical period for action on male genital dysfunction, **139,** 23
effect on enzyme levels, **139,** 29
effect on female human genital system, **139,** 21
effect on gonadal steroidgenesis, **139,** 10–11
effect on sexually dimorphic brain and sexual behavior, **139,** 26–28
general considerations for, **139,** 44–45
genital abnormalities and
in guinea pigs, **139,** 13–14
in mice, **139,** 17–21, 22–23
in rats, **139,** 16, 22–23
history, **139,** 1–2
IGF-I and, **139,** 42–43
K^+ channels of neuroblastoma cells, **139,** 42
neonatal treatment with, mouse pelvis and, **139,** 36–39
pregnancy and, **139,** 43
protein kinase C and, **139,** 42
T-antigen
coordination with SV40 DNA, **162A,** 517–518
NLS, **162B,** 259–260
nuclear subclasses, **162A,** 501–508
regulated SV40 DNA replication, **162A,** 497–501
transforming potential, **162A,** 532–533
Tarichagranulosa, **138,** 92–93, 97–99, 112, 124
TA series, anther-specific genes, **140,** 31
Tau proteins
microtubule
bundling, **165,** 252–253
interaction, **165,** 252
as microtubule-associated proteins
gene organization, **171,** 170
splicing, **171,** 170–171
structure
MT assembly domain, **171,** 171–172
projection domain, **171,** 171
in nervous system
axonal compartments, **151,** 74–77
intertubule spacing, **151,** 92–93
somatodendritic compartments, **151,** 74–77
O-glycosylation, **171,** 174
pathological
as biochemical marker for Alzheimer's disease
Tau aggregation, **171,** 191–195
Tau phosphorylation, **171,** 188–191
in glial cells, **171,** 207
isoform distribution in neuron population, **171,** 206
in muscles, **171,** 207
in neurodegenerative disorders
dystrophic neurites, **171,** 187
neurofibrillary tangles, **171,** 185–187
Pick bodies, **171,** 187–188
in neurofibrillary degeneration subtypes
corticobasal degeneration, **171,** 200–201
disorders with PHF-Tau, **171,** 195–199
Gerstmann–Straussler–Scheinker disease, **171,** 202–203
neuronal injuries, **171,** 203
Tau 55, **171,** 201–202
Tau 55 and 64, **171,** 201
Tau 64 and 69, **171,** 199–200
Tau triplet, **171,** 199
role in microtubule assembly, **171,** 204–206
as phosphoprotein, **171,** 172–174
role in microtubule assembly
isoforms in neurons, **171,** 184–185
microtubule assembly domain, **171,** 176
other interactions, **171,** 181–182
projection domain, **171,** 175–176
proteins in nonneuronal cells, **171,** 185
Tau phosphorylation
associated kinases, **171,** 178–179
associated phosphatases, **171,** 179–180
regulations, **171,** 180–181
Tau sorting in nerve cells, **171,** 182–184
Tau–tubulin interactions, **171,** 176–177
role in pathological events, **171,** 207–208
structure, **165,** 252

Taurine
leak pathway, LTD_4 role, **161,** 223–224
neurotransmission, **149,** 257–258
regulation, **161,** 198–199
swelling-activated channel, **161,** 199–201
Taxol
cytoskeleton, **166,** 15
environmental stimuli in plant cells and, **145,** 285, 287
mitosis and, **144,** 229, 236–237, 246
myofibrillogenesis and, **143,** 177
stabilization of microtubules, **157,** 22
Taxonomy
cell wall components, **171,** 85
in whole-chromosome hybridization studies, **153,** 200
Tax protein, transcriptional regulation and, **143,** 42–44
Taxus brevifolia, mitosis and, **144,** 236
T cell leukemia virus-1, human, mapping of NOS, **162B,** 308–310
T cell mitogens, **156,** 170–172, 306
T cell receptor
antigen binding site, **157,** 182–183
antigen recognition, **157,** 189–191, **167,** 96–97
antigen recognition activation motif, **157,** 184, 231–233, 239, 240
components, **157,** 188–189
coreceptors, **157,** 183, 190–191, 199, 214–216
MHC class restriction, **157,** 190–191
phosphorylation, **157,** 241
role
apoptosis, **157,** 250–256
helper T cell differentiation, **157,** 202
T cell selection, **157,** 198–199
signal transduction
cell cycle entry, **157,** 248–250
defects, **157,** 183
phosphatidylinositol 3-kinase pathway, **157,** 227–231
phospholipase C pathway, **157,** 216–221
Ras pathway, **157,** 221–226
regulation of gene expression, **157,** 243–248
tyrosine kinases, **157,** 213, 235–236
structure, **157,** 188–189
T cell activation, **157,** 252–256
transcriptional regulation and, **143,** 10
T cells
apoptosis, **157,** 183, 198–199
cytokine production, **157,** 201–202
dendritic cell interactions
antigen-specific sensitization, **153,** 77
costimulation, **153,** 78–82, 84–86
differentiation
cytolytic T cells, **157,** 203–204
helper T cells, **157,** 201–203
effects on IFN-γ-induced inhibition of GM-CFC, **167,** 231–232
energy, **157,** 20
in gut wall
in dome epithelium, **167,** 134
as intraepithelial lymphocytes, **167,** 96–97
lamina propria, **167,** 97–98
memory
in dome epithelium, **167,** 134
lamina propria lymphocyte expression of CD45RO antigen marker for, **167,** 98
microgravity, **156,** 306–311, 316
migration through mucosal system, **167,** 93–94
MIP-1α gene expression in activated, **167,** 199
negative selection, **157,** 198–201
positive selection, **157,** 198–201
role in B cell activation, **157,** 206–207
systems for antigen-specific stimulation, **157,** 191
TGF-β production by, **167,** 210
tolerance, **157,** 181–182, 198–201
transcriptional regulation and, **143,** 7, 10, 11–17, 22, 28, 39–41
genes, **143,** 12–14, 16–17
viruses, **143,** 39–43
types, **157,** 182, 188–190
TCR, *see* T cell receptor
TdT, *see* Terminal deoxynucleotidyltransferase
Tectal cortex, lamination, **175,** 267
Tectum mesencephali, amphibian development, **169,** 117–118, 120–121
Teeth
enamel proteins, **158,** 251–252
human development, **158,** 221
regenerative cementogenesis and, *see* Cementogenesis, regenerative
Teichoic acids, bacterial cell wall, **161,** 270–272
Teleosts
avian salt glands and, **144,** 129

chromatophores and, **143,** 246
iridophores, **143,** 209–210
motility control, **143,** 225, 228
signal transduction, **143,** 240, 245
ependymm
inner endomeningeal layer, **165,** 133–134
intermediate endomeningeal layer, **165,** 129–132
meninges, **165,** 128–134
outer endomeningeal layer, **165,** 129
gonadotropin-releasing hormone gene, *see* Gonadotropin-releasing hormone
karyosphere and, **144,** 25
Telomeres
ciliates and, **141,** 10
intermediate filament protein interactions with, **154,** 33
TEM, *see* Transmission electron microscope
Temperature
and biological clocks in retina, **146,** 95
cytoplasmic streaming affected by, **155,** 125–127
effect on *in vivo* reproductive process, **140,** 392–394
control of male and female quality, **140,** 396–397
in vitro pollination in maize, **140,** 394–396
effect on phases of *in vitro* reproductive process
kernel development, **140,** 399
pollen selection, **140,** 399–400
postpollination, **140,** 397–398
stress on reproductive tissues, **140,** 398–399
and extracellular matrix, **146,** 301
high, stress responses to, **140,** 391–392
intracellular calcium messenger system and, **142,** 305
low, environmental stimuli in plant cells and, **145,** 251–252, 288–290, 291
and melanocytes, **146,** 203, 205
regulation of chromoplast development, **163,** 199–200
sensitive mutants, mitosis and, **144,** 226–227, 229–233, 238, 255
vertebrate cell culture technology and, **142,** 164, 174–175, 191, 196, 200
Template prepattern hypothesis, diatom shell morphogenesis, **150,** 258–260
Tenascin, synthesis in cartilage, **159,** 291
Tenebrionid beetles, karyosphere and, **144,** 19–25
Teniposide, *see* VM-26
Tenotomy, effect on muscle fiber type transitions, **170,** 180–181
Tensegrity
cellular, cytoskeletal regulation
actin cytoskeleton
models, **150,** 189–196
tension molding, **150,** 196–201
cell function control, **150,** 179–185
cell shape control, **150,** 179–185
development, implications for, **150,** 211–214
mechanical signaling system
focal adhesion, **150,** 207–208
long-range signal transfer, **150,** 208–210
structural inheritance, **150,** 210–211
mechanical stress response, **150,** 185–189
mechanisms, **150,** 173–175
structural integration
intermediate filaments as tensile stiffeners, **150,** 205–206
microtubules as compression-resistant struts, **150,** 201–205
nuclear tensegrity, **150,** 206–207
tissue pattern establishment, **150,** 175–178
transmembrane coupling, **150,** 178–179
nuclear, **150,** 206–207
Tensin
cytoskeletal dynamics and, **144,** 103
expression in chondrocytes, **159,** 291–292
phosphorylation by focal adhesion kinase/Src complex, **168,** 91, 93
as potential FAK substrate, **167,** 172
tyrosine phosphorylation in FAK-deficient cells, **167,** 177–178
vinculin *in vitro* interaction with, **167,** 162, 163
Tension
cortical, cytoskeletal dynamics, **144,** 115–117
superhelical, complete relaxation, **162B,** 407–408
Tension hypothesis, nematocyst discharge, **156,** 279
Tentyrianomas taurica, karyosphere and, **144,** 19, 21

Teratogen, effect on fetus
  dose response, **158,** 259
  time of exposure, **158,** 259
TERM, animal mtDNA and, **141,** 229
Terminal deoxynucleotidyltransferase, marker of B cell differentiation, **157,** 140
Terminal web
  M cells *versus* enterocytes, **167,** 111
  variations in development of, **167,** 129
Termination
  animal mtDNA and, **141,** 185, 188–189
    regulation, **141,** 226, 229
    replication, **141,** 224
    transcription, **141,** 221–222
  endosymbiont hypothesis and, **141,** 298, 303–305, 307
  higher plant mitochondrial genomes and, **141,** 161
  kinetoplastid mtDNA and, **141,** 83
  regions, ciliates and, **141,** 9
Terpenoids, chalcone synthase and, **147,** 247
Test grids
  for boundary length, **161,** 118
  for length density, **161,** 120–121
  for volume density, **161,** 117–118
Testicular cancer, **156,** 128
Testicular macrophages
  cellular structure, **149,** 102–106
  characteristics, **149,** 99, 129
  clinical implications
    AIDS, **149,** 126
    autoimmune disease, **149,** 124–125
    cryptorchidism, **149,** 126–127
    infections, **149,** 125–126
    vasectomy, **149,** 127
  endocrinology
    follicle-stimulating hormone, **149,** 118–119
    human chorionic gonadotropin, **149,** 119–120
    hypophysectomy, **149,** 121
    photoperiod, **149,** 121–122
    steroids, **149,** 120
  future considerations, **149,** 130
  immunology, **149,** 122–124
  isolation procedure, **149,** 127–129
  and Leydig cells, interactions
    functional, **149,** 109–111
    structural, **149,** 105–109
  ontogeny in mammals, **149,** 100–102
  paracrinology, **149,** 104–105
    comparisons with ovarian macrophages, **149,** 116–117
  phagocytic cell evolution, **149,** 99–100
  secretory products, **149,** 111–115
  steroid metabolism, **149,** 115–116
Testins, Sertoli-germ cell communication network and, **147,** 33, 56, 64, 76, 78
Testis
  blood barrier
    glucose transport, **172,** 31–32
    structure, **172,** 31–32
  determination, mammalian, **154,** 221–222
  parathyroid hormone-related protein, **166,** 250
Testis-determining factor
  function, **154,** 222
  regional localization, **154,** 226–229
  SRY gene
    characterization, **154,** 233–237, 247
    cloning, **154,** 233–237
    demonstration of sex determination, **154,** 236, 238
    evolution of function, **154,** 241–245
    evolution of organization, **154,** 241–245
    function mechanism, **154,** 238–241
    genetic research, **154,** 229–233
    identification, **154,** 246
  Y chromosome role, **154,** 223–226
Testosterone
  control of amphibian metamorphosis and, **145,** 119, 121
  effect on muscle fiber-type transitions, **170,** 185–186
  effect on peroxisome proliferator response, **160,** 329–330
  β-endorphin effect on, **149,** 115
  GnRH-encoding gene in teleosts and, **147,** 175
  gonadal differentiation role, **154,** 218
  neonatal exposure to sex hormones and antihormones and, **139,** 10–11
  production by Leydig cells in response to interleukin-l, **149,** 113
  role in cartilage formation, **159,** 316
  seminal vesicle epithelium response, **160,** 284–286, 294

Sertoli-germ cell communication network and, **147,** 35, 70, 77, 80
function, **147,** 39, 44–45, 58
stage-dependent changes, **147,** 60–61
sterility induction in neonatal rats, **160,** 3–4, 11
Testosterone-repressed prostate message-2
encoding gene, cell death-related expression, **163,** 116–117, 152
Sertoli-germ cell communication network and, **147,** 76–78
Tetradecanoyl phorbol acetate
fish chromatophores and, **143,** 240–241
melanocytes and, **146,** 236–237, 241–243, 248
transdifferentiation in medusae and, **142,** 247–249, 257
transformation of retinal pigment epithelial cell, **150,** 124–125
Tetraethylammonium
avian salt glands and, **144,** 175–176
biological clocks in retina and, **146,** 111, 127
5,6,7,8-Tetrahydrofolate, **156,** 111
4-Tetrahydroxychalcone in pollen extracts of plants, **140,** 47, 49
Tetrahymena
dynein
cross-bridge cycle, **155,** 168–169
inner row *in situ* localization, **155,** 164
microtubule translocation, **155,** 168
outer row
composition, **155,** 147
identification, **155,** 146
structure of outer arms, **155,** 157–160
endosymbiont hypothesis and, **141,** 310, 314
filament protein, *see* Filament protein, *Tetrahymena*
hormone receptor imprinting
liver cells of higher mammals compared to, **155,** 18
mechanism, **155,** 19–21
hormone receptors
antibody treatment, **155,** 11
downregulation, **155,** 12
FITC-insulin binding
in offspring generation, **155,** 17
of plasma membranes, **155,** 13–14
histamine-induced phagocytosis, **155,** 3
insulin imprinting, **155,** 9–10
insulin treatment of nuclei, **155,** 15
mating receptor role in development, **155,** 4
model for evolutionary studies, **155,** 5
recognition of hormones, **155,** 6–7
second messengers, **155,** 9
specificity, **155,** 7–8
kinetoplastid mtDNA and, **141,** 72
microgravity experimennt, **156,** 343
mitochondrial genomes of, **141,** 3–4, 58–60
DNA sequences, **141,** 13, 35–36, 38, 45
genetic code, **141,** 53, 55
rRNA genes, **141,** 46–48
structure, **141,** 6, 8–10
mitosis, **144,** 228
replication band
chromatin replication models, **153,** 165
immunochemical studies, **153,** 157–159
transcription factors related to, **154,** 17
Tetrahymena pyriformis
dynein ATPase, **173,** 13–14
endosymbiont hypothesis and, **141,** 295, 297, 314
mitochondrial genomes of, **141,** 1–2
DNA sequences, **141,** 12–13, 35–36, 38, 40–43
genetic code, **141,** 52–53, 56
rRNA genes, **141,** 46–50, 52
structure, **141,** 4, 7–9
Tetrahymena thermophila, mitochondrial genomes, **141,** 1, 6–8, 10, 43–45
Tetrameric protein transthyretin, RPE as source, **171,** 236–237
Tetramethylrhodamine isothiocyanate, in FISH diagnosis, **153,** 11–12
Tetrasaccharides, **156,** 2
Tetraselmis, eyespot reflection in, **148,** 259, 261
Tetrodotoxin
avian salt glands, **144,** 157, 165
calcium channels and, **147,** 205
control of rhythmic behaviors and, **147,** 12
proteins in striated muscle, **142,** 87, 98
t-Force
in geometric clutch design, **173,** 46–47
role in geometric clutch model, **173,** 53
TGF, *see* Transforming growth factor
TGN, *see* Trans-Golgi network

TH1 cells, cytokine and TNFβ production, **167,** 98
TH2 cells, cytokines produced, **167,** 98
Thalassiosira bulbosa, morphogenesis, **150,** 261
Thalassiosira eccentrica, morphogenesis
areolae formation, **150,** 257
scale vesicle formation, **150,** 269
Thalassiosira mediterranea, morphogenesis
colloidal silica formation, **150,** 325
microfilaments, **150,** 333
Thalassiosira trifulta, morphogenesis involving microfilaments, **150,** 333
Thalassiosira tumida, valve formation, **150,** 269
Thalidomide, inhibition of endothelial cell growth, **161,** 22
Thermobia domestica
antennae, **174,** 6–7
complex eye, **174,** 5–6
epidermis, **174,** 8
labrum, **174,** 7
mouthparts, **174,** 7–8
Thermodynamics
actin filament bundle formation *in vitro,* **175,** 56–58
bacterial flagellar motor, **147,** 98–99, 159
model comparisons, **147,** 146, 150
motor energetics, **147,** 118, 121–123, 125–127
in drug pump model, **171,** 135
Thermokinetic balancing, bacterial flagellar motor, **147,** 149
Thermostability
general cell, **148,** 172–173
primary, **148,** 172–173
increase, **148,** 182, 184–186
proteins
heat shock-induced increase, **148,** 179–182
in tolerant zone, **148,** 211
response to extreme heat, **148,** 172–175
in tolerant zone, **148,** 211–212
Thiamine, and neurotransmission, **149,** 272–274
Thick-filament proteins, in striated muscle
cardiac muscle, **142,** 103–107
skeletal muscle, **142,** 62–83
Thickness, glomerular basement membrane, **161,** 161–162
Thin-filament proteins, in striated muscle
cardiac muscle, **142,** 107–110
skeletal muscle, **142,** 83–93
6-Thioguanine, mechanisms, **156,** 121–122
Thiolactomycin, higher plant plastids and, **145,** 184
Thionins, extracellular matrix and, **146,** 269
THP, *see* Thymus hormonal products
Three-dimensional analysis, NPCs, **162B,** 230–231, 294–295, 238–241
Threonine
flowering plants, **143,** 318–321, 323
proteins rich in, extracellular matrix, **146,** 268
Threonine dehydratase, flowering plants and, **143,** 323
Thrombin
effect on APP synthesis, **170,** 257
effect on arterial smooth muscle cell culture proliferation, **169,** 213–214
Thrombocytopenia, in chemotherapy with growth factors, **167,** 240
Thrombospondin
effect on arterial smooth muscle cell culture phenotype, **169,** 203–204
inhibition of angiogenesis, **159,** 143–144
inhibition of endothelial cell growth, **161,** 18–19
synthesis in arterial smooth muscle cell culture, **169,** 226
Thuja occidentalis, **140,** 323, 325
Thy 1 antigen, density on thymocyte surface, **156,** 167–168
Thylakoids
chloroplasts, **144,** 261, 263, 266, 273
dinoflagellates, **144,** 285, 288–291
euglenoids, **144,** 283–284
heterokont algae, **144,** 277–278
higher plant plastids
metabolism, **145,** 168, 172, 174, 197
structure, **145,** 153, 161, 166
Thymidine
avian salt glands and, **144,** 143
transdifferentiation in medusae, **142,** 252
Thymidine kinase, cell cycle dependence, **162B,** 388–391
Thymidylate synthase, cell cycle dependence, **162B,** 388–391
Thymidylate synthetase
5-fluorouracil metabolism, **156,** 117–118
methotrexate metabolism, **156,** 111, 116

Thymocytes
apoptosis, **163,** 117–118
differentiation and maturation, **156,** 162–174
rosette-forming capacity, **156,** 173–174
transcriptional regulation, **143,** 7
Thymosin β_4, **156,** 168, 226–227, 229, 254, **167,** 223
Thymus
blood barrier
glucose transport, **172,** 32–33
structure, **172,** 32–33
dendritic cells
isolation, **153,** 68
properties, **153,** 68
hormonal products, **156,** 159, 175–180
Thyroid, **166,** 22–23, 28–29
Thyroid hormone
avian salt glands, **144,** 199
control of amphibian metamorphosis, **145,** 105–107, 110–113, 115–117, 119, 131, 141
adrenocortical hormones, **145,** 124
feedback, **145,** 117–119
growth hormone, **145,** 138
hypothalamus, **145,** 115–117
pituitary, **145,** 112–115
prolactin, **145,** 130, 133–136
effect on Harderian gland secretion, **168,** 50
effect on muscle fiber-type transitions, **170,** 183–184
homology with gonadotropic hormone, **159,** 68
peroxisome response, **160,** 327
proteins in striated muscle, **142,** 105–106
RBF-like proteins, **162B,** 355
synthesis, **159,** 72
Thyroid hormone receptor
accessory factors, **162B,** 360–361
homology with related receptors, **155,** 27
Thyrosine kinase, stimulation by endothelin, **171,** 280
Thyrotrophs, regulation, **159,** 72
Thyrotropin-releasing hormone
analogs, **159,** 40
conformation effect on activity, **159,** 39
control of amphibian metamorphosis, **145,** 115–116, 136–137, 140
effect on neurohormone release, **159,** 39–40
effect on prolactin release, **159,** 63
immunohistochemical localization, **138,** 101–103
neuroendocrine effects, **138,** 103–106
neuronal distribution, **159,** 39
of THR, **138,** 101–103
transport, **159,** 40
Thyroxine
amphibian brain development role, **169,** 134–136, 138
control of amphibian metamorphosis and, **145,** 107, 110, 112–113, 115–116, 118–119
adrenocortical hormones, **145,** 120–124
prolactin, **145,** 129–131, 136
effect on Harderian gland secretion, **168,** 51–52
proteins in striated muscle, **142,** 105–106
TIBA, **156,** 44
Tight junctions
gap junction association, **157,** 330–332
integral membrane proteins, **157,** 332
in lens epithelial cells, **163,** 337
Tine proteins
in *Difflugia* movement, **150,** 62
in *Hyalodiscus* movement, **150,** 62
membrane raking function, **150,** 48–49, 51, 63
shelled amoebae locomotion, **150,** 54
Tipula, karyosphere, **144,** 18
Tissue factors
characteristics, **152,** 50–51
as cytokine growth factor superfamily member, **152,** 51–53
cytokine-mediated induction, **152,** 59–64
endothelial cell, regulation of, **152,** 64–73
on fibroblasts, **152,** 73–74
function, **152,** 53–55
localization, **152,** 55–57
monocyte-macrophage
adhesion receptors, induction by, **152,** 75–76
properties, **152,** 57–59
Tissue inhibitor of metalloproteinase II, **147,** 54
Tissues
analysis, grease in plant cells, **145,** 92
ascidian
brain induction, **176,** 278–279

Tissues (*continued*)
differentiation, autonomy, **176,** 255–257
notochord induction, **176,** 272–278
sensory pigment cell induction, **176,** 279–281
trunk lateral cell induction, **176,** 281
calcified, *see* Calcified tissues
cell cultures, locomotion
ameboid locomotion in relation to, **150,** 35–36, 61–62
future research, **150,** 62–65
mechanisms, **150,** 38–41
methods of study, **150,** 35–37
nomenclature, **150,** 38–41
rearward traction forces, **150,** 41–43
retrograde surface transport
actin treadmilling, **150,** 51–54
hypothesis, alternative, **150,** 46–48
mechanism, **150,** 43–46
membrane raking hypothesis, **150,** 48–51
cholesteric supramolecular order, **166,** 70–84
collagen fibril diameters, **173,** 109–110
connective, *see* Connective tissue
extracellular matrix, **166,** 70–81
FAK effects in, **167,** 175–178
fibrillar collagen architecture, **173,** 131–132
fixation, polyamines, **138,** 65–66
fluidity, **166,** 60–61
lymphoid, *see* Lymphoid tissue
MIP-1α effects on nonhemopoietic, **167,** 208
NMR analysis, **145,** 45, 50
imaging, **145,** 25–26, 30–32
magnetization transfer, **145,** 33–34
spectroscopy, **145,** 20–24
order, **166,** 60–61
peripheral and central, vertebrate, POMC gene expression, **170,** 83
plant, culture, effect of microgravity, **171,** 42–44
remodeling, collagen III during, **173,** 86
specification of gene expression in RPE, **171,** 228–232
TGF-β effects on nonhemopoietic, **167,** 216, 220
Tissue specificity
Arabidopsis thaliana , **144,** 70
in development, **148,** 105–106
Drosophila muscles and, **143,** 65, 80, 111, 117
in evolution, **148,** 105–106
extracellular matrix and, **146,** 287, 295
flowering plants and, **143,** 308
growth factor-induced retinal regeneration and, **146,** 60
higher plant plastids and, **145,** 184, 186
myofibrillogenesis and, **143,** 159
nuclear magnetic resonance and, **145,** 31
proteins in striated muscle and, **142,** 66, 106
transcriptional regulation and, **143,** 13
transdifferentiation in medusae and, **142,** 236
Titin
in cytoplasm, **156,** 245–247
Drosophila muscles and, **143,** 124, 140
myfibrillogenesis and, **143,** 176, 182–183
in striated muscle, **142,** 62
cardiac muscle, **142,** 109–112
skeletal muscle, **142,** 91, 93–96, 99, 101–102
T lymphocytes, *see* T cells
T-lymphotropic virus type 1, human, tumorigenesis in transgenic mice bearing gene for, **164,** 63
TMTX, *see* Trimetrexate
TNF, *see* Tumor necrosis factor
Tobacco
chalcone synthase, **147,** 255–256
chromoplast differentiation
chemical regulation, **163,** 204
stress regulation, **163,** 203
endosymbiont hypothesis and, **141,** 263, 265, 267, 272
environmental stimuli in plant cells and, **145,** 291, 293
flowering plants and
amino acids, **143,** 311, 323–324
ammonium, **143,** 307–309
nitrate, **143,** 299, 301
higher plant mitochondrial genomes and, **141,** 157–158
intracellular calcium messenger system and, **142,** 319–32O, 322–324
microgravity experiments, **156,** 333
plant-pathogen interactions in, **144,** 71
plastids and
metabolism, **145,** 175–176, 193
structure, **145,** 153, 155, 157–159
transgenic, **156,** 50
urease in plant cells and, **145,** 80, 90

Tolerance
Arabidopsis thaliana, **144,** 71, 76
drought, stomatal opening regulation, **165,** 35
immunotolerance, induction by antigens in gut epithelium, **167,** 142–143
Toll receptor, signal transduction, **164,** 201–202
Tomato
chromoplast differentiation
chemical regulation, **163,** 204
genetic factors, **163,** 191–193
hormonal regulation, **163,** 201–203
light regulation, **163,** 199
temperature regulation, **163,** 199–200
transcription activity, **163,** 195–197
flowering plants and, **143,** 311–313, 322
plant-pathogen interactions in, **144,** 72, 74
plastid envelope membrane budding, **163,** 206
plastids and, **145,** 161–163
pollen-specific genes, **140,** 23
symplast and, **142,** 297
urease in plant cells and, **145,** 81
Tongue, human development, **158,** 219
Tonicity, avian salt glands and, **144,** 157, 177–185
Tonoplast proton ATPase
function, **174,** 92–95
molecular cloning
16- and 17-kDa subunit gene, **174,** 97–98
55- and 62-kDa subunit gene, **174,** 97
67- to 73-kDa subunit gene, **174,** 96–97
95- to 116-kDa subunit gene, **174,** 98–99
molecular structure, **174,** 92–95
peripheral studies, **174,** 99–100
physiological functions, **174,** 86–87
physiological role
in metabolic processes of cells, **174,** 89–92
in space filling of cells, **174,** 87–89
purification, **174,** 104–105
reconstitution into liposomes, **174,** 105–113
solubilization, **174,** 103–104
Tonoplasts
environmental stimuli in plant cells, **145,** 283, 290
intracellular calcium messenger system, **142,** 313, 315–317, 324–325
Tonsils
dendritic cells
isolation, **153,** 66–67
properties, **153,** 66–67
M cells, **167,** 135–136
Tooth, regenerative cementogenesis and, *see* Cementogenesis, regenerative
Topography
functional, *in situ* nuclear matrix, **162A,** 8–10
structural, SV40 DNA replication, **162A,** 495–519
Topoisomerase I
associated with nuclear matrix, **162A,** 306–307
drug resistance mediated by, **156,** 106–107
Topoisomerase II
active drugs, **162A,** 559–564
activity in cells, **156,** 130–131
association with SARs, **162A,** 128
binding sites on MAR/ORI elements, **162A,** 343
cellular levels, **162A,** 306–307
consensus cleavage sites, **162A,** 292
DNA
biological activities, **162A,** 556–559
in chromosome formation, **162B,** 23–26
nucleoskeletal component, **162B,** 200–202
and drug resistance, **156,** 97–108
interaction with anthracyclines, **156,** 80–81
kinetoplastid mtDNA, **141,** 66, 69
localization to internal matrix network, **162B,** 107
mediated DNA cleavage, **162B,** 425–432
mitosis and, **144,** 227
phosphorylation, **156,** 105
Topoisomerase I inhibitors, **156,** 105–107
Topoisomerase II inhibitors, **156,** 98–105
Topotecan, **156,** 105
Torpedo, calcium channels, **147,** 201
Torpedo marorata, **138,** 112
Torque
generation, bacterial flagellar motor and, **147,** 97, 115
models, **147,** 134, 136–137, 143, 145
structure, **147,** 107–110
stall, bacterial flagellar motor, **147,** 110
Torque-velocity curve, bacterial flagellar motor and, **147,** 128–129, 152

Totipotency, ability of plants, **158,** 65–66, 121–123
Toxicity
osmotic, salt tolerance, **165,** 4–6
vertebrate cell culture technology and, **142,** 174, 198
Toxicology, genetic, whole-chromosome hybridization applications, **153,** 208–213
Toxins, *see also specific toxin*
Gram-positive bacteria, **161,** 274–275
host-specific, in plant-parasite interactions, **172,** 61
intracellular calcium messenger system and, **142,** 309–311
microbial, superantigen, **161,** 277
vertebrate cell culture technology and, **142,** 191
TPA, *see* Tetradecanoyl phorbol acetate
Tracers
experiments on sporopollenin, **140,** 58–61
M cell transport
description, **167,** 117–122
transcellular route of, **167,** 126–128
Tracheary elements, lignification, **176,** 98–100
Tracheloraphis totevi, karyosphere and, **144,** 36
Tracking polymerases, localization, **162A,** 131–137
Traction
cytoskeletal dynamics and, **144,** 89–91
theories, **139,** 168–170
in tissue cells
actin treadmilling role, **150,** 63
definition, **150,** 41
mechanism, **150,** 45–46
rearward forces, **150,** 41–43
Tradescantia
division apparatus of generative cell, **140,** 172–175
environmental stimuli in plant cells, **145,** 291, 293
hair cells, **138,** 247, 249–251, 256–257, 259
leaf response to heat shock, **148,** 173–174, 184
acquired tolerance, **148,** 190, 192, 194–196, 198
recovery, **148,** 187–189
repair, **148,** 208, 210
Tradescantia paludosa
generative organ cells, effect of microgravity, **171,** 42
pollen, **140,** 4–5, 175
Tradescantia virginiana, **140,** 152–154, 156–158, 162
Trafficking, molecular, through NPC, **162B,** 243–249
trans-acting factors
animal mtDNA and, **141,** 219, 229
chalcone synthase and, **147,** 259–260, 271–277
endosymbiont hypothesis and, **141,** 281, 301
Transcellular transport, antigens and tracers, **167,** 117–122, 126–127
Transcinnamic acid, chalcone synthase and, **147,** 268–269
Transcoumaric acid, chalcone synthase and, **147,** 268–269
Transcription
activity of isolated cells, **140,** 225
animal mtDNA and, **141,** 174–175, 207–208, 217–222, 229–231
control region, **141,** 201–202
protein genes, **141,** 183, 187, 189
regulation, **141,** 225–229
replication, **141,** 223–225
tRNA, **141,** 193, 198
Arabidopsis thaliana and, **144,** 59, 73, 77
association with repair replication, **162A,** 139–140
avian salt glands and, **144,** 192
biological clocks in retina and, **146,** 108, 122–123
chalcone synthase gene, control, **147,** 252, 254–277
during chromoplast development, **163,** 195–197
ciliates and, **141,** 35, 37, 50, 52
coordinate, tissue-specific genes, **148,** 87
coordination with RNA splicing, **162A,** 39
dependent nucleosomal changes, **162A,** 403–407
Drosophila muscles and, **143,** 64–65, 137
actins, **143,** 109–111
contractile protein genes, **143,** 94, 100–101, 103
development, **143,** 77–83, 85
myosin light chains, **143,** 105, 107

paramyosin, **143,** 108
tropomyosin, **143,** 115–117
troponin, **143,** 119–121
effects of cell differentiation, **148,** 101
endosymbiont hypothesis and gene expression, **141,** 298–306
gene transfer, **141,** 325
genome organization, **141,** 289, 294–297
lineages of life 240, **141,** 246
mitochondrial, **141,** 329
molecular biology, **141,** 311, 313, 325
nuclear genome, **141,** 247
plastids, **141,** 266–267
enhancers
cohabitation with MARs, **162A,** 345–346
MARs as, **162A,** 309–312
extracellular matrix and
modification, **146,** 295, 297, 300
pathogenesis, **146,** 309, 311
stress, **146,** 303–304
flowering plants and, **143,** 302, 307, 309, 312, 324–325
gene, MAR sequence role, **162B,** 356–359
GnRH gene in teleosts, **147,** 169, 172, 175, 185–186
growth factor-induced retinal regeneration and, **146,** 69
higher plant mitochondrial genomes and, **141,** 149, 158, 160–162, 164
abnormal phenotypes, **141,** 131, 133, 142–143, 146
higher plant plastids and, **145,** 161–164
hormonal control of amphibian metamorphosis and, **145,** 112
intracellular calcium messenger system and, **142,** 328, 337
karyosphere and, **144,** 44–45
kinetoplastid mtDNA and, **141,** 65–66, 80–84
maxicircle genes, **141,** 72–74
maxicircle transcripts, **141,** 74–79
mdrl, **156,** 79
melanocytes and, **146,** 195
genetics of melanoma, **146,** 228, 240–241, 245–247, 249
pigment-cell genetics, **146,** 209, 213–214
mitochondrial, in *Saccharomyces cerevisiae*
mechanism, **152,** 167–168
promoters, **152,** 146–149
RNA polymerase role, **152,** 149
transcription factors, **152,** 149–150
mitochondrial genomes in fungi and, **141,** 93, 103, 107–108, 115
myofibrillogenesis and, **143,** 154, 173
nuclear matrix role, **162A,** 100–103,157–159
oocytes of insects, **144,** 18–19, 23
pollen-expressed genes, specificity, **140,** 12–16
regulation, and S/MARs, **162A,** 416–425
regulation by NF-KB, **143,** 1–2, 47–48
biochemistry, **143,** 22–25
function, **143,** 2–7
genes, **143,** 11–12
inflammation, **143,** 17–19
T Iymphocytes, **143,** 12–17
inducible activation, **143,** 7–11
molecular cloning, **143,** 25–27
ankyrin repeat proteins, **143,** 36–39
cDNA, **143,** 29–33
Drosophila, **143,** 35–36
modes of regulation, **143,** 34–35
p65 subunit, **143,** 33–34
Rel, **143,** 27–29
proteins, **143,** 45–47
structure of sites, **143,** 19–22
viruses, **143,** 39, 44
HIV, **143,** 39–42
HTLV-I, **143,** 42–44
regulators, interactions among, **162B,** 365–366
restriction by AT-rich sequences, **148,** 91–87
retinal pigmented epithelium and, **146,** 77
retinal regeneration and, **146,** 57
reverse
higher plant mitochondrial genomes and, **141,** 149, 158–159
mitochondrial genomes in fungi and, **141,** 107–109
RNA polymerase II
foci, **162A,** 136–137
transcript domains, **162A,** 71
self-incompatibility and, **143,** 270, 283, 290

Transcription (*continued*)
Sertoli-germ cell communication network and, **147,** 69, 74, 77
sites, **162A,** 154–156,166
and size of chromatin domain, **162A,** 430–431
specific genes during pollen development, **140,** 5–7
transdifferentiation in medusae and, **142,** 227, 229
tyrosinase gene, regulation, **171,** 233–235
vertebrates, **144,** 26–27, 38
Transcriptional activator, mtTF1, endosymbiont hypothesis, **141,** 299–301
Transcription factors
AP-1 complex, **157,** 244, 253
basic helix-loop-helix, role in chalcone synthase regulation, **147,** 272–273
binding sites on MAR/ORIs, **162A,** 325–333
binding steroid receptors, **162B,** 366
and chromatin, **162A,** 228–230
Drosophila muscles and, **143,** 78–79, 81, 83
enriched on nuclear matrix, **162A,** 293–303
heat shock TF, **165,** 27–28
instability of homogeneous state
post-translational modification, **154,** 351
recox regulation, **154,** 347–348
spatially differentiated modulation of activity, **154,** 354
intermediate filament proteins related to, **154,** 13–18
nuclear distribution, **162A,** 156
nuclear localization signal masking, **151,** 203–206
and nuclear matrix, **162A,** 230–231
orphan receptor family, **164,** 202–203
Pit-1, **162A,** 102–103
initiation of mammotroph differentiation, **157,** 38
repression of prolactin expression, **157,** 40
regulation and, **143,** 1, 7, 17, 19, 42–44
role in apoptotic processes, **163,** 119–120
steroid hormone receptors as, **162B,** 338–340
steroid receptor family, **164,** 202–203
SW15, phosphorylation effects, **151,** 207–208
Transcripts, in transgenic plants, transcript localization, **140,** 25–28
Transcytosis
antigens by dome epithelium, **167,** 126–128
polymeric immunoglobulins, **172,** 231
Transdifferentiation
amphibian RPE, **171,** 248–249
chick RPE *in vitro,* **171,** 249
growth factor-induced retinal regeneration and, **146,** 49, 52–55, 62, 68–69
in medusae, **142,** 213–214, 256–257
animals, **142,** 216–218
hydromedusae
DNA replication, **142,** 252–256
initiation, **142,** 236–251
isolation of tissues, **142,** 222–223
potential, **142,** 223–236
tissue organization, **142,** 218–222
plants, **142,** 215–216
terminology, **142,** 214–215
rat RPE *in vitro,* **171,** 249–254
and substratum mechanics
cellular mechanics, **150,** 128–132
cytomechanics, **150,** 120–125
retinal pigment epithelial cells, **150,** 119–120
Transducin
growth factor-induced retinal regeneration and, **146,** 59
retinal precursor cells and, **146,** 147–148
Transduction, *see* Phototransduction
Transfection
associated studies, **156,** 83
melanocytes and, **146,** 203, 205, 245, 251
myofibrillogenesis and, **143,** 174
transcriptional regulation and, **143,** 5, 41
Transferrin
chondrocyte differentiation, **159,** 314–315
endocytosis by germ cells, **160,** 187
neurogenesis, **159,** 17
secretion by Sertoli cells, **160,** 201–202
Sertoli-germ cell communication network and
function, **147,** 45, 50, 59, 64
germ cell *in* vitro, **147,** 67–71
mechanisms, **147,** 72, 75–78

sperm maturation role
endocytosis, **154,** 135–137
receptor-mediated endocytosis at base of Sertoli cells, **154,** 120, 122
Transfer RNA
animal mtDNA and, **141,** 174–175, 201, 203, 207–209
anticodon-codon interactions, **141,** 191–193
gene evolution, **141,** 199–201
genome content, **141,** 179–180
modified bases, **141,** 199
protein genes, **141,** 185, 189
regulation, **141,** 228
replication, **141,** 223–224
rRNA, **141,** 189–191
transcription, **141,** 218, 221–222
variant structural forms, **141,** 193–199
ciliates and, **141,** 58–59
genetic code, **141,** 52–56
Paramecium aurelia, **141,** 12, 35–36
rRNA genes, **141,** 48, 50–52
cyanelle genome organization, **151,** 41–42
endosymbiont hypothesis and
gene expression, **141,** 298, 300–304, 306–307
gene structure, **141,** 311–316
genome organization, **141,** 289–290, 292–297
lineages of life, **141,** 238, 240, 244
mitochondria, **141,** 281
nuclear genome, **141,** 247, 266–268
Plasmodium, **141,** 326
rRNA sequences, **141,** 323
higher plant mitochondrial genomes and, **141,** 129, 161, 164
gene location, **141,** 159
structure, **141,** 137, 149–150
higher plant plastids and, **145,** 156–157, 163
kinetoplastid mtDNA and, **141,** 72, 75
mitochondrial genomes in fungi and, **141,** 120
DNA diversity, **141,** 93, 96–98, 100, 103
generation of mtDNA diversity, **141,** 110, 112, 116–118
Transformations
malignant, and feedback regulation, **167,** 232–233
results and controls, **140,** 435–438
in situ, female sexual organs, **140,** 441–442
strategies, **140,** 432–435
Transforming growth factor-α
melanocytes and, **146,** 240–241
migration induction by, **155,** 68–69
tumorigenesis in transgenic mice bearing gene for, **164,** 51–52, 75
Transforming growth factor-β
angiogenic activity, **159,** 134
antiproliferative effect on mucosa in gut, **167,** 212
autoregulation, **159,** 309
bifunctionality, **161,** 15
clinical potential of polyfunctional nature of, **167,** 239
CML progenitor cells and, **167,** 195–196
description, **167,** 208–210
diversity of effects of, **167,** 224
effect on chondrocyte differentiation, **159,** 270, 307–309, 331
effects on hemopoietic progenitor cells, **167,** 210–214
inhibition of
anchorage-dependent growth of tumor cells, **167,** 237
EGF-stimulated keratinocytes, **167,** 193
in LTBMC system, **167,** 203–204
malignant B-cell excess production of, **167,** 237–238
mechanisms of inhibition, **167,** 215–216
melanocytes and, **146,** 237, 239–240
migration induction by, **155,** 74–75
MIP-1α receptors and, **167,** 207
RB gene pathway and, **167,** 233
receptors for, **167,** 214–215
retinal pigmented epithelium and, **146,** 78, 80
retinal regeneration and, **146,** 56, 59
role in bone fracture repair, **159,** 324
Sertoli-germ cell communication network and, **147,** 47
superfamily members, **159,** 308–309
synergism with basic fibroblast growth factor, **159,** 308–309
Transforming growth factor-β1
effect on arterial smooth muscle cell culture proliferation, **169,** 211–212
role in apoptotic processes, **163,** 118
synthesis in arterial smooth muscle cell culture, **169,** 222

Transforming growth factors
discovery, **155,** 49
effect on pituitary cell proliferation, **157,** 49–50
and neuronal plasticity
neurite initiation and elongation role, **153,** 254–255
phenotypic expression, **153,** 274
retinal precursor cells and, **146,** 178
Transgenic mouse
γ5 gene deletions, **157,** 192
gene knockout studies, **158,** 238
generation, **158,** 238
system for antigen-specific stimulation of lymphocytes, **157,** 191, 193, 196–197
tumorigenesis
chromosomal changes
karyology, **164,** 73
recombinations, **164,** 73
co-oncogenic factors, **164,** 64–69
embryonal abnormalities, **164,** 74–76
methods using, **164,** 40–41
multiple hyperplastic and neoplastic changes, **164,** 64–65
multistep nature of, **164,** 76–79
proto-oncogene-bearing
bcr/abl, **164,** 47–49
c-fos, **164,** 46–47
c-myc, **164,** 45–46
E2A-PBX homeobox gene, **164,** 52–54
growth factor/growth factor receptor genes, **164,** 49–52
pathologic changes, table, **164,** 42–44
ras, **164,** 47
tumor morphology, **164,** 60–72
tumors of rare localization, **164,** 72
viral gene-bearing
human hepatitis B virus, **164,** 61–62
human immunodeficiency virus, **164,** 63–64
human T-lymphotropic virus type 1, **164,** 63
papillomavirus, **164,** 62–63
polymomavirus, **164,** 62
simian virus 40, **164,** 54–60
Transgenic organisms, *Drosophila* muscles and, **143,** 64–65
Transgenic plants, plastids and, **145,** 175–175
Transillumination-assisted microdissection, **147,** 59–60
Transitions
conformational, ependymin, **165,** 127–128
muscle fiber-type
factors
aging, **170,** 186–188
denervation, **170,** 176–177
development, **170,** 158–162
electrostimulation-induced fast-to-slow transitions, **170,** 163–170
electrostimulation-induced slow-to-fast transitions, **170,** 177–178
exercise training, **170,** 171–172
hindlimb suspension, **170,** 181–182
hormone action, **170,** 184–186
immobilization in shortened position, **170,** 180–181
innervation, **170,** 162–163
mechanical overloading, **170,** 172–175
microgravity, **170,** 182–183
myotonia, **170,** 170–171
spinal cord transection and isolation, **170,** 177
thyroid hormone action, **170,** 183–184
mechanism, **170,** 188–192
related to energy metabolism, **170,** 152–153
related to myosin isoform, **170,** 153–155
phase, *see* Phase transitions
zone in lens, cell migration, **163,** 310
Translation
activity of isolated cells, **140,** 225–226
animal mtDNA and, **141,** 174, 183, 187–189, 191, 208
regulation, **141,** 226
transcription, **141,** 222
ciliates and, **141,** 12, 55, 59
Drosophila muscles and, **143,** 112, 122, 139
endosymbiont hypothesis and gene expression, **141,** 306–308
genome organization, **141,** 292–294, 296
mitochondrial 280, **141,** 323, 329
plastids, **141,** 367
flowering plants and, **143,** 302

higher plant mitochondrial genomes and, **141,** 164
higher plant plastids and, **145,** 161
kinetoplastid mtDNA and, **141,** 76
mitochondrial, in *Saccharomyces cerevisiae,* **152,** 159–162
mitochondrial genomes in fungi and, **141,** 93, 118
myofibrillogenesis and, **143,** 167, 177
in organello, during chromoplast differentiation, **163,** 197–199
posttranslational modification
collagen I, **173,** 82–83
role in MT–microfilament interactions, **173,** 230
proteins in striated muscle and, **142,** 107
transdifferentiation in medusae and, **142,** 229
Translation apparatus proteins, endosymbiont hypothesis and, **141,** 311–313
Translation factors, mitochondrial, in *Saccharomyces cerevisiae,* **152,** 166–167, 169
Translesion synthesis, DNA repair, **156,** 125–126
Translocation
cytoskeletal dynamics and, **144,** 89, 100, 110, 118
Drosophila muscles and, **143,** 66
endosymbiont hypothesis and, **141,** 289
environmental stimuli in plant cells and, **145,** 284, 287, 290
explaining nucleocytoplasmic transport, **162B,** 202–204
fish chromatophores and, **143,** 199, 247
flowering plants and, **143,** 298
higher plant plastids and
amino acid synthesis, **145,** 193, 199
fatty acid synthesis, **145,** 189–190
metabolism, **145,** 167–171
starch synthesis, **145,** 179–180
insulin to cytoplasm and nucleus, **173,** 258–260
intracellular calcium messenger system and, **142,** 317, 332
kinetoplastid mtDNA and, **141,** 71
latex bead, **140,** 304–305
macromolecules into cytoplasm, mechanisms, **173,** 260–261
matrix-driven, in protein fiber networks
definition, **150,** 140
physical mechanisms, **150,** 143–146
percolation, **150,** 145–146
wetting, **150,** 143–145
process, **150,** 141–143
mitochondrial genomes in fungi and, **141,** 108
mitosis and, **144,** 220
myofibrillogenesis and, **143,** 168
Robertsonian, effect on gametes, **158,** 226
transcriptional regulation and, **143,** 22, 26, 30, 38
urease in plant cells and, **145,** 77, 94
Translocation proteins
across mitochondrial membranes, **174,** 145–146
inner membrane components, **174,** 144–145
outer membrane components, **174,** 142–144
Transmembrane domains, G-protein-coupled receptors, **164,** 205–207
Transmembrane linkage proteins, in cytoskeleton–membrane connections, **176,** 150–154
Transmembrane loop, biological clocks in retina and, **146,** 129–133
Transmembrane protein, R8-specific boss, **139,** 148–149
Transmembranes, glycine receptor, components
GlyR assembly, **176,** 220–223
GlyR-rich postsynaptic microdomains, **176,** 219–220
Transmission electron microscopy
in analysis of calcifications, **172,** 138
avian salt glands, **144,** 138
eggshells and biomineralization and
fabrication, **145,** 239–240
structure, **145,** 220–221, 224, 232–233
extracellular matrix in animal systems, **140,** 299
retinal precursor cells and, **146,** 157, 182
scanning, mass analysis of NPCs, **162B,** 228–230
Transpiration, by plant cell walls, effect of microgravity, **170,** 52–54
Transplantation
biological clocks in retina and, **146,** 84

Transplantation (*continued*)
brain
effect on behavior, **175,** 291
effect on pathology
avian reflex epilepsy, **175,** 291–297
Fepi/JA chimera construction, **175,** 294
quail–chick experiments
regional specifications, **175,** 271–272
tectal and cerebellar structures, **175,** 266–271
transplantation effects, **175,** 264–266
closed-eye technique, **138,** 9–10
cytokine production and, **146,** 79
dendritic cell role in
clinical applications, **153,** 90–91
process, **153,** 87
growth factor-induced retinal regeneration and, **146,** 51, 61
melanocytes and, **146,** 209–210, 215, 220–221, 233
NMR analysis of transplants, **145,** 41
retinal pigment epithelium, **171,** 247
retinal precursor cells and, **146,** 166
testis as site, **149,** 123
transcleral and superior-lesion approach, **138,** 10–11
transplant rejection, proteins in striated muscle and, **142,** 115–116
Transport
active, Ca^{2+}, **176,** 175–176
amino acids, role of V-ATPase, **174,** 90–91
antigens by M cells
as central characteristic, **167,** 100, 112
histochemical marker correlation with, **167,** 105
maturation stages *versus* capacity for, **167,** 117
and tracers, **167,** 117–122
ATP
effect of MDR protein expression, **171,** 155–156
hypothesis for ABC transporters, **171,** 156
axonal, by microtubules, **171,** 169
axonal transport, **165,** 238
cell wall precursors
callose, **176,** 172
cellulose, **176,** 171–172
pectins, **176,** 170–171
chloride, **165,** 23
drug, *see* Drug transport
electron
higher plant plastids and, **145,** 150, 191, 193
intracellular calcium messenger system and, **142,** 329, 331–332
fast axonal transport, **165,** 238
axonal assembly, **165,** 269–273
dynein, **165,** 271
intracellular
diffusionist theory, **162B,** 186–189
role of nonmuscle tropomyosin, **170,** 18–19
intra-Golgi, role of Rab6, **176,** 31–32
membrane
methotrexate, **156,** 109–110, 113, 116
Pgp, **156,** 82
membrane machinery, linkage to Rab GTPases, **176,** 57–58
mitochondrial, in axons, **173,** 223–225
nucleocytoplasmic
import pathways, **162B,** 258–260
perspectives, models, and evidence, **162B,** 193–210
organelle
different motors, **173,** 225–226
model systems, **173,** 222–223
paracellular, antigens and tracers in gut epithelium, **167,** 117–118
protein, plant cell coated vesicles and, **167,** 6–8
receptor intracellular transport, **166,** 11–12
retrograde surface
actin treadmilling, **150,** 51–54
definition, **150,** 41
future research, **150,** 62–65
hypothesis, alternative, **150,** 46–48
mechanism, **150,** 43–46
membrane raking hypothesis, **150,** 48–51
slow axonal, **165,** 238
axonal assembly, **165,** 274–279
cytoskeleton, **165,** 276, 277–278
surface, retrograde, *see* Retrograde surface transport
water, *see* Water transport
Transporters, Pgp, **156,** 82
Transport proteins, **156,** 27–28, 82–83; **172,** 218

Transport vesicles
formation, role of Rab proteins, **176,** 40–41
in polarized epithelial cells, **173,** 225
role of SNARE, **176,** 57–58
Transscleral approach, to RPE, **138,** 10
trans-splicing, endosymbiont hypothesis, **141,** 267, 281, 304
Transverse tubules, proteins in striated muscle and, **142,** 119–122, 126–127
TRAP
avian retina and, **146,** 24
T3 receptor auxiliary protein, **162B,** 361
Traumatotactic migration, environmental stimuli in plant cells, **145,** 292, 294, 300
Treacher Collins syndrome, *see* Mandibulofacial dystosis
Tree diagram, **139,** 201, 203
Trehalose
intracellular concentrations, **165,** 10
metabolic cycle, **165,** 7, 9
TPS1 gene, **165,** 9–10
TPS2 gene, **165,** 9–10
Trematomus, fish chromatophores and, **143,** 198
Trematomus bernacchii, fish chromatophores and, **143,** 193
TRH, *see* Thyrotropin-releasing hormone
Tribolium castanaeum, larva, genetics of development, **174,** 43–45
Tribolium confusum, aging, effect of altered gravity, **171,** 55–56
Tricarboxylic acid cycle
γ-aminobutyric acid metabolism and, **147,** 286, 311, 316
glutamate metabolism and, **147,** 300
Tricellular pollen development, diagrammatic representation of, **140,** 206
Trichoplusia ni, kinetic chromosome organization, **152,** 9
Trichostatin A, supporting S/MAR functions, **162A,** 423–425
Tridentiger, fish chromatophores and, **143,** 227
Tridentiger obscurus, fish chromatophores and, **143,** 237
Tridihexethylchloride, avian salt glands and, **144,** 162
Trifluoperazine
in calmodulin inhibition, **139,** 42
effect on clofibrate-induced peroxisomes, **160,** 328
inhibition of ecto-ATPases, **158,** 191
Trifluoromethane, nuclear magnetic resonance and, **145,** 32
Trifolium repens, **156,** 39–40
Trigonellin, **156,** 9
Triiodobenzoic acid, **152,** 269; **156,** 44
Triiodothyronine
control of amphibian metamorphosis and, **145,** 107, 110, 121–122, 129–130
effect on Harderian gland secretion, **168,** 551–52
Trimetrexate, **156,** 109, 112, 115–116
Triose phosphate, higher plant plastids and, **145,** 168–169, 179–180, 201
Triskelions, extraction from coated vesicles
assembly into cages of bovine brain, **167,** 13–14
binding of uncoating ATPase at vertices of, **167,** 22
description, **167,** 10–11
structure and size among plants, animals, and yeasts, **167,** 26
Trisomies, *see also* Aneuploidy; Mosaicism, chromosomal
association with clinical syndromes, **167,** 264
chromosome-specific differences in mosaicisms and, **167,** 287
deliveries with karotypically-normal cells in placenta, **167,** 289
determining parent of origin, **167,** 272–274
FISH diagnosis, **153,** 26–29
mosaicism and, distribution of aneuploidy and euploid cells, **167,** 288
paternal age and incidence of, **167,** 281–282
sex ratios in, **167,** 286
Trisomy 13, *see* Patau syndrome
Trisomy 16
maternal age and, **167,** 276–277
as most common trisomy in
from gametogenesis onward, **167,** 270
preimplantation embryos, **167,** 267
spontaneous abortions, **167,** 267
Trisomy 18
associated craniofacial abnormalities, **158,** 223
incidence in live births, **167,** 265

Trisomy 21, *see* Down's syndrome
Triticum
environmental stimuli in plant cells and, **145,** 259
mitosis and, **144,** 225
Triticum aestivum
higher plant mitochondrial genomes and, **141,** 142
microgravity experiments, **156,** 321, 322, 324, 326
Triticum durum, cell wall, effect of weightlessness
epidermal cells, **170,** 51–52
vascular bundles, **170,** 50–51
Triton X-100
inhibition of T-tubule ATPase, **158,** 195–196
solubilization of E-type ATPases, **158,** 177–178
Triturus cristatus, hormonal control of metamorphosis and, **145,** 124–125
Triturus cristatus carnifex, **138,** 99, 101–102, 105, 108, 127–130, 132, 143, 147, 149, 156–158, 172–173
Triturus vulgaris, DNA fiber radioautography analysis, **154,** 272–273
tRNA, *see* Transfer RNA
Trochanter muscles, in *Drosophila*
contractile protein genes, **143,** 100–101, 103, 123–124
development, **143,** 71, 86
large and small cells, **143,** 138
tergal depressor, **143,** 64
Trophic factors
cytokine production and, **146,** 79
growth factor-induced retinal regeneration and, **146,** 64
melanocytes and, **146,** 218
neurite initiation and elongation role
acetyl-L-carnitine, **153,** 255
brain-derived neurotrophic factor, **153,** 247, 251–252
ciliary neurotrophic factor, **153,** 252–253
epidermal growth factor, **153,** 254–255
fibroblast growth factors, **153,** 253–254
midkine, **153,** 255
nerve growth factor, **153,** 247–252
neurotrophins, **153,** 248–252
retinoic acid, **153,** 255
soluble neurite-promoting factor, **153,** 255
transforming growth factors, **153,** 254–255
Trophoblast
confined mosaicism and, **167,** 288–289
field vole
cell morphology, **165,** 63–65
secondary trophoblast giant cells, **165,** 63
trophoblast supergiant cells, **165,** 63–65
fox, cell morphology, **165,** 65–66
giant cell, **165,** 54–56
chromatin, **165,** 87
cytoplasmic compartmentalization membranous structures, **165,** 98–102
dense fibrillar component, **165,** 87
depolytenization, **165,** 90–102
differentiation, terminal stage, **165,** 90–102
fibrillar center, **165,** 87
fragmentation, **165,** 90–102
genome distribution of chromosomal material in nuclear fragments, **165,** 90–92
granular component, **165,** 87
nuclear fragmentation membranous structures, **165,** 98–102
nucleolar lacunar space, **165,** 87
nucleolus components, **165,** 87–89
nucleus differentiation, **165,** 86
perinucleolar chromatin, **165,** 87
polytene chromosome
differentiation, **165,** 103–104
genome multiplication, **165,** 103–104
placenta, **165,** 102–103, 105–106
significance, **165,** 102–106
polytenic nucleus, ultrastructural organization, **165,** 81–89
granulated metrial gland cell modification by, **153,** 124–126
mink, cell morphology, **165,** 65–66
mouse, cell morphology, **165,** 58–61
polytene chromosome, **165,** 54–75
cell morphology, **165,** 56–69
cellular reproduction, **165,** 54–75
classic *vs.* nonclassic, **165,** 79–81
Diptera-rat similarity, **165,** 76–77
mammal-Diptera differences, **165,** 77–79

morphology, **165,** 54–75
nondassic features, **165,** 77–79
trophoblast giant cell, **165,** 54–56
rabbit, cell morphology, **165,** 61, 62
rat, cell morphology, **165,** 56–68
Trophoectoderm material, diagnosis, **153,** 4–5
Tropomodulin
effect on short actin filaments, **166,** 5
function, **170,** 15
role in tropomyosin function, **170,** 15–16
Tropomyosin
binding of actin filaments, **166,** 5
Drosophila muscles and
contractile protein genes, **143,** 94–95, 98, 114–117
function, **143,** 135–136
interactions, **143,** 126
troponin, **143,** 117–118, 121–122
effect on tumorigenicity, **170,** 9–10
encoding gene
in *Drosophila,* **170,** 6
in invertebrates, **170,** 6
in vertebrates, **170,** 6–7
in yeast, **170,** 5–6
expression in transformed cells, **170,** 7–9
function, role of tropomodulin, **170,** 15–16
general structure, **170,** 5
isoform-specific properties, **170,** 25
mRNA and protein expression patterns, **170,** 7
muscle isoform, sequence diversity, **170,** 4–5
myofibrillogenesis and, **143,** 153
assembly, **143,** 165, 168, 172, 175–176
isoform change, **143,** 154
sarcomers formation, **143,** 183
nonmuscle isoforms
characterization, **170,** 5
exon-specific function, **170,** 22–25
expression, **170,** 2
function, role of caldesmon, **170,** 13–15
generation, **170,** 3–4
regulatory functions, **170,** 2, 17–18
role in actin filament stabilization, **170,** 16–17
role in cytokinesis, **170,** 20–22
role in intracellular granule movement, **170,** 18–19
role in muscle development, **170,** 10–11
as stress fiber-like structure component, **170,** 10
structure and properties, **170,** 3–5
proteins in striated muscle, **142,** 84–85, 108, 111, 119
role in cell shape change, **170,** 19–20
role in cytokinesis, **170,** 23–24
role in neuron development, **170,** 12–13
Troponin
Drosophila muscles and
contractile protein genes, **143,** 95, 98–99, 105, 117–122
function, **143,** 136
interactions, **143,** 132
tropomyosin, **143,** 116
myofibrillogenesis and, **143,** 153, 172, 175–176, 183
in striated muscle, **142,** 62
cardiac muscle, **142,** 108, 111, 117–118
skeletal muscle, **142,** 84–87, 101
Troponin-C
Drosophila muscles and, **143,** 117
myofibrillogenesis and, **143,** 175
Troponin-H, *Drosophila* muscles and, **143,** 115, 117, 120–122, 136, 138
Troponin-I
Drosophila muscles and, **143,** 119, 121, 131, 116117
myofibrillogenesis and, **143,** 175
Troponin-T
Drosophila muscles and, **143,** 116–119, 131
muscle fiber type-specific distribution, **170,** 157
myofibrillogenesis and, **143,** 154–155, 175–176
Trunk-derived NC cells, development restriction, **138,** 292–293
Trunk lateral cells, induction in ascidian tissue formation, **176,** 281
Trypanosoma, chloroplasts and, **144,** 284
Trypanosoma brucei
binding to sialic acids, **175,** 182–183
endosymbiont hypothesis and, **141,** 252, 305
kinetoplastid mtDNA and, **141,** 67, 82–84
developmental control, **141,** 80–82
maxicircle genes, **141,** 69–74
maxicircle transcripts, **141,** 74–76, 78–80

Trypanosoma congolense, binding to sialic acids, **175,** 182–183
Trypanosoma cruzi
binding to sialic acids, **175,** 182–183
kinetoplastid mtDNA and, **141,** 74–75, 82–83
Trypanosoma equiperdum, kinetoplastid mtDNA and, **141,** 68, 83–84
Trypanosoma evansi, kinetoplastid mtDNA and, **141,** 68
Trypanosoma lewisi, kinetoplastid mtDNA and, **141,** 67
Trypanosomes
ciliates and, **141,** 55, 60
endosymbiont hypothesis and, **141,** 294–297, 305, 310
kinetoplastid mtDNA and, **141,** 66–68, 83
Tryphine, self-incompatibility and, **143,** 261
Trypsin
intracellular calcium messenger system and, **142,** 33O, 332
vertebrate cell culture technology and, **142,** 149, 172
Trypsin-modulating oostatic factors, insect, characterization, **164,** 230
Tryptophan, flowering plants and, **143,** 309, 311, 313–318
Tryptophan synthase β, flowering plants and, **143,** 313–315, 318
TS, *see* Thymidylate synthetase
TSH, *see* Thyroid-stimulating hormone
Tubal conceptus transfer, **164,** 167
Tubularia, instability of homogeneous state, reduction fields, **154,** 349
Tubules, transverse, proteins in striated muscle and, **142,** 119–122, 126–127
Tubulin
basicity in cold adaptation, **157,** 11
biochemical characteristics, **176,** 141–143
in cytoplasmic cytoskeleton, **166,** 5
cytoskeletal dynamics and, **144,** 109
–dynein interaction, role of t-force, **173,** 46–47
–dynein sliding, dynamics, **173,** 17–19
encoding genes, **139,** 225–226, 228–230, 251–253
environmental stimuli in plant cells and, **145,** 284, 286–287
fish chromatophores and, **143,** 199, 216, 221–224, 242, 246
GTP, role in dimerization, **157,** 3
isotypes, **157,** 6, 11, 18–19
and microtubules
biochemical characteristics of pollen tubules, **140,** 91–92
methods to visualize microtubules in pollen, **140,** 92–93
structure of molecules and microtubules, **140,** 90–91
mitosis and, **144,** 227, 235, 238
polymers, comparison with actin and 14nm filament protein, **164,** 4
posttranslational modification, **157,** 6, 11–12, 17
species variability, **157,** 12
structure, **157,** 2–3
–Tau protein interactions, in microtubule assembly, **171,** 176–177
temperature-induced conformational change, **157,** 15
α-Tubulin, introns, **139,** 248
β-Tubulin, *Drosophila* muscles and, **143,** 84–85
γ-Tubulin
assembly, **169,** 45–46
binding proteins, **169,** 46–47
discovery, **169,** 43
divergent form, **169,** 47–48
immunolocalization, **169,** 43, 47
microtubule nucleation role, **169,** 44–46, 71
overexpression in mammalian cells, **169,** 44–45
plants, **169,** 62
stoichiometry of microtubule binding, **169,** 44
Tubulin-binding proteins, regulation of MT–microfilament interactions, **173,** 226–230
Tubulobulbar complex, Sertoli cells, **160,** 183, 189–191
Tulipa, pollen, **140,** 61
Tumor angiogenesis, placenta growth factor role, **161,** 13–14
Tumor cells
adenocarcinoma
drug resistance, **156,** 119
MDR reversal, **156,** 91
cadherin expression, **165,** 178–179
HeLa cells
cytoskeletal dynamics and, **144,** 95

nuclear magnetic resonance and, **145,** 35, 37
vertebrate cell culture technology and, **142,** 148, 166, 195–196
human trophoblast, **165,** 68–69
as migration modulators, **155,** 49, 55
NMR analysis, **145,** 37, 39, 42
reverse transformation, cytoskeleton role, **154,** 44
trophoblast, **165,** 68–69
Tumorigenesis, *see also* Carcinogenesis
and changes in nuclear matrix, **162A,** 108–110
Drosophila muscles and, **143,** 114
in transgenic mice
chromosomal changes, **164,** 73
co-oncogenic factors, **164,** 64–69
embryonal abnormalities, **164,** 74–76
multiple hyperplastic and neoplastic changes, **164,** 64–65
multistep nature of, **164,** 76–79
proto-oncogene-bearing
bcr/abl, **164,** 47–49
c-fos, **164,** 46–47
c-myc, **164,** 45–46
E2A-PBX homeobox gene, **164,** 52–54
growth factor/growth factor receptor genes, **164,** 49–52
pathologic changes, table, **164,** 42–44
ras, **164,** 47
tumor morphology, **164,** 60–72
tumors of rare localization, **164,** 72
viral gene-bearing
human hepatitis B virus, **164,** 61–62
human immunodeficiency virus, **164,** 63–64
human T-lymphotropic virus type 1, **164,** 63
papillomavirus, **164,** 62–63
polymomavirus, **164,** 62
simian virus 40, **164,** 54–60
Tumorigenicity, effect of tropomyosin levels, **170,** 9–10
Tumor necrosis factor
bifunctionality, **161,** 16
dendritic cell interactions, **153,** 83–84
endothelial cell tissue factor induction by, **152,** 69–70, 72
hemopoietic progenitor cells and, **167,** 228–230
microgravity, **156,** 310
neurotransmission, **149,** 269
retinal pigmented epithelium and, **146,** 78
SDKP sequence in, **167,** 223–224
Tumor necrosis factor-α
angiogenic activity, **159,** 129
clinical potential of polyfunctional nature of, **167,** 239
diversity of effects of, **167,** 224
as effective antitumor agent, **167,** 237
effect of adrenocorticotropin hormone, **170,** 89
influence on Leydig cells, **149,** 112–113
in lens fiber DNA degradation, **163,** 142–143
macrophage production of, **161,** 312–313
migration modulation by, **155,** 80
in reducing toxicity of 4-hydroxy-cyclophosphamide, **167,** 241
role in apoptotic processes, **163,** 118–119
role in immune response, **170,** 105–107
transcriptional regulation and, **143,** 2, 45
genes, **143,** 14, 16–18
inducible activation, **143,** 7, 10
molecular cloning, **143,** 30
viruses, **143,** 40–42
Tumor necrosis factor-β
intestinal lymphocytes producing, TH1 cells, **167,** 98
transcriptional regulation and, **143,** 7, 14
Tumors, *see also* Cancer; Carcinoma; Malignancies
aldosterone-producing, role of endothelin, **171,** 293–294
coagulation effects, **152,** 81, 83
DNA, as whole-chromosome hybridization probe, **153,** 192
effects on glucose transporter expression, **172,** 36–37
feedback regulators, *see also* Malignant tissues; Tumor necrosis factor
growth inhibitory factors, **167,** 234–235
growth modulation by proliferation inhibitors, **167,** 235–237
malignant transformation, **167,** 232–238
resistance mechanisms, **167,** 237–238
genetic, *see* Genetic tumors
growth factor-induced retinal regeneration and, **146,** 60
hypotetraploid ascites, growth regulation, **167,** 235

Tumors (*continued*)
melanocytes and, **146,** 196, 210, 228, 241
melanoma and, **146,** 225
biology, **146,** 236
clinical evidence, **146,** 228, 230, 232
fish, **146,** 233–234
growth factors, **146,** 241
oncogenes, **146,** 241–245
tumor suppressor genes, **146,** 247–249
nuclear magnetic resonance and, **145,** 1, 3, 50
applications, **145,** 35, 39, 41–44
techniques, **145,** 17, 19–21, 31
proteins in striated muscle and, **142,** 101–102
retinal pigmented epithelium and, **146,** 77
role of endothelin, **171,** 295
Sertoli-germ cell communication network and, **147,** 70
spheroidal aggregate cultures, **155,** 186
stem cell cultures, **155,** 184
transcriptional regulation and, **143,** 2, 14, 16, 47
transdifferentiation in medusae and, **142,** 247
vertebrate cell culture technology and, **142,** 147–148
whole-chromosome hybridization analysis
comparative genomic hybridization, **153,** 218
lymphoid tumors, **153,** 216
methods, **153,** 218–219
Tumor suppressor genes
enhanced growth stimulation in loss of, **167,** 238
transcriptional regulation and, **143,** 20
tumor cell control by, **167,** 233
TUNEL technique, for DNA fragmentation, **163,** 121–123
Tunicamycin, proteinase B and, **139,** 68
Turgor pressure
effect on pollen tube tip growth, **176,** 178–179
symplast, **142,** 278
Turing biochemical mechanism
characteristics, **154,** 322–335
phosphofructokinase effects, **154,** 365
Turing–Child metabolic system
metabolic field, **154,** 356–359
metabolic pattern, **154,** 344
phosphorylation field, **154,** 345–346
principle, **154,** 310
Turner syndrome, associated craniofacial abnormalities, **158,** 224
Turning, amoeboid cells, **163,** 80–83
Turnip crinkle virus, *Arabidopsis thaliana,* **144,** 70
Turnip yellow mosaic virus, *Arabidopsis thaliana,* **144,** 70–71
Turnover
eggshells and biomineralization and, **145,** 243
higher plant plastids and, **145,** 177
urease in plant cells and, **145,** 66–68, 98
Turtles, avian salt glands and, **144,** 147, 161, 171, 188
Tweezers, optical, bacterial flagellar motor, **147,** 117, 119–121, 124–126, 128, 156
Twisted liquid crystalline supramolecular arrangement
cell machinery implications, **166,** 94–95
cholesteric twist origin, **166,** 93
hierarchical levels, **166,** 93–94
cytoskeleton, **166,** 95
fluidity at time of secretion, **166,** 94
helicoid, **166,** 95
morphogenesis, **166,** 59–96
order misknowledge, **166,** 92–93
Twistor mechanism, bacterial flagellar motor and, **147,** 140–141
Twitchin
Drosophila muscles and, **143,** 124–125, 140
myofibrillogenesis and, **143,** 183
Two-dimensional analysis
nuclear matrix proteins, **162A,** 24–29
replication initiation, **162A,** 469–473
Two-headed myosin, **156,** 218–219, 222–225, 254
Two-phase thin-layer countercurrent distribution
analysis of electromagnetic field effects on cells, **158,** 300–302
cell partitioning, **158,** 300
Typhoid fever vaccinations, Peyer's patch M cells and oral, **167,** 140–141
p-Tyramine, uptake systems, **164,** 216
Tyrosinase
gene transcription, regulation, **171,** 233–235

melanocytes and
biology, **146,** 193
genetics of melanoma, **146,** 234, 243
pigment-cell genetics, **146,** 197, 202, 205–208, 212–214
synthesis in retinal pigment epithelium, **171,** 232–233
Tyrosinase-related protein-1, melanocytes and, **146,** 205–208, 212–213
Tyrosinase-related protein-2, melanocytes and, **146,** 205–208, 214
Tyrosine
flowering plants and, **143,** 309, 311
melanocytes and, **146,** 193, 195–196, 198, 212
in platelet-derived growth factor receptor binding, **172,** 100, 101–102
Tyrosine ammonia lyase, in phenylpropanoid pathway, **172,** 252
Tyrosine kinase
autophosphorylation modulation of enzymatic activity of, **167,** 168
melanocytes and
genetics of melanoma, **146,** 234, 243, 246–247
pigment-cell genetics, **146,** 215, 217–218
platelet-derived growth factors and, **172,** 99
$pp60^{STC}$ in discovery of $pp125^{FAK}$, **167,** 164
in signal transduction systems, **139,** 326–327
Src family, **167,** 174
transcriptional regulation end, **143,** 6, 10, 37
vertebrate receptor, **139,** 149
Tyrosine phosphatase inhibitors, **167,** 177
Tyrphostin, **167,** 176

U

Ubiquitin
Drosophila muscles and, **143,** 136
encoding gene, **139,** 227, 234–237; **163,** 116–117, 132
Ubiquitination
CFTR protein, **172,** 204–205
histone, **162A,** 211–212
role in Tau protein aggregation, **171,** 194
Ubiquitous urease, in plant cells, **145,** 84–86, 92
Ubisch bodies, **140,** 52
Udotea, environmental stimuli in, **145,** 294
Ulcerative stomatitis–mucositis, as chemotherapeutic dose-limiting factor, **167,** 241–242
Ulcers, healing, therapeutic strategies, **161,** 28
Ulex europaeus agglutin in M cell studies, **167,** 114
Ultrafiltration theory, leaky pores, **163,** 8
Ultrasonography, prenatal diagnosis of craniofacial dysmorphologies, **158,** 262–263
Ultrastructure
lens
elongating fibers, **163,** 326–330
epithelium, **163,** 323–326
mature fibers, **163,** 326–330
senescent fiber cells, **163,** 330–334
M cell
in identification of, **167,** 100–102
site-specific variations in, **167,** 114–115
Ultraviolet light
chalcone synthase and, **147,** 248, 253, 256, 261, 263–264
environmental stimuli in plant cells and, **145,** 256, 282, 299
protectants, chalcone synthase and, **147,** 233, 245–247
Umbrella, transdifferentiation in medusae and, **142,** 232–233
Uncoating, coated vesicles in plant cells, **167,** 22–23
Undaria pinnatifida, intracellular calcium messenger system and, **142,** 336
Undulipodia, endosymbiont hypothesis and, **141,** 330–331
Unicellular organisms, microgravity experiments, **156,** 339–343
Unidentified reading frames
animal mtDNA and, **141,** 174, 181
higher plant mitochondrial genomes and, **141,** 133, 148, 164
mitochondrial genomes in fungi and, **141,** 96, 98, 117
Uniparental disomy
chromosomal deletions, **160,** 84–85

Uniparental disomy (*continued*)
maternal cytoplasmic inheritance mechanism, **140,** 379–382
syndromes associated with, **167,** 289
Unit evolutionary period, GnRH-encoding gene in teleosts and, **147,** 184
Unmanned spaceflight, **156,** 349, 354–355
UPD, *see* Uniparental disomy
Uptake, tracers and antigens by Peyer's patch M cells, **167,** 117–122
Urea, in plant cells
biochemical genetics, **145,** 88, 90
elimination, **145,** 81, 83–84
metabolic origins, **145,** 71–72, 74–76, 79
nickel metabolism, **145,** 82–97
Urease, in plant cells
biochemical genetics
accessory genes, **145,** 86–88
gene products, **145,** 88–92
structural genes, **145,** 84–87
elimination, **145,** 79–80
germination, **145,** 81–82
loss of protection, **145,** 82–84
protein deposition, **145,** 80–81
metabolic origins, **145,** 66
arginine, **145,** 66–72
purines, **145,** 73–79
nickel metabolism, **145,** 92–97
urea substrate, **145,** 97–99
Urechis caupo, sperm-binding proteins, **163,** 278–279
Ureides, urease in plant cells
derived urea, **145,** 66
elimination, **145,** 81, 83
metabolic origins, **145,** 73–79
nickel metabolism, **145,** 95–97
as nitrogen sources, **145,** 98
Uricase, peroxisome content, **160,** 312–313
Uridines, kinetoplastid mtDNA and, **141,** 75–78, 80–81
Urinary calculus, mineral phases in, **172,** 148
Urogenital system, parathyroid hormonerelated protein, **166,** 247–249
Uromastyx acanthinurus, avian salt glands, **144,** 139, 147
Urotensins, secretion in fish, **159,** 7
Urubu, avian salt glands and, **144,** 130
Ustilago cynodontis, mitochondrial genomes in fungi, **141,** 104–105
Uterus
aging, related hormone receptor changes, **160,** 26
androgen effects in neonatal rats, **160,** 2, 10, 12–13
development, ovariectomy effects, **160,** 14–15, 31
endometrial epithelial cells, surface changes, **160,** 34–37
endometrial fibrosis and decidual cell reaction, **160,** 28
estrogen
growth effects, **160,** 31–33
uptake, **160,** 8, 13, 24
pregnant
decidualization, **153,** 105–106
granulated lymphoid cells, *see* Lymphoid cells, granulated, of pregnant uterus
progesterone
growth effects, **160,** 31–33
uptake, **160,** 8, 13–14, 24
progesterone receptor levels, **160,** 33–34
prostaglandin synthesis, **160,** 17–18, 20, 29
protein synthesis, ovarian hormone effects, **160,** 37–38

V

cis-Vaccenic acid, **156,** 19, 24
Vaccines
M cells as potential gateway for oral
description, **167,** 139–142
membrane binding properties and, **167,** 122
overview, **167,** 138
oral, M cell antigen transport potential for delivery, **167,** 138, 139–142
vertebrate cell culture technology and, **142,** 146, 202
bioreactors, **142,** 172, 174, 176
traditional cultures, **142,** 151, 153
Vacuolar biogenesis, in *Saccharomyces cerevisiae*
endocytosis in, **139,** 79–81
fructose 1,6-bis-phosphatase degradation and, **139,** 77–79
general considerations for, **139,** 59–63, 110–114

genes required for, **139,** 93–97, 102–110
mutants, **139,** 93–97
in vitro transport system, **139,** 109–110
VPS1 5 gene, **139,** 103–106
VPS1 gene, **139,** 106–107
Golgi apparatus and, **139,** 76–77
mutants defective for, **139,** 89–102
eukaryotic secretary pathway, **139,** 89–93
genes required for, **139,** 93–97
vacuolar morphology, **139,** 97–102
protein biosynthesis and, **139,** 63–75
aminopeptidase I, **139,** 67, 69–70
carborypeptidase Y, **139,** 64–66, 70
α-mannose, **139,** 74–75
membrane proteins, **139,** 70–74
proteinase A, **139,** 67–68, 70
proteinase B, **139,** 67, 68–69
sorting pathways, **139,** 63–64
protein degradation and, **139,** 75–76
protein traffic to, **139,** 60–63
vacuolar H^+-ATPase and, **139,** 81–89
Vacuolar H^+-ATPases, **167,** 24–26
Vacuoles
chloroplasts and, **144,** 261, 266, 270, 283, 292
cytoskeletal dynamics and, **144,** 98
environmental stimuli in plant cells, **145,** 269, 277, 292, 295, 298
fusion, in yeast, **176,** 63–64
intracellular calcium messenger system and, **142,** 316–317, 319–32O, 324
pea cotyledon, origins of protein storage, **167,** 7–8
plant cell coated vesicles
receptors in sorting of proteins destined for, **167,** 20
synthesis and translocation of proteins destined for, **167,** 19–20
salt tolerance, **165,** 24–25
Sertoli-germ cell, communication network and, **147,** 64, 68
symplast and, **142,** 263–265, 268
Vacuolization, plant cell, effect of microgravity, **171,** 49–50
Vairimorpha nectarix, endosymbiont hypothesis and, **141,** 247, 251
Valine, **143,** 323–324; **167,** 131
Valinomycin, intracellular calcium messenger system, **142,** 316
Vallisneria, environmental stimuli
chemicals, **145,** 297–298
chloroplasts, **145,** 268, 271, 275–277
mesophyll cells, **145,** 299–301
photodinesis, **145,** 252–254, 256–257, 260–261, 263–265, 268
Vallisneria gigantea, intracellular calcium messenger system and, **142,** 308
Valonia, symplast, **142,** 289
Vanadate
dynein activity affected by, **155,** 150–151
effect on flagellar motility, **173,** 27–28
environmental stimuli in plant cells and, **145,** 266, 283
fish chromatophores and, **143,** 222
inhibition of ecto-ATPases, **158,** 191
intracellular calcium messenger system and, **142,** 313–316, 318, 328
in studies of FAK activity in epithelial cells, **167,** 177
Variable region, kinetoplastid mtDNA, **141,** 73–74
Variant I protein, mitochondrial genomes in fungi and, **141,** 93, 98, 105, 109, 120
Vascular bundles
cell wall, effect of weightlessness, **170,** 50–51
symplast and, **142,** 27O, 264, 299
Vascular endothelial growth factor
antibodies, for vascular disease control, **161,** 31
homology with platelet-derived growth factor, **159,** 136
induction of angiogenesis, **159,** 136–137
stimulation of vascular cells, **161,** 7–8
synergism with basic fibroblast growth factor, **161,** 3
Vascularization
avian retina, **146,** 4
growth factor-induced retinal regeneration and, **146,** 60
neovascularization
retinal pigment epithelium, **138,** 34
in retinas of RCS dystrophic rats by RPE cell transplantation, **138,** 33–34
serine protease role, **161,** 4
Vascular permeability factor, *see* Vascular endothelial growth factor
Vascular volume, avian salt glands and, **144,** 185–187

Vascular wall, parathyroid hormone-related protein, **166,** 243, 244
Vasectomy, testicular macrophage involvement, **149,** 127
Vasoactive factor, parathyroid hormone related protein, **166,** 243, 244
Vasoactive intestinal peptide
astrocyte reaction to, **148,** 142–144
effect on prolactin release, **159,** 63–64
and related peptides, **138,** 169–171
seminal vesicle epithelium response, **160,** 289, 294
size, **159,** 63
Vasoactive intestinal polypeptide
avian salt glands and, **144,** 162–167, 197, 201
control of amphibian metamorphosis and, **145,** 127, 138
Vasoconstriction, avian salt glands and, **144,** 154
Vasopressin
astrocyte reaction to, **148,** 138–139
biological clocks in retina and, **146,** 84
cellular localization, **159,** 34
cytoskeleton, **166,** 17–21
gene, **159,** 33
messenger RNA levels, **159,** 9, 35–36
precursor for smaller hormones, **159,** 33–34
role in hormone release, **159,** 36
Vasotocin
immunohistochemical localization of AVT and MT, **138,** 120–124
neuroendocrine and behavioral effects of AVT and MT, **138,** 124–126
Vastatins, modulation of arterial smooth muscle cell behavior, **169,** 228
V-ATPase, *see* Tonoplast proton ATPase
Vaucheria, environmental stimuli in, **145,** 268–271, 283
Vectors, transforming DNA, pollen grains as, **140,** 431–440
Vegetative cells
cytoplasm and somatic cells GSP comparison with MTs in, **140,** 155–158
cytoplasm of pollen grains and tubes
calcium gradients and calcium-cytoskeleton interactions, **140,** 80–83
dry pollen, **140,** 76–77
growing pollen tubes, **140,** 78–80
ionic compounds and intracellular pH in pollen tubes, **140,** 83
movements of organelles and nuclei, **140,** 84–86
pollen hydration, activation and germination, **140,** 77–78
interactions with, **140,** 178–179
interface between, **140,** 193–197
nucleus, functional significance, **140,** 144–145
in pollen and pollen tubes, **140,** 190–193
cell walls, **140,** 192, 193, 194, 196, 198–200
exine, **140,** 193
intine, **140,** 193
external plasma membranes, **140,** 190–191
organelles, **140,** 193
vegetative nucleus, **140,** 191–193
in volvocine algae, **139,** 192–194
Vegetative segregation, mitochondrial, **161,** 63–64
Veiled dendritic cells, human, **153,** 65–66
Velocity
amoeboid cell, **163,** 77–80
cell, computation, **163,** 58–60
sliding, flagellar dynein regulation, **155,** 170–171
Venules, high endothelial, in Peyer's patches, **167,** 95
Verapamil
calcium channels and, **147,** 209–211, 220
intracellular calcium messenger system and, **142,** 320
MDR reversal, **156,** 89–91
in P-glycoprotein-mediated drug resistance, **156,** 79
as second messenger, **156,** 289
Veratridine, avian salt glands and, **144,** 165
Vertebrate cells
culture technology, **142,** 145–147, 201–204
bioreactors
airlift reactors, **142,** 169–170
cell propagators, **142,** 170–171
fiber-bed reactors, **142,** 185–187
fluidized-bed reactors, **142,** 178–180
hollow-fiber reactors, **142,** 180–183
membrane-based reactors, **142,** 183–185

microcarriers, **142,** 171–177
microencapsulation, **142,** 177–178
packed-bed reactors, **142,** 187–192
perfusion, **142,** 167–169
production system, **142,** 162–167
two-stage reactors, **142,** 169
optimization, **142,** 192–193
bioreactor environment, **142,** 199–201
cellular metabolism, **142,** 193–195
mixing in bioreactors, **142,** 195–197
oxygen transport, **142,** 197–200
three-dimensional cultures
Bergenholtz's system, **142,** 160, 162
hydrated collagen lattices, **142,** 159–160
sail-sheet cultures, **142,** 160–161
silica-fiber cultures, **142,** 156, 158–159
spider web cultures, **142,** 153–155
sponge, **142,** 154, 156–157
traditional cultures
anchorage-dependent cultures, **142,** 149–150
background, **142,** 147–149
growth environment, **142,** 151–153
suspension cultures, **142,** 150–151
parathyroid hormonerelated protein, **166,** 231–268
Vertebrates
cells and organs
cytokines, **170,** 102
proopiomelanocortin gene, **170,** 82–85
role of nitric oxide, **170,** 110–113
gonadal differentiation, **154,** 220–221
head formation, **175,** 260–261
immune response, role of POMC products, **170,** 87–88
karyosphere and, **144,** 25–27
anurans, **144,** 27–36
birds, **144,** 37–38
mammals, **144,** 38–39
reptiles, **144,** 36–37
skeletal structure, extracellular matrix, **166,** 74–77, 78
stress response
evolution, **170,** 97–98
role of POMC products, **170,** 87–88
tropomyosin gene, **170,** 6–7
Verticillium, **167,** 19
Vesicles
avian salt glands and, **144,** 144
biological clocks in retina and, **146,** 89
clathrin-coated, *see* Clathrin, coated vesicles
cnidarian neuronal dense-cored, **167,** 40
cytoskeletal dynamics and, **144,** 97–98, 100, 111
eggshells and biomineralization and, **145,** 230
electron-dense, at preprophase band, **175,** 108
encephalic, fate map, **175,** 246, 261–262
environmental stimuli in plant cells and
light, **145,** 263, 277–278, 281
low temperature, **145,** 289
wounding, **145,** 295
exocytic, regulated, Rab3 GTPase in, **176,** 22–23
germinal
breakdown, mitosis and, **144,** 247
karyosphere and, **144,** 44, 46–47
innervation, **144,** 159, 162–163, 165, 167
karyosphere and, **144,** 1, 19, 38, 44, 46–47
M cell
description, **167,** 111
versus enterocytes in rabbits, **167,** 113*f*
transport of endocytic, to endosomal compartment, **167,** 127
melanocytes and, **146,** 193–195
mitosis and, **144,** 219, 234, 245, 247
phospholipid-artificial membrane, in inducing immunity against streptococci, **167,** 141–142
secretory
aggregation, **176,** 173–174
pathway, **176,** 173
in pollen tube, **176,** 140
secretory mechanism, **144,** 169, 172
silica deposition, in diatom morphogenesis
definition, **150,** 253
formation, **150,** 269
shell, modeling, **150,** 275–277, 279, 283
structure, **150,** 254
synaptic, *see* Synaptic vesicles
transport
in polarized epithelial cells, **173,** 225
role of Rab proteins in formation, **176,** 40–41
role of SNARE, **176,** 57–58

Vespa tropica, **138,** 174
Vetch, nodule development, **156,** 3, 18, 27–28, 30–31, 46–48
Viability, and maintenance of isolated cells, **140,** 221–224
Vibrio alginolyticus, bacterial flagellar motor and, **147,** 154
Vibrio cholerae
binding to enterocytes by toxin produced by, **167,** 139
M cells and
fucose-specific adhesin in binding to, **167,** 124
preferential adherence to, **167,** 123
uptake of entire, **167,** 120*t,* 122
Vicia faba
intracellular calcium messenger system and, **142,** 309, 311, 320–321, 327
photosynthetic cells, effect of microgravity, **171,** 38
plastids and, **145,** 155
stylar matrix of, **140,** 305–307
urease in plant cells and, **145,** 67
Vicia sativa, **156,** 19, 28, 47
Vigna, symplast and
auxin, **142,** 280–281
electrophysiological structure, **142,** 266, 268, 270–273
integration of activity, **142,** 287, 291–292, 295
proton pumps, **142,** 274, 276
Vigna mungo, microgravity experiments, **156,** 321–322, 325–327, 331–332
Villi, *see* Microvilli
Villin, **144,** 107; **156,** 235–236, 254–255
Villus enterocytes, dome epithelium enterocytes, comparison, **167,** 131
Vimentin
binding to human erythrocyte inside-out vesicles, **164,** 108
double nuclear membrane barrier, **154,** 28–31
leukemia cell differentiation by, **154,** 52
M cell expression
as cytoplasmic marker in rabbits, **167,** 104, 129
in undifferentiated crypt cells, **167,** 108
myofibrillogenesis and, **143,** 182
nucleic-acid binding potential, **154,** 5
nucleic acid interactions, **154,** 6
presence in palantine tonsil crypt epithelium, **167,** 136
regenerative cementogenesis and, **142,** 40
in reverse transformation model of gene regulation, **154,** 39, 41, 44–45
in striated muscle, **142,** 90, 96–98, 112–114
in vitro binding to nuclear components, **164,** 99–100
Vinblastine, **144,** 235; **156,** 83; **166,** 15
Vinca alkaloids, **156,** 79, 80, 96, 100
Vincristine, **144,** 235; **156,** 83–84
Vinculin
binding partners, **167,** 162–163, 173–174
myofibrillogenesis and, **143,** 159
in striated muscle, **142,** 126
cardiac muscle, **142,** 108–109, 114–115, 117
skeletal muscle, **142,** 92–93, 99
Vinyl acetate, copolymer with ethylene, **146,** 60–62, 66
VIP, *see* Vasoactive intestinal peptide
Viral genes, in transgenic mice
human hepatitis B virus, **164,** 61–62
human immunodeficiency virus, **164,** 63–64
human T-lymphotropic virus type 1, **164,** 63
papillomavirus, **164,** 62–63
polymomavirus, **164,** 62
simian virus 40, **164,** 54–60
Viral infection, chromoplast-like plastids in tissue, **163,** 185–186
Virions
preferential adherence to M cell apical membranes of, **167,** 123
SV40
assembly stages, **162A,** 520–528
association with chromatin, **162A,** 530–531
during infection, **162A,** 522–523,525–528
Virulence, *Arabidopsis thaliana* and, **144,** 54, 57, 67, 69, 73, 75
Viruses
Arabidopsis thaliana and, **144,** 69–71
binding to mucins, **175,** 157–162
coronaviruses, **175,** 171–172
ecto-ATPase levels, **158,** 175
–host interaction
LLC-MK_2 as model system, **162A,** 531–532
and nuclear matrix, **162A,** 486–488
influenza A virus, **175,** 162–169

influenza B virus, **175,** 162–169
influenza C virus, **175,** 169–170
paramyxoviruses, **175,** 170–171
Peyer's patch dome epithelium uptake of, **167,** 117–122
reoviruses, **175,** 172–175
transcriptional regulation and, **143,** 2, 6, 17, 46
HIV, **143,** 39–42
HTLV-I, **143,** 42–44
molecular cloning, **143,** 27–29
vertebrate cell culture technology and, **142,** 148, 169, 172, 174, 176, 196
Vischeria, eyespot reflection in, **148,** 259, 269
Viscin threads, **140,** 53
Viscosity
environmental stimuli in plant cells and, **145,** 257, 259
microviscosity, in seedlings, effect of microgravity, **171,** 40–41
Viscous drag, effect on flagella movement, **173,** 35–38
Vision, growth factor-induced retinal regeneration and, **146,** 50, 52, 60
Visualization software, in multidimensional computer image analysis, **162A,** 50–53
Visual pigment, retinal precursor cells and, **146,** 148, 175, 180
Visual signaling, insect, **164,** 232
Visual system
amphibian development
comparison of amphibian orders, **169,** 113
lens, **169,** 110–111
life history effects, **169,** 113–115, 120–121
retina, **169,** 112–113
tectum mesencephali, **169,** 117–118, 120–121
visual projections, **169,** 115, 117
cell death during development, **163,** 149–150
Vitamin A
retinal precursor cells and, **146,** 180
Sertoli-germ cell communication network and, **147,** 59
Vitamin C
regenerative cementogenesis and, **142,** 38
role
in cartilage mineralization, **159,** 297
in chondrocyte differentiation, **159,** 292–293
in collagen synthesis, **159,** 295
Vitamin D
eggshells and biomineralization and, **145,** 236
metabolite role in cartilage formation, **159,** 316–317
responsive element, enhancer function, **162A,** 261–263, 266–269
Vitamin D receptor, Harderian gland, **168,** 42
Vitamin K, eggshells and biomineralization and, **145,** 234
Vitellin, **166,** 107
Vitelline coat, sperm-binding proteins
abalone, **163,** 262–263
ascidian, **163,** 258–259, 258–260
Vitellogenesis, **159,** 173; **166,** 104, 105–106
Vitellogenin
endosome, **166,** 110–111
evolution, **159,** 179
key features, **166,** 108, 109
oocyte, **166,** 107–109
genes, **166,** 108–109
precursor-product relationships, **166,** 107
receptor binding, **166,** 108–109
serum, **166,** 107
structure, **166,** 108–109
yolk, **166,** 107
structure, **159,** 173
synthesis, **159,** 173, 179
Vitellogenin receptor
biochemical studies, **166,** 128–131
chicken oocyte
homologous receptors, **166,** 123–124
molecular biological studies, **166,** 120–123
cockroach, **166,** 130–131
fish, **166,** 129–130
frog, **166,** 129–130
locust, **166,** 130–131
molecular biological studies, **166,** 120–128
mosquito, **166,** 130–131
oocyte, future trends, **166,** 131–132
Vitreoretinopathy, *see* Proliferative vitreoretinopathy
Vitreous
growth factor-induced retinal regeneration and, **146,** 53, 55, 57, 59
retinal pigmented epithelium and, **146,** 78–80
Vitronectin, plant analog, **174,** 271–272

Vitronectin-like proteins, in cytoskeleton–membrane connections, **176,** 152–154
Vittellogenesis
hormonal control of amphibian metamorphosis and, **145,** 112
karyosphere and, **144,** 15, 21
VM-26, **156,** 102–104; **162A,** 559–561
Vocalization, species-specific, effect of brain transplant, **175,** 291
Volatiles, pollen, and surface lipids, **140,** 36–41
Voltage-gated ion channels, **156,** 278, 285, 290
Voltage sensitivity, bacterial flagellar motor and, **147,** 144–145
Volume
average
cell types in renal glomeruli, **161,** 151–154
endothelial cells, **161,** 151–154
glomerular epithelial cells, **161,** 151–154
mesangial cells, **161,** 151–154
parietal epithelial cells, **161,** 151–154
cell, computation in 3D motion analysis systems, **163,** 75
interstitial fluid, avian salt glands, **144,** 185, 187–188
mean glomerular, renal glomeruli, **161,** 141–143
stereological estimation
embedded kidney, **161,** 141
glomerular, **161,** 141–143
indirect methods, **161,** 129–130
kidney, **161,** 141–143
point-sampled intercepts, **161,** 130–131
selector method, **161,** 131–312
vascular, avian salt glands and, **144,** 185–187
Volume density
glomerular capillary lumina, **161,** 158–159
glomerular urinary space, **161,** 158–159
mean glomerular, **161,** 141
parameter specification, **161,** 116–117
stereological methods, **161,** 141–143
Volume regulation, *see also* Regulatory volume decrease; Regulatory volume increase
amino acids as cellular osmolytes, **161,** 197–202
coordinated control, **161,** 236–238
ion transport systems in, **161,** 177–181
lipoxygenase products in, **161,** 222–223
polysaturated fatty acid effects, **161,** 224–225
protein phosphorylation/dephosphorylation effects, **161,** 230–236
requirements, **161,** 177
Volume sensing
agonist-induced Ca^{2+}-dependent volume changes, **161,** 218–219
eicosanoid role, **161,** 220–222
and free $Ca^{2+}{}_i$ changes, **161,** 217–218
and inositol phosphate cycle, **161,** 219–220
macromolecular crowding effects, **161,** 211–214
stretch-activated channels, **161,** 207–208
nonselective cation channels, **161,** 208–210
selective channels, **161,** 210–211
Volume-sensitive organic osmolyte/anion channel, **161,** 184, 200
Volvocine lineage hypothesis, **139,** 199–204
Volvox, see Algae, volvocine
VP-16 drug, **156,** 101, 104, 132; **162A,** 547
VT, *see* Vasotocin

W

Wall
cell, *see* Cell walls
pollen as living structure, **140,** 568–569
stylar
role in pollination, **140,** 303–304
structure and chemical composition, **140,** 301–303
Water
D_2O, cytoskeleton, **166,** 15
effects on protein conformation, **160,** 103–105
intracellular environment, **160,** 101–102, 116
macromolecular interactions in, entropic influences, **160,** 102
membrane permeability, **160,** 110
symplast and
absorption
auxin, **142,** 277–279, 282–283
integration of activity, **142,** 285–286, 289–290, 296

depletion, **142,** 280
potential, **142,** 264, 285
transport, **142,** 287–292, 296–297
uptake, **142,** 285–287, 291–297, 300
Water channels
in ADH-sensitive epithelia, **163,** 36–37
membrane-spanning proteins, **163,** 37–38
Water stress, extracellular matrix and, **146,** 301, 303
Water suppression, nuclear magnetic resonance and, **145,** 45
Water transport, *see also* Osmosis
cotransporter proteins
conformational change in water transport, **160,** 134–136, 145
evolution, **160,** 113
permeation, **160,** 111–113
structure, **160,** 136–137, 145
cotransport thermodynamics, **160,** 131–134
isotonic transport mechanism, **160,** 142–144
measurement
fluorophore quenching, **160,** 109
ionoselective microelectrode, **160,** 108–109, 125
optical methods, **160,** 109
osmometer, **160,** 99–100
mechanism
ion-coupled active transport, **160,** 100
passive, **160,** 100, 105–106
phospholipid permeation, **160,** 111–112
potassium/chloride cotransporter, **160,** 124–125
dark cell epithelia, **160,** 127–129
kidney proximal tubule, **160,** 129–130
leaky epithelia, **160,** 125–126
mechanism, **160,** 140–142, 145
rate of transport, **160,** 132–133
thermodynamics, **160,** 132–133
small intestine, **160,** 115–116, 130–131
sodium/alanine cotransporter, **160,** 130–131
sodium/glucose cotransporter, **160,** 130–131, 134
uphill transport, mechanism, **160,** 140–142
Weightlessness, *see also* Clinorotation; Clinostation; Microgravity
as active factor in space flight, **171,** 2
definition, **171,** 4
effect on epidermal plant cell wall structure, **170,** 51–54
effect on plant cell wall polysaccharide synthesis, **170,** 54–56
effect on vascular bundle cell walls, **170,** 50–51
Welwitschia, potential double fertilization in, **140,** 342–348
Wheat
effect of microgravity, **171,** 41
endosymbiont hypothesis and, **141,** 303–304, 316
higher plant mitochondrial genomes and, **141,** 157–158, 161–163
intracellular calcium messenger system and, **142,** 307, 322–323, 329–330, 332
microgravity experiments, **156,** 321, 322, 324, 326–328
plastids and, **145,** 179, 197, 203
symplast and, **142,** 284
urease in plant cells and, **145,** 67
Wheat germ agglutinin, nuclear protein import affected by, **151,** 199–200
Whitlockites, in pathological calcifications, **172,** 157–158
Whole-chromosome hybridization, *see* Hybridization, whole-chromosome
Windowing, cell, 2D analysis systems, **163,** 68
Wings, separated, *Pinus* pollen, sporopollenin isolation, **140,** 55
WL 28325, induction of resistance in mouse, **168,** 326
Wolffian duct, programmed cell death in, **163,** 141
Wolff–Parkinson–White syndrome, gap junction role, **157,** 340
Woloszynskia coronata, photoreceptive apparatuses, **148,** 250
Wortmannin, inhibition phosphatidylinositol 3-kinase, **157,** 227–228
Wound assay, *in vitro,* as migration analysis method, **155,** 61–62
Wound healing
acute inflammation
duration, **169,** 151
platelet role, **169,** 152–153, 155
polymorphonuclear leukocyte role, **169,** 155–156
apoptosis, **169,** 165

Wound healing (*continued*)
burns, **169,** 174–175
epithelialization
duration, **169,** 152, 165
keratinocyte role, **169,** 165–168, 170–172
fetal surgery, **169,** 175–176
granulation tissue formation
endothelial cell role, **169,** 172–174
fibroblast role, **169,** 160–165
macrophage role, **169,** 156–157, 159–160
monocyte role, **169,** 157, 159
growth factors in, **161,** 27
phases, **169,** 151–152
regenerative cementogenesis and, **142,** 2, 37
established root surfaces *in vivo* 43-45, **142,** 47, 49–50
retinal pigmented epithelium and, **146,** 75, 78, 80
role of APP, **170,** 239–240
tissue remodeling, duration, **169,** 152
transdifferentiation in medusae and, **142,** 245
Wounding, environmental stimuli in plant cells, **145,** 252, 290–297, 299

X

Xanthine dehydrogenase, urease in plant cells, **145,** 77, 95
Xanthomonas, plant-pathogen interactions in, **144,** 57
Xanthomonas campestris, plant-pathogen interactions, **144,** 66, 69, 75
Xanthophores, cytophysiology
as chromatophore, **143,** 192, 194, 197–200
motility control, **143,** 228–229, 231, 234
signal transduction, **143,** 241–242
Xanthophylls, chloroplasts and, **144,** 265, 282, 289–290
Xanthophytes, chloroplasts and, **144,** 277–278, 282
Xanthosomes, fish chromatophores and, **143,** 228, 242
X chromosome, *see also* Gonosomes
age-related increase in nondisjunction of, **167,** 279
heterochromatic regulation, **160,** 82–84
imprinting and inactivation, **160,** 54, 63, 79–80, 83
mammalian, *see* Sex chromosomes, mammalian
methylation, **160,** 63
monosomy of, **167,** 267
Xist gene and regulation, **160,** 80–82
Xenogenous origin of cells, endosymbiont hypothesis, **141,** 233–234, 237
Xenopus
animal mtDNA, **141,** 224–225
endosymbiont hypothesis and, **141,** 289, 298
mitochondrial genomes of, **141,** 12
biological clocks in retina and, **146,** 85
control of metamorphosis
adrenocortical hormones, **145,** 121, 123–127
growth hormone, **145,** 139
prolactin, **145,** 129
thyroid hormone, **145,** 107, 110, 112
egg extracts, **138,** 263
embryos, **138,** 247, 262
FAK expression
in embryonic brain and spinal cord, **167,** 168
in myotendinous junctions of skeletal muscle, **167,** 175
growth factor-induced retinal regeneration and, **146,** 52
instability of homogeneous state
localized activity, **154,** 344
metabolic field, **154,** 359
proliferation, metabolism, **154,** 362–363
karyosphere and, **144,** 45
keratin mRNA depletion in oocytes, **154,** 61
microgravity experiments, **156,** 334
mitosis and, **144,** 218, 250
M phase MAP kinase, **138,** 228–229, 231–232
oocytes, **138,** 227–228; **151,** 190; **164,** 233
pore complex glycoproteins, **151,** 186–187
proteins in striated muscle and, **142,** 125
retinal precursor cells and, **146,** 168
signal-mediated protein import through, **151,** 191
transcriptional regulation and, **143,** 25
Xenopus laevis
animal mtDNA and
control region, **141,** 201–202

genome content, **141,** 176–177
nucleotide bias, **141,** 204–206
protein genes, **141,** 183, 185, 188
rRNA, **141,** 190
sequences, **141,** 203–204
transcription, **141,** 218
tRNA, **141,** 196
carnosine localization, **138,** 171
control of metamorphosis and, **145,** 106, 108–109
adrenocortical hormones, **145,** 119, 125
growth hormone, **145,** 140
thyroid hormone, **145,** 107, 111, 113
CRF-ir fibers, **138,** 108
development, **176,** 292
endosymbiont hypothesis and, **141,** 289, 299–300
GnRH content, **138,** 98–99
intermediate lobe, **138,** 130
karyosphere and, **144,** 30, 40
kidney and interrenal gland, **138,** 137
lateral line organ, **138,** 166
notochord induction, **176,** 275–276
NPY mechanism, **138,** 132–133
POMC cDNA clones, **138,** 139
preoptic nucleus, **138,** 147
skin, **138,** 176
thyrotropin-releasing hormone, **138,** 99–101
ventral infundibular wall, **138,** 92–93
Xeroderma pigmentosum, **156,** 128
Xiphophorus
fish chromatophores, **143,** 199, 227, 240
melanocytes and, **146,** 232–234, 241, 247, 251
Xiphophorus helleri, fish chromatophores and, **143,** 228
X-protein
myofibrillogenesis, **143,** 176
in striated muscle, **142,** 82
X-ray crystallography, influenza A virus, **175,** 162–165
X-ray diffraction
in analysis of calcifications, **172,** 135–136
eggshells and biomineralization and, **145,** 230, 239
of enamel apatites, **172,** 131
X-ray microradiography, in analysis of calcifications, **172,** 135
XRP, *see* Xyloglucan-related proteins
Xylan, extracellular matrix and, **146,** 265–266, 274, 276, 285, 300
Xylem
extracellular matrix, **146,** 295, 297, 300
symplast and, **142,** 299–300
auxin, **142,** 279–280, 285
electrophysiological structure, **142,** 267–271, 273
integration of activity, **142,** 285–287, 290–292, 294, 296–299
proton pumps, **142,** 275–277
urease in plant cells and, **145,** 77, 79
Xylogenesis, extracellular matrix and, **146,** 294, 300
Xyloglucan endotransglycosylase, activity of EXGT, detection, **173,** 177
Xyloglucan-related proteins
amino acid sequence, **173,** 182–184
catalytic site, **173,** 184–186
gene expression
environmental signals, **173,** 191–192
other hormones, **173,** 191
regulation by auxin, **173,** 188–189
regulation by brassinosteroids, **173,** 189–191
spatial and temporal regulation, **173,** 186–188
gene family, **173,** 182–184
Xyloglucans
–cellulose framework, **173,** 162–165
disproportioning reaction for EXGT, **173,** 179
extracellular matrix
characterization, **146,** 264–266
immunology, **146,** 284–285
modification, **146,** 292–295, 301, 304
interaction with cellulose microfibrils, **173,** 195–196
fucosylation, **173,** 165–166
pectic polysaccharide network, **173,** 167
xyloglucan forms, **173,** 166–167
Xylose, extracellular matrix and, **146,** 276, 293
Xylosyl transferases, extracellular matrix and, **146,** 274–275

Y

YAC, *see* Yeast artificial chromosome
Yariv phenylglycosides, as AGP probe, **174,** 204–210

Y chromosomes, mammalian, *see* Sex chromosomes, mammalian
Yeast
- animal mtDNA and, **141,** 181, 198, 230–231
- ciliates and, **141,** 11
- clathrin in internalization of mating pheromones, **167,** 18–19
- cytoskeletal dynamics and, **144,** 97
- endosymbiont hypothesis and future prospects, **141,** 332
  - gene expression, **141,** 300–301, 306
  - gene structure, **141,** 310–313
  - genome organization, **141,** 291
- heat tolerance of cell, **165,** 28
- higher plant mitochondrial genomes and, **141,** 136, 144
- kinetoplastid mtDNA and, **141,** 68
- in M cell studies
  - uptake by porcine Peyer's patches, **167,** 121*f*
  - variations in uptake of, **167,** 117
- meiotic recombination errors and nondisjunction, **167,** 282
- mitochondria, DNA escape, **165,** 222–225
- mitochondrial genomes in fungi, **141,** 89, 121
  - DNA diversity, **141,** 93, 96–104
  - generation of mtDNA diversity, **141,** 110, 112, 115–119
- mitosis and
  - chromosomes, **144,** 245, 247–248, 256
  - dissociation of events, **144,** 218, 226–228, 232, 237
- NPC membrane proteins, **162B,** 290
- NPCs, bulk isolation, **162B,** 263–264
- nucleoporins, **162B,** 275–290
  - GLFG family, **162B,** 282–286
  - XFXFG family, **162B,** 279–282
- precursor protein processing, **167,** 81
- transcription, **141,** 219–220
- urease in plant cells and, **145,** 74–75
- vacuolar function in physiology and growth, **139,** 59–114

Yeast artificial chromosome
- *Arabidopsis thaliana* and, **144,** 63, 67
- fluorescence *in situ* hybridization, **158,** 233
- human libraries, **158,** 232
- overlap mapping, **158,** 232
- rearrangements, **158,** 232
- screening of other libraries, **158,** 233, 235
- sequence-tagged site content mapping, **158,** 232–233
- in whole-chromosome hybridization, **153,** 195

Yellow flavonols in various pollen types, occurrence, **140,** 46
Yersinia enterocolitica, **167,** 120*t,* 122

Z

Zacco, fish chromatophores and, **143,** 230
Zamia umbrosa, **140,** 329–348
Z-bands
- *Drosophila* muscles and, **143,** 85
- myofibrillogenesis and, **143,** 174

Z-bodies, *Drosophila* muscles and, **143,** 75, 85–86
Z-disc, myofibrillogenesis and, **143,** 174–175, 181–182
Z-disc proteins, *Drosophila* muscles and, **143,** 97–98, 118, 122–125, 135
Zea mays
- cell wall, effect of weightlessness
  - cellulose content, **170,** 55–56
  - meristem cell regeneration, **170,** 47
  - polysaccharide content, **170,** 57
- chalcone synthase and
  - control of activity, **147,** 254, 256, 259–260
  - control of gene, **147,** 267–268, 271–272, 274–276
  - functions of products, **147,** 250–252
- chromoplast differentiation
  - chemical regulation, **163,** 204
  - genetic factors, **163,** 193–194
  - nutritional regulation, **163,** 203
- ciliates and, **141,** 37, 39–40, 58
- endoplasmic reticulum, **140,** 202
- endosymbiont hypothesis and, **141,** 272, 303–304, 306, 315
- environmental stimuli in, **145,** 299
- fixed pollen, **140,** 191, 194, 195
- flavonoids, **140,** 43
- flowering plants and, **143,** 299, 314, 316–317
- higher plant mitochondrial genomes and, **141,** 157–161
  - abnormal phenotypes, **141,** 130–132, 134

recombination, **141,** 149–150, 152, 154, 156
structure, **141,** 136–137, 139, 141–142, 145–146, 148
impairment of entry into meiosis, **158,** 93–94
intracellular calcium messenger system and, **142,** 308, 311, 320, 333, 337
microgravity experiments, **156,** 322, 324–330, 332–333
mutants, chromoplast-like plastids, **163,** 187
plastids and
amino acid synthesis, **145,** 192, 194, 201
metabolism, **145,** 170, 172, 180, 185
structure, **145,** 159, 167
pollen
bursting, effect of pH of incubation medium on, **140,** 208
composition of classes of compounds comprising lipids from, **140,** 40
replicas, **140,** 203
root statocytes, effect of microgravity, **171,** 20–22
sperm cell, **140,** 199–200, 201
urease in plant cells and, **145,** 94
in vitro pollination in, **140,** 394–396
Zearalenone, effects of perinatal exposure, **139,** 12–13
Zebrafish, ependymin gene, structure, **165,** 123
Zeiosis, **139,** 175–176
Zephyranthes grandifflora, **140,** 55
Zeugmatin, in striated muscle, **142,** 91, 95
Zinc
intracellular calcium messenger system and, **142,** 319
urease in plant cells and, **145,** 94
Zinc-finger proteins
Drosophila muscles and, **143,** 79, 81, 137
transcriptional regulation and, **143,** 2, 19, 45–46
Zinnia, extracellular matrix and, **146,** 297, 300
Zinniol, intracellular calcium messenger system and, **142,** 308
Z-line, *Drosophila* muscles and, **143,** 66–67
Zona fasciculata reticularis, glucocorticoid secretion, **171,** 277–278
Zona glomerulosa, mineralocorticoid secretion, effect of ET, **171,** 275–277
Zona pellucida
lysins, *in vitro* fertilized conceptuses, **164,** 165–166
sperm-binding proteins
oligosaccharyl moieties, **163,** 281–282
ZP2, **163,** 267–268
ZP3, **163,** 264–267
Zone of polarizing activity, **163,** 145–146
Zonulae adherentes, avian salt glands and, **144,** 135, 138–139
Zonulae occludentes, avian salt glands and, **144,** 135, 138–139
ZP2 sperm-binding protein, **163,** 267–268
ZP3 sperm-binding protein, **163,** 264–267
Zucchini hypocotyls, coated vesicles
AP-like proteins, **167,** 18
isolation of, **167,** 9
Zygotes
artificial, attempts to obtain, **140,** 579
production, **140,** 409–411
volvacacean, **139,** 199
Zyxin, binding partners, **167,** 163

Contributor Index

Boldface numerals indicate volume number

A

Adachi, Eijiro, **173**, 73
Adler, Ruben, **146**, 145
Aebi, Ueli, **162B**, 225
Agutter, Paul S., **162B**, 184
Ainis, Luigi, **157**, 277
Akin, Debra, **151**, 183
Alexandrov, Vladimir Ya., **148**, 171
Andersen, Ann C., **138**, 89
Andersen, Oivind, **147**, 165
Arias, Jose L., **145**, 217
Ataliotis, Paris, **172**, 95

B

Baba, Takeshi, **166**, 182
Bade, Ernesto G., **155**, 49
Balczon, Ron, **169**, 25
Baluska, Frantisek, **175**, 91
Banchereau, Jacques, **166**, 140
Barber, Mark S., **172**, 243
Barlow, Peter W., **175**, 91
Barnstable, Colin J., **171**, 225
Baroffio, Anne, **138**, 269
Bastos, Ricardo, **162B**, 257
Batini, Cesira, **175**, 242
Belloni, Anna S., **171**, 267
Beloussov, L. V., **150**, 1
Bennett, D. C., **146**, 191
Berezney, Ronald, **162**, 2
Bernstein, Sanford I., **143**, 63
Bignold, Leon P., **139**, 157
Birchmeier, Carmen, **160**, 222
Björklund, Natalie K., **150**, 373
Blencowe, Benjamin J., **162**, 67
Block, G. D., **146**, 83
Bode, J., **162**, 389
Bohnert, Hans J., **151**, 29
Bolwell, G. Paul, **146**, 261
Borg, Thomas K., **150**, 69
Bouck, G. Benjamin, **169**, 268
Bouffler, S. D., **153**, 171
Bouler, J.-M., **172**, 129
Boulikas, Teni, **162**, 279
Bouvier, Florence, **163**, 175
Brasch, Klaus, **159**, 161
Brière, C. **150**, 225
Brinkley, B. R., **162B**, 1
Brix, Klaudia, **149**, 145
Brodland, G. Wayne, **150**, 95
Buee, Luc, **171**, 167
Bugenings, Markus, **155**, 181
Buire, Michel, **156**, 1
Burke, Brian, **162B**, 257

C

Camara, Bilal, **163**, 175
Campochiaro, Peter A., **146**, 75
Cancedda, Fiorella Descalzi, **159**, 265
Cancedda, Ranieri, **159**, 265
Caplan, Arnold I., **145**, 217
Caplan, S. Roy, **147**, 97
Cardoso, M. Christina, **162B**, 303
Carroll, Sean B., **139**, 121

Carver, Wayne, **150**, 69
Castagnola, Patrizio, **159**, 265
Catala, Martin, **175**, 242
Catapano, C. V., **162**, 539
Chui, Jason J.-S., **143**, 1
Claassen, Dale E., **156**, 301
Clark, Margaret S., **174**, 127
Clark-Walker, G. D., **141**, 89
Clayton, David A., **141**, 217
Cockerill, Gillian W., **159**, 113
Cohen, Jack S., **145**, 1
Conlon, J. Michael, **138**, 89
Cook, Peter R., **162**, 125
Cosio, Eric G., **148**, 1
Crabbe, M. J. C., **140**, 525
Cresti, Mauro, **140**, 73
Cripps, Richard M., **143**, 63
Csaba, G., **155**, 1
Csillik, Bertalan, **147**, 193
Cummings, Donald J., **141**, 1

D

Daculsi, G., **172**, 129
Dangl, Jeffrey L., **144**, 53
Danks, Janine A., **166**, 231
Davie, James R., **162**, 191
Davies, S. P., **140**, 19
De Boni, Umberto, **150**, 149
de Bouteiller, Odette, **166**, 140
de Jong, Luitzen, **162**, 151
de la Espina, Susana Moreno Diaz, **162B**, 75
De Loof, Arnold, **166**, 1
Delacourte, Andre, **171**, 167
Deppert, W., **162**, 485
Dickinson, H. G., **140**, 525
Dieckmann, Carol L., **152**, 145
Digonnet, Catherine, **140**, 407
Dijkwel, P. A., **162**, 455
Dike, Laura, **150**, 173
Drevet, Joel R., **140**, 425
Drum, Ryan W., **150**, 243
Dulac, Catherine, **138**, 269
Dumas, Christian, **140**, 565
Dupin, Elisabeth, **138**, 269
Dupuis, Isabelle, **140**, 391

E

Ebel, Jurgen, **148**, 1
Eberharter, Anton, **162B**, 377
Egner, William, **153**, 41
Emes, M. J., **145**, 149
Ermilov, Alexander N., **164**, 37
Eugene, Daniel, **159**, 196

F

Fabry, Stefan, **139**, 189
Fasolo, Aldo, **138**, 89
Fasulo, Salvatore, **157**, 277
Faure, Jean-Emmanuel, **140**, 407
Feagin, Jean E., **141**, 65
Fechheimer, Marcus, **175**, 29
Feldherr, Carl M., **151**, 183
Fernandes, D. J., **162**, 539
Fifkova, Eva, **139**, 267
Fink, David J., **145**, 217
Flickinger, Reed A., **148**, 81
Folkerts, Otto, **141**, 129
Foltz, Kathleen R., **163**, 249
Forgacs, Gabor, **150**, 139
Foster, Jamie W., **154**, 191
Francheschi, Claudio, **170**, 79
Franchini, Antonella, **170**, 79
Franklin, F. C. H., **158**, 1
Franklin-Tong, V. E., **158**, 1
Fraser, Lynn R., **149**, 1
Friedman, William E., **140**, 319
Fry, Stephen C., **151**, 229
Fujii, Ryozo, **143**, 191
Fujii, Yasuhisa, **166**, 182
Fujita, Tomomichi, **152**, 265
Fukui, Yoshio, **144**, 85
Furukawa, Ruth, **175**, 29

G

Gamble, Jennifer R., **159**, 113
Gaude, T., **140**, 525
Gaur, Vinod P., **138**, 1
Gavin, R. H., **173**, 207
Geczy, Carolyn L., **152**, 49
Georgatos, Spyros D., **164**, 91
Gerstberger, Rudiger, **144**, 129
Geusz, M., **146**, 83
Ghosh, Sibdas, **144**, 217
Giraud-Guille, Marie-Madeleine, **166**, 59
Gold, Michael R., **157**, 129
Goldman, Robert D., **162B**, 141

Goodwin, B. C., **150**, 225
Goodman, Eugene M., **158**, 279
Gordon, Richard, **150**, 243, 373
Gotoh, Yukiko, **138**, 211
Grande, Marjolein A., **162**, 151
Graves, Jennifer A. Marshall, **154**, 191
Gray, David A., **144**, 129
Gray, Michael W., **141**, 233
Grebecki, Andrzej, **148**, 37
Greenebaum, Ben, **158**, 279
Griffin, Darren K., **153**, 1
Grilli, Mariagrazia, **143**, 1
Gromova, Irina I., **162B**, 405
Grouard, Geraldine, **166**, 140
Gruzova, Mira N., **144**, 1
Gubatz, S., **140**, 35
Guraya, Sardul S., **160**, 164

H

Haimo, Leah T., **164**, 269
Hamlin, J. L., **162**, 455
Hansen, Linda, **150**, 173
Hanson, Maureen R., **141**, 129
Harada, Shuko, **173**, 243
Harris, Albert K., **150**, 35
Hart, Derek N. J., **153**, 41
Hassell, Thomas, **142**, 145
Hatano, Sadashi, **156**, 199
Hato, Fumihiko, **156**, 159
Hayashi, Toshihiko, **173**, 73
He, D., **162B**, 1
Heidemann, Steven R., **165**, 235
Heizmann, Philippe, **140**, 485
Hepler, Peter K., **138**, 239
Hermo, Louis, **154**, 105
Heuer, Arthur H., **145**, 217
Hill, Adrian E., **163**, 1
Hinata, Kokichi, **143**, 257
Hiramoto, Yukio, **157**, 99
Hirano, Hiroshi, **172**, 1
Hoffman, Werner, **165**, 121
Hoj, Peter B., **174**, 127
Holland, Mark A., **145**, 65
Hollenberg, Martin J., **146**, 49
Hoogenraad, Nicholas J., **174**, 127
Hopkinson, Ian, **173**, 73
Hougaard, David M., **138**, 51
Howald, Isabelle, **139**, 59
Huber, Joan L., **149**, 47
Huber, Steven C., **149**, 47
Hugueney, Phillippe, **163**, 175
Hutson, James C., **149**, 99

I

Iarovaia, Olga V., **162B**, 405
Ichinose, Yuki, **172**, 55
Iguchi, Taisen, **139**, 1
Ingber, Donald E., **150**, 173
Ingleton, Patricia M., **166**, 231
Ishida, Sarahmi, **152**, 109
Isogai, Akira, **143**, 257

J

Jackson, Dean A., **162**, 125
Janssen, Ine, **166**, 1
Jarett, Leonard, **173**, 243
Jaynes, C. D., **138**, 1
Jegou, B., **147**, 25
Jiang, Zhi-Gang, **153**, 233
Jilling, Tamas, **172**, 193

K

Kamitsubo, Eiji, **157**, 99
Kaneko, Toyoji, **169**, 1
Kanous, Kathleen S., **173**, 1
Kao, Teh-Hui, **140**, 449
Kara-Ivanov, Michael, **147**, 97
Karp, Seth, **150**, 173
Kasahara, Michihiro, **172**, 1
Kasamo, Kunihiro, **174**, 85
Katou, Kiyoshi, **142**, 263
Kaufman, Thomas C., **174**, 1
Kawamura, Kousuke, **145**, 105
Kay, V., **162**, 389
Kee, Barbara L., **157**, 129
Kelm, Sorge, **175**, 137
Khalsa, S. B. S., **146**, 83
Kiba, Akinori, **172**, 55
Kikuyama, Sakae, **145**, 105
Kinoshita, Yoshihiro, **156**, 159
Kirk, David L., **139**, 189
Kirk, Kevin L., **172**, 193
Klehr-Wirth, D., **162**, 389
Klungland, Helge, **147**, 165
Knox, J. Paul, **171**, 79

Knox, R. B., **140**, 19
Kolon, Birgit, **155**, 181
Kondorosi, Adam, **156**, 1
Kondorosi, Eva, **156**, 1
Kopf-Maier, Petra, **155**, 181
Kordyum, Elizabeth J., **171**, 1
Korte, Gary E., **152**, 223
Kranz, Erhard, **140**, 407
Kreimer, Georg, **148**, 229
Krisch, Brigitte, **148**, 119
Kugler, Peter, **147**, 285
Kuntz, Marcel, **163**, 175
Kuszak, J. R., **163**, 306

L

Last, Robert L., **143**, 297
Latham, Keith E., **160**, 53
Lauber, Andrea, **162B**, 337
Lawrence, M. J., **158**, 1
Layer, Paul G., **146**, 1
Layer, Paul G., **151**, 139
Le Douarin, Nicole M., **138**, 269
Le Douarin, Nicole M., **175**, 242
LeGeros, R. Z., **172**, 129
Leonhardt, Heinrich, **162B**, 303
Li, Linxi, **138**, 1
Lian, Jane B., **162**, 251
Liaspunova, Natalia A., **154**, 261
Liley, Helen, **150**, 173
Lin, Jenny L.-C., **170**, 1
Lin, Jim J.-C., **170**, 1
Lindemann, Charles B., **173**, 1
Liu, Chau-Ching, **153**, 105
Liu, Yong-Jun, **166**, 140
Loffelhardt, Wolfgang, **151**, 29
Loidl, Peter, **162B**, 377
Lord, Elisabeth M., **140**, 297
Lorz, Horst, **140**, 407

M

Ma, Hong, **162**, 2
Maare, Christian, **156**, 77
Mackiewicz, Andrzej, **170**, 225
Maison, Christele, **164**, 91
Makita, Takashi, **160**, 304
Maniotis, Andrew, **150**, 173
Manske, Marianne, **155**, 49
Marron, Michael T., **158**, 279
Marrs, James A., **165**, 159
Martin, C. R., **147**, 233
Martinez-Murillo, R., **149**, 217
Martinez-Rodriguez, R., **149**, 217
Mascarenhas, Joseph P., **140**, 3
Mata, Lucinda R., **160**, 268
Matsuuchi, Linda, **157**, 129
McGrath, James, **160**, 53
McMahon, D. G., **146**, 83
McMichael, Jr., Robert W., **149**, 47
McNamee, Helen, **150**, 173
Mentlein, Rolf, **148**, 119
Mercola, Mark, **172**, 95
Meyer, Dirk, **160**, 222
Michel, S., **146**, 83
Mielke, C., **162**, 389
Mire-Thibodeaux, Patricia, **156**, 275
Mitchell, David R., **155**, 141
Mitchell, Heidi J., **172**, 243
Mogensen, H. Lloyd, **140**, 129
Moir, Robert D., **162B**, 141
Moloney, Maurice M., **140**, 425
Moneger, Rene, **163**, 175
Montecino, Martin, **162**, 251
Moore, Gudrun E., **158**, 215
Moore, Karen E., **139**, 59
Morales, Carlos R., **154**, 105
Morales, Marisela, **139**, 267
Mortillaro, Michael J., **162**, 2
Muggleton-Harris, Audrey L., **151**, 1
Muto, Shoshi, **142**, 305

N

Nagai, Reiko, **145**, 251
Nagata, Toshiyuki, **152**, 109
Nanney, Lillian B., **169**, 151
Naumidi, I. I., **150**, 1
Naylor, Dean J., **174**, 127
Nedukha, Elena M., **170**, 39
Nelson, W. James, **165**, 159
Newman, Stuart A., **150**, 139
Ngo, Huan, **169**, 268
Nickerson, Jeffrey A., **162**, 67
Nielsen, Dorthe, **156**, 77
Nieuwkoop, Pieter D., **150**, 373
Nishida, Eisuke, **138**, 211

Nishitani, Kazuhiko, **173**, 157
Nothnagel, Eugene A., **174**, 195
Novoselov, V. V., **150**, 1
Numata, Osamu, **164**, 1
Nussdorfer, Gastone G., **171**, 267

O

O'Donnell, Patrick T., **143**, 63
Obar, Robert A., **151**, 67
Obinata, Takashi, **143**, 153
Ochs, Robert L., **159**, 161
Ohno, Shinichi, **166**, 182
Ohta, Yasuhiko, **160**, 1
Okamoto, Hisashi, **142**, 263
Oko, Richard, **154**, 105
Olins, Ada L., **153**, 137
Olins, Donald E., **153**, 137
Opas, Michal, **150**, 119
Orenic, Teresa V., **139**, 121
Ottaviani, Enzo, **170**, 79
Ozil, Jean-Pierre, **152**, 183

P

Paige, Christopher J., **157**, 129
Pante, Nelly, **162B**, 225
Pante, Nelly, **162B**, 257
Pantic, Vladimir R., **159**, 1
Parfenov, Vladimir N., **144**, 1
Park, Carol M., **146**, 49
Parr, Earl L., **153**, 105
Parulekar, Satish J., **142**, 145
Paweletz, Neidhard, **144**, 217
Pelletier, Georges, **138**, 89
Penman, Sheldon, **162**, 67
Perlman, Jay I., **152**, 223
Pette, Dirk, **170**, 143
Pierson, Elisabeth S., **140**, 73
Plesner, Liselotte, **158**, 141
Plopper, George, **150**, 173
Polacco, Joseph C., **145**, 65
Pollack, Ayala, **152**, 223

R

Ratet, Pascal, **156**, 1
Raymond, Christopher K., **139**, 59
Razin, Sergey V., **162B**, 405
Riethmacher, Dieter, **160**, 222
Rios-Ramirez, M., **162**, 389
Rizzolo, Lawrence J., **171**, 225
Roberts, Christopher J., **139**, 59
Roeckel, Patricia, **140**, 425
Roepe, Paul D., **171**, 122
Rogers, Bryan T., **174**, 1
Rohlmann, Astrid, **157**, 315
Ronchi, Vittoria Nuti, **158**, 65
Rossi, Gian Paola, **171**, 267
Roth, Gerard, **169**, 83
Ruiz-Cabello, Jesus, **145**, 1
Russell, Scott D., **140**, 357
Russell, Scott D., **140**, 565
Ryan, Michael T., **174**, 127

S

Samarabandu, Jagath, **162**, 2
Sanders, E. J., **163**, 105
Sanders, Luraynne C., **140**, 297
Sandhu, Nicole P., **162B**, 337
Saveliev, S. V., **150**, 1
Schaffer, Christopher J., **169**, 151
Schauer, Roland, **175**, 137
Schiffmann, Yoram, **154**, 309
Schirmbeck, R., **162**, 485
Schlake, T., **162**, 389
Schmid, Volker, **142**, 213
Schmidt, Andrea, **169**, 83
Schmitt, Rudiger, **139**, 189
Schneider, Wolfgang Johann, **166**, 103
Schoenfeld, Thomas A., **151**, 29
Schroeder, Hubert E., **142**, 1
Schuchard, Mark, **162B**, 337
Schul, Wouter, **162**, 151
Schultze, Michael, **156**, 1
Schwarz, Heinz, **165**, 121
Seaton, Anthony D., **138**, 1
Selverston, Allen I., **147**, 1
Serrano, Ramon, **165**, 1
Sheedlo, Harold John, **138**, 1
Shimmen, Teruo, **155**, 97
Shiraishi, Tomonori, **172**, 55
Shoeman, Robert L., **154**, 1

Shvemberger, Irina N., **164**, 37
Simpson, David G., **150**, 69
Sims, John, **150**, 173
Singh, Anuradha, **140**, 449
Singh, M. B., **140**, 19
Skovsgaard, Torben, **156**, 77
Smith, Robert A., **153**, 233
Smith, Robert M., **173**, 243
Soll, David R., **163**, 43
Solter, Davor, **160**, 53
Sonobe, Seiji, **175**, 1
Spann, Timothy P., **162B**, 141
Spelsberg, Thomas C., **162B**, 337
Spooner, Brian S., **156**, 301
Staples, Robin R., **152**, 145
Staron, Robert S., **170**, 143
Stein, Gary S., **162**, 251
Stein, Janet, **162**, 251
Stengert, M., **162**, 389
Stevens, Tom H., **139**, 59
Stockem, Wilhelm, **149**, 145
Stovall, S. V., **138**, 1
Stromberg, Elisabeth, **157**, 1
Stromer, Marvin H., **142**, 61
Stuart, Kenneth, **141**, 65
Subramaniam, M., **162B**, 337
Swann, Karl, **152**, 183
Syono, Kunihiko, **152**, 265

T

Takahashi, Sumio, **157**, 33
Takahashi, Yohsuke, **152**, 109
Takata, Kuniaki, **172**, 1
Tanaka, Shigeyasu, **145**, 105
Taxi, Jacques, **159**, 196
Terada, Nobuo, **166**, 182
Terracio, Louis, **150**, 69
Thaler, Catherine D., **164**, 269
Thorsness, Peter E., **165**, 207
Thyberg, Johan, **169**, 184
Tobin, A. K., **145**, 149
Tonon, Marie-Christine, **138**, 89
Toriyama, Kinya, **143**, 257
Toyoda, Kazuhiro, **172**, 55
Traub, Peter, **154**, 1
Trick, Martin, **140**, 485
Tripathi, Satish C., **142**, 145
Turner, James E., **138**, 1

U

Ueda, Hideho, **166**, 182
Uno, Isao, **139**, 309

V

Vadas, Mathew A., **159**, 113
van Driel, Roel, **162**, 151
van Steensel, Bas, **162**, 151
van Wijnen, Andre J., **162**, 251
Vanden Broeck, Jozef, **166**, 1
Vanden Broeck, Jozef J. M., **164**, 189
Vaudry, Hubert, **138**, 89
Volkmann, Dieter, **175**, 91

W

Wadkins, Randy M., **171**, 122
Wallace, Graham, **151**, 229
Wallin, Margareta, **157**, 1
Wamboldt, Dawn D., **170**, 1
Wang, Ning, **150**, 173
Wang, Tao, **170**, 1
Wansink, Derick G., **162**, 151
Warren, Kerri S., **170**, 1
Wassermann, Karsten, **156**, 77
Watanabe, Masao, **143**, 257
Watson, Glen M., **156**, 275
Weber, Eric R., **165**, 207
Wei, Xiangyun, **162**, 2
Whatley, Jean M., **144**, 259
Wiermann, R., **140**, 35
Willbold, Elmar, **146**, 1
Willbold, Elmar, **151**, 139
Williams, Lisa A., **153**, 41
Winston, Nicola J., **164**, 139
Wolburg, Hartwig, **157**, 315
Wolf, Klaus Werner, **152**, 1
Wolstenholme, David R., **141**, 173
Wride, M. A., **163**, 105

X

Xiao, Si-Qun, **145**, 217

Y

Yamada, Tetsuji, **172**, 55
Yamamoto, Kazutoshi, **145**, 105
Yamanishi, Hiroyasu, **174**, 85
Yokoya, Etsuo, **155**, 97
Young, John Ding-E, **153**, 105
Young, Richard W., **138**, 1

Z

Zaccone, Giacomo, **157**, 277
Zeng, C., **162B**, 1
Zeuthen, T., **160**, 99
Zhao, Shulei, **171**, 225
Zybina, Eugenia V., **165**, 53
Zybina, Tatiana G., **165**, 53